ABIOTIC STRESS TOLERANCE IN CROP PLANTS

Breeding and Biotechnology

books for everyone
PAPERBACKS

NIPA® GENX ELECTRONIC RESOURCES & SOLUTIONS P. LTD.
New Delhi-110 034

Abiotic Stress Tolerance in Crop Plants

Breeding and Biotechnology

Dr. BIDHAN ROY
Scientist (Genetics and Plant Breeding)
Uttar Banga Krishi Viswavidyalaya, Pundibari, Cooch Behar 736 165
West Bengal (India)

Dr. ASIT KUMAR BASU
Professor (Seed Science and Technology)
Bidhan Chandra Krishi Viswavidyalaya, Mohanpur, Nadia 741 252
West Bengal (India)

nipa® books for everyone
PAPERBACKS

NIPA® GENX ELECTRONIC RESOURCES & SOLUTIONS P. LTD.
New Delhi-110 034

nipa books for everyone
PAPERBACKS

NIPA® GENX ELECTRONIC RESOURCES & SOLUTIONS P. LTD.

101,103, Vikas Surya Plaza, CU Block
L.S.C. Market, Pitam Pura, New Delhi-110 034
Ph : +91 11 27341616, 27341717, 27341718
E-mail: newindiapublishingagency@gmail.com
www: www.nipabooks.com

For customer assistance, please contact
Phone: + 91-11-27 34 17 17
Fax: + 91-11- 27 34 16 16
E-Mail: feedbacks@nipabooks.com

ISBN 978-81-96053-68-0

Composed & Designed by NIPA

Dedicated to

The Sacred Memory of My (Bidhan Roy) Parents

Late Balaram Roy and Late Sonalakshmi Roy

PREFACE

The load of abiotic stresses on crop production is being incremented gradually by the directed demands of human beings for their food and luxuries. The ground water is depleting fast due to both intensive and extensive cultivation during off-monsoon periods as well as supplementation of water through irrigation during monsoon. Poor quality irrigation water and depletion of ground water increased salinity in arid and semi-arid zones. Intensive cultivation and heavy feeder crops lead deficiency of some nutrients and toxicity of others. The industrial growth and increase in vehicles created heavy metal toxicity in many industrial areas. In combination of industrial growth and improvement of transportation systems facilitate aggressive air pollution, which subsequently forcing climate change around the globe. All of these exert greater influence on plant growth and crop productivity.

Growing population in both developing and underdeveloped countries already has alarmed to increased food grain production. The productivity of major staple food crops have reached to a plateau. There is very little scope to increase crop production area too. Therefore, more emphasis is required to use problem soils effectively. The soil reclamation is a costly affair and it is temporary. Development of crop genotypes tolerant/resistant to the adverse conditions is the only solution of such problem. To develop tolerant/resistant plant genotypes, the plant breeder or plant biotechnologist should have knowledge regarding the injury and tolerance mechanisms in plants for specific stress and plant symptoms to know the nature of abiotic stress, breeding methods and biotechnological approaches. This book deals with those above mentioned requirements. Apart from breeding and biotechnology, each chapter deals with crop management under specific stress environment.

Involvement of the first author in development of metal-toxicity (aluminium and iron) tolerant rice genotypes using biotechnological tools (somaclonal variations), gene tagging for salt tolerance and transgenic development using

SOD gene is continuing since his Ph. D. programme. This book has been divided into seven major parts: physical stress (salt), water stresses (drought and waterlogging), temperature stresses (heat and cold), metal toxicities (aluminium, iron, cadmium, lead, nickel, molybdenum etc), non-metal toxicities (boron and arsenic), oxidative stress, and finally atmospheric stresses (air pollution, radiation and climate change). Emphases have been given to include latest development in the field of abiotic stresses with appropriate citations and achievements. It is our hope that this book will adequately serve the students, teachers and plant scientists in the field of agriculture and botany.

It is our pleasure to acknowledge the help of Shri Rajeeb Lochan Maharana, Ph. D. scholar, Department of Seed Science and Technology, Bidhan Chandra Krishi Viswavidyalaya, Mohanpur, Nadia, West Bengal, for helping in collection of writing materials and abstracts from Central Library of BCKV.

We are extremely greateful to staff of Genetics and Plant Breeding Department, Uttar Banga Krishi Viswavidyalaya, Pundibari, West Bengal, and staff of Cooch Behar Krishi Vigyan Kendra, UBKV, Pundibari, particularly Shri S.K. Bose (Office Superintendent) for continuous encouragement for preparation of this manuscript. We are especially indebted to the Vice-Chancellors of both the universities (UBKV and BCKV) for allowing us in preparation of this book. We would like to place on record our gratefulness to Dr. C. P. Suresh and Dr. Pankaj Panwar for their help and support in promoting the development of this book on abiotic stresses. Any suggestion from the readers for further development of this book will be highly respected.

August, 2008
West Bengal

BIDHAN ROY
ASIT KUMAR BASU

CONTENTS

Unit-II
WATER STRESSES

Unit-III
TEMPERATURE STRESSES

Unit-IV
OXIDATIVE STRESS

Unit-V
METAL TOXICITY TOLERANCE

Unit-VI
NON METAL TOXICITY TOLERANCE

Unit-VII
ATMOSPHERIC STRESSES

1

Introduction

Abiotic stresses are major constraints for many crop plants in specific areas over the globe which limits the crop production. Dudal (1976) estimated that only 10% of world's arable land may be categorized as free from stress. The rapid change in environmental conditions likely to override the adaptive potential of plants, this environmental changes mainly originated from anthropogenic activities, which has caused soil and air pollution, plant are exposed to natural climatic or edaphic stresses, such as, high irradiation, heat, chilling, freezing, drought, excess water and waterlogging and nutrient imbalance. Among abiotic stresses drought is the main abiotic factor as it affects 26% of arable area (Table 1.1). Water stress is a single most severe, limitation to the productivity of rice in the rainfed ecosystem (Widawgky and O'Toole, 1990). Mineral toxicity/deficiencies are second in importance. Among mineral toxicity, salinity is wide spread and is estimated to affect 10% of the world land surface (Richards, 1995). Increased salinization of arable land is expected to have devastating global effects, resulting in 30% land losses within the next 25 years and up to 50% by the year 2050 (Wang *et al.*, 2003). Singh and Bandyopadhyay (1996) reported that salt affected soils occupy about 2.5 ´ 10^6 ha in India. Soil acidity is another major problem common to tropical regions, which constitutes about 3.95 billion ha of land (FAO, 1991). Acid soils caused by combination of aluminium (Al) and manganese toxicity are major constraints to soil fertility and crop productivity. Al toxicity problems are of enormous importance in production of rice, maize, and sorghum. The yield losses caused by various abiotic stresses in South and Southeast Asia have been compiled in Table 1.2 (Dey and Upadhyay, 1996).

Table 1.1. The fraction of world arable land affected by abiotic stresses

Abiotic stress	Fraction (%) of arable land
Drought	26
Mineral (Toxicity / deficiency)	20
Freezing	15
No stress	10

Source: Dudal, 1976

Table 1.2. Estimated yield loss caused by various abiotic factors (Dey and Upadhyaya, 1996)

Country/ Region	Yield loss (kg/ha)				
	All technical constraints	Abiotic factors	Drought	Cold	Submergence
Southern India	468	117	17	4	0
Eastern India	658	306	144	18	24
Bangladesh	635	284	93	10	140
North Eastern China	1350	1156	153	194	95
Central China	1515	1444	250	317	163
Nepal	1422	406	236	0	13
Southern China	1091	990	143	159	112
Northern China	1288	1033	169	160	79

There is serious concern for food security of developing countries, which demands a conscious effort to improve production from areas commonly exposed to abiotic stresses. As mindless urbanization and industrialization swallow fertile lands and overuse of pesticides puts the environment jeopardy, researchers are designing crops that could tolerate abiotic stresses. The limitations to increase crop yield on existing cropped areas can be partially overcome with greater production inputs, which are costly, laborious and beyond the reach of small and marginal farmers.

The research on the genetic control mechanisms and the heritability of plant responses to stress should lead to the development of new crop cultivars specially breed for adaptation to the stress situations. Crop varieties bred to provide a genetically controlled environmental resiliency can contribute markedly to yield stability (Christiansen, 1982). The new breeding strategy for abiotic stresses consists of six phases (Fig. 1.1).

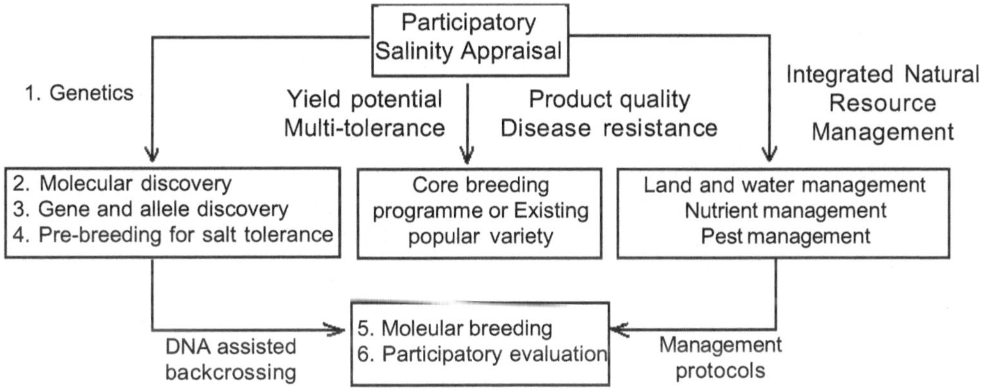

Fig. 1.1. Diagrammatic representation of six-phase breeding strategy

In conventional breeding programme, genetic markers are being used by plant breeders, which are morphological traits and controlled by single locus. The morphological markers are not always useful for selection for abiotic stresses. Therefore, molecular breeding (biochemical and/or DNA markers) is a new chapter for quick selection to improve the crops for problem soils. Transgenic development is another straight forward technology to improve crop yield in abiotic stress affected land. The advent of plant transformation may have placed within the grasp the possibility of engineering greater abiotic stress tolerance in plants. Molecular control mechanisms for abiotic stress tolerance are based on the activation and regulation of specific stress-related genes. These genes are involved in the whole sequence of stress response, such as signaling, transcriptional control, protection of membranes and proteins and free-radical and toxic compound scavenging. Genes from biological species can be transferred to crop species by genetic engineering methods. Introduction of molecular change by genetic engineering takes lesser time compared to plant breeding methods. Only desired gene(s) can be transferred, whereas, in conventional breeding approach associated with simultaneous transfer of undesired gene(s). For re-cultivation of degraded soils and reclamation of industrial sites, stress tolerant plants are required. Biotechnological efforts are under way to improve plant stress tolerance and the ability to extract pollutants from the soils with the aim of using plants for soil clean up (Salt *et al.*, 1995). In order to device new strategies for abiotic stress tolerance for phytoremidation and improved tolerance, it is important to understand the basic principles as to how stress produce injuries and its tolerance mechanisms, and subsequent development of tolerant genotypes.

REFERENCES

Christiansen MN. 1982. World environmental limitations to food and fiber culture. In: Breeding Plants for Less Favourable Environments, (eds.), Christiansen NM, Lewis CF John Wiley, USA, pp. 1-11.

Dey MM, Upadhyay HK. 1996. Yield loss due to drought, cold and submergence tolerance. In: Rice Research Asia: Progress and Priorities, (eds.), Evenson RE, Herdt RW, Hossain M, International Rice Research Institute in collaboration with CAB International, UK. pp. 291-303.

Dudal R. 1976. Inventory of major soils of the world with special reference to mineral stress. In: Plant Adaptation of Mineral Stress in Problem Soils, (ed.), Wright MJ, Cornell University, Agriculture Experiment Station, Ithaca, New York.

FAO. 1991. FAOSTAT agriculture data. Food and Agriculture Organization of the United Nations.

Richards RA. 1995. Improving crop production on salt-affected soils: by breeding or management? Expl Agri. 31(4): 395-408.

Salt DE, Blaylock M, Kumar NPBA, Dushenkov V, Ensley BD, Chet I, Raskin I. 1995. Phytoremediation: a novel strategy for the removal of toxic metals from the environment using plants. Biotechnol. 13: 468-474.

Singh NT, Bandyopadhyay AK. 1996. Chemical degradation leading to salt-affected soils and their management for agriculture and alternate uses. Technical Bulletin No. 17, New Delhi, India, Indian Society of Soil Science.

Wang WX, Vinocur B, Altman A. 2003. Plant responses to drought salinity and extreme temperatures: towards genetic engineering for stress tolerance. Planta. 218(1): 1-14.

Windawsky, DA, O'Toole JC. 1990. Prioritizing rice biotechnology research agenda for Eastern India. New York (USA); The Rockfeller Foundation.

Unit-I
PHYSICAL STRESS

2

Salt Tolerance

2.1. INTRODUCTION

Salt stress has become an ever increasing threat to food production. It is a major factor limiting the crop productivity in the arid and semi-arid areas of the world (Ashraf, 1994; Foolad and Jones, 1992) and it affects about 10% of the total global land area (Richards, 1995). Increased salinization of arable lands is expected to have devastating global effects, resulting in 30% land losses within the next 25 years and up to 50% by the year 2050 (Wang *et al.,* 2003).

Soluble salts can cause harm to plant, if they are in high concentration in water or soils and it limit crop cultivation world wide. Generally an array of stresses interplay in saline soils and reduces productivity of salt sensitive crops. In India, about 12 mha of land has been affected with salinity and alkalinity (Yadav and Gupta, 1984); an area of nearly 4 mha of land suitable for rice affected with salinity (Paul and Ghosh, 1986). The optimum salt concentration for the growth of halophytes is found to be about 0.5 M NaCl (Flower and Yeo, 1981). Most of the crop plants are salt sensitive glycophytes. Although the tolerance to saline conditions to plant is variable, crop species are generally intolerant of $1/3^{rd}$ of the concentration of sea water. The quantum of diversity in respect of salt-tolerance, which often exists among the cultivars of any glycophyte species, indicates that stable genetic mechanism exist even in glycophytic plants that may provide incremental levels of protection against salt stress.

Salt affected soils are mainly of two types: (i) saline and (ii) alkaline. A comparative summary of the various properties of saline and alkaline soils has been presented in Table 2.1.

Table 2.1. A summary of comparative characteristics of saline and alkaline soils

Feature	Saline soils	Alkaline soils
pH	< 8.2	> 8.2
Electrical conductivity	> 4 dSm^{-1}	< 4 dSm^{-1}*
Exchangeable sodium (%)	< 15	> 15
Gypsum	Present in significant quantities	Nearly always absent
Soil structure	Stable	Unstable
Visible symptom	White layer of salt on surface	Dark brown or black/ ash coloured clay crust

* Conductivity may exceed 4 dSm^{-1} if appreciable Na_2CO_3 and $NaHCO_3$ are present

This problem is pronounced in 8 out of 15 agro-climatic zones of India, particularly in the Indo-Gangetic plains, arid and semi-arid region of Rajasthan and Gujarat, black soil belts covering Madhya Pradesh, Gujarat, Maharastra, Karnataka, Andhra Pradesh, Tamil Nadu, and coastal belt (Yadav, 1993).

Before proceeding towards the selection of crops and their varieties, it is necessary for the grower to have the knowledge on salinity and alkalinity classes as well as salt tolerance of different crops. These classes indicate the extent of salinity. The salinity classes and their interpretation in terms of plant response (salinity scale) as recommended by Richards (1954) and has been depicted in Table 2.2.

Table 2.2. Classification of saline soils based on their effects on growth of most of the crops

Class	Conductivity dS/m	Plant response
Non-saline	0-2	Negligible
Slightly saline	2-4	Yield of sensitive crops may be restricted
Moderately saline	4-8	Yield of many crops are restricted
Strongly saline	8-16	Only tolerant crops yield satisfactorily
Very strong saline	> 16	Only very tolerant crops yield satisfactorily

2.2. SALT TOXICITY SYMPTOMS

2.2.1. Salt Affected Soils

The saline and alkali (sodic) soils can be identified in the field by some visual symptoms. The saline soil is covered by white layer of salts on the surface. This white layer of salts disappears on irrigation/rain and reappears during dry period. The alkali soils are covered with dark brown or black/ash coloured clay crust due to dissolution of humus in alkali soils (sodium humate). During rains or irrigation water will not stagnate for a longer period on normal or saline soils whereas it will continue to stagnate on alkali/sodic soils for a longer period. The water on the soil surface of saline/normal soil is very clear whereas in alkali soil it is dark coloured, soapy and muddy. The physical condition of saline soil is always better than the alkali soil. The alkali soils are dispersed, fluffy and slake when they are wetted by rain or irrigation and may develop hard crust or cracks on drying.

In the coastal area, the winds carry salt spray and leave salt deposits on plants. In addition, plants can be damaged by salt in the irrigation water. Most salt damage will show up during dry season, when plants have to be watered with well water that contains salt. In this case, salts build up, and there is not rain to rinse salt off the foliage. Alkaline soil may lock up vital mineral nutrients needed for plant growth. Sand along the coasts generally lacks organic matter or any other nutrients and moisture holding material. This lackuna can be partly overcome by the addition of organic matter.

2.2.2. Plant Symptoms

1. Tip and edge burn of leaves, slow growth, nutrient deficiencies, wilting and eventually death of the plant can occur.
2. Increase salinity decrease and delay in seed germination (Jain *et al.*, 2003).
3. At seedling stage, cause substantial reduction in plant stand, shoot length (Jain *et al.*, 2003), root and shoot dry weight and finally grain yield (Scardaci *et al.*, 1996; Shannon *et al.*, 1998).
4. Salinity stress decrease chlorophyll content, total sugar, starch and potassium in seedling stage (Trivedi *et al.*, 2004).
5. Early reproductive stages such as panicle initiation and pollination are found to be highly sensitive to salinity that effects the formation of grain components and ultimately the grain yield (Khatun and Flower, 1995; Zeng *et al.*, 2001).

6. Several reports indicate that Na and Cl^{-1} contents in shoot increase and plant growth decreases with increasing levels of soil salinity (Shannon *et al.*, 1998).

The detrimental effect, as death of plants or decrease in productivity and suppression of growth occurs in all plants, but their tolerance levels and rates of growth reduction at lethal concentrations of salt vary widely among different plants species.

2.3. INJURY MECHANISMS

Salt can cause damage to the plants when it is absorbed by the roots and accumulated in the plants to toxic levels. Salt stress has mainly three fold effects; it reduces water potential, causes ion imbalance or disturbance in ion homeostasis and toxicity. The major causes of salt injury have been briefly discussed below.

2.3.1. Reduced Water Potential

The altered water status leads to initial growth reduction and limitation of plant productivity. The plant-water relationship parameters were found to be directly affected by salinity. High levels of salinity not only impose osmotic and ionic stresses on plant, but also impair uptake and transport of essential nutrients. Since salt stress involves both osmotic and ionic stress (Hayashi and Murata, 1998; Munns, 2002; Benlloch-Gonzalez *et al.*, 2005), growth suspension is directly related to total concentration of soluble salts or osmotic potential of soil water (Flowers, 2004).

2.3.2. Increased Energy Utilization and Inhibition in Translocation

Extra expenditure of energy for osmotic adjustment or in repair system under salinity causes growth reduction (Pasternak, 1987). Seedlings in salinity stressed condition require relatively high energy for performing normal metabolism and, therefore, available sugar gets oxidized for release of energy for prolonging its survival. Decrease in dry weight of shoot under salt stress may be due to inhibition of hydrolyses of reserve/synthesized food or/and its translocation to the growing axis. Munns *et al.* (1995), Singh *et al.* (2001a) also indicated that salt within the plant reduces growth by causing premature senescence of old leaves and hence a reduced supply of assimilates to the growing regions. According to Narayana and Rao (1987), deleterious effect of salinity might be due to its adverse effect on translocation and partitioning of assimilates towards sink and other metabolic processes. The poor seed set in rice under saline condition is primarily due to reduced translocation of soluble carbohydrate into primary and secondary spikelets, accumulation of more Na and less K in the floral parts and inhibition of specific activity under salt stress due to reduced soluble carbohydrate content in different floral parts.

2.3.3. Ionic Imbalance

In saline soil, cation (Na) and anions (Cl, SO_4, HCO_3) are generally predominant either in the form of solution or exchange complexes. According to specific ion theory, soil salinity exerts a detrimental effect on plants through toxicity of one or more specific ions in the salts present in excess. Increased salinity level decreases K, Mg, Ca, P, Zn, Fe, B and Mn contents in shoot (Hassanien, 2000). Ca stabilizes the membrane integrity and permeability and high Na^+ concentration reduces exogenous Ca activity and availability to root cell membrane and further increases Na^+ uptake into root xylem. Salinity stress affects metabolism of plant cells leading to severe crop damage. Accumulation of Na and Cl in leaves reduces photosynthetic activity and brings ectopic ultra-structural and metabolic changes.

2.3.4. Inhibitions of Enzymes Activities

Salinity inhibits seedling germination by lowering water uptake and inhibition of activities of hydrolytic enzymes such as d-amylase, protease and ribonuclease in the endosperm (Dubey and Rani, 1990; Kumar *et al.*, 1996). Similarly, the activity of peroxidase is related to maintenance of cell membrane integrity through its environment in detoxifying H_2O_2 to water (Levitt, 1980), thus, modify the effect of free radicals under stress.

Singh *et al.* (2001a) found an increase in starch content with rise in salt stress levels chickpea seedlings. This may be the fact that with increase in salinity, breakdown of starch is less due to low amylase activity. Similar findings was recorded by Dubey (1982, 1983) that the amylase activity was lower at higher salt stress and activity was higher tolerant as compared to susceptible genotypes. The rate of starch degradation by amylase was found to be less in NaCl treated seeds of sorghum (Ogra and Baijal, 1982). Rathert (1983) also found that starch contents were higher in plants grown in saline habitats, because starch was the predominant carbohydrates which are affected by salinity. The starch synthesis activity in developing rice grains was found to be significantly inhibited under salinity stress (Abdullah *et al.*, 2001).

2.3.5. Change in Physiological Parameters

Root physiology is drastically affected by higher concentration of salt. Among the physiological parameters, leaf temperature increased with increasing salinity while osmotic potential, stomatal conductance and transpiration rate decreased with increase in soil salinity (Manjunatha *et al.*, 2003).

Photosynthesis drastically reduced as the chlorophyll production is severely affected under salinity environment. Root dehydration is increased due to the fact that the solute concentration is higher in the outside of the cell than it is inside of the cell. When salt goes into solution it interferes with water uptake by the plant. Water leaves the cell faster than it may be absorbed due to the fact that the sodium ions in the salt tend to replace both P and K ions in the soil, making the unavailable to the plant. With a high accumulation of salt around the roots, plant growth is adversely affected and the root die of drought.

In grain legumes, (*Vigna radiata, V. mungo* and *Cyamopsis tetragonoloba*), germination and length of seedlings were significantly reduced. The reduced physiological and metabolic activity in shoot as a result of high osmotic pressure in the vicinity of the seed is responsible for retardation of plumule and radicle growth (Maliwal and Paliwal, 1982).

Salt stress affects all the major processes such as growth, photosynthesis and energy and lipid metabolism (Ramoliya *et al.*, 2004; Parida and Das, 2005).

2.4. TOLERANCE MECHANISMS

Salt tolerance denotes the ability of a plant to prevent, reduce or over come the possible injurious effects caused directly or indirectly by excessive presence of soluble salt in the soil liquid phase or growth medium. The tolerance ability of plants against salinity may arise from tolerance to water stress or osmotic effects of salinity and tolerance to salinity induce ionic toxicity or toxic effects of salinity or a combination of both. The possible mechanisms of salinity tolerance by plant have been detailed in this section.

2.4.1. Cell Membrane Stability

Cell membrane stability is an indicator of degree of salt damage. The plant cell wall is a highly organized composite of many different polysaccharides, proteins and aromatic substances. The percentage of cell membrane electrolyte leakage increases with increasing salt concentrations. Hoque and Arima (2000) reported significant differences among the water chestnut species based on cell membrane stability and ranked them as:

Italian I (*Trapa natans*) > Japanese (*Trapa japonica*) > Chinese (*Trapa bicornis*) > Indian (*Trapa bispinosa*) > Italian II (*Trapa quadrispinosa*).

2.4.2. Osmotic Adjustment

Salt tolerance of plant is characterized by its capacity to endure the effects of excess salt in the medium of root growth i.e., plant can withstand a precise amount of salt without adverse effect. Protection from osmotic stress injury is accomplished by the accumulation of organic osmolytes. Tolerant plants accumulate osmoprotectants and osmolytes, which have been shown to accumulate in response to drought, salinity and cold stress. Osmolytes are molecules with low molecular mass and they could be quaternary amines, amino acids, or sugar alcohol. Enhanced accumulation of these molecules is known to increase the osmotic potential of the cell, which could help in combating water stress eventually they could stabilize membrane and other macromolecular structures. Not all higher plants accumulate the same substances, and even when they are able to synthesize a given substance not all of the accumulates in the quantities necessary to diminish the osmotic potential, thereby increasing the driving force for the entry of water inside the cells. A brief description of different organic osmolytes have been given below.

2.4.2.1. Glycinebetaine

The quaternary ammonium compound (QAC) glycinebetaine is an important osmoprotectant synthesized by many plants in response to environmental stresses. The QAC are considered to act as non-toxic cytoplasmic osmodium which maintains the intercellular osmotic balance between the cytoplasm and the NaCl in the vacuole. Some preliminary reports on cereals indicate that once synthesized, betaine may not further be metabolized and may be mobile within the plant. Because of these properties the glycinebetaine content could serve as a cumulative index of the internal water status of the plant with potential application in both plant breeding and crop management. Some evidences suggest that salt resistance may be correlated with the accumulation of QAC, particularly glycinebetaine, choline in a number of plant species.

Enhanced salt resistance, associated with an increased synthesis of glycinebetaine, was reported by Saneoka *et al*. (1995) and Colmer *et al*. (1995).

Colmer *et al.* (1995) showed that the enhanced salt resistance of amphidiploid *Triticum aestivum, Lophopyrum elogatum* relative to its wheat parent correlates with enhanced accumulation of glycinebetaine in the youngest leave. Thus, accumulation of betaine in youngest leaves could be used as an indicator for selection criterion for salt tolerance. Furthermore, foliar application of glycinebetaine improved water stressed soybean plants (Agboma *et al.*, 1997).

The biosynthetic pathway of glycinebetaine in dicotyledonous plants is probably and ancient origin (Weretilnyk *et al.*, 1989). All plants do not accumulate betaine under salt stress or in normal soil environment. Spinach, sunflower, maize, wheat and wheat relatives etc. are glycinebetaine accumulators, whereas lettuce, tobacco, soybean, rice etc. are glycinebetaine non-accumulators. Synthesis of glycinebetaine in most plants is a two-step reaction. In first step, choline is converted into betaine-aldehyde, and in second step, betaine-aldehyde is oxidized to glycinebetaine.

2.4.2.2. Manitol

Manitol is sugar alcohol and found play an important role in osmotic adjustment (OA) and provides enhanced tolerance in response to high salinity or water stress. Accumulation of manitol depends on its synthesis, its degradation by enzyme and long-distance transport. In celery, degradation of manitol occurs through the activity of a manitol dehydrogenase (*mtD*), and reduction of this enzyme leads to enhanced manitol accumulation, which take place during acclimation to salinity. *E. coli* (*mt1D*) gene has been tailored to tobacco plants resulted in enhanced accumulation of manitol, thereby led to increased tolerance of transgenic plants in response to salinity (Tarezynski *et al.,* 1993). Enhanced manitol biosynthesis in the chloroplast of transgenic plants resulted in increased tolerance of oxidative stress (Shen *et al.,* 1997). Thomas *et al.* (1995) germinated seeds of transgenic *Arabidopsis* transformed with *mt1D* in a medium supplemented with high amount of NaCl (250 mM); it showed that this gene confers high salt tolerance.

2.4.2.3. Proline

Another important class of organic solutes is amino acid, proline is among of them. Proline accumulation is found to be positively correlated with salt tolerance (Igarashi *et al.,* 1997). Salt stress is known to inhibit protein synthesis in various plant tissues (Everse *et al.,* 1997) due to the inhibition of amino acids incorporation into protein. Protein hydrolysis in salanized plants is always associated with increase in proline and free amino acids (Irigoyen *et al.,* 1992). It is positively correlated with salt tolerance in crop plants. Although the precise role of proline accumulation is still debated, proline is often considered to act as a compatible solute involved in osmotic adjustment. Accumulation of proline may be through an increase in its synthesis concomitantly with inhibition of its catabolism (Yoshida *et al.,* 1997). Some researchers reported that proline accumulation in plants may be a symptom of stress in less salinity tolerance species and suggested that it plays multiple roles in plant stress tolerance. Proline may act as a mediator of osmotic adjustment (Yoshida *et al.,* 1997), protect macromolecules during dehydration (Sanchez *et al.,* 1998) and serve as hydroxyl radical scavenger (Alia *et al.,* 1995). However, its role in imparting resistance to salt stress is controversial. In contradictory, many authors described proline as injury symptom rather than a salt tolerant sensor (Lutts *et al.,* 1996; Azooz *et al.,* 2004).

Biosynthesis of proline from glutamic acid involves two enzymes, the first is Δ^{-1} pyrroline-5-carboxylate synthetase (*P5CS*), and second is Δ^{-1} pyrroline-5-carboxylate reductase (*P5CR*). A backup pathway of proline biosynthesis uses ornithine as precursor.

2.4.2.4. Other amino acids

Gururaja Rao *et al.* (2001) found that kito-acids like a-kitoglutaric acid, pyruvic acid and oxaloacetic acid increased with salinity which play important role in synthesis of amino acids like glutamic acid, alanine, glutamine, aspartic acid and asparogine. These amino acids along with other inorganic ions play an important role in osmotic adjustment, thereby enabling plant species to exhibit good growth and canopy spread even under high salinity. Joshi *et al.* (2002) also found that the asparagines, aspartic acid, glycine, phenylalanine and proline increased in salt treated plants.

2.4.2.5. Other osmoprotectants

Wild plants that tolerate salt growth in saline environment have high inter cellular salt concentration. A major component of the osmotic adjustment in these cells is accomplished by ion uptake. The utilization of inorganic ions for osmotic adjustment would suggest that salt tolerant plant must be able to tolerate high levels of salts within their cells. However, enzyme extracted from these plants show high sensitivity to salt, suggesting that the plants are able to keep Na^+ ions away from the cytosol. Plants can use three strategies for the maintenance of a low cytosolic sodium concentration: sodium exclusion, compartmentation and secretion. One mechanism for sodium transport out of the cell is through operation of plasma membrane bound Na^+/H^+ antiports, as confirmed by the characterization of *S0S1*, a putative plasma membrane Na^+/H^+ antiport from *Arabidopsis thaliana* (Shi *et al.*, 2000). The efficient compartmentation of sodium is likewise accomplished through operation of vacuolar Na^+/H^+ antiports that are more potentially harmful ions from cytosol into large, internally acidic, tonoplast bound vacuoles (Apse *et al.*, 1999). These ions, in turn, act as an osmoticum within the vacuole to maintain water flow into the cell (Glenn *et al.*, 1999). Antiports use the protomotive force generated by vacuolar H^+ translocating enzymes, H^+ adenosine triphosphatare (*ATPase*) and H^+ inorganic phytophosphatase (PPiase), to couple downhill movement of H^+ (down its electrochemical potential) with uphill movement of Na^+ (against its electrochemical potential).

Carbohydrate accumulates in various plants under salinity condition (Balibrea *et al.*, 1997; Azooz *et al.*, 2004).

In succulent, Chenopodiceae, osmotic adjustment is achieved by accumulation of high levels of sodium and chloride in the shoots, accompanied by synthesis of substantial amounts of the compatible solute glycinebetaine (Gorham *et al.*, 1985).

2.4.3. Phytohormones

The major effect of salinity in the root environment has been attributed to reduce hormone delivery from root to leaves, which could induce an inhibition of crop growth. Hence, various growth promoting substances such as GA_3, IAA or kinetin have been used to overcome the drastic effect of salt stress (Singh *et al.*, 1994; Aldesuguy *et al.*, 1998). Presoaking of soybean seeds with IAA or GA_3 significantly mitigated the adverse effects of salinity on growth, tissue ion concentrations, and ionic balance (Zaidi and Singh, 1996). The most effective dose was 200 ppm of IAA to confer salt tolerance in soybean. Angrish *et al.* (2001) also reported that presoaking of wheat seeds with kinetin or GA_3 or their combination alleviated the deleterious effect of salinity. Further, it had been observed that the plant growth regulators-induced salinity stress alleviation was due to a concomitant increase in the tissue N content and nitrate reductase activity. Treatment with BA (Benzyladenine) improved photosynthetic pigment content in maize under salt stress (Shadi *et al.*, 2001), which indicated that BA increase salt tolerance in rice. Pretreatment with ABA resulted in appearance of several new proteins. ABA responsive proteins were heat-stable and their accumulation increased with an enhanced tolerance to the lethal levels of salinity. ABA treatment increased plant height and kinetin increased germination under saline environment in rice (Kim *et al.*, 1997). Higher concentration of ABA and kinetin also increased seedling starch and proline content of rice seedlings. Spraying with IAA in most cases resulted in pronounced increase of total free amino acids in root and shoot system in sorghum (Azooz *et al.*, 2004). They also observed that IAA markedly retarded the accumulation of proline where a significant increase was observed as compared with the corresponding salinized plants. Pre-treatment of ABA also enhance the survival of seedlings of *Eleusine coracana* at a lethal stress (Uma *et al.*, 1993). Exogenous application of salicylic acid counter acted on the deleterious effects of NaCl on shoot and root length, fresh and dry weight, leaf area, ribulosebisphosphate carboxylase activity, photosynthetic activity, pigment and sugar content of maize (Khodary, 2004). Salicylic acid enhances salt tolerance of crop by improving its growth.

Brassinosteroids have emerged as sixth group of phytohormones in plants (Rao *et al.*, 2002). It is considered as phytohormones with pleiotropic effect (Sasse, 1997), as it influences varied physiological process like growth, germination, flowering and senescence. The ability of these hormones to confer resistance to plants against abiotic stress is gaining much attention. The involvement of brassinosteroids to increase resistance to drought stress have been reported in sugar beat (Schilling *et al.*, 1991) and wheat (Sairam, 1994). The ability of epibrassinolide to confer resistance to bromegrass (*Bromus inermis*) cell cultures against high and low temperature has also been reported (Wilen *et al.*, 1995). The role of brassinosteriods in protecting plants against environmental stress will be an important research, therefore, clarifying the mode of action of brassinosteroids and may contribute greatly to the usage of brassinosteroids in agricultural production. Anuradha and Seeta Ram Roa (2002) found that Brassinolide reverse the inhibitory effect on germination and seedling growth of rice under saline environment. The activation of seedling growth by brassionlide under salinity stress was found to be associated with enhanced levels of nucleic acids, soluble proteins and free proline.

2.4.4. Enzymes

There are variable changes in enzyme activities in salt susceptible and tolerant genotypes under salinity stress. Higher activities of amylase, protease, peroxidase and catalase had been observed in tolerant genotypes of chick pea (*Cicer arientinum*) than in susceptible ones (Singh *et al.*, 2001a). Mandal and Singh (2000) also found higher amylase and peroxidase activities in tolerant cultivars of rice than the susceptible ones, but protease activity was lower in tolerant than susceptible cultivars. Salt stress stimulated the catalase activity in lentil (*Lens culinaris* Medic.) genotypes (Singh *et al.,* 2001a). Dubey (1994) also reported differences in salt tolerance when seedling of rice under increasing levels of salinity and distinct morphological differences as well as enzymatic activities. The changes in enzyme activities and their possible role in salt tolerance of *Suaeda nudiflora,* a succulent halophyte have been reported by Cherian and Reddy (2000). They found that the relative salt tolerance of *S. nudiflora* is characterized by massive accumulation of inorganic ions, succulency, stimulated activities of peroxidase, superoxide dismutase and altered response of ATPase and acid phosphatase to increase salinity. Dash and Panda (2000) also observed an increase in activity of phosphatase and ATPas at higher salt concentration in germinating wheat seeds.

Peroxidase activity increases in salinity stress, thus its increase may be due to increased enzyme synthesis and is useful for adoption under conditions requiring prevention of peroxidation of membrane lipids (Kalir *et al.,* 1984). Similar increase in peroxidase activity in salt tolerant varieties of *Pisum sativum* was observed by Olmos *et al.* (1994). Increase in peroxidase activity could be due to formation of H_2O_2 which could release the enzyme from the membrane structure (Zhang and Kirkhan, 1994).

Catalase activity showed an increasing trend with rise in salt concentration level (Singh, 2001a and b). Induction of several antioxidantation enzymes in development of salt tolerant cell lines of pea was reported by Olmos *et al.* (1994). Many salt tolerant species are reported to reduce the membrane damage by increasing enzymatic defense against active oxyen radicales (Smirnoff, 1993). Enhanced activity of catalase was reported to be essential for the survival of the halophyte *Halimione portulacoides* (Kalir and Poljakoff-Mayber, 1981).

2.4.5. Vitamins in Salt Tolerance

Vitamins also play a major role in salinity stress in plants. Radi *et al.* (2001) found that the rate of transpiration significantly reduced in vitamin B6 (pyridoxine) or PP (nicotinamide) treated plants at all salinity levels compared with untreated salinized plants. However, vitamin treatments generally exhibited a significant increase in the leaf area and dry matter yield of salt stressed bean lines resulting in a complete alleviation of the adverse effect of the high levels of salinity. Further, Ismail and Azooz (2002) found that seed soaking in either vitamin B6 or vitamin PP alleviated the accumulation of carbohydrates in the beans. Vitamin treatment also increased the protein content in bean. Soaking of seeds in one of the two vitamins exhibited variable response in the production of the free amino acids and proline in seedlings of bean. They also noticed that vitamin B6 was more effective than PP. Wang *et al.* (1991) also reported increased fat and protein metabolism within the vitamin B6 treatment of coton seeds during germination. They also found that vitamin B6 (10 ppm) treated seeds increased respiration by 33.5% during germination in saline and alkali soils. Consequently, the seedling emergence rate was 14.1-20.0% higher for treated seeds, and subsequently cotton yields were increased by 13.0-16.5%.

2.4.6. Ion Accumulation and Ion Balance

Plants accumulate sodium ion in the cytoplasm by transporting it to the vocule using older leaves as sink, thus avoid salt accumulation in sensitive tissues, such as meristems. Ion transport into the vacuole works against a concentration gradient and thus requires energy. This is achieved by coupling the transport the protein to a proton pump, transporting H^+ ions in the opposite direction, hence it is known as *'antiport'* protein. Rush and Epstein (1976, 1981) found ion accumulation in the wild tomato species, *Lycopersicon choesmanii* to be reflection of the halophytic nature. Whereas, Sacher *et al.* (1982, 1983) could not established relationship between ion accumulation and salt tolerance in crosses of *Lycopersicon peruvianum* and its cultivated descendent *L. esculentum*. When halophytes and non-halophytes are compared ion accumulation appears to be superior tolerance mechanism for growth in a saline habitat (Greenway, 1965). Indica varieties are generally more tolerant to salinity than japonica rice varieties (Lee *et al.*, 2003). Tolerant indica varieties are good Na^+ excluder, absorb high amount of K and maintain a low Na : K ratio in shoot. Plants growing under salt stress show marked imbalance in their ionic composition (Greenway and Munnus, 1990). There exists an inverse relationship between salt tolerance and sodium accumulation in non-halophytes and it may be used as an index for characterizing salt tolerant genotypes. Tolerant genotypes not only absorb less Na but also show minimum imbalance in K, Ca and Mg status of the shoot. High K/Na, Ca/Na and Mn/Na ratios (Balasubramanian and Rao, 1977) or low Na^+/K^+ (Won *et al.*, 1992; Chen *et al.*, 2007), Na^+/Ca^+ and Na/Mg ratios (Joshi *et al.*, 1980) may be used as criteria for salt tolerance. Yilmaz *et al.* (2004) also reported negative association with leaf K^+, Na^+ and K^+/Na^+ with increasing NaCl concentrations. Pandey and Sharma (2002) found that the salt tolerant genotypes of rice had relatively lower K^+/Na^+, K^+/Ca^+ and K^+/Mg^+ ratio as compared to salt sensitive genotypes which have medium to high ratios in the leaves.

2.4.7. Ion Exclusion

Ion exclusion rate also play an important role in salt tolerance in some crop species. Differences have been observed among the crops as well as among the genotypes within the species for ion exclusion. Tsuchiya *et al.* (1995a) studied ion exclusion in excised nodal roots of rice. They reported highest exclusion rate in maize followed by sorghum and rice. The exclusion rate of different ions were in the order of $Mg^{2+} > Ca^{2+} > Na^{2+} > Cl^{-} > K^{+}$. Further, Tsuchiya *et al.* (1995b) reported that at the high pressure, varieties with higher amount of lignin in roots showed lower Na^{2+} and Cl^{-} concentrations in the exudate. Their findings suggested that the presence of negative pressure in roots which can be provided by the transpiration or osmotic adjustment, thus it will increase the salt exclusion rate.

Bicellular leaf salt glands were observed in all species of the genus *Zoysia*, lying recumbent to the leaf surface in parallel rows at top intercostals ridges. Marcum *et al.* (1998) observed negative correlation of salinity tolerance with the leaf sap Na^{2+} concentration and positive correlation with salt gland Na^{2+} secretion rate and leaf gland density. It is conventionally assumed that an ability to exclude Na^{2+} correlates with plant salt tolerance (Munns and James 2003; Garthwaite *et al.*, 2005). The findings indicated that salinity tolerance in the *Zoysia* genus is associated with the shoot saline ion exclusion via leaf salt gland secretion, which in turn is related to leaf gland density.

2.4.8. Silicon Supply

Influence of silicon (Si) supply on salt stress in tomato has been studied by Al-Aghabary *et al.* (2004). They reported that Si partially offset the negative impact of NaCl stress, increasing tolerance in tomato plants to NaCl by increasing superoxide dismutase (SOD) and catalase (CAT) activities, chlorophyll content and photochemical efficiency of Photosystem II. Salt stress decreases SOD and CAT activities and soluble protein in leaves. Enhanced activities of SOD and CAT by Si addition may protect the plant tissue from oxidative damage induced by salt, thus, mitigating salt toxicity and improving the growth of tomato plants.

Silicon has been found to minimize the salinity effects and transpiration rate in plants (Tisdale *et al.*, 1993; Epstein, 1994; Marschner, 1995) and Na^{+} uptake in rice under salinity (Yeo *et al.*, 1998; Guruaja Rao, 2001). It protects leaves from UV radiation damage by filtering the harmful UV rays (Tisdale *et al.*, 1993) and increases the leaf erectness and seed yield in paddy (Yoshida *et al.*, 1969). Silicate supplementation has been found to increase the stomatal conductance of salt treated plants, photosynthesis and reduce Na uptake by the partial blockage of the transpirational bypass flow, the pathway by which large portion of the uptake of Na occurs in rice (Yeo *et al.*, 1998; Gururaja Rao, 2001). Trivedi *et al.* (2004) found that the silicon application lowered sodium transport into the shoots resulted in higher chlorophyll (by minimizing sodium toxicity) and tissue tolerance that in turn help in carbohydrate synthesis.

2.4.9. Leaf Characteristics

2.4.9.1. Xerophytic nature of leaves

Phonix dactylifera is a halophyte and adapted to salt stress condition by the evaluation of an effective osmoregulation for salt tolerance only to a limited range in the range of salt stress. Ramoliya and Pandey (2003) suggested that the ability of this species to thrive in dry regions is conferred by the xerophytic feature of its leaves.

2.4.9.2. Epicuticular wax

Rao (1989) screened *Cajanus cajan* for salt tolerance based on the amount and composition of leaf epicuticular wax and cuticular transpiration. He found that epicuticular wax accumulated in all the cultivars under salinity. The cultivar ICP 7035 and ICP 7065 showing higher amounts in thin layer chromatography analysis revealed the presence of high amount of primary and secondary alcohols, aldehydes and β-diketones under salinity, whereas control plants showed high fatty acid content. It was also noted that the cuticular transpiration was reduced in all the cultivars under salinity when compared to their respective controls. Accumulation of epicuticular wax causing an additive reduction in cuticular transpiration may play a vital role in inducing tolerance to salinity in those cultivars.

2.4.10. Crop Stages

There are considerable variations for salt tolerance based on different stages of the same genotype or crop. *Brassica* exhibited susceptibility to salinity at seedling emergence and early seedling growth, but are relatively more tolerant at later growth stages, particularly at flowering to formation of siliqua (Kumar, 1995).

Some salt tolerant crop plant find difficult to germinate in presence of high salt concentration, otherwise those plants are salt tolerant. Germination is the first critical phase on which success of crop depends. Therefore, understanding the mechanism during seed germination and at early growth stages is essential for introduction of salt tolerant species (Leith, 1999). Somehow, a rice variety is intolerant during germination and vegetative growth, but very tolerant during the reproductive development (Heenan *et al.,* 1988). IR8 was found to be sensitive during germination, otherwise relatively tolerant at the seedling stage (Akbar and Yabuno, 1977). Joshi et al. (2002) reported that the seeds of *Helechoa setulosa* failed to germinate even at moderate salinity (8 dSm^{-1}) of sea water but the young plant could withstand salinity upto 40 dSm^{-1} in growth medium. So, if the species is to be introduced on saline wasteland, the young plants are to be transplanted on saline soils with matching salinity levels. *Suaeda nudiflora* Moq is highly tolerant to salinity, and it is very sensitive during the germination stage of development. To overcome this problem, Cherian and Reddy (2002) developed micropropagation protocol, which after hardening can be directly transplanted in the field.

2.4.11. Biological Inoculums

Various soil reclamation technologies for salt tolerance, such as chemical and engineering methods have been developed. Those methods are expensive and temporary. Biological amelioration is the concern of the days, which includes blue-green algae (Cyanobacterium), *Thiobacilli* and mycorrhizas. Application of organic amendments and crop residues growing salt tolerant crops and planting N-fixing trees overcome the effects salinity of salt effected soils. Mycorrhizal root colonization occurs irrespective of salt tolerant or intolerant cultivars under stressed or non-stressed soils.

Application of bioinoculants Arbuscular mycorrhizal (AM) fungi and one of the plant growth promoting rhizobacteria such as *Azospirillum, Agrobacteria, Pseudomonas,* several Gram positive *Bacillus* are environment friendly, energy efficient and economically viable approach for reclaiming wastelands and increasing biomass production. The beneficial effects of these bacteria in combination with fungi have been reported by a number of workers (Tian *et al.,* 2004; Patreze and Cordeiro, 2004; Domenech *et al.,* 2004). Many studies have demonstrated that inoculation with AM fungi improves growth of plants under a variety of salinity stress conditions (Diallo *et al.,* 2001; Burke *et al.,* 2003; Tain *et al.,* 2004). Preinoculation of transplants with AM fungi can help to alleviate deleterious effects of saline soils on crop yield (Cantrell and Linderman, 2001; Al-Karaki and Hammad, 2001; Rabie and Almadini, 2005). Thus, mycorrhizal inoculated plants acquire greater tolerance ability to salt stress. Mycorrhizae were involved in protection against salt stress via better access to nutritional status (Zandavalli *et al.*, 2004), and modification of plant physiology i.e., osmotic modifications (Rao and Tak, 2002) and photosynthesis (Meroguihur *et al.*, 2002). To some extent, these AM fungi have been considered as bio-ameliorators of saline soils (Yano-Melo *et al.*, 2003; Tain *et al.*, 2004).

In the biological inoculation technology, bacterial-mycorrhizal legume tripartite symbiosis in saline condition had been documented (Rabie and Almadimi, 2005) as well as the effects of dual inoculation with *Azospirillum brasilense* and Arbuscular mycorrhizal fungus *Glomus clarum* on the host plants (*Vicia faba*). *Glomus clarum* significantly increased tolerance to salinity, mycorrhizal dependency, phosphorus level, phosphatase enzymes, nodule number, nitrogen level, protein content and nitrogenase enzymes of all salanized faba plants in comparison with control and non-*G. clarum* inoculated plants either in absence or presence of *A. brasilense.* In *G. clarum* inoculated plants, K^+ / Na^+, Mg^+/Na^+ and Ca^+/Na^+ ratios were higher.

2.5. TSUNAMI IN ANDAMAN AND NICOBAR ISLANDS

The 2004 Indian Ocean earthquake, known by the scientific community as the *Sumatra-Andaman earthquake* (Lay *et al.*, 2005), was an undersea earthquake that occurred on December 26, 2004, with an epicentre of the west coast of Sumatra, Indonesia. The earthquake triggered a series of devastating tsunamis along the coasts of most landmasses bordering the Indian Ocean, killing large numbers of people and inundating coastal communities across South and Southeast Asia, including parts of Indonesia, Sri Lanka, India (Andaman and Nicobar Islands, coastal areas of Tamil Nadu and Kerala), and Thailand.

The magnitude of the earthquake was originally recorded as 9.0, but has been increased to between 9.1 and 9.3. At this magnitude, it is the second largest earthquake ever recorded on a seismograph. This earthquake was also reported to be the longest duration of faulting ever observed, lasting between 500 and 600 seconds (8.3 to 10 minutes), and it was large enough that it caused the entire planet to vibrate as much as half an inch, or over a centimeter (Walton, 2005). It also triggered earthquakes in other locations as far away as Alaska (West *et al.*, 2005).

2.5.1. Immediate Effect of Tsunami on Soil and Plants

The devastating wave of tsunami trounced the standing crops, particularly paddy (Fig. 2.1C). The coconut trees along coastal line uprooted/damaged heavily. It damaged more than 3900 ha of agricultural land of these islands. Land lost its productive potential due to salt waterlogging in embankment areas and/or low lying areas; even possibility of getting permanently water logged due to large scale embankment erosion and lost fertility of coastal lands due to sediment deposits. The tsunami tidal waves has transported large volume of sea water into inland water bodies and also created large tidal pools of sea water which has percolated into coastal fresh water aquifers (aquifers are the major sources of drinking water in coastal areas) and salinize them. It also led sedimentation of drainage channels, creeks' mouths, lagoons and water ways. In many areas, sediments of fine grey layer to grayish brown layer deposits of varying depths (5 cm - 35 cm) in the low lying areas was observed by the visiting team from IARI. Residual high content of salt is in the layers of clay and silt left behind by the tsunami waves. These layers can be easily identified by cracks that spread across the surface of the soil.

In deep brown coastal soil zones the quality of shallow ground water has deteriorated. EC of shallow ground water (25m below ground level) changed from the pre-tsunami value of 0.5 dS/m to the post tsunami value of 4.8 dS/m.

2.5.2. Impact of Soil Health and Crop Growth

The top silt litter over the salt affected soil reduced the permeability of the soil and thus water infiltration into deeper layer of crop extraction. Cracks are formed high salt content in the soil. Once the rain water flow through the cracks and by pass the whole soil surface, water is lost and the salt content is also reduced due to water application. It will reduce higher amount of irrigation water for leaching as well as growing good crops stand and yield. During dry season, salt appears on the surface in the tsunami effected areas.

In order to improve such situation removal of salt by surface scraping or flush or leach by huge water application is alternative to quick reclamation. If tillage operation is done through mechanical methods the salt will further mix and this will take long time to reclaim the soil. Traditional tillage practices and controlled irrigation water applications will help to keep the salt away from crop root zone.

In most of the affected areas water intrusion changed the soil conditions and turned it to be unfavourable for immediate crop cultivation. The deep-brown coastal-fertile alluvial soil, which has slightly higher amount of clay content, normal pH and EC, and better drainage facilities have been altered after tsunami. The EC of such silty clay tsunami deposits vary between 6.7 dS/m and 23.7 dS/m. A saline soil has an EC of the soil saturated extract more than 4 dS/m. Water containing EC of 1 dS/m is approximately 640 ppm (9.5 ton/ha salt in 6 inch soil layer). The primary effect of total salinity is a reduction of water availability to root through osmotic effects. Yields of most crops are substantially reduced as EC increases above 4.0 dS/m. Post tsunami soil EC level have increased by 5-15 times compare to normal season, while soil pH levels increased marginally. However, the sub-soil EC of the silty clay deposited is remarkably less than that of the surface soil. In few areas there was unique impact of stagnation of sea sediments, debris and sea water. The thick slushy black deposit on the soil surface causing heavy damage to the soil structure and standing crop.

Fig. 2.1. Effect of tsunami on crops in Andaman and Nicobar Islands. **a)** Coconut grove before tsunami, **b)** Coconut grove one year after tsunami, **c)** Rice field at the next day of tsunami. *(For colour version of this figure see page 533)*

2.6. GENETIC VARIABILITY OF SALINITY TOLERANCE

Differences in salt tolerance exist among different genera and species as well as within the same species. Tolerance ability of crop plants varies greatly in their ability to grow and yield under saline conditions. Plant growth is affected by the salinity level of the soil, which is a function of the soils and initial salt content as well as of the salinity of the water. Some crops are very sensitive to salinity while others are very tolerant. Genetic variability of some crops has been detailed in this section.

2.6.1. Rice (*Oryza sativa*)

Rice is generally salt sensitive crop; however, there are large numbers of varieties which show variation in salt tolerance (Datta, 1972). A reliable estimation of magnitude of this variability, especially within the species, is critical to the success of breeding programmes for salinity resistance. Blum (1988) has suggested convincingly that the most reliable approach will be to evaluate the extent of yield of genetically different materials under a range of salinity levels. Reduction in grain yield due to salinity is found to be positively correlated with Na^+ contents in panicles after booting stage (Asch *et al.*, 1999). Both additive and dominance effects govern by the most characters associated with salt tolerance (Gregorio and Senadhira, 1993; Lee, 1995). Characters such as shoot length, Na^+ and K^+ content in shoots and roots showed significant and high additive effects at seedling stage. At maturity stage, plant height and yield per plant showed significantly high additive effects indicating importance of additive gene action (Moeljopawiro and Ikehashi, 1981; Akbar *et al.*, 1985; Mishra *et al.*, 1998; Narayanan *et al.*, 1990). Salinity tolerance measured as low Na/K ratio in the shoot was reported to be governed by both additive and dominance effects (Gregorio and Sendhira, 1993). The population derived from Basmati 370, CSR 10 displayed skewed distribution whereas, other crosses showed a non-skewed distribution indicating the involvement of a few major genes along with many minor genes governing salinity tolerance (Mishra *et al.*, 1998).

Genetic studies at IRRI indicated that both additive and dominance effects are important in the inheritance of almost all characters associated with salt tolerance in rice (Mishra *et al.*, 1990; Gregorio and Senadhira, 1993; Lee, 1995). At the seedling stage under saline condition, characters associated with salinity tolerance such as shoot and root showed highly significant additive effects as well as the heritability in these characters is low. Characters at maturing stage, such as plant height and yield per plant showed highly significant additive effects suggesting the greater importance of additive gene action in the inheritance of these characters (Moeljopawiro and Ikehashi, 1981; Akbar *et al.*, 1986; Mishra *et al.*, 1990).

Jones and Stenhouse (1984) reported that the maternal factors play an important role in determining root growth of rice hybrids under saline soil. In most of the rice hybrids salt tolerance is controlled by dominant genes, ranging in action from partial dominance to over dominance (Lavrichenko, 1985). Akbar *et al.* (1985) and Mosina (1986) also have reported additive and dominance effects to be important in the inheritance of quantitative and qualitative characters determining salt tolerance. Highly significant additive effects were observed for plant height and yield per plant with high heritability value. Jones (1986) reported that a simple additive dominance model with maternal components adequately describes the variation in swamp rice.

Akbar and Yabuno (1977) reported that panicle sterility in rice under salt stress is a dominant character controlled by a small number of genes. Moeljopawiro and Ikehashi (1981) reported over dominance for tolerance despite of the high environmental fluctuation; many lines were more tolerant than the presents.

The study of Gregorio and Senadhira (1993) using a 9-parent complete diallel analysis of N/K ratio in shoots in rice showed that both additive and dominant gene-effect govern salinity tolerance (low Na/K ratio). The trait exhibited over dominance and is conrolled by at least two groups of dominant genes. Environmental effects were large and narrow sense heritability was low. Genetic combining ability study by Gregorio and Senadhira (1993) showed the significance of both general combining ability (GCA) and specific combining ability (SCA) affect the inheritance of salinity tolerance in rice.

Narayanan and Sree Rangaswamy (1991) reported additive dominant gene effects for days to flowering, plant height, tiller number, panicle length, number of spikelets/panicle, 1000 grain weight and dry matter accumulation under both normal and salinized conditions. However, the additive effect was significant for grain yield only in the salinized situation. Salinization suppressed to a greater extent the dominance effect, contributing to the expression of grain yields, suggesting that varieties with predominantly mere additive genes for grain yield would perform better in saline soils. The genetic component analysis revealed that a low Na/K ratio is governed by both additive and dominant gene effects (Gregorio and Senadhira, 1993). The environmental effects were large and the heritability of trait was low. Their findings suggested that when breeding for salt tolerance, selection must be done in a later generation and under controlled conditions in order to minimize environmental effects. Combining ability analysis revealed that both GCA and SCA effects were important in the genetics of salt tolerance. The presence of reciprocal effects among crosses necessitates the use of susceptible parents as male in hybridization programmes. Lee *et al.* (1997) reported higher mean square of GCA than those for SCA in *japonica* rice seedlings parameters, indicating the predominance of the additive gene effect. Among the tolerant varieties, Gaori and Namyang 7 was good combiner for improving salinity tolerance at the seedling stage.

In a complete diallel cross, Mahmood *et al.* (2004) reported significant contribution of additive and dominant gene effects to the total heritable variation for plant height, panicle length, sodium and potassium concentrations in shoot. Additive genetic effects were important for expression of variation of productive tillers, number of primary branches per panicle, while dominant genetic effects were important for the expression of variation of number of days to maturity. Zheng (2004) observed higher broad-sense and realized heritability of tiller per plant suggesting that the character is genetically controlled. Moderately high generalized heritability indicates the effectiveness of early selection for salt tolerance if tiller number is used as selection criteria.

2.6.2. Wheat (*Triticum* sp.) and Rye (*Secale cereale*)

Genetic studies on wheat species by Gorham (1988) and Gorham *et al.* (1990) demonstrated a major effect of gene(s), located on the long arm of chromosome *4D*, on the ability to discriminate between Na and K uptake and transport to the roots, reducing the Na content of the plant and increasing its salt tolerance. Rye and triticale accessions grown in saline hydroponic culture exhibited low Na and high K concentrations in their leaves which are characteristic features of the enhanced K/Na discrimination trait found in *D* genome of wheat (Gorham, 1990). This trait was not consistently improved by the presence of *D* genome in octaploid triticale or in *D* genome substitution lines of hexaploid triticale. The presence of rye genome did not significantly affect anion concentrations within the leaves. Cocco *et al.* (1976) reported that with increase in ploidy level the uptake efficiency of excised roots of wheat (2n, 4n and 6n) increased in the maximum uptake velocity.

Some triticales inherit salt resistance from its parent while others are susceptible similar to the wheat parent. Bisnoi and Pancholy (1980) determined the effects of salinity on germination of triticale and rye showed similar trend in reduction of germination as salinity increased from 0 to 1.5% NaCl and exhibited fairly high salt tolerance. However, triticale 6TA131 and wheat parent showed a significant drop in germination even at 1.5% NaCl concentration and were more susceptible to salt injury.

2.6.3. Sorghum (*Sorghum bicolor*)

Salt tolerance in sorghum appeared to be controlled by dominant genes and non-allelic gene interaction was found to be important during early growth stage (Daddur, 1977). Ratanodilok *et al.* (1978) observed moderate to high broad sense heritability estimates for salt tolerance in sorghum. The parent, midparent and F_2 values revealed that salt tolerance might be controlled by complementary gene action, incomplete dominance or additive effects of several genes. Additive and non-additive effects were important for the expression of variation under low and high salinity levels. Data on root lengths of two weeks old seedlings of 5 accessions germinated in 50-200 mM NaCl solutions were used to estimate broad sense heritability for salinity tolerance (Azhar and McNeilly, 1988) and they suggested 'selection' for improvement of tolerance in crop plants. Spivakov (1990) reported that in sorghum salt resistance is inherited mainly along the male line and is controlled by several non-allelic genes which determine the dominant or over dominant expression of resistance.

2.6.4. Barley (*Hordium* sp.)

Salt tolerance in barley is a heritable character (Epstein, 1977). Foster *et al.* (1990) located the genes influencing salt tolerance to specific chromosomes of the *H* and *Hch* genomes of *Hordeum valgare* and *H. chilene,* respectively. Genes with positive effects for salt tolerance were located on chromosomes *4H* and *5H* of *H. valgare* and *1Hch, 4Hch* and *5Hch* of *H. chilense.* Mahgoub and Sayed Ahmed (1999) reported significant differences for Na and K levels in root, stem and leaves of the parents and hybrids in sorghum. The expression of these physiological traits in F_1 hybrids indicated that these physiological traits were controlled by dominant gene(s).

2.6.5. Maize (*Zea mays*)

Hybrid maize varieties showed significant variation under salt stress environment (Eker *et al.,* 2006). The higher salt tolerance in maize varieties based on the severity of leaf symptoms was associated with significantly lower Na concentrations in shoots. The K/Na and Ca/Na ratios were significantly greater in the most tolerant varieties. Under salt treatment significant correlations were found between K/Na ratios and shoot dry matter production (r = 0.444***). K/Na ratios and leaf damage (r = 0.411***) and Ca/Na ratios and shoot dry matter production (r = 0.444***). Their result indicated existence of a large genotypic variation in tolerance to NaCl toxicity in maize.

2.6.6. Tomato

Lyon (1941) reported that single gene controls the presence and absence of tolerance in tomato. He found that *Lycopersicon pimpinellifolium* was less sensitive than *L. esculantum* to Na_2SO_4 and F_1 hybrids between these two species were identical in salt tolerance to sensitive parent. Survival and yield data for selections from a cross between tolerant wild tomato *L. cheesmani* and a domestic sensitive cultivar of *L. esculentum* indicated that salt tolerance is genetically controlled and can be transferred between species (Epstein, 1977; Rush and Epstein, 1976). They also reported that salt tolerance in tomatoes is heritable. Generation mean analysis of an interspecific cross in tomato, indicated that there were no significant embryo (additive, dominance or epistatic) effects on germination of seed under salt stress (Foolad and Jones, 1991). Variance component analysis indicated a large genetic variation with additive gene action as the predominant component. That variance was attributed to endosperm additive effects on germination ability under salt stress. Narrow senses of heritability estimates were moderately high. Further, Foolad and Jones (1992) found similarity of the estimates obtained in both parent-offspring regression suggesting the absence of significant dominant genetic effects and when considered with the moderately high heritability estimates, reasonably rapid response to selection in the early segregating generations would be expected for the important seed characters in tomato.

2.6.7. Other Crops

Seed germination of sunflower under salinity appeared to be dominantly controlled (Miller, 1995). Shao *et al.* (1994) reported that salt tolerance of soybean was controlled by a pair of genes. Shoot length, Na and Ca content of the shoots and shoot and root dry weights showed significant additive effects with a high degree of heritability. Ashraf and Waheed (1998) observed significant additive and non-additive effects with the prevalence of additive gene action over dominance for both yield components. Estimates of broad and narrow sense of heritability were high, indicating potential for improvement in salinity tolerance in lentil (*Lens culinaris*) through selection.

2.7. SOURCES OF TOLERANCE

Salt causes water to be removed out of plants through exosmosis, especially on young leaves. This often results in marginal burning and loss of leaves in plant that are salt tolerant. Salt tolerance of a plant relates to the resistance and ability to grow under the condition of high winds, salt spray, alkaline soils, infertile sandy soils, and salt in irrigation water. The tolerance of a given plant to salt may be affected if any of those five conditions become extreme. The salt tolerant ability varies among the species as well as among the genotypes within the species, and may be classified as follows:

2.7.1. High salt tolerance

The plant that can withstand most salt conditions and exposure to ocean or high salt in irrigation water (up to 3500 ppm salt). Salt tolerant plants are highly resistant to salt drift.

2.7.2. Medium salt tolerance

The plant that can withstand some salt in the irrigation water (up to 2500 ppm salt); the yields of plants are restricted by salinity. Moderately salt tolerant plants tolerate some salt spray.

2.7.3. Low salt tolerance

The plant that can withstand salt in irrigation water less than 2000 ppm salt. The yield of crop is restricted to a certain limit.

Table 2.3 is a guide to select salt tolerant crops in salt effected areas. Effect of salt concentrations in irrigated water on landscape plants have been given below (Rockledge Gardes Information Sheet, 2153 So. US Hwy#1, Rockledge, FL32955.321.636.7662):

- A reading of 300 ppm will injure orchids
- A reading of 1000 + ppm will injure *azaleas* (a shrub with brightly coloured flowers)
- A reading of 1500 + ppm will damage many foliage plants
- A reading of 1800 + ppm will injure many woody plants (including rose and citrus)
- A reading of 2200 + ppm make the water unsuitable for most of the plants

Table 2.3. List of crop species based on their degree of salt tolerance in saline soils

Scientific name	Common name	Thresh hold level dS/m	Rating
I. Cereal crops			
Avena sativa	Oats	-	MT
Hordium valgare	Barley (e)	8.0	T
Oryza sativa	Rice	3.0	S
Secale cereale	Rye	11.4	T
Setaria italica	Foxtail Millet	-	MS
Sorghum bicolor	Sorghum	6.8	MT
Triticum aestivum	Wheat	6.0	MT
Triticum aestivum	Wheat (Semi-dwarf)	8.6	T
Triticum turgidum	Wheat, Durum	5.9	T
Zea mays	Maize	1.7	MS
II. Vegetable crops			
Abelomoschus esculentus	Okra	-	S
Allium cepa	Onion	1.2	S
Asparagus officinalis	Asparagus	4.1	T
Beata valgaris	Beet red (h)	4.0	MT
Brassica napa	Turnip	0.9	MS
Brassica oleraceae botrytis	Cauliflower	-	MS
Brassica oleraceae botrytis	Broccoli	2.8	MS
Brassica oleraceae capitata	Cabbage	1.8	MS
Brassica oleraceae germifera	Brussel sprouts	-	MS
Capsicum annum	Pepper	1.5	MS
Citrullus lanatus	Water melon	-	MS
Cucumis melo	Musk melon	-	MS
Cucurbita pepo	Pumkin	-	MS
Cucurbita pepo melopepo	Squash, scallop	3.2	MS
Cucurbita pepo melopepo	Squash, guchini	4.7	MT
Daucus carota	Carrot	1.0	MS
Ipomea batatas	Sweet potato	1.5	MS
Latuca sativa	Lettuce	1.3	MS
Lycopersicon esculentum	Tomato	2.5	MS
Lycopersicon esculentum var. cerasiforme	Tomato, cherry	1.7	MS

Contd...

Table 2.3. Contd...

Phaseolus valgare	Bean	1.0	S
Pisum sativum	Garden pea	-	S
Raphanus sativa	Radish	1.2	MS
Solanum melongena	Brinjal/ Egg plant	1.1	MS
Solanum tuberosum	Potato	1.7	MS
Spinacia olericeae	Spinach	2.0	MS
Vicia faba	Broad bean	1.6	MS
Vigna radiata	Bean (Mung)	1.8	S
Vigna unguiculata	Cowpea	4.9	MT
III. Oil seed crops			
Arachis hypogaea	Peanut	3.2	MS
Cantharus tinctorius	Safflower	-	MT
Glycine max	Soybean	5.0	MT
Heliathus annus	Sunflower	-	MS
Ricinus communis	Castor	-	MS
Sesamum indicum	Sesame (m)	-	S
IV. Fiber crops			
Gossypium hirsutum	Cotton	7.7	T
Hibiscus cannabinus	Kenaf	8.1	MT
Linum usitassimum	Flax	1.7	MS
VI. Fruit crops			
Ananas comosus	Pineapple	-	MT
Annola cherimola	Cherimoya	-	S
Carica papaya	Papaya	-	MT
Citrus limon	Lemon (e)	-	S
Citrus maxima	Pummelo	-	S
Citrus sinensis	Orange	1.7	S
Citrus sp.	Lime	-	S
Citrus paradise	Grapefruit	1.8	S
Diospyros virginiana	Perssimon	-	S
Eriobotrya japonica	Loquat	-	S
Ficus carica	Fig	-	MT
Mangifera indica	Mango	-	S
Malus sylvestris	Apple (e)	-	S

Contd...

Table 2.3. Contd...

Olea europeaea	Olive	-	MT
Parthenium argentatum	Guayul	15.0	T
Pasflora edulis	Passion fruit	-	S
Persea americana	Avacado (e)	-	S
Phonix dactylifera	Date palm	4.0	T
Prunus ameniaca	Apricot	1.6	S
Prunus avium	Cherry, sweet	-	S
Prunus besseyi	Cherry, sand	-	S
Prunus communis	Pear	-	S
Prunus domestica	Plum (e)	1.5	S
Prunus ductalis	Almond	1.5	S
Prunus persica	Peach	1.7	S
Punica granatum	Pomegranate	-	MT
Rubus indacus	Raspberry	-	S
Ribes sp.	Currant	-	S
Ribes sp.	Gooseberry	-	S
Rubus sp.	Blackberry	1.5	S
Rubus ursuilis	Boysenberry	1.5	S
Simmondsia chinensis	Jojoba (e)	-	T
Vitis sp.	Grape (e)	1.5	MS
Ziziphus jujuba	Jujube	-	MT
V. Forage Crops			
Agropyron cristatum	Wheatgrass, fairway crested	7.5	T
A. elonga	Wheatgrass, tall	7.5	T
A. intermedium	Wheatgrass, intermediate	-	MT
A. smithii	Wheatgrass, Western	-	MT
A. sibiricum	Wheatgrass, standard crested	3.5	MT
A. trachycaulum	Wheatgrass, slender	-	MT
Agrostis stolonifera	Bentgrass	-	MS
Alopecurus pratensis	Foxtail, meadew	1.5	MS
Astregalus	Milkvetch, Cieer	-	MS
Arrhenatherum danthonia	Oatgrass, tall	-	MS
Avena sativa	Oat (forage)	-	MS
Bouteloua gracilis	Grama, blue	-	MS

Contd...

Table 2.3. Contd...

Brassica napus	Rape	-	MT
Bromus inermis	Brome, smooth	-	MS
B. marginatus	Brome, mountain	-	MT
B. unioloides	Rescuegrass	-	MT
Cenchrus ciliaris	Buffelgrass	-	MS
Cloris gayana	Rhodegrass	-	MT
Cynodon ductylon	Bumudagrass	6.9	MS
Dactylis glomerata	Orchardgrass	1.5	MS
Dicanthium aristatum	Bluestem, Angleton	-	MS
Distichlis stricta	Saltgrass, dessert	-	T
Diplachne fusca	Kallargrass	-	T
Elymus angustus	Wildrye, Altai	-	T
E. canadensis	Wildrye, Canadian	-	MT
E. junceus	Wildrye, Russian	-	T
E. triticoides	Wildrye, breadless	2.7	MT
Eragrostis sp.	Lovegrass	2.0	MS
Festuca elatior	Fescue, tall	3.9	MT
F. pratensis	Festuca, meadew	-	MT
Hordium vulgare	Barley, forage	6.0	MT
Lolium italicum multiflorum	Ryegrass, Italian	-	MT
Lolium perenne	Ryegrass perennial	5.6	MT
Lotus uliginosus	Trefoil, big	2.3	MS
Macroptilium atroperpureum	Sirata	-	MS
Medicago sativa	Alfalfa	2.0	MS
Melilotus alba	Clover, Hubam	-	MT
Paspalum dilatatum	-	-	MS
Phalaris arundinacea	Canarygrass, reed	-	MT
Phalaris tuberosa	Harding grass	4.6	MT
Phleum pratense	Timothy	-	MS
Poterium sanguisorba	Burnetgrass	-	MS
Puccinellia airroides	Alkaligrass, Nuttall	-	T
Secale cereale	Rye (forage)	-	MS
Sesbania exaltata	Sesbania	2.3	MS
Sorghum sudanense	Sudangrass	2.8	MT
Sporobolus airroides	Alkai Sacaton	-	T
Trifolium alexandrium	Clover, Berseem	1.5	MS

Contd...

Table 2.3. Contd...

Trifolium fiagiferum	Clover, strawberry	1.5	MS
Trifolium hybridum	Clover, alsike	1.5	MS
Trifolium pratense	Clover, red	1.5	MS
Trifolium melilotus	Clover, sweet	-	MT
Trifolium repens	Clover, White Dutch	-	MS
Trifolium repens	Clover, ladino	1.5	MS
Vicia angustifolia	Vetch, common	3.0	MS
Vigna unguiculata	Cowpea (forage)	2.5	MS
Zea mays	Corn (forage) (f)	1.8	MS

T: Tolerant; **MT:** Medium tolerant; **S:** Susceptible; **MS:** Medium susceptible

Variability differences in salinity tolerance are known in many crops. For any crop improvement programme, the first requirement is one would enquire about the natural variability in the crop tolerance behaviour and whether the existing cultivated and wild resources have been thoroughly screened. In paddy, coarse varieties are more tolerant than the finer one; again *indica* rice are comparatively tolerant than *japonica* types. Within a crop varieties can be designated as highly tolerant, medium tolerant and low tolerant. Traditional cultivars are the most tolerant to abiotic stresses. Rice cultivars, Pokkali, Getu, Cheriveruppu, Nona Bokra, SR26B, Damodar are tolerant of salinity, but possess poor agronomic characters. Salt tolerant indica rice cultures seen to have originated or been selected in coastal areas of India (states of Kerala and West Bengal). Generally, the salinity tolerant varieties are adapted to regions having salinity or alkalinity problems, and have lower yield potential under non-stress conditions than normal varieties. It is the sensitivity of a crop rather than the soil parameter *per se* that essentially determines the occurrence as well as the magnitude of the actual soil problem. Therefore, impairing crop varieties for salt tolerance offers a more effective, less costly, non-polluting and long lasting strategy to manage salt affected soils. The tolerant varieties of different crops have been summarized in Table 2.4, which can be used as donor plant in crop improvement.

Tolerant oil seed crop, *Brassica* is greatly modified by cultural, climatic and biological factors, and by the degree of heterogeneity in saline crop lands. Kumar (1995) reported that amphidiploids, *Brassica napus*, *B. carinata*, and *B. juncea* are more tolerant to salinity and alkalinity than their respective diploid progenitors. Early maturing bold seeded cultivars tolerant to salinity are available in *B. juncia*. Thus, the salt tolerant ability of *Brassica* differs among the species as well as among the the varieties within the species.

Table 2.4. Salt tolerant genotypes of different crops

Crop	Tolerant variety	Reference
Rice *Oryza sativa*	Arya-33, AU-1, Bhura Rata, Bilekagga, BR4-10, CSR-1, CSR-2, CSR-3, CSR-6, Damodar, Dasal, Getu, Hamilton, Jhona 349, Jhona 341, Kala Rata, KR1-24, Karekagga, Matla, MCM1, MCM2, Mo1, Mo2, Mo3, Nona Bokra, Nona Sail, Ormundhan, Orpandy, Patnai 23, Pokkali, PVR1, SR3-9, SR22B, SR26B, Vytilla, AU1	Chopra and Paroda, 1986
	CSC1, Co 43	Krishnamurty *et al.*, 1988
	CSR-10, CSR-11, CSR-12, CSR-13, CSR-19, CSR-20, CSR-21, CSR-24, CSR-26	Mishra, 1996
	Pulot Daeng Maradka, Kautik Serai, Kautik Putih, IR2153-26-3-5,	Gupta, 1997
	TNHR16	Ali *et al.*, 1998
	Ketumbar, Khao Seetha, Soc Nau, Pat Cheriviruppu	Gregorio *et al.*, 2002
	Bicol	Senadhira *et al.*, 2002
	Satha, IR8-SR-268, Puspal, CSR-4	Chhipa, 2003
	TRY(R)2	Rajagopalan *et al.*, 2004
	V1, V2, V3, V4, V6, BR11, BR23, BR29, BR31	Alam *et al.*, 2004
	Marvik	Eker *et al.*, 2006
Wheat *Triticum astium*	*HD-2285, HD-2329, WH-542,* C-306	Mishra, 1996
	Kharchia-65, KRL-1-4, Job-666, WH-147, Raj-3077, WG-357, C-306, HD-2009	Chhipa, 2003

Contd...

Table 2.4. Contd...

Barley *Hordium vulgare*	CSB-1, CSB-2, CSB-3, DL-200, Ratna, BH97, DL348	Mishra, 1996
	Amber, DL-88, DL-120, jyoti, DL-48, RD-137, BL-2	Chhipa, 2003
Pearl millet Pennisetum americanum	PHP-14	Chhipa, 2003
Rye/Mustard *Brassica* spp.	Pusa Bold, Kranti, CS-52, CS-416, CSTR-330-1, CSTR-609-B10, CSTR-610-10-1-1	Mishra, 1996
	RL-18, RLM-1, Prakash, T-59 (varuna)	Chhipa, 2003
	Kranti	Verma *et al.*, 2003
Cotton *Gossipium* sp.	Acala-1517, Stoneville, Coker 100-6, G-27, J-34, Krishna	Chhipa, 2003
	NAIB-999	Ali *et al.*, 2004
Sugarcane *Saccharum officinarum*	Co-453, Co-1341, Co-6801, Co-62329, Co-1111	Mishra, 1996
	Co-205, Co-286, Co-210, Co-331, Co-312, Co-321, Co-513, Co-453	Chhipa, 2003
Berseem	Fahli, Mescaw, BL-1	Chhipa, 2003
Chickpea *Cicer arientinum*	SG-11, DHG-84-11	Singh *et al.*, 2001a
Green bean *Phaseolus* sp	Gevas Sirik 57 (GS57)	Yasar *et al.*, 2006
Sesame *Sesamum indicum*	Sesaco-7, Sesaco-8	
Sun flower *Helianthus annus*	GP336, SF91, SF30 GP255, GP336, Modern	Sassikumar *et al.*, 2003 Sassikumar *et al.*, 2004
Soybean *Glycine max*	SL-432, JS 94-67	Jain *et al.*, 2003

2.8. SCREENING FOR SALT TOLERANCE

The success of breeding programmes would depend upon a reliable estimation of the salinity resistance of various genotypes. For crop improvement programme, the foremost requirement is to gather information on natural variability in the crop tolerance behaviour and whether the existing cultivated and wild resources have been thoroughly screened. So far, different methods and protocols have been developed for successful screening of crop genotypes have been discussed here.

2.8.1. Germination Test

Germination test in a saline medium is the most common way of screening genotypes. Several techniques can be employed for assessing germination potential under salinity stress (Singh, 2003). The simplest technique for assessing germination under stress is to plant seeds on filter papers wetted with the proper salt solution kept in a container. Pattern of response to salt stress may be better distinguished when seed is germinated at numerous osmotic potentials (Robinson *et al.* 1986). A gradient of osmotic potential can be easily created by making a series of small dilutions of a concentrated salt solution. Seed may be plated on an agar medium containing suitable fungicides and the desired concentration of salt solution. Carlson *et al.* (1983) germinated seeds of 15 cultivars of alfalfa on agar containing different concentrations of NaCl. Significant differences among the cultures were observed for ability to germinate under salt stress.

2.8.2. Sand Culture

In this approach pots are filled with sand and subsequently irrigated with saline water. Sharma and Kumar (1985) used polyethylene bags filled with thoroughly washed river sand. The bags were wrapped with a polyethylene sheet to protect roots from exposure to light and to prevent algal growth. The bags were then flooded with Hogland nutrient solution, with salt mixture required to be added for salinization and those bags were used to screen for salinity resistance. They found that the wheat var. C306 is more resistant during the vegetative stage while the other var. Khachia 65 is more resistant during the reproductive stage.

2.8.3. Hydroponics

Jana and Slinkard (1979) screened *Lens culinaris* for salt tolerance in plastic tanks. The tanks contained 16 bottomless growth pouches arranged vertically in lines. The tanks were filled with a known quantity of salt solution, half Hoagland nutrient solution added and aerated continuously. The growth pouches permitted the seeds to remain above the salt solution while imbibing salt water from the tank. Some fungicide (captan) may be used to retard fungal invasion of the seed. Germination and seedling growth were recorded after three weeks.

Marconi *et al.* (2001) made a comparative analysis of regenerated potato (*Solanum tuberosum*) in hydrophonic culture under NaCl stress. The relative growth rate, relative accumulation rate of water, K^+, and Na^+ content had been considered as selection criteria to isolate salt tolerant mutant.

2.8.4. Natural Selection

Generally, the saline soils are associated with acid, acid sulfate, and/or peat soils with nutrient deficiencies and mineral toxicities. Because of these multiple stresses, breeding materials must be evaluated under naturally occurring saline soil and climatic conditions to improve selection efficiency. This method is the best method for screening against abiotic stresses. But, it is based on season and involve more cost.

2.8.5. Pollen Grains Screening

Identification and evaluation of salt tolerant genotypes are important to improve production of crops in salt affected land. The criteria used in the past for screening of crop plants have failed to prove their legitimacy as the final economic yield is not positively correlated with any of the those parameters. Thus, need of the hour is to have a simple, efficient, economically viable and dependable technique for screening of salinity tolerant genotypes of crop plants that is reflected in economic yield as well.

Pollen grains, like microbial organisms, are characterized by large population, general simplicity and easy amenability for *in vitro* studies. Those features make pollen a suitable system for the study of adaptability to stressful environment. Studies have provided sufficient evidences to believe that a substantial complements of structural genes (60-80%) of sporophyte are superimposed in pollen also (Ottaviano and Mulcahy, 1989; Frova *et al.*, 1991). Therefore, culturing of pollen in a suitable germination medium fortified with a stressing agent should permit germination of tolerant pollen more successfully than the non-tolerant ones and thus selection pressure can be applied to pollen independent of the parent plant. Sacher and Mulcahy (1981) recorded differences in *in vitro* pollen tube growth of salt sensitive and salt-tolerant tomato cultivars. Pollen tube growth of a sibling mated with heterogeneous species, *Lycopersicon peruvianum*, exhibited a significantly wider range of inhibition in salinized media, while in the control medium; growth was uniformly inhibited in saline media. Thus, pollen tube growth rate serve as a sensitive screening criterion.

Sureena and Dhingra (2003) reported positive correlation between pollen and plant under salt stress. Sodium chloride, in general, inhibits both pollen germination and pollen tube growth in different species of *Brassica*, and inhibition varied with the species and the cultivar studied. Based on overall pollen performance among 4 tested species, inhibition was more in *B. juncea* cv. Varuna and hence adjusted as the most salt sensitive. The inhibition, however, decreased from *B. alba, B. carinata* to *B. napus*, thereby, indicating that *B. nupus* is the most salt tolerant. Some tolerant varieties of *B. juncea* as identified are- RH-7846, RH-8701, Domo-4 and Pusa Bold.

2.8.6. Young Seedling Screening

Salinity affects more the rate of germination than the ultimate germination percentage, and suppresses elongation of roots than of shoots (Datta and Pradhan, 1981), and in some other crops, more of shoots than of roots (Cruz and Cuartero, 1990). Minimum suppression in coleoptile elongation has been suggested as a better index for screening sorghum varieties (Ogra and Baijal, 1978).

A simple method for screening salt tolerant genotypes had been developed by Sukarin *et al.* (1993) to improve the efficiency of the selection in breeding procedure. They have transplanted young seedlings of rice, maize, *Coix laceryma-jobi*, buckwheat, *Cajanus cajan*, *Phaseolus aureus* (*Vigna radiata*), cotton, rosette, and *Jatropha carcus* on petri dishes filled with sand and impregnated with salt solution. After 10 days of transplanting, the height of the cultured seedlings was measured and linear regression equation of plant height of the crop tested on the salt concentration index, were derived. A smaller co-efficient indicated a high salt tolerance. This method of screening requires small number of seeds, rapid screening and relatively high accuracy of the result.

2.8.7. Seedling Root Dip

For laboratory screening, roots of 30 days old seedlings can be dipped in 0.5% salt solution for 72 hours. The salt tolerant varieties are characterized by lower absorption of salt. While breeding for rice varieties suitable for saline-alkaline situations, one should select for number leaf tip burn, better root system and plant free from dark band at the juncture of root and shoot.

2.8.8. Cell Membrane Stability (CMS)

The general protocol involves the application of stress to the leaf after it has been subjected to hardening, followed by the measurement of electrolyte leakage using the conductometric method. The most common application is for salt tolerance and therefore the initial detail is given for heat stress.

1. The plant must be exposed to salt stress for at least 72h before the test in order to allow for hardening (acclimation). The capacity for hardening is a major component of the capacity for tolerance. Hardening can be achieved in the natural field environment, if salt stress occurs, or in the greenhouse or a programmed chamber.
2. Leaf discs or pieces of leaf tissue cut with scissors or even whole small leaves are detached and placed in standard glass vials that can accommodate a conductivity electrode. The total area of leaf material per vial is about 15 to 25 cm^2. The sample is then washed for 2-3 times with de-ionized water. The water is drained off but samples remain wet so that they would not desiccate. In the case of screening, at least 10 vials (samples) are prepared for each genotype. In that case 5 pairs are taken from five different plants (replicates). For each pair, one vial is designated as treatment (T) and the other as control (C).
3. The treatment vials are subjected to the heat stress treatment *in vitro*. They are placed in racks and covered (not stoppered) with 'Saran' wrap so as to drying the samples. Racks are placed in thermostated water bath so that the leaf samples will be completely below the water surface level. Temperature is set to a predetermined stress (treatment) temperature and the samples remain in the bath for 1h. The control vials are placed in a rack, covered with Saran wrap and placed at room temperature (18-25 ^{0}C). The treatment temperature should be such that it will result in average population CMS values around 50%-60% for full separation of the accessions.
4. After treatment 20cc of deionized water is added to each vial making certain that all leaf materials are submerged. All vials are then placed for incubation at about 10 ^{0}C (typically, on the lowest refrigerator shelf) for 24h. After incubation samples are equilibrated for 1h to room temperature and the conductivity of the medium is measured by inserting a conductivity electrode into each vial.
5. All vials covered with Saran wrap are placed in an autoclave for 15 min to kill all tissues. Conductivity of all samples is measured after samples

are equilibrated to room temperature. Calculation: where T1 and T2 are treatment conductivities before and after autoclaving and C1 and C2 are the respective control conductivities.

CMS % = [1-(T1/T2)]/[1-(C1/C2)] X 100%

Injury = 100-CMS

2.8.9. Other Methods of Screening

Many other methods, such as, tetrazolium test, paper towel methods etc. have been recorded to be utilized for screening against salinity. A general reduction in shoot and root length, vigour index and dry matter production with increased salt concentration was observed in the plants using paper towel method (Sassikumar *et al.*, 2004).

2.9. CONVENTIONAL BREEDING FOR SALT TOLERANCE

Sustainable food production puts a high demand on breeding more salt-tolerant crops (Cuartero *et al.*, 2006). Breeding of crops tolerant to soil salinity and yield well in salt affected soils, and employing crop management practices to counter salinity, have been proposed to maintain crop productivity on increasing area of salt-affected lands, because reclamation of salt prone land, the other possible way out, is an expensive proposition. The generalized steps for conventional breeding approaches for salt tolerance are:

1. Screening germplasm collections for donors of salt tolerance
2. Crossing a donor with an elite line and advancing the F_1 hybrid to about $F_{7/8}$ generation while selecting the elite characters
3. Selecting for tolerance starting about F_4 generation

Most of the best specific cross-combinations for salt-tolerance in rice had susceptible but high-yielding parents such as IR 28 and IR 29. The presence of reciprocal effects among the crosses necessitates the use of high-yielding susceptible parents as males in a hybridization programme. Further, the large heterotic effects observed suggest the potential of a hybrid rice programme for salt affected lands. Several modern plant types developed at IRRI using traditional salt tolerant parents such as Nona Bokra, Pokkali, SR 26B and Kalarata do not possess the level of tolerance because donors had too many undesirable traits, most of them are linked to salinity tolerance, preventing transfer to an improved plant type.

Smith *et al.* (1994) improved the performance elite populations of alfalfa (*Medicago sativa*) by mass selection under salinity stress. The wild perennial *Glycine* species have been suggested as potential source of germplam to improve soybean for salt tolerance (Pantalone and Kenworthy, 1989). Wild soybean *G. soja* Ieb. & Zucc. is the unique close relative of cultivated soybean and has been efficiently used in soybean breeding. Hu and Wang (1997) reported variation for salt tolerance in natural population of *G. soja*.

2.9.1. Limitations in Conventional Breeding for Salt Tolerance

Most crop plants are sensitive to salinity throughout the ontogeny of the plant. Despite considerable research on salinity tolerance in plants, there are only a few instances where salt tolerant cultivars have been developed; this is due to the complexity of the character. Plant responses to salt stress are modulated by many physiological and agronomical characteristics, which may be controlled by the action of several to many genes whose expressions are influenced by various environmental factors. Further, salinity tolerance is developmentally regulated, stage-specific phenomenon; tolerance at one stage of the plant development is often not correlated with tolerance at other stages. Specific ontogenic stage should be evaluated separately for assessment of tolerance and, identification, characterization and utilization of useful genetic components. The major limitations in development of salt tolerant genotype has been listed below.

a) Creation of reliable, dependable and controlled salinity environment for selection of salt tolerant genotypes in crop species is tedious, costly and beyond reach of many breeders.
b) Salt tolerance is greatly modified by the culture, climate and biological factors and by the degree of heterogeneity in saline crop lands.
c) Lack of simple method to score reliable and dependable selection criterion for salinity resistance.
d) Salt tolerance is complex genetically and physiologically. The genetic control of salinity resistance is generally polygenic, which makes transfer from germplasm lines, especially, related species a very difficult task.
e) Salt tolerance often shows the characteristics of a mutagenic trait with quantitative trait loci associated with tolerance and with ion transport under saline condition.
f) High strength of genotypic 'χ' environment interaction.

The basis of salinity resistance is poorly understood. This makes genetic analysis and breeding efforts considerably difficult.

2.10. SOIL RECLAMATION

Liu *et al.* (2004) investigated the effects of regulatory measures, such as fertilizer application, irrigation and chemical soil amendment on distribution of soil salt and ion uptake by crops. Their result showed that fertilizer application increased electrical conductivity of the surface soil layer. Electrical conductivity of the surface soil increased with the soil amendment with $CaSO_4$ at an early stage. Fertilizer application and irrigation improved the growing environment for barley and enhanced its salt tolerance ability. They also reported higher Na^+ content in stem than in the leaves demonstrating the function of barley retaining Na^+ in the stem and mitigate Na^+ damage to the leaves. Thus, the soil amendment showed positive effect in increasing K uptake and decreasing Na^+ content in plant and alleviating stress of the plant from salt ions.

Haloxylon recurvum (locally know as *Khar*) is a drought and salt tolerant plant of Thar Desert in India. This plant is a major biomass producer and has economic and ecological importance for the region (Dagla and Shekhawat, 2005). Its genetic improvement for utilization in reclamation of soils of the desert is essential.

2.11. BIOTECHNOLOGY IN BREEDING FOR SALT TOLERANCE

Cell and tissue culture techniques and genetic engineering are considered to be the main potential genetic approaches for screening against environmental stresses. Marker aided selection also gaining momentum in pyramiding salt tolerant genes in desirable genotypes. The achievements in biotechnology to develop salt tolerant crop plants have been discussed in this section.

2.11.1. Somaclonal Variation

Production of salt-tolerant plants through *in vitro* technology is one of the ways to utilize the waste saline lands. Selection of favourable mutant cells and somaclonal variant strains from callus culture is supplementary tools to traditional breeding for production of stress-resistant plants (Larkin and Scowcroft, 1981; Dix, 1993; Ashraf, 1994). The recent development in the *in vitro* technology offers a meaningful tool for determining the tolerance and also in screening and developing salt tolerant genotypes. The introduction of a specific genotype in *in vitro* selection programme depends on potentiality of callus induction and embryogenic callus induction, subsequently green plantlet regeneration. Studies have shown that genotype affects *in vitro* culture response (Van Sint Jan *et al.,* 1990; Arzani and Mirodjagh, 1999; Mandal and Roy, 2003; Roy and Mandal, 2005). *In vitro* cell culture may be effectively used in genetic modification of crop plants. One approach of such genetic modification involves the isolation of stable variant cell lines from established cell culture either directly or following mutagen treatment; regeneration of plants from such lines requires that the cells be subjected to appropriate selection pressure, i.e., exposure to high salt concentration and subsequently test for retention of the tolerance. *In vitro* selection of salt-tolerant cell lines and regenerated plants has been reported in several species, such as, potato (Sabbah and Tal, 1990), rice (Lutts *et al.,* 1999; Aditya and Baker, 2005), *Hordeum* (Sibi and Fakiri, 2000), wheat (Barakat and Abdel-Latif, 1996), sunflower (Alvarez *et al.,* 2003), sugarcane (Gandonou *et al.,* 2005).

NaCl tolerant calli (variants) were selected from the cotyledonary embryo of *Arachis hypogea* L. on agar solidified medium supplemented with 1% NaCl, a concentration otherwise lethal to non-selected calli. This suggests that tissue culture selection can be used to improve salt tolerance of plants. *In vitro* cell selection and somaclonal variation offer an alternative to tradition breeding methodology for generating improved breeding lines. Callus cultures are initiated from chosen variety of concerned crop. The callus should have high potential for regeneration of complete plants. The proliferated calli are then subjected to NaCl supplemented medium. The level of NaCl varies depending upon genotypes and species; in general, salt concentration should be high enough to kill over 95% of the cells. NaCl concentrations in the culture medium have reverse effect on the plantlet regeration from callus (Basu *et al.,* 1997; Lutts *et al.,* 1999). Puspalatha and Padnabhan (1998) observed reduced fresh weight of calli with increasing concentration of NaCl in the medium and no callus growth was observed by many workers (Shankhdhar *et al.,* 2000; Aditya and Baker, 2005) in rice. The salt tolerant selected lines required the presence of 0.5% NaCl for optimal growth and successfully grew at 1.5% NaCl, a concentration that was lethal to the unselected lines.

2.11.1.1. Length of selection period and selection cycles

The surviving cells one sfressed medium may be subjected to one or more selection cycle(s). Salt resistant clones are then cultured on salt-free medium. Prolonged callus of sorghum cultured on a NaCl containing selection medium grew better when transferred to a fresh NaCl containing medium than the control callus, which was not selected on the NaCl selection medium (Bhaskaran *et al.,* 1983). Prolonged *in vitro* selection of embryo-derived calli of rice in a toxic concentration of NaCl led to arrested growth in most of the calli except a few which maintained healthy and stable growth throughout the three selective progenies (Reddy and Vaidyanathan, 1985). Barakat and Abdel-Latif (1995) suggested stepwise method for high efficiency of plantlet regeneration for salt tolerant cell line. They have reported decrease in the relative growth rate of callus in general. But, the selected cell lines revealed significantly higher relative growth responses compared to unselected cell lines. They also found more proline content in selected lines. A generalized selection cycles and length have been presented in Fig. 2.2.

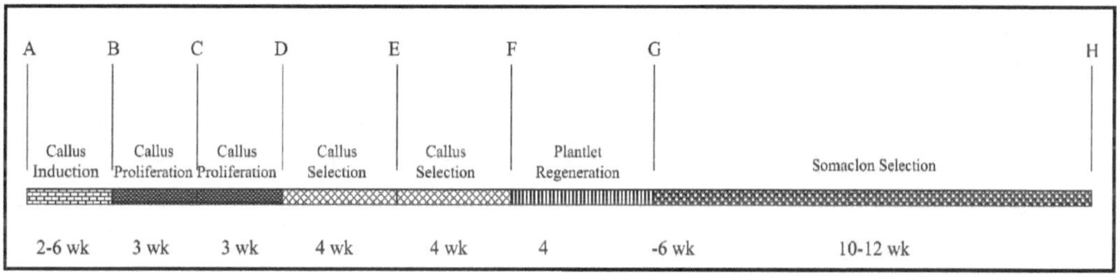

Fig. 2.2. Gereralized diagramatic representation of *in vitro* selection procedure for salt tolerance. (A) Explant inoculation on Callus Induction Medium, (B) First sub-culture on Callus Maitenance Medium, (C) Secon sub-culture on Callus Maitenance Medium, (D) First sub-culture on stressed clallus maintenance medium, (E) Second sub-culture on stressed clallus maintenance medium, F) Transferred to stress free medium, (G) *In vivo* selection, (H) Selected somaclones.

2.11.1.2. Acheivments

Some of the successes in development of salt tolerant genotypes and creation of variability through *in vitro* screening have been discussed in this section.

Rice (Oryza sativa): Subhashini and Reddy (1989) grew regenerated plants from calli of rice cultivars Basmati 370, Gopal Bhog, Pakistan Basmati and Chittimityalu, selected from medium containing 1% NaCl or 50% sea water up to maturity. Three weeks old seedlings raised from such grains were transplanted to pots and grown to maturity with NaCl stress along with controls. The regenerated plants were superior in saline soil for all the agronomic characters in comparison of control.

Prolonged *in vitro* selection of embryo-derived calli of rice cv. 27814 in high concentration of NaCl led to arrested growth in most of the calli except a few, which maintained healthy and stable growth (Reddy and Vaidyanathan, 1985). Wong *et al.* (1986) developed salt tolerant rice plants from cv. Taichung 67 by *in vitro* methods. Tolerant calli were isolated and regenerated on medium containing 0.6, 1.0 or 1.5% NaCl. Plantlets were obtained on medium containing < 1.0% NaCl. The regenerated plants were maintained until tillering, when a single tiller was removed and subjected to 0.6% NaCl-salinized solution for 30 days. Approximately half of the regenerants showed tolerance.

Irradiated anthers of indica rice at the uninucleate pollen grain stage with 0.5-3.0 kRad of γ-rays were cultured on N6 (Chu *et al.*, 1975) medium (Kuang and Dong, 1988). About 30% of regenerated plantlets showed salt tolerance ability, of which about 50% transmitted their resistance to the progeny.

Mandal *et al.* (1999), Pramanik and Mandal (2000) had exploilted somaclonal variation in crop improvement. They isolated salt tolerant clones of Pokkali (a salt-tolerant rice cultivar). Among the somaclones, they found BTS24 as better yielder both in saline and non-saline soils. At IRRI (Gregorio *et al.*, 2002), somaclonal variants of Pokkali with improved agronomic traits were identified, the characters for which the variants differed significantly from Pokkali. The variant, TCCP 266-2-49-B-B-3 had desirable levels of all tested characteristics and retained salinity tolerance equal to Pokkali. The variant showed vigourous growth and, unlike the parent Pokkali, is semi-dwarf, a character essential in increasing yield potential without lodging. This somaclone became a popular donor parent and has produced new high yielding salinity tolerant lines, some of which have been released as cultivars.

Sorghum (Sorghum bicolor): In vitro screening of *Sorghum bicolor* and *S. halepense* by Yang *et al.* (1990a, b) revealed that the accumulation of NaCl accounted for three fourth of the solutes involved in osmotic adjustment in the calli of both species. The growth and Na : K ratio in salinized callus was correlated with whole plant responses, which can be used as indicators of whole plant salt tolerance in sorghum. They grew callus of those two species for two weeks in a liquid Murashige and Skoog (1962) medium to which NaCl was added to concentration of 0.05, 0.10 and 0.15 M.

Wheat (Triticum sp.*):* Embryogenic calli isolated from immature embryos of four wheat cultivar subjected to three *in vitro* selection methods for salt tolerance (Barakat and Abdet-Latif, 1996). The relative growth rate of callus decreased as the concentration of NaCl increased in the selected and unselected callus lines. *In vitro* selection for NaCl tolerance showed that the stepwise method of increasing NaCl in the medium was more effective for plant regeneration.

Sugarcane (Saccharum sp.*):* Salt tolerant cell lines were selected in sugarcane by Fitch and Moore (1981). The differentiated sugarcane plantlets were transferred to a solid MS basal medium containing 20% coconut water and 18 or 20 g/L of NaCl. Survivor plantlets were grown for an additional 5-18 months on media containing 12-20 g/L of NaCl. Three sugarcane cultivars (CP 43/33, BF-162 and IM-61) exhibited salt tolerant potential for callus formation (Tanvir *et al.*, 2002). Those plantlets were multiplied on non-salinised field and then planted on saline sodic soil with pH 8.5 and EC 4.9 dSm^{-1}. Somaclones of BF-162 and IM-61 showed dominance over the parents in cane yield, height, girth and brisk. Gandonou *et al.* (2005) found that sugarcane genotypes CP70-321 and NC0310 were superior for salt tolerance together with their high potential for embryogenic callus induction and they suggested that those varieties as a good model to study physiological mechanisms associated with *in vitro* salt tolerance and *in vitro* selection for salt tolerance in sugar cane.

Alfalfa (Medicago sativa): Croughan *et al.* (1978) screened salt tolerant alfalfa cells on agar-solidified media containing 1% (w/v) NaCl. Within two months of exposure, more than 99% of the planted cells discolourized. A small number of cells remained healthy, these cells were sub-cultured in every 3-4 weeks for 8 months. The salt selected cell lines grew better than the unselected cultures at high levels of NaCl.

Tobacco (Nicotiana tabaccum): Regenerant tobacco cell lines obtained from three varieties by sub-culturing on medium with sea water or sodium sulphate or manitol could tolerate salt concentrations toxic to parent plants (Levenko *et al.*, 1989). Dridze *et al.* (1991) also developed 11 cell lines with increased tolerance to salinity; these selected cell lines retained the aquired character even after six months of subculture under unstressed conditions.

Mothbean: Bhargava and Chandra (1989) grew calli of moth bean (*Vigna aconitifolia*) on MS medium with the addition of 0.5, 1.0 or 1.5 NaCl. A decrease in growth during initial subculture was followed by a gradual recovery with 0.5% and 1.0% NaCl. They observed a shift towards salt tolerance in the line cultured on 1.0% NaCl solution.

Mustard (Brassica juncea): In vitro selection of salt tolerant plant of Indian mustard (*Brassica juncea* L.) cv. Prakash has been accomplished on high NaCl media (Jain *et al.*, 1990). The salt selected shoot retained salt tolerance following three month growth and multiplication on control medium. While two of these somaclones flowered and set seeds. In green house, selected salt tolerant somaclones performed better for plant growth, yield and other agronomic traits at higher salt treatments, indicating thereby that salt tolerance character selected *in vitro* was expressed in the whole plant, and is genetically stable and transmitted onto the progeny.

Black Gram (Vigana radiata): Geetha and Rao (1997) isolated a NaCl tolerant callus line of blackgram from callus culture medium supplemented with increasing concentrations of NaCl. The growth of the tolerant line, in the presence of NaCl (300 mM) was comparable to that of a sensitive callus line growing in the absence of NaCl.

Potato (Solanum tuberosum): Ochatt *et al.* (1999) selected a stable salt tolerant potato cell lines, able to grow on medium containing 60-450 mM NaCl. Regenerated plantlets from salt tolerant callus, exhibited salt tolerance and produced more tuber/plant under salt stress.

Banana (Musa sp.*):* C-Ventura *et al.* (2002) screened banana on M-3 medium supplemented with different concentrations of NaCl. Cultivar Gran Enano showed tolerance up to 4500 ppm with a 96.7% survival rate and 25% leaf damage. Whereas, FHIA 03 showed tolerance up to 8000 ppm of NaCl with 80.9% survival rate and 72.5% leaf damage.

Cotton (*Gossypium hirsutum* L.): An endeavour was under taken by Ganesan and Jayabalan (2005) to develop the protocol for consistent production of salt-tolerant cotton plants through direct organogenesis. NaCl at concentrations of 25-200 mM were tested for shoot tip, cotyledonary node and leaf node cultures. The salt-tolerant plants were selected between 100 and 175 mM of NaCl-treated cultures. The survival percentage of selected salt-tolerant plant was 85% in the field.

2.11.2. Androclonal Variation

Mostly single crop is grown in a year in salt-affected soils, therefore at least 8-10 years is required to develop a promising cultivar through conventional breeding under natural selection method. The F_1 another culture (AC) technique is one strategy for reducing evaluation process. AC shortens the breeding cycle, producing homozygous lines in one generation. In traditional breeding, breeder has to establish 3000-5000 plants in F_2 generation alone, whereas, in AC only 60-70 plants are required (Khush and Senadhira, 1994). Snape (1989) reported on increase in selection efficiency using double haploid breeding compared to conventional practices. Wong *et al.* (1986) regenerated anthes culture in NaCl - stressed medium. Anthers from plants grown from seeds treated with ethylmethane sulphate were cultured on a medium containing 0.6 or 1.0 % NaCl. Thus, it should be easier to identify superior genotypes in a cross and produce new cultivars.

The AC method is based on gamete instead of sporophyte selection and the probability to obtain the described genotype is higher in haploid than diploids. The AC strategy should also enable a great use of minor genes for resistance, the cumulative effect of which may provide more desirable resistance than that provided by major genes (Toenniessen and Khush, 1991). The possibility of generating recombinant with desirable characteristics is high (Zapata *et al.,* 1991) in AC.

Cross was made between IR28 (high yielding salt-susceptible) and Pokkali (low yielding salt-tolerant) Roy and Mandal (2004). The F_1 and parents were subjected to anther culture on organic adjuvant and sucrose to obtain high green plantlets regeneration. Maltose (6%) was found to be most suitable for androgenic callus development (Table 2.5) and subsequently regenerate good number of green plantlets (14.0%) of F_1 plants. The regenerated A_1 generation in the housed condition showed immense variability among themselves viz. plant height, number of panicles per plant, panicle length, number of filled grains per panicle and spikelet sterility (Table 2.6). In AC, development of undifferentiated calli masses before regeneration into whole plant brings about the possibility of somaclonal variation; such *de novo* synthesized variation may be used in genetic improvement (Schaeffer, 1983; Mandal *et al.,* 2000). Two hundred nine plantlets (DH lines) from anther culture of F_1 have been developed. It is more than the suggesion made by Khush and Senadhira (1994) that is 60-70 androgenic plants are required for selection of desired line (Fig. 2.3).

Table 2.5: Anther culture response of F_1 hybrids and the parents

Treatment	Genotype	Callus induction (%)	Regeneration (%)		
			Green plantlets	Albino plantlet	Total
Maltose	IR28	25.85	16.16	26.88	43.04
	IR28 × Pokkali	5.96	14.00	26.00	40.00
	Pokkali	21.83	15.28	27.84	43.12
15% Coconut water	IR28	11.19	10.25	21.27	31.52
	IR28 × Pokkali	0.00	0.00	0.00	0.00
	Pokkali	7.39	12.24	34.69	46.93
Sucrose (5%)	IR28	1.52	0.00	12.50	12.50
	IR28 × Pokkali	0.00	0.00	0.00	0.00
	Pokkali	2.57	3.45	3.45	6.90

Table 2.6. Performance of parent and F_1 under net house condition

Character	F_1 hybrid	Parent	
		IR28	Pokkali
Plant height (cm)	125.6	87.2	136.5
No. of tillers/ plant	7.25	10.52	4.32
Panicle length (cm)	22.06	23.92	18.95
No. of filled grains/ panicle	60.16	96.44	47.05
Spikelet sterility (%)	44.71	28.61	33.17
Grain yield/ plant (g)	12.21	24.11	5.69

Fig. 2.3. Anther culture and plantlet regeneration in F_1 hybrids of IR28 ´ Pokkali. **a)** Plated anthers on N6 medium supplemented with 2 mg L^{-1} 2,4-D + 6% maltose. **b)** Emerging androgenic calli N6 medium. **c)** Regenerating androgenic green plantlet on MS medium fortified with 1 mg L^{-1} BAP + 1 mg L^{-1} kinetin + 0.5 mg L^{-1} NAA. **d)** Rooted androgenic plantlets on MS basal medium. **e)** Anther culture in rice- IR 28, F_1 and Pokkali; **f)** and roclonal variants- A_0 generation. *(For colour version of this figure see page 534)*

Some high-yielding salt tolerant AC-derived lines viz. IR51500-AC11-1, IR51500-AC17, IR51485-AC6534-4, IR72132-AC6-1, IR69997-AC1, IR69997-AC2, IR69997-AC3 and IR69997-AC4 have been developed at IRRI in 1996 (Gregorio *et al.,* 2002). Most of these lines had been used as donor parents in breeding programmes in Bangladesh, Dominican Republic, Egypt, Mexico, Mayamnar, Philippines and Thailand (ISTORN, 1991).

Senadhira *et al.* (2002) developed 79 di-haploid lines in rice through AC the cross of IR5657-33-2 x IR4630-22-2-5-1-3. Among those, di-haploids, IR51500-AC9-7 and IR51500-AC11-1 were selected for saline prone areas. IR51500-AC11-1 performed better than the other cultivars grown in saline prone lands. In 1995, this line was released as a salt tolerant cultivar in Philippines with the name PSBRc50 or 'Bicol' (Fig. 2.4). This is the first AC derived line from *indica* crosses released as variety. In India, IR51500-AC-17 and IR51485-AC6534-4 were named as commercial cultivars CSR21 and CSR28, respectively, for cultivation in saline soils (IRRI, 1997).

Fig. 2.4. Development, identification, and release of AC-derived salt tolerant rice cultivar IR51500-AC11-1 (Senadhira *et al.*, 2002).

DS: Dry Season; **WS:** Wet Season; **NCT:** National Cooperative Testing; **RVIG:** Rice Varietal Improvement Group; **NSIC:** National Seed Industry Council (Philippines)

Schum and Peril (1998) selected salt tolerant lines in *Chrysanthemum* treating callus with EMS. Somaclonal variations lead to the selection of salt-tolerant plants in *Coleus blumei* (Ibrahim *et al.*, 1992).

2.11.3. Limitations in Development of *in vitro* Culture

Although all the above cited experimental results described the tremendous potential of cell culture as a powerful technique for plant improvement, this approach has certain limitations and problems as:

1. Only a certain type of characters can be selected.
2. Till today there is no way of direct selection for increased yield through the cell culture techniques.
3. Metabolic adaptation by cultured cells, genotype-dependent characters, a dramatic decrease in the regeneration of embryogenic cultures (Lutts *et al.*, 1999) are characteristic features of anther culture.
4. Regeneration from long term callus culture is difficult and some time not possible. For instance, after 5 months of subculture, only 10% of subcultured callus differentiated into shoots (Ozias-Akins and Vasil, 1982).
5. The cultured cells frequently contain nuclei showing various degrees of polyploidy, aneuploidy and chromosome abnormalities. This chromosomal variation often leads to the loss of morphogenetic potentiality.
6. Usually many selected resistant clones become susceptible after passages on salt free medium.
7. Phenotypic changes in regenerated plants and non-expression of salt-tolerant phenotypes are frequently observed in regenerated plantlets
8. The difficulty of conserving other characteristics of the original varieties in plantlets selected for their tolerance to salt.

2.11.4. Enzyme Analysis and Gel Electrophoresis

Abdel-Tawab *et al.* (2003) developed some biochemical markers, such as SDS protein, esterase and acid phospatase isozymes for sugarcane in relation to salt and drought stress. The electrophoretic pattern of esterase isozymes of all cultivars in the sand culture experiment under control, salt and drought treatments exhibited nine bands. While acid phosphatase isozymes of all cultivars in the sand culture experiments showed a maximum number of five bands. Based on these findings they have suggested that the *East-8* and *Act-5* bands may be used as positive biochemical markers for salt and drought tolerance in sugarcane. SDS-PAGE banding pattern of maize inbreds showed differential responses with respect to salt tolerance and ABA or BA treatments (Shadi *et al.*, 2001, 2003), in which esterase and peroxidase isozyme profiles were weakly or negatively correlated with salt tolerance and ABA treatment.

Marambe and Ando (1995) found five isozyme patterns of α-amylase in germinating sorghum seeds. The additive nature of the respective isozymes was enhanced with the increasing duration of imbibition in distilled water. They observed variation in sensitivity of the isozymes to the salinity treatments, and isozyme 1 was the most tolerant. Their results revealed that the salinity tolerance could be attributable to the compensation of loss in activity of certain isozymes of α-amylase. Chen *et al.* (1997) examined electrophoresis effects of salinity of soybean on different SOD isozymes. They observed 80, 11, and 9% of total SOD activity occurred in the cytosol, mitochondria and chloroplast, respectively at seedling stage. The difference between the sensitivity and tolerant cultivars was mainly in the chloroplast SOD, and it was suggested that *SODc1c2c3* bands are closely related to salt tolerance.

2.11.5. Marker Assisted Selection

Degradation of agricultural land and water supplies is a result of intensive agricultural practices employed in developed and developing countries for a long time. Ideally these practices should be changed to a more sustainable use of land and water resources. For example, mixed cropping with perennials and tree would alleviate the accumulation of sodium and other salts in upper soil layers. Nonetheless, this type of change in farming systems and the development of new products is likely a long and difficult process, since it will require the use of new land will not address the problem of growing crops in land that is already compromised. Development and use of crops that can tolerate high levels of soil salinity is a practical solution. Although conventional breeding for salt tolerance has been attempted for a long time, the lack of success in generating tolerant varieties would suggest that conventional breeding practices are not enough. Conventional breeding involving use of tolerant lines through *in vitro* or *in vivo* screening methods, their subsequent through phenotypic characterization are found to be inadequate for improving salt tolerance in rice. Advantage of genomics are making possible and alternative approach in which a prebreeding phase is used to pyramid several known genes and finally to map major quantitative trait loci (QTLs) for complementary aspects of salt tolerance. Recent development in molecular marker assisted selection (MAS) helps to speed up the process and can assists in pyramiding to achieve abiotic stress tolerance in less time. It is also mentioned that molecular physical maps are important source for tagging gene(s) as well as in facilitating cloning of characters of economic importance, sequencing of genomic DNAs and analysis of chromosomes and genome structure in detail. Thus, physical maps would occupy control place in genetic and molecular enquiries including genome analysis, gene cloning and crop genetic improvement. DNA based selection protocols that are used to pyramid genes are again employed during the breeding phase to transfer the entire set of genes for salt tolerance into any elite lines by backcrossing. These increase the breeding efficiency (Bennett and Khush, 2003) as below:

1. Elite characters are more easily recovered by back crossing than the pedigree approach.
2. The genetic complexity of salt tolerance is reduced to a small set of well-defined genes and loci of large effect.
3. The confounding effects of genotype x environment interactions are eliminated from the breeding phase.
4. The investment in the gene discovery and QTL mapping can be recovered by transferring the same set of genes to a range of recipient lines that require salt tolerance.

2.11.5.1. Sources of gene

Gene discovery is first step for molecular breeding for abiotic stresses. There are four broad methods for gene discovery (Table 2.7). The identified/ discovered gene(s) can be used for many other purpose also, such as molecular breeding and cloning and transgenic development etc.

Table 2.7. Sources of gene for abiotic stress tolerance

Type	Resources
Genotype	Germplasm collection Breeding lines Segregating populations Insertional and deletional mutants Specific transformants
Phenotyping	Field-based screens Screening in managed environment Physiological measurement Biochemical assay
Molccular	Large- and small-insert genomic libraries cDNA libraries Cloned DNA markers Antibodies Metabolomics
Bioinformatic	Data base of DNA and protein sequences Genetics and physical maps Proteomics database Databases on mutant libraries Database on phenotypic and molecular characterization Software

2.11.5.2. Achievements in marker assisted selection

A remarkable development has been achieved in the field of MAS in different crops. Some of the achievements have been briefed here for few crops.

Rice (Oryza sativa): Rapid and reliable techniques have been developed at IRRI to detect tolerance to salinity stress. The genes governing tolerance in Pokkali were mapped through the use of F_8 recombinant inbred lines (RIL) derived from the cross between Pokkali and the salinity sensitive variety IR29. A major gene for salinity tolerance was mapped on chromosome 1 (IRRI, 1997). According to Gregorio *et al.* (2002), this gene may be the same as *SalT* gene in the Cornell map (Causse *et al.*, 1994) which code for protein in 140 amino acid residues. This protein accumulates very rapidly in sheaths and roots of mature plants and seedlings when subjected to salt (Claes *et al.*, 1990).

For high K absorption, in addition to the major gene in chromosome 1, QTLs were detected on chromosomes 4 and 12. For Na absorption, there were QTLs on chromosomes 1, 10 and two on chromosome 3. For Na/K ratio, three QTLs were mapped to chromosomes 1, 10 and 12. Three QTLs for several days of seedling under salt stress were detected on chromosomes 1, 6 and 7, respectively and explained 13.9% and 18.0% of the total phenotypic variance (Lin *et al.*, 2004). Based on the correlations between survival days of seedling and other physiological characters, it was considered that the damage of leaves was attributed to accumulation of Na^+ in the shoot by transport of Na^+ from the root to the shoot in external high concentration. They found 8 QTLs including 3 for three characters of the shoot, and 5 for 4 characters of root at 5 chromosomal regions controlled complex physiological characters related to salt tolerance. A common major QTL was detected on chromosome 1 for the three traits putatively associated with salinity tolerance (IRRI, 1998). In a different population (Agami ´ IR3352-AC202) developed at IRRI, a major QTL was also mapped on the same segment of chromosome 1. A major gene controlling salt tolerance in a mutant line was mapped on chromosome 7 near RFLP marker RG4 (Zhang *et al.*, 1995).

QTLs or gene(s) have been mapped for salt tolerance using improved methods of identifying and measuring its physiological components. Major and minor QTLs for morphological attributes, ion accumulation and other traits associated with salinity tolerance were reported (Flower *et al.*, 2000; Koyama *et al.*, 2001; Lang *et al.*, 2001). Flower *et al.* (2000) have identified four putative markers for sodium and potassium ion transport selectivity through AFLP analysis using a mapping population involving IR 4630 and IR15324. The same has also been used for the identification of QTLs, which govern Na^+ and K^+ uptake, and Na selectivity using AFLP, SSR and RFLP markers (Koyama *et al.*, 2001). Four QTLs were identified for time of seedling survival in salt solution, one QTL for shoot dry weight, two QTLs for root dry weight, one QTL each for Na^+ and K^+ absorption and four QTLs for the Na/K ratio by RFLP and SSR marker analysis of 108 F_8 of Tasamai2 x CBRILs (Lang *et al.*, 2001).

Seven QTLs that control various rice seedling traits conferring salt tolerance were mapped by RFLP analysis using a doubled haploid population derived from IR64 and Azucena (Prasad *et al.*, 2000). Tripathy *et al.* (2000) identified QTLs for cell membrane stability under drought stress in rice using RFLP, AFLP and microsatellite markers. A major QTL for salinity tolerance was located on a 30 cM region of rice chromosome 1, using AFLP markers in a recombinant inbred population between traditional salt tolerant Pokkali and sensitive high yielding IR29. Flanking microsatelites to this QTL viz. *SALTOL*, were also identified subsequently viz. RM23 and RM140 which are approximately 26 cM apart.

Wheat (Triticum aestivum): Nemoto *et al.* (1999) isolated five cDNAs for salt stress responding genes from *Triticum aestivum* cv. Chinese spring. The clones were collectively named *WESR* (Wheat Early Stress Responding) genes *WESR1*, *WESR2* and *WESR3* were isolated from leaves and *WESR4* and *WESR5* from roots. Induction of the gene expression by NaCl treatment for 2h was confirmed by Northern blot analysis. WESR gene transcripts accumulated from 1.3-2.6 folds over the control. Based on nucleotide and deduced amino acid sequence analysis, WESR showed homology to a cDNA clone isolated from barley containing the zinc-finger motification. The deduced amino acid sequence of *WESR5* indicated homology to the glucose-6-phosphate dehydrogenase (*G5PDH*) reported in potato and alfalfa. Other *WESR* clones did not show homology with any genes of known function and all *WESR* clones are novel salt stress responding genes.

Tomato (Solanum esculentum): In tomato, genetic resources from salt tolerance have been identified largely within the related wild species, and considerable efforts have been made to characterize the genetic control of tolerance at various developmental stages (Foolad, 2004). The inheritance of several tolerance related characters have been determined and QTLs associated with tolerance at individual developmental stages have been identified and characterized. It has been confirmed that each stage of salt tolerance is largely controlled by a few QTLs with major effects and several QTLs with smaller effects, which suggest the absence of genetic relationship among the stages in tolerance to salinity. Zhang *et al.* (2003) identified 9 QTLs for salt tolerance during seed germination and 8 QTLs during vegetative stage. Mostly different QTLs were identified for salt tolerance during those two stages suggesting different genetic and physiological mechanism contributing to the salt tolerance.

Soybean (Glycine max): Guo *et al.* (2000) used PCR markers for tagging salt tolerance gene in soybean. They have identified two codominant PCR markers, a 600 bp in salt susceptible genotypes and a 700 bp or 700/600 bp bands in tolerant genotypes. The markers were closely related with salt tolerance or salt susceptible alleles. The findings suggested that those markers can be applied in the identification of salt tolerant genotypes in soybeans.

2.11.6. Transgenic Development

Salt tolerance is an important trait that requires overcoming salinity induced reduction in plant productivity. The genetic response of plants to the abiotic stresses is complex involving simultaneous expression of a number of genes. Plant genetic engineering techniques could be effectively utilized to exploit some of the untapped potentials to increase the harvestable crop yield. It involves specific gene manipulations either through over expression or silencing of alien/native genes. If a salt tolerant gene is identified which can lead to betterment of the crops, it is possible to transfer that progress in transgenic research for inclusion salinity stress tolerance, which has been presented in Table 2.8. Genes from any biological species can be transferred to crop species by genetic engineering methods. Introduction of molecular changes by genetic engineering takes lesser time compared to plant breeding methods. Only desired gene(s) can be transferred, whereas, conventional breeding approach associates with simultaneous transfer of undesired gene(s).

Transgenic research has made significant progress in crop genetic improvement with the advent of modern *rDNA* technologies. A large number of transgenics in diverse crops are on large-scale cultivation. Unlike classical breeding genetic engineering for transgenic development is faster and selectants are made available precisely with more confidence. Moreover, multiple genes can be stacked or transformed to a stock of interest. There are a large number of genes found to be instrumental and there are many functional targets for engineering tolerance to salinity. Few of the genes of importance have been briefed below.

2.11.6.1. Antiport genes

A salt concentration of 200 mM is equivalent to 40% of the salt concentration of sea water and will inhibit growth of almost all crop plants. It is observed that the wild type plants are severely inhibited by the presence of 200 mM NaCl in the growth solution and most of the plants died or were severely stunted. On the other hand, the transgenic plant grew, flowered and produced fruit. The high sodium and chloride content in the leaves of transgenic plants grown in salty water demonstrated that enhanced vacuolar accumulation of Na^+ ions, mediated by the Na^+/H^+ antiport, allowed transgenic plants to ameliorate the toxic effects of Na^+. Most notable was the production of fruit by these transgenic plants grown in the presence of 200 mM NaCl. While transgenic leaves accumulated Na^+ to almost 1% of their dry weight, the fruit displayed only a marginal increase in Na^+ content and a 25% increase in K^+ content.

A construct containing the *DtNHX* gene, coding for a vacuolar Na^+/H^+ antiport from *Arabidopsis thaliana,* was introduced into the genome of *Brassica napus* cv. Westar by Zhang *et al.* (2001). Over expression of the vacuolar $Na^+/$

H^+ antiport did not affect the growth of transgenic plants since similar growth was observed when wild type and transgenic plants were grown in the presence of 10 mM NaCl. While growth of wild type plants was severely affected by the presence of 200 mM NaCl in the growth solution, transgenic plants grew, flowered and produced seeds. They noted that the transgenic plants grown at 200 mM NaCl produced number of seeds similar to those of wild-type plants grown at low salinity. Moreover, qualitative and quantitative analysis of oil content showed no significant differences between seeds from wild type plants grown at low salinity and transgenic plants grown at high salinity. They also observed that the transgenic plants accumulated up to 6% Na without altering the yield and oil content.

2.11.6.2. CodA gene

The *adi* (Choline oxidase) gene isolated from the soil bacterium *Arthrobacter globiformis* converts choline to glycine betaine via betain aldehyde (Deshnium *et al.,* 1995). This gene encodes choline oxidase, the enzyme that converts choline to glycinebetaine. Transgenic potato plants expressing the oxalate oxidase enzyme were produced by Turhan (2005) using *Agrobacterium*-mediated genetic transformation. His findings revealed a relatively higher salt tolerance ability of transgenic than the non-transgenic genotypes *in vitro.* However, the glasshouse results were less consistent, but some transgenic genotypes showed superior yield characteristics to the non-transgenic under salinity.

The ability to synthesize and accumulate glycinebetaine contributes to salt and drought tolerance (Grumet and Hanson, 1986). Transformation of *Arabidopsis thaliana* with *adi* gene resulted in enhanced salt tolerance of transgenic *Arabidopsis* plants (Hayashi *et al.,* 1997). Thus synthesis of glycinebetaine in transgenic plants *in vitro,* as a result of the expression of *adi* gene, might be very useful in improving the ability of crop plants to tolerate salt stress (Hayashi *et al.* 2001). Sakamoto *et al.* (1998) reported transgenic rice expressing the *adi* gene in the chloroplast and the cytosol recovered to normal growth at a faster rate than the wild type after an initial growth inhibition under salt and low temperature stress.

Glycinebetaine is an extremely efficient osmoprotectant widely distributed among plants. Lilius *et al.* (1996) transformed tobacco plants with *adi* gene, which encodes for choline dehydrogenase protein (responsible for conversion of choline to betaine aldehyde) from *E. coli*. The *adi* gene introduction rendered the transgenic tobacco plants tolerant to high concentration of salt. Betaine aldehyde dehydrogenease (BADH) is the second enzyme in this pathway responsible for conversion of betaine aldehyde to glycinebetaine.

Table 2.8. List of transgenic research for salt stress tolerance

Gene	Gene action	Species	Phenotypic expression by transgenic plant	Reference
Mt1D	Manitol-1-phosphate dehydrogenase (manitol systhesis)	Tobacco	Increased plant height and fresh weight under salinity stress	Tarezynski *et al.*, 1993
Mt1D	Manitol-1-phosphate dehydrogenase (manitol systhesis)	Arabidopsis	Increased germination under salinity stress	Thomas *et al.*, 1995
MtlD	-	Tobacco	Increased salt tolerance	Liu *et al.*, 1995
p5cs	Pyrroline carboxylase synthetase (proline synthesis)	Tobacco	Enhanced biomass and flowering under salt stress	Kavi-Kishore *et al.*, 1995
betA	Choline dehydrogenase (glycine betaine synthesis)	Tobacco	Increased tolerant of salinity stress	Lilius *et al.*, 1996
HVA1	Group 3 LEA protein gene	*Oryza sativa*	Increased tolerance of drought and salinity	Xu *et al.*, 1996
codA	Coline oxidase	Arabidopsis	Seedling tolerant of salinity stress and increased germination under cold	Hayashi *et al.*, 1997
IMT1	Myo-inositol-O-methyl transferase (D-ononitol synthesis)	Tobacco	Performed better under drought and salinity stress	Sheveleva *et al.*, 1997
Nt107	Glutathione S-transeferase	Tobacco	Sustained growth under cold and salinity stress	Roxas *et al.*, 1997
codA	Coline oxidase	*Oryza sativa*	Increased tolerance of salinity and cold	Sakamoto *et al.*, 1998
p5cs	Pyrroline carboxylate synthetase (proline synthesis)	*Oryza sativa*	Increased biomass production under drought and salinity stress	Zhu *et al.*, 1998
codA	*Glycine oxidase*	Arabidopsis	Increased tolerance to salt and cold	Hayashi and Murata, 1998
DREB	Transcription factor	Arabidopsis	Increased tolerance to cold, drought and salinity	Kasuga *et al.* 1999
NHX1	-	Arabidopsis	Increased salt tolerance	Apse *et al.*, 1999
DnaK	-	Tobacco	Increased salt tolerance	Sugino *et al.*, 1999
MsPRP2	Transcription factor	Alfalfa	Increased salinity tolerance	Winicov and Bastola, 1999
Osmotin	-	*Solanum tuberosum*	Increased salt tolerance	Evers *et al.*, 1999
proDH	Proline dehydrogenase	Arabidopsis	Tolerant to freezing and high salinity	Nanjo *et al.*, 1999a
P5CS	-	Arabidopsis	Salt tolerance (hypersensitive)	Nanjo *et al.*, 1999b
AtHAL3a	Phosphoprotein phoapahatase	Arabidopsis	Regulates salinity, osmotic tolerance and plant growth	Espinosa-Ruiz *et al.*, 1999
Glycolase	-	Tobacco	Increased salt tolerance	Veena and Sopory., 1999

Contd...

Table 2.8 Contd...

AtNHX1	Over expression of Na^+/H^+ antiport	Arabidopsis	Transgenic plants grow in 200 mM NaCl	Glenn *et al.*, 1999
Mn-SOD	Expression of superoxide dismutase	*Oryza sativa*	Increased salt tolerance	Tanaka *et al.*, 1999
GS2	Chloroplastic glutamine synthetase	*Oryza sativa*	Increasede salinity resistance and chilling tolerance	Hosida *et al.*, 2000
HAL1	-	*Solanum esculentum*	Increased salt tolerance	Gisbert *et al.*, 2000
OsCDPK7	Transcription factor	*Oryza sativa*	Increased tolerance of cold, salinity and drought	Saijo *et al.*, 2000
codA	*Glycine betaine synthesis*	Brassica	Increased salt tolerance	Prasad *et al.*, 2000
p5csF	Proline synthesis (feed back inhibition removed)	Tobacco	Survived 200 mM NaCl, reduced oxidative stress induced by osmotic stress	Hong *et al.*, 2000
Apolnv	Invertase (Sucrose break down)	Tobacco	Salt tolerance high osmotic pressure increase in cell sap	Fukushima *et al.*, 2001
AtNHX1	Over expression of Na^+/H^+ antiport	Tomato	Transgenic plants grow in 200 mM NaCl	Zhang and Bhumwald, 2001
DtNHX	Over expression of Na^+/H^+ antiport	Brassica	Transgenic plants grow in 200 mM NaCl	Zhang *et al.*, 2001
Osmotin	-	*Oryza sativa*	Incrased salt and drought tolerance	Barthakur *et al.*, 2001
Mt1D	Manitol-1-phosphate dehydrogenase (manitol systhesis)	Rice	Increased salt tolerance	Li *et al.*, 2004
TSase	Trehalose synthase	Tobacco	Increased tolerance to drought and salt	Zhang *et al.*, 2005

Tobacco has been transferred with the *adi* and *betB* genes, involved in the glycinebetaine (betaine) pathway of *Escherichia coli* (Holmstrom, 1998). The transgenic plants produced and accumulated osmolyte and exhibited enhanced stress tolerance as shown by enhanced growth under salt stress and improved recovery of PS-II after salt stress.

2.11.6.3. mt1D gene

Osmotically shocked cells synthesize and accumulate massive amounts of osmoprotectory compounds. Such compounds possibly help the cells to lower their osmotic potential and to draw water from the outside medium. Manitol as an osmoprotectory compound is primarily found in microbes. By introducing manitol-1-phosphate dehydrogenease gene (*mt1D*) isolated from *E. coli* (Tarezynski *et al.*, 1993) showed over-expression of manitol in tobacco plants. These transgenic plants showed tolerance to high NaCl levels (250 mM). Seeds of transgenic *Arabidopsis* transformed with *mt1D* gene under control of CaMV 35 promoter over produced manitol and germinated in a medium supplemented with high amount of NaCl (Thomas *et al.*, 1995). Li *et al.* (2004) introduced *mt1D* gene into upland rice (*Oryza sativa* var. *japonica*) by microprojectile bombardment. Growth rate of transgenic plants was significantly higher than the control on MS medium containing 1% NaCl. Non-transgenic plants died after 35 days. They reported less membrane damage and low Na^+/K^+ ratio than the control under salt stress.

2.11.6.4. AtNHX1 gene

Glenn *et al.* (1999) have engineered transgenic *Arabidopsis* plants that overexpress *AtNHX1*, a vacuolar Na^+/H^+ antiport, which allowed the plants to grow in 200 mM NaCl. Zhang and Bhumwald (2001) reported the genetic modification of tomato plants to overexpress the *Arabidopsis thaliana AtNHX1* antiport, which likewise allowed those plants to grow in the presence of 200 mM NaCl. Besides providing farmers with a cash crop for salted lands, such, plants may also pull salt out of soils, enabling other crops to thrive again. Wheat productivity is severely affected by soil salinity due to Na^+ toxicity to plant cells. Xue *et al.* (2004) generated transgenic wheat expressing a vacuolar Na^+/H^+ antiport gene *AtNHX1*. The transgenic wheat lines exhibited improved biomass production. The field trial revealed that the transgenic wheat lines produced higher grain yield and heavier and larger grains in the field of saline soil. The transgenic rice accumulated a lower level of Na^+ and higher level of K^+ in the leaves than the non-transgenic plants under saline environment.

2.11.6.5. nahA gene

The *Escherchia coli 'nahA'* gene encodes a Na^+/H^+ antiporter, which plays critical role in ion homeostasis has been transferred into rice (*Oryza sativa* L. ssp. *Japonica*) by Wu *et al.* (2005). The transgenic plants showed better germination rate, growth and average yield per plant than control. They also reported higher sodium and proline content in transgenic lines, implying that *nhaA* over-expression enhance osmoregulation by activating the bio-synthesis of proline.

2.11.6.6. LEA gene

LEA (Late Embryogenesis Abundant protein) proteins represent a category of characteristic proteins which become abundant during the late embryogenesis. Xu *et al.* (1996) found that the *hva1* gene, which encodes for a specific class of LEA proteins, when overexpressed in rice leads to increased salt tolerance. This study demonstrated that subcellular compartmentalization of the biosynthesis of glycinebetaine was a critical step in attaining enhancement of tolerance for salinity and water stress.

Trehalose is a non-reducing disaccharide of glucose that functions as a protectant in the stabilization of biological structure and enhances the tolerance of organisms to abiotic stress. Zhang *et al.* (2005) transformed tobacco plants with trehalose synthetase (*Tsase*) gene for manipulating abiotic stress tolerance. They reported higher trehalose accumulation in transgenic plants as compared to non-transgenics. The finding suggested that the transgenic plants transformed with *Tsase* gene can accumulate higher levels of trehalose and have enhance tolerance to drought and salt stresses.

Ellul *et al.* (2003) developed transgenic water melon [*Citrullus lanatus* (Thunb.) Matsun. & Nakaj] with *HAL1* gene related to salt tolerance using *Agrobacterium*. Integration of gene was confirmed by Northern Blot analysis, and the diploid level of transgenic plants was confirmed by flow cytometry.

2.11.6.7. H^+-pyrophosphatase (H^+-Ppase) gene

An H^+-Ppase gene named TsVP involved in basic biochemical and physiological mechanisms was cloned from *Thellungiella halophila*. Transgenic tobacco overexpressing TsVP had 60% greater dry weight than wild-type tobacco at 300 mM NaCl (Gao *et al.*, 2006). Their findings suggested that over expression of H^+-Ppase causes the accumulation of Na^+ in vacuoles instead of in the cytoplasm and avoids the toxicity of excess Na^+ in plant cells.

2.11.6.8. Others

Transgenic canola plants (*Brassica napus*) developed by Zhang *et al.* (2001) grown in high salinity conditions accumulated sodium upto 6% of total dry weight. Taking consideration that a nature *Brassica* plant in the field can weight 2 kg fresh weight of 300 g dry weight, each plant could accumulate 18 g of sodium when grown in the presence of 200 mM NaCl (Epstein, 1983). This significant amount of sodium taken up by transgenic plants would suggest that in addition to value as an agronomic crop, these plants could be used as one component needed to reclaim saline soils.

2.11.7. Direct injection of genomic DNA

There are various methods of direct gene transfer into the host plants. In case of microinjection, the DNA solution is injected directly inside the cell using capillary glass micropipettes with the help of micromanipulators of microinjection assembly. The injection of DNA into lumen of developing inflorescence and it is considered that the DNA is taken up by the microspores during some specific stage of development. Transfer of DNA from the same or a different species of plant by direct injection into floral tillers may find wide application in the generation of transgenic plants with specific qualities. Prakash and Padayatty (1989) directly injected DNA of Pokkali into developing floral tillers of variety IR20, produced transgenic seeds that were similar to Pokkali in husk colour, germinated well in 0.2M NaCl and have a 4-6 folds higher proline content.

2.12. REFERENCES

Abdel-Tawab FM, Fahmy EM, Allam AI, Rashidy HA, Shoaib RM. 2003. Marker-assisted selection for abiotic stress tolerance in sugarcane (Saccharum spp.). Egyptian J Agril Res. 81(2): 635-646

Abdullah Z, Mustaq AK, Flowers TJ. 2001. Causes of sterility in seed set of rice under salinity stress. J Agron Crops Sci. 187: 25-32.

Aditya TL, Baker DA. 2005. The effect of continuous NaCl stress on the retoration of high callus induction frequency, regeneration potential and *in vitro* growth of somaclones of four Bangladeshi indica rice genotypes. Oryza. 42(2): 103-108.

Agboma PC, Sinclair TR, Kokinen P, Peltom-Sainio P, Pehu E. 997. An evaluation on the effect of exogenous glycinebetaine on the growth and yield of soybean: Timing of application, watering regimes and cultivars. Field Crop Res. 54: 51-64.

Ahsraf M. 1994. Breeding for salinity tolerant plants. Critical Rev Plant Sci. 13: 17-42.

Akbar M, Khush GS, Hille Ris Lambers D. 1985. Genetics of salt tolerance in rice. In: Rice Genetics I, IRRI, Philippines. pp. 399-409.

Akbar M, Khush GS, HilleisLambers D. 1986. Genetics of salt tolerance in rice. In: Proceedings of the International Rice Genetics Symposium. IRRI, Los Banos, Philippines. pp. 399-409.

Akbar M, Yabuno T. 1977. Breeding for saline resistant varieties of rice. Inheritance of delayed type panicle sterility induced by salinity. Jap J Bred. 27: 237-240.

Al-Aghabary R, Zhu ZJ, Shi QH. 2004. Influence of silicon supply on chlorophyll content; chlorophyll fluorescence, and antioxidative enzyme activities in tomato plants under salt stress. J Plant Nutrition. 27(12): 2101-2115.

Alam MZ, Stuchbury T, Naylor REL, Rashid MA. 2004. Effects of salinity on the growth of some modern rice cultivars. J Agron. 3(1): 1-10.

Aldesuguy SA, Badawy AM, El-Dohlolo SM. 1998. Physiological studies on adaptation of wheat plants to irrigation with sea water. J Union Arab Biol (Cairo Botany). 6(B): 267-279.

Ali AJ, Rangaswamy M, Rajagopalan R, Mohamed SEN, Manickam TS. 1998. TNRH16: a salt tolerant rice hybrid. Intrl Rice Res Notes. 23(2): 22.

Ali Y, Aslam Z, Asif M. 2004. A high yielding and slat tolerant new cotton variety -NIAB-999. Intrl J Biol Biotechnol. 1(2): 181-185.

Alia KV, Prasad SK, Saradhi PP. 1995. Effect of zinc on free radicals and proline in Brassica and Cajanus. Phytochem. 39: 45-47.

Al-Karaki GN, Hamad R. 2001. Mycorrhizal influence on fruit yield and mineral content of tomato grown under salt stress. J Plant Nutr. 24(8): 1311-1323.

Alvarez I, Tomaro LM, Benavides PM. 2003. Changes polyamines, proline and ethylene in sunflower calluses treated with NaCl. Plant Cell Tiss Org Cult. 00: 1-9.

Angrish R, Kumar B, Datta KS. 2001. Effect of gibbrellic acid and kinetin on nitrogen content and nitrate reductase activity in wheat under saline conditions. Indian J Plant Physiol. 6(2): 172-177.

Anuradha S, Seeta Ram Rao S. 2002. Alleviating influence of brassinolide on salinity stress induced inhibition of germinating and seedling growth of rice. Indian J Plant Physiol. 7(4): 384-387.

Apse MP, Aharon GS, Snedden WS, Blumwald E. 1999. Salt tolence conferred by overexpression of a vacuolar Na^+/H^+ antiporter in Arabidopsis. Science. 285: 1256-1258.

Arzani A, Mirodjagh SS. 1999. Response of durum wheat cultivars to immature embryo culture, callus induction and in vitro salt stress. Plant Cell Tissue Org Cult. 58: 67-72.

Asch F, Dingkuhn M, Wiltstock C, Dorffling K. 1999. Sodium and potassium uptake or rice panicles as affected by salinity and season in relation to yield and yield components. Plant Soil. 207: 133-145.

Ashraf M, Waheed A. 1998. Genetic basis of salt (NaCl) tolerance in lentil. Lens Newsl. 25(1-2):15-22.

Ashraf M. 1994. Breeding for salinity tolerance in plants. Critical Review. Plant Sci. 13: 17-42.

Azhar FM, McNielly T. 1988. The genetic basis of variation for salt tolerance in Sorghum bicolor L. seedlings. Plant Breed. 101: 114-121.

Azooz MM, Shaddad MA, Abdel-Latef AA. 2004. The accumulation and compartmentation of proline in relation to salt tolerance of three sorghum cultivars. Indian J Plant Physiol. 9(1): 1-8.

Balasubramanian V, Rao S. 1977. Physiological basis of salt tolerance in rice. Piso. 26: 291-294.

Balibrea ME, Rus-Alvarez AM, Bolarin MC, Alfocea FP. 1997. Fast changes in soluble carbohydrate and proline content in tomato seedlings in response to ionic and non-ionic iso-osmotic stress. J Plant Physiol. 151: 221-226.

Barakat MN, Abdel-Latif TH. 1995. In vitro selection for salt tolerant lines in wheat. II. In vitro characterization of cell lines and plant regeneration. Alexandria J Agril Res. 40(3): 139-165.

Barakat MN, Abdel-Latif TH. 1996. In vitro selection of wheat callus tolerant to high levels of salt and plant regeneration. Euphytica. 91: 127-140.

Barthakur S, Babu V, Bansal KC. 2001. Over-expression of osmotin induces proline accumulation in transgenic tobacco. J Plant Biochem Biotechnol. 10: 31-37.

Basu S, Gangopadhyay G, Mukherjee BB, Gupta S. 1997. Plant regeneration of salt adopted callus of indicia rice (var. Basmati 370) in saline condition. Plant Cell Tiss Org Cult. 57: 3-11.

Benlloch-Gonzalez M, Fournier J, Ramos J, Benlloch M. 2005. Strategies underlying salt tolerance in Halophytes are present in Cynara carduncullus. Plant Sci. 168 (3): 653-659.

Bennett J, Khush GS. 2003. Enhancing salt tolerance in crop through molecular breeding: a new strategy. J Crop Prod. 7(1-2): 11-65.

Bhargava S, Chandra N. 1989. Effect of sodium chloride on callus cultures of moth bean Vigna aconitofoloa (Jacq.) Marechal cv. IPCMO 926. Indian J Exp Biol. 27: 83-84.

Bhaskaran S, Smith RH, Schertz K. 1983. Sodium chloride tolerant callus of Sorghum bicolor. Zeitschuft fur Pflazenphysiol. 112: 459-463.

Bisnoi UR, Pancholy DK. 1980. Comparative salt tolerance in triticle, wheat and rye during germination. Plant Soil. 55: 491-493.

Blum A. 1988. Plant Breeding for stress eninvironments. CRC Press, IRC, Boca Raton, Florida.

Burke DJ, Hamerlynek EP, Hahn D. 2003. Interactions between the salt sarsh grass season, Spartina patens arbuscular mycorrhizal fungi and sediment bacteria during the growing. Soil Biol Biochem. 35: 501-511.

Cantrell IC, Linderman RG.2001. Preinoculation of lettuce and onion with VA mycorrhizal fungi reduces deleterious effects of soil salinity. Plant Soil. 233(2): 269-281.

Carlson JR, Jr Ditterline RL, Martin JM, Sands DC, Lund RE. 1983. Alfalfa seed germination in abiotic agar containing NaCl. Crop Sci. 23: 882-885.

Causse MA, Fulton TM, Cho YG, Ahn SN, Chunwongse J, Wu K, Xiao, Yu Z, Ronald PC, Harrington SE, Second G, Mc Couch SR, Tangsley SD. 1994. Saturated molecular map of rice genome based on an interspecific back cross population. Genetics. 138: 1251-1274.

Chen YW, Shao GH, Chang RZ. 1997. The effect of salt stress on superoxide dismutase in various organelles of cotyledons of soybean seedlings. Acta Agronomica Sinica. 23(2): 214-219.

Chen Z, Zhou M, Newman IA, Mendham NJ, Zhang G, Shabala S. 2007. Potassium and sodium relations in salinised barley tissue as basis of differentioal salt tolerance. Functional Plant Biology. 34: 150-162.

Cherian S, Reddy MP. 2000. Salt tolerance in the halophyte Sueda mudiflora Moq.: Effect of NaCl on growth, ion accumulation and oxidative enzymes. Indian J Plant Physiol. 5(1): 32-37.

Cherian S, Reddy MP. 2002. Micropropagation of the halophyte Suaeda mediflora Moq. through axillary bud culture. Indian J Plant Physiol. 7 (1): 40-43.

Chhipa BR. 2003. Management of salt affected soils through cultural manipulation. In: Abiotic Stresses and Crop Production, Maloo SR (ed.), Agrotech Publishing Academy, Udaipur, Indian. pp. 143-192.

Chopra VL, Paroda RS. 1986. Approaches for incorporating drought and salinity resistance in crop plants. Oxford and IBH Publ. Co. Pvt. Ltd., New Delhi.

Chu C, Wang CC, Sun CS, Hsu C, Yin KC, Chu CY, Bi FY. 1975. Establishment of an efficient medium for anther culture of rice through comparative experiments on the nitrogen sources. Scientia Sinica. 5: 659-668.

Claes B, Dekeyser R, Villaroel R, VanDen Bulcke M, Bauwi G. 1990. Characterization of a rice gene showing organ specific expression in response to salt stress and drought. Plant Cell. 2: 19-27.

Cocco G, Ferrari G, Lucci GC. 1976. Uptake efficiency of roots in plants at different ploidy levels. J Agric Sci, UK. 87: 585-589.

Colmer TD, Epstein E, Dvorak J. 1995. Differential solute regulation in leaf blades of various ages in salt-sensitive wheat and a salt-tolerant wheat ´ Lophopyrum elongatum (Host) A. Love amphidiploid. Plant Physiol. 108: 1715-1724.

Croughan TP, Stavarek SJ, Rains DW. 1978. Selection of a NaCl tolerance line of cultured alfalfa cells. Crop Sci. 18: 959-963.

Cruz V, Cuartero J. 1990. Effects of salinity at several developmental stages of six genotypes of tomato (Lycopersicon spp.). In: Pric XI Eucarpia Meeting on Tomato Genetics and Breeding, Malaga, Spain. pp. 81-86.

Cuartero J, Bolarin MC, Asins MJ, Moreno V. 2006. Increasing salt tolerance in the tomato. J Exp Bot. 57: 1045-1058.

C-Ventura J dala, Rinz L, Lopez J, Medero V, Garcia M, Rodriguez S, Cabera M, Milian M, Montano N, Reynaldo D, Torres M, cazvajal D, Galvez JR, Garica J, Albert J, Dela C-Ventura J. 2002. Effect on sodium chloride on in vitro growth of Grande naine and FHIA 03 cultivars. Biotechnologia Vegetal. 2(2): 89-93.

Daddur AM. 1977. The inheritance of salt tolerance the barley (Hordeum vulgare L.). Diss Abstr Int B. 37: 5911B.

Dagla HR, Shekhawat NS. 2005. In vitro multiplication of Haloxylon recurvum (Moq.)- a plant for saline soil reclamation. J Plant Biotechnol. 7(3): 1-6.

Dash M, Panda SK. 2000. Influence of sodium chloride salinity on some hydrolytic enzyme activity in germinating wheat seeds. Indian J Plant Physiol. 5(4): 403-404.

Datta SK, Pradhan SS. 1981. A screening method for salt tolerance of rice varieties at seedling stage. Sci Cult. 47: 444-446.

Datta SK. 1972. A study of salt tolerance of twelve varieties of rice. Curr Sci. 41: 456-457.

Deshnium, Los DA, Hayashi H, Mustard L, Murata N. 1995. Transformation of Synechococcus with a gene for choline oxidase enhances tolerance to salt stress. Plant Mol Biol. 29: 897-907.

Diallo A, Samb P, Macauley H. 2001. Water status and stomatal behaviour of cowpea Vigna unguiculata (E) Walp plants inoculated with two Glomus species at low soil moisture levels. Eur J Soil Biol. 37: 187-196.

Dix PJ. 1993. The role of mutant cell lines in studies on environmental stress tolerance: an assessment. Plant J. 3: 309-313.

Domenech J, Rasmos-Solano B, Probaza A, Lucas-Garcia JA, Juan JC, Gutierrez-Manero FJ. 2004. Bacillus spp. And Pisolithus tinetorius effects on Quercus ilex ssp. Battota: a study on tree growth, rhizosphere community structure and mycorrhizal infection. Forest Ecol Mangt. 196: 293-303.

Dridze IL, Khadeeva NV, Maisuryan AN. 1991. Production of tobacco cell lines tolerant to certain stress factors. Soviet Plant Physiol. 38: 121-125.

Dubey RS, Rani M. 1990. Influence of NaCl salinity on the behavior of protease, aminopeptidase and carboxypeptidase in rice seedling in relation to salt tolerance. Aust J Plant Physiol. 17: 215-221.

Dubey RS. 1982. Biochemical changes in germinating rice seeds under saline stress. Bio Chemine and Physiologie der Pflangen. 177: 523-525.

Dubey RS. 1983. Hydrolic enzymes of rice seeds differing in salt tolerance. Plant Physiol Biochem. 10: 168-175.

Dubey RS. 1994. Protein synthesis by plants under stressful conditions. In: Plant and Crop Stress, Pessarkli M (ed.), Marcel Dekker, Inc. New York. pp. 277-299.

Eker S, Comertpay G, Konuskan O, Ulger AC, Ozturk L, Cakmak I. 2006. Effect of salinity stress on dry matter production and ion accumulation in hybride maize varieties. Turkish J Agri Forestry. 30(5): 365-373.

Ellul P, Rios G, Atares A, Roig LA, Serrano R, Moreno V. 2003. The expression of the Saccharomyces cerevisal HAL 1 gene increase salt tolerance in transgenic watermelon [Citrullus lensetus (Thunb.) Matsun. & Nakai.]. Theor Appl Genet. 107(3): 462-469.

Epstein E. 1977. Genetic potential for solving problems of soil mineral stress: adaptation of crops to salinity. In: Proc workshop on Plant Adaptation to Mineral Stress in Problem Soil, Wright MJ (ed.) A special publication of Cornel Univ Agric Exp Sta, Ithaca, New York. pp. 73-82.

Epstein E. 1983. Crops tolerant to salinity and other mineral stress. In: Better Crops for Food, Ciba. Foundation Symposium, Nugent J, O'connor M (eds.), London: Pitman. pp. 97.

Epstein E. 1994. The anomaly of silicon in plant biology. Proc Nat Acad. Sci, (USA). 91: 11-17.

Espinosa-Ruiz A, Belles JM, Serrano R, Culianez Macia FA. 1999. Arabidopsis thaliana ALHAL: A flavoprotein related to salt and osmotic tolerance and plant growth. Plant J. 20: 529-539.

Evers D, Overney S, Simon P, Greppin H, Hausman. 1999. Salt tolerance of Solanum tuberosum L. overexpressing an heterologous osmotin-like protein. Biol Plant. 42: 105-112.

Evers D, Schmit C, Mailliget Y, Hausman TF. 1997. Growth characteristics and bio-chemical changes of polar shoots in in vitro under sodium chloride stress. J Plant Physiol. 151: 748-753.

Fitch MM, Moore PH. 1981. The selection resistance in sugarcane (Saccharum sp. Hybrid) tissue culture. Plant Physiol. 67 (4 Suppl): 26.

Flower TJ, Yeo AR. 1981. Variability in the resistance of sodium chloride salinity within rice (Oryza sativa L.) varieties. New Delhi Phytol. 88: 363-373.

Flowers TJ. 2004. Improving crops salt tolerance. J Exp Bot. 55: 307-319.

Flowers TJ, Koyama ML, Flowers SA, Sudhakar C, Singh KP, Yeo AR. 2000. QTL their place in engineering tolerance of rice to salinity. J Exp Biol. 51: 99-106.

Foolad MR, Jones RA. 1992. Parent-offspring regression estimates of heritability for salt-tolerance during germination in tomato. Crop Sci. 32(2): 439-442.

Foolad MR. 2004. Recent advances in genetics of salt tolerance in tomato. Plant Cell Tiss Org Cult. 76(2): 101-119

Fooland MR, Jones RA. 1991. Genetic analysis of salt tolerance during germination of Lycopersicon. Theo Appl Genet. 81: 321-326.

Foster BP, Phillips MS, Miller TE, Baird E, Powell W. 1990. Chromosome location of genes controlling tolerance to salt (NaCl) and vigour in Hodeum valgare and H. Chilense. Heredity 65: 99-107.

Frova C, Taramino G, Bimelli G. 1991. Sporophytic and gametophytic heat shock protein synthesis in Sorghum bicolar. Plant Sci. 73: 35-44.

Fukushima E, Arata Y, Endo T, Sonnewald U, Sato F. 2001. Improved salt tolerance of transgenic tobacco expressing apoplastic yeast-derived invertase. Plant Cell Physiol. 42: 245-249.

Gandonou CH. Abrini J, Idaomar M, Senhaji S. 2005. Response of sugarcane (Saccharum sp.) varieties to embryo callus induction and in vitro salt stress. African J Biotechnol. 4(4): 350-354.

Ganesan M, Jayabalan N. 2005. Salt-tolerant adaptation development in cotton cultivars (Gossipium hirsutum L.) through in vitro regeneration. Plant Cell Biotechnol Mol Biol. 6(3/4): 127-132.

Gao F, Gao Q, Duan XG, Yue GD, Yang AF, Zhang JR. 2006. Cloning of an H^+-Ppase gene from Thellungiella halophila and its heterologous expression to improve tobacco salt tolerance. J Exp Bot. 57(12): 3259-3270.

Garthwaite AJ, von Bothmer R, Colmer TD. 2005. Salt tolerance in wild Hordeum species is associated with restricted entry of Na^{2+} and Cl- into the shoots. J Exp Bot. 56: 2365-2378.

Geetha N, Rao GR. 1997. Growth, proline accumulation and ion contents of NaCl-tolerant and sensitive callus lives of blackgram (Vigna radiata L. Hopper). Adv Plant Sci. 10: 137-143.

Gisbert C, Rus AM, Bolarin MC, Lopez-Coronado JM, Arrillaga I, Montesinos C, Caro M, Serrano R, Moreno V. 2000. The yeast HAL1 gene improves salt tolerance of transgenic tomato. Plant Physiol. 123: 393-402.

Glenn E, Brown JJ, Blumwald E. 1999. Salt tolerant mechanisms and crop potential of halophytes. Critical Rev Plant Sci. 18: 227-255.

Gorham J, Jones RGW, Bristol A. 1990. Partial characterization of the trait for enhanced K^+ - Na^+ discrimination in the D genome of wheat. Planta. 1980: 590-597.

Gorham J, Wyn Jones RG, McDounell E. 1985. Some mechanisms of salt tolerance in crop plants. Plant Soil. 89: 15-40.

Gorham J. 1988. Genetics of sodium uptake in wheat. In: Proc 7th Intrl Wheat Genetics Symp, Miller TE, Koelsner RMD (eds.), Cambridge, UK, 13-14 July, 1988. pp. 817-821.

Gorham J. 1990. Salt tolerance in the triteceae: ion discrimination in rye and triticle. J Exp Bot. 41: 609-614.

Greenway H, Munnus RA. 1990. Mechanism of salt tolerance in non-halophytes. Ann Rev Plant Physiol. 31: 149-190.

Greenway H. 1965. Plant response to saline substrates. IV. Chloride uptake by Hordeum vulgare as affected by inhibitors, transpiration, and nutrients in the medium. Aust J Biol Sci. 18: 763-779.

Gregorio GB, Senadhira D, Mendoza RD, Manigbas NL, Roxas JP, Guerta CQ. 2002. Progress in breeding for salinity tolerance and associated abiotic stresses in rice. Field Crop Res. 76: 91-101.

Gregorio GB, Senadhira D. 1993. Genetic analysis of salinity tolerance in rice. Theor Appl Genet. 86: 333-338.

Grumet R, Hanson AD. 1986. Genetic evidence for an osmoregulatory function of glycinebetaine accumulation in barley. Aust J Plant Physiol. 13: 353-364.

Guo B, Qin LJ, Shao GH, Chang RZ, Liu LH, Xu ZY, Li XH, Sun JY. 2000. Tagging salt tolerant gene using PCR markers in soybean. Scientia Agricultura Sinica. 33(1): 10-16.

Gupta US. 1997. Crop Improvement, Vol. 2. Stress Tolerance, Oxfor & IBH Publishing Co Pvt Ltd, New Delhi. pp. 1-32.

Gururaja Rao G, Ravindra Babu V, Abhay Nath, Raj Kumar. 2001. Salt tolerance in Salvadora persica Osmotic constituents and growth during immature phase. 6(2): 131-135.

Gururaja Rao G. 2001. Studies on salt tolerance in rice: Role of silicon. J Plant Biol. 27: 57-60.

Hassanein AA. 2000. Physiological responses induced by shock and gradual salinization in rice (Oryza sativa L.) seedlings and the possible roles played by glutathione treatment. Acta Botanica Hungamica. 42: 139-159.

Hayashi H, Alia, Sakamoto A, Nonaka H, Chen THH, Murata N. 2001. Enhance germination under high salt condition of seeds of transgenic Arabidopsis with a bacterial gene (cod A) for choline oxilase. J Plant Res. 111: 357-362.

Hayashi H, Murata N. 1998. Genetically engineered enhancement of salt tolerance in higher plants. In: Stress Response of Photosynthetic Organisms: Molecular Mechanisms and Molecular Regulation, Sato Murata N (ed.), Elsevier, Ansterdam. pp. 133-148.

Hayashi HA, Mustardy L, Deshnium P, Ida M, Murata N. 1997. Transformation of Arabidopsis thaliana with the adi gene for choline oxidase: accumulation of glycine betaine and enhanced tolerance to salt and cold stresses. Plant J. 12: 133-142.

Heenan DP, Lewin LG, McCaffery DW. 1988. Salinity tolerance in rice varieties at different growth stages. Aust J Exp Agric. 28: 343-349.

Holmstrom KO. 1998. Engineering plant adaptation to water stress. Acta Univ Agric Sueciae, Agaria. No. 84: 49.

Hong Z, Lakkineni K, Zhang Z, Verma DPS. 2000. Removal of feedback inhibition of Δ^{-1} pyrroline-5-carboxylate synthatase results in increased proline accumulation and protection of plants from osmotic stress. Plant Physiol. 122: 1129-1136.

Hoque MA, Arima S. 2000. Evaluation of salt damage through cell membrane stability monitored by electrolyte leakage in water chestnut (Trapa sp.). Bulletin of the Faculty of Agriculture, Saga University. No. 85: 141-146.

Hoshida H, Tanaka Y, Hibino T, Hayashi Y, Tanaka A, Takabe T. 2000. Enhance tolerance to salt stress in transgenic rice over expression chloroplast glutamine synthetase. Plant Mol Biol. 43: 103-111.

Hu ZA, Wang HX. 1997. Salt tolerance of wild soybean (Glycine soja) in natural populations evaluated by a new method. Soybean Genet Newsl. 24: 79-80.

Ibrahim KM, Collins JC, Collins HA. 1992. Characterization of progeny of Coleus blumei following an in vitro selection for salt tolerance. Plant Cell Tiss Org Cult. 28: 139-145.

Igarasi Y, Yoshida Y, Sanada Y, Wada K, Shinazoki KY, Shinozaki K. 1997. Characterization of the gene for pyrroline-5-carbohydrate synthetase and correlation between the expression of the gene and salt tolerance in Oryza sativa L. Plant Mol Biol. 33: 857-865.

Irigoyen JJ, Emerich DW, Daiz S. 1992. Water stress induced changes in concentrations of proline and total soluble sugars in nodulated alfalfa (Medicago sativa) plants. Physiol Plant. 84: 55-60.

IRRI. 1997. Programs Report for 1996. International Rice Research Institute, Los Banos, Philippines.

IRRI. 1998. Program Report for 1997. International Rice Research Institute, Los Banos, Philippines.

IRSTON. 1991. Final Report of the 15th International Rice Salinity Tolerance Observation Nursery (14th IRSTON), 1990. In: Final Report of the 1990 INGER Nurseries. International Rice Research Institute, Los Banos, Phillippines.

Ismail AM, Azooz MM. 2002. Response of Vicia faba to salinity and vitamins. Indian J Plant Physiol. 7(3): 298-301.

Jain A, Punia MS, Dhari R. 2003. Studies on physiological parameters related to salt tolerance in soybean (Glycine max). National J Plant Improvement. 5(2): 68-70.

Jain RK, Jain S, Nainawatee HS, Chowdhury JB. 1990. Salt tolerance in Brassica juncea L. I. In vitro selection, agronomic evaluation and genetic stability. Euphytica. 48: 141-152.

Jana HK, Slinkard AE. 1979. Screening for salt tolerance in lentils. LENS Newsl. 6: 25-27.

Jones MP, Stenhouse JW. 1984. Inheritance of salt tolerance in mangrove swamp rice. Int Rice Res Newsl. 9(9): 9-10.

Jones MP. 1986. Genetic analysis of salt tolerance in mangrove swamp rice. In: Rice Genetics Proc Int Rice Genet Symp, 27-31 May, 1985, Manila, Philippines. pp. 411-422.

Joshi AJ, Sagar Kumar A, Hinglajia H. 2002. Effects of sea water on germination, growth, accumulation of organic compounds and inorganic ions in halophytic grass Heleochloa setulosa (Trin) Blatt Et Me Cann. Indian J Plant Physiol. 7(1): 26-30.

Joshi YC, Quadar A, Bal AR, Rana RS. 1980. Selective absorption and exclusion process of toxic ions. Intern Symp Salt Affected Soil, Karnal (Indian). pp. 457.

Kalir A, Omri G, Poljakoff-Mayber A. 1984. Peroxidase and catalase activity in leaves of Halminione porttulacoides L. exposed to salinity. Physiol Plant. 62: 238-244.

Kalir A, Poljakoff-Mayber A. 1981. Changes in activity of malate dehydrogenase, catalase peroxidase and superoxide dismutase in leaves of Halmione portulacoides (L.) exposed to higher sodium chloride concentrations. Annal Bot. 47: 75-85.

Kasuga M, Liu Q, Miura S, Yamaguchi Shinozaki K, Shinozaki K. 1999. Improving plant drought salt and freezing tolerance by gene transfer of a single stress inducible transcription factor. Nature Biotechnol. 17: 287-291.

Kavi Kishor PB, Hong Z, Miao GH, Hu CAA, Verma DPS. 1995. Over expression of Δ^{-1} proline-5-carboxylate synthetase increases proline production and confers osmotolerance in transgenic plants. Plant Physiol. 108: 1387-1394.

Khatun S, Flowers TJ. 1995. Effect of salinity on seed set in rice. Plant Cell Environ. 18: 61-67.

Khodary SEA. 2004. Effect of salicylic acid on growth, photosynthesis and carbohydrate metabolism in salt stressed maize plants. Intrl J Agri Biol. 6(1): 5-8.

Khush GS, Senadhira D. 1994. Strategies for breeding improved rice varieties with tolerance of adverse soil conditions. In: Rice and Problem soils in South and South-East Asia, Senadhira D (ed.), IRRI Discussion Paper Series. No. 4. pp. 145-154.

Kim SK, Lee SC, Won JG, Min GG, Lee SP, Choi BS. 1997. Effect of ABA and kinetin on NaCl injury during rice germination. Korean J Crop Sci. 42(2): 182-188.

Koyama ML, Levesley A, Koebner RMD, Flowers TJ, Yeo AR. 2001. Quantitative trait loci for component physiological traits determining tolerance in rice. Plant Physiol. 125: 406-422.

Krishnamurthy R, Anbazhagan M, Bhagawat KA. 1988. Glycinebetaine accumulation and varietal adaptability to salinity as a potential as a potential metabolic measure of salt tolerance in rice. Curr Sci. 57(5): 259-261.

Kuang VD, Dong NK. 1988. Mutagenesis and direct selection of salt resistant in the anther culture of rice. In: Biologiya Kul"tiviruemykh kletok I biotekhnologiya, 1. pp. 181.

Kumar D. 1995. Salt tolerance in oilseed Brassica- present status and future prospects. Plant Bred Abstr. 65(10): 1439-1447.

Kumar SA, Muthukumarsamy M, Paneerselvam R. 1996. Nitrogen metabolism in blackgram under NaCl stress. J Indian Bot Soc. 75: 69-71.

Lang NT, Yanagihara S, Buu BC. 2001. QTL analysis of salt tolerance in rice (Oryza sativa L.). SABRAO J Bredd Genet. 33: 11-20.

Larkin PJ, Scowcroft WR. 1981. Somaclonal variation a novel source of variability from cell culture for plant improvement. Theor Appl Genet. 60: 197-214.

Lavrichenko VG. 1985. Inheritance of salt tolerance in rice hybrids of the first two generations. Byulleten Nauchnotekhnicheskoi Informatsii Vsesoyuznogo Nauchno-issledovalerskogo Instituta Risa No 34: 3-6.

Lay T, Kanamori H, Ammon C, Nettles M, Ward S, Aster R, Beck S, Bilek S, Brudzinski M, Butler R, DeShon H, Ekström G, Satake K, Sipkin S. 2005. The Great Sumatra-Andaman Earthquake of December 26, 2004. Science. 308: 1127–1133.

Lee KS, Choi WY, Ko JC, Kim TS, Gregorio GB. 2003. Salinity tolerance of japonica and indica rice (Oryza sativa L.) at the seedling stage. Planta. 216: 1043-1046.

Lee KS, Park NK, Yang SJ. 1997. Combining ability of Japanica rices for salinity tolerance at seedling stage. Korean J Crop Sci. 42(3): 270-274.

Lee KS. 1995. Variability and genetics of salt tolerance in japonica rice (Oryza sativa L.). Ph. D. Thesis, University of the Philippines. Los Banos, Philippines.

Leith H. 1999. Sustainable Lalophytes Utilization in the Mediterran and Subtropical Dry Regions, University of Osnabrueck. pp. 1-47.

Levenko BA, Serveeva LE, Vinogradov VA, Larkina NI, Mironov EK. 1989. Selection and analysis of tobacco cell lines and their regenerants resistant to salt and water stress. In: Eksperimantalnaya genetika V uskorenii seleksionnogo prostsessa. Kev, Ukraisian SSR. pp. 101-110.

Levitt J. 1980. Response of plants to environmental stress, vol 2. Water radiation, salt and other stresses (2nd edition), Academic Press, New York. pp. 1-507.

Li ZC, Zhang XC, Zhang L, Thuang BC, Zhang CL, Wang GY, Fu YC. 2004. Expression of MtD1 gene in transgenic rice leads to enhanced salt-tolerance. J China Agril Univ. 9(6): 38-43.

Lilius G., Holmberg N, Bulow L. 1996. Enhanced NaCl stress tolerance in transgenic tobacco expression bacterial choline dehydrogenase. Bio/Technology. 14: 177-180.

Lin HX, Zhu MZ, Yano M, Gao JP, Liang ZW, Su WA, Hu XH, Ren ZH, Chas DY. 2004. QTLs for Na^+ and K^+ uptake of the shoots and roots controlling rice salt tolerance. Theor Appl Genet. 108(2): 253-260.

Liu CQ, Yan JS, Chen DM. 2004. Effects of regulatory measures on distribution of soil salt and ionic uptake by crop. Acta Pedologica Sinica. 41(2): 230-236.

Liu J, Huang S, Peng X, Liu W, Wang H. 1995. Studies on high salt tolerance of transgenic tobacco. Chinese J Biotechnol. 11: 275-280.

Lutts S, Bouharmont J, Kinet JM. 1999. Physiological characterization of salt resistance rice (Oryza sativa L.) somaclones. Aust J Bol. 47: 835-849.

Lutts S, Kinet JM, Bouharmant J. 1996. Effect of salt stress on growth, mineral nutrition and proline accumulation in relation to osmotic adjustment in rice (Oryza sativa L.) cultivars differing in salinity resistance. Plant Growth Regul. 19: 207-218.

Lyon CB. 1941. Response of two species of tomatoes and their F_1 generation to sodium sulphate in the nutrient medium. Bot Gaz. 103: 107-122.

Mahgoub EMI, Sayed-Ahmed MS. 1999. The genetics of sodium and potassium uptake and translocation in barley (Hardeum vulgare L.). Arab Univ J Agric Sci. 7: 129-143.

Mahmood T, Turner M, Stoddard FL, Javed MA. 2004. Genetic analysis of quantitative traits in rice (Oryza sativa L.) exposed to salinity. Aust J Agril Res. 55(11): 1173-1181.

Maliwal GL, Paliwal KV. 1982. Salt tolerance of some mungbean (Vigna radiata), urdbean (Vigna mungo) and guar (Cyamopsis tetrogonoloba) varieties at germination and early growth stages. Leg Res. 5: 23-30.

Mandal AB, Pramanik SC, Bikash Chowdhury, Bandopadhyay AK. 1999. Salt tolerant Pokkali somaclones performance under normal and saline soils in Bay Islands. Field Crops Res. 61: 13-21.

Mandal AB, Roy B. 2003. In vitro selection of mature seed derived calli for increased tolerance toward Fe-toxicity in rice and their isozyme profiling. J Genet & Breed. 57: 325-340.

Mandal AB, Sheeja TE, Roy Bidhan. 2000. Assesment of androclonal variation in an indica rice PTB28. Indian J Expt Biol. 38: 1054-1057.

Mandal MP, Singh RA. 2000. Effect of salt stress on amylase, peroxidase and protease activity in rice (Oryza sativa L.) seedlings. Indian J Plant Physiol. 5(2): 183-457.

Manjunatha H, Rajakumar GR, Ravishankar G, Raghaviah CV. 2003. Effect of salinity stress on seed yield through physiological parameters in sunflower genotypes. Helia. 26(39): 155-160.

Marambe B, Ando T. 1995. Physiological basis of salinity tolerance of sorghum seeds during germination. J Agro Crop Sci. 174(5): 291-296.

Marconi PL, Benavides MP, Caso OH. 2001. Growth and physiological characterization of regenerated potato (Solanum tuberosum) plants affected by NaCl stress. New Zeland of Crop Hort Sci. 29(1): 45-50.

Marcum KB, Anderson SJ, Engelke MC. 1998. Salt ion recreation: a salinity tolerance mechanism among five zoysiagrass species. Crop Sci. 38(3): 806-810.

Marschner H. 1995. Mineral Nutrition of Higher Plants, Academic Press, London.

Meroguihur AE, Burity HA, Tabosa JN, Maia FL. 2002. Salt stress response and proline accumulation in Brachiaria humidicola. Revista Argentina de Microbiologia, Argentina A. 34(2): 77-82.

Miller JF. 1995. Inheritance of salt tolerance in sunflower. Helia. 18(23): 9-16.

Mishra B, Akbar M, Seshu DV. 1990. Genetic studies on salinity tolerance in rice towards better productivity in salt-affected soils. In: Proceedings of the Paper Presented at the Rice Research Seminar, July 12, 1990. IRRI, Los Banos, Philippines.

Mishra B, Singh RK, Jetly V. 1998. Inheritance pattern of salinity tolerance in rice. J Genet Bred, (Romi) 92: 325-331.

Mishra B. 1996. Highlights of Research on Crops and Varieties for Salt Affected Soils. Central Soil Salinity Research Institute, Karnal, India. Research Bulletin. pp. 1-28.

Moeljopawiro S, Ikehashi H. 1981. Inheritance of salt tolerance in rice. Euphytica. 30: 291-300.

Mosina SB. 1986. Breeding rice for salt resistance. Truly. No. 226-294: 5-15.

Munns R, James RA. 2003. Screening methods for salinity tolerance: a case study with tetraploid wheat. Plant and Soil. 253: 201-218.

Munns R, Schachtman DP, Condon AG. 1995. The significance of two phase growth response to salinity in wheat and barley. Aust J Plant Physiol. 22: 561-569.

Munns R. 2002. Comparative physiology of salt and water stress. Plant Cell Environ. 20: 239-250.

Murashige T, Skoog, F. 1962. A revised medium for rapid growth and bioassays with tobacco tissue culture. Plant Physiol. 15: 473-497.

Narayan KK, Sree Rangasamy SR. 1991. Genetic analysis for salt tolerance in rice. In: Rice Genetics. II. Proceedings of the Second International Rice Genetics Symposium, International Rice Research Institute, Manila, Philippines. pp. 167-173.

Narayana KC, Rao CGP. 1987. Effect on salinity on seed germination and early seedling growth of Cajana cajan L. J Indian Bot Soc. 68: 451-452.

Narayanan KK, Krishnaraj S, Sree Rangaswamy SR. 1990. Genetic analysis of salt tolerance in rice. In Rice Genetics II, IRRI, Philippines. pp. 167-173.

Nemoto Y, Kuwakani N, Sarakuma T. 1999. Isolation of novel early salt responding genes for wheat (Triticum astivum L.) by differential display. Theor Appl Genet. 98: 673-378.

Ochatt SJ, Marconi PL, Radice S, Arnozis PA, Caso OH. 1999. In vitro recurrent selection of potato: Production and characterization of salt tolerant cell lines and plants. Plant Cell Tiss Organ Cult. 55: 1-8.

Ogra RK, Baijal BD. 1978. Tolerance of some sorghum varieties to salt stress at early seedling stage. Indian J Agri Sci. 48: 713-717.

Olmos E, Hemandez JA, Sevilla F, Hellin E. 1994. Induction of several antioxidant enzymes in the selection of salt tolerant cell lines of Pisum sativum. J Plant Physiol. 144: 594-598.

Orga RK, Baijal BD. 1982. Physiological studies on the effect of salinity of sorghum I. Changes in alpha-amylase and acid prefease during seedling growth. Indian J Plant Physiol. 25: 133-140.

Ottaviano E, Mulcahy DI. 1989. Genetics angiosperm pollen. Adv Genet. 26: 1-64.

Ozias-Akins Vasil IK. 1982. Plant regeneration from cultured immature embryos and inflorescence of Triticum aestivum (L.) (wheat): Evidence for somatic embryogenesis. Protoplasma. 110: 95-105.

Pandey UK, Sharma AP. 2002. Effect of salinity on potassium, calcium and magnesium content in rice varieties. Indian J Plant Physiol. 7(3): 302-304.

Pantalone VR, Kenworthy WJ. 1989. Salt tolerance in Glycine max and perennial Glycine. Soybean Genet Newsl. 16: 145-146.

Parida SK, Das AB. 2005. Salt tolerance and salinity effects on plants. Ecotoxicol Environ Safety. 60 (3): 324-349.

Pasternak D. 1987. Salt tolerance and crop production– A comprehensive approach. Ann Rev Phytopathol. 25: 271-291.

Patreze CM, Cordeiro L. 2004. Nitrogen fixing and vasicular-arbuscular mycorrhizal symbiosis in some tropical legume trees of tribe Mimosease. Forest Ecol Mongt. 196: 275-285.

Paul NK, Ghosh PD. 1986. In vitro selection for NaCl tolerant cell culture in Oryza sativa L. Current Sci. 55: 568-569.

Praksah KS, Padayatty JD. 1989. Transfer of saline tolerance from one strain of rice to another by injection of DNA. Curr Sci. 58(17): 991-993.

Pramanik SC, Mandal AB. 2000. Response of salt-tolerant rice somclones to different levels of nitrogen. Intrl Rice Res Notes. 25(3): 32-33.

Prasad SR, Bagali PG, Hittalmani, Shashidhar HE. 2000. Molecular mapping of quantitative trait loci associated with seedling tolerance to salt stress in rice (Oryza sativa L.). Curr Sci. 78: 162-164.

Pushpalatha T, Padmanabhan C. 1998. Studies on in vitro testing for salt tolerance in callus cultures of rice (Oryza sativa L.). Curr Agri. 22: 109-113.

Rabie GH, Almadine AM. 2005. Role of bioinculants in development of salt tolerance of Vicia faba plants under salinity stress. African J Biotechnol. 4(3): 210-222.

Radi AE, Ismail AM, Azooz MM. 2001. Interactive effect of some vitamins and salinity on the rate of transpiration and growth of some broad bean lines. Indian J Plant Physiol. 6 (1): 24-29.

Rajagopalan R, Robin S, Sivasubramanian P, Mohammad SEN, Ali AJ, Kandasamy M, Sivanantham M. 2004. TRYI2: a short duration, salt tolerant rice variety. International Rice Res Notes. 29(1): 28.

Ramoliya P, Patel H, Pandey AN. 2004. Effect of salinization of soil on growth and macro and micro nutrient accumulation in seedling of Salvadora persica (Salvadoraceae). Forest Ecol Mangt. 202 (1-3): 181-193.

Ramoliya PJ, Pandy AN. 2003. Soil salinity and water status affect growth of Phoenix dactylifera seedlings. New Zealand J Crop Hort Sci. 31(4): 345-353.

Rao AV, Tak R. 2002. Growth of different tree species and their nutrient uptake in limestone minspoil as influenced by arbuscular mycorrhizal (AM) fungi in India arid zone. J Arid Environ. 51: 113-119.

Rao GG. 1989. Studies on salt tolerances of pigeon pea cultivars. Leg Res. 12: 22-26.

Rao SSR, Varshini BV, Sujata E, Anuradha S. 2002. Brassinosteroids-New class of phytohormones. Curr Sci. 82: 1239-1245.

Ratanodilok N, Marcarrian V, Schmalzd C. 1978. Salt tolerance in grain sorghum. In: Agron Abstr, A.S.A., USA. p. 160.

Rathert G. 1983 Carbohydrate status in response to ion regulation of two rice cultivars (Oryza sativa L.) growing in saline medium. J Plant Nutr. 6: 817-829.

Reddy PJ, Vaidyanathan K. 1985. In vitro selection for salt tolerance in Basmati rice. Indian J Plant Physiol. 28: 88-91.

Richards LA. 1954. Diagnosia and improvement of saline and alkali soils, USDA HBK 60, p. 160.

Richards RA. 1995. Improving crop production on salt-affected soils: by breeding or management? Expl Agri. 31(4): 395-408.

Robinson DL, Dobrenz AK, Smith SE. 1986. Evaluating the genetic grains for germination salt tolerance in alfalfa using a sodium chloride gradient. Agron J. 78: 1099-1103.

Roxas VP, Smith RK Jr, Allen ER, Allen RD. 1997. Over expression of glutathione-S-transferase / glutathione peroxidase enhances the growth of transgenic tobacco seedlings during stress. Nature Biotechnol. 15: 988-991.

Roy B, Mandal AB. 2004. Towards development of mapping populations through anther culture and conventional recombination breeding for molecular tagging of salt-tolerant gene/s involving IR28 and Pokkali. 9th National Rice Biotechnology Network Meeting. pp. 183-185.

Roy B, Mandal AB. 2005. Towards development of Al-toxicity tolerant lines in indica rice by exploiting somaclonal variation. Euphytica. 145(3): 221-227.

Rush DW, Epstein E. 1976. Genotypic to salinity. Differences Between salt sensitive and salt tolerant genotypes of tomato. Plant Physiol. 57: 162-166.

Rush DW, Epstein E. 1981. Breeding and selection for salt tolerance by the incorporation of wild germplasm into a domestic tomato. J Am Soc Hortic Sci. 106: 699-704.

Sabbah S, Tal M. 1990. Development of callus and suspension cultures of potato resistant to NaCl and manitol and their response to stress. Plant Cell Tiss Org Cult. 21: 119-128.

Sacher RE, Mulcahy DL. 1981. Gametophytic selection for salt tolerance in tomato and related species. Plant Physiol. 67 (4 suppl): 19.

Sacher RF, Staples RC, Robinson RW. 1982. Salt tolerance in hybrids of Lycopersicon esculentum, Solanum pennellii and selected breeding lines. In: Biosaline Research a Look to the Future, San Pietro A (ed.), Plenum Press, New York. pp. 325-336.

Sacher RF, Staples RC, Robinson RW. 1983. Ion regulation and response of tomato to sodium chloride a homeostatic system. J Am Soc Hortic Sci. 108: 566-569.

Sairam RK. 1994. Effect of homobrassinolide application on plant metabolism and grain yield under irrigated and moisture stress conditions of two wheat varieties. Plant Growth Regul. 14: 173-181.

Sakamoto A, Alia HH, Murata N. 1998. Metabolic engineering of rice leading to biosynthesis of glycinebetaine and tolerance to salt and cold. Plant Mol Biol. 38: 1011-1019.

Sanchez FJ, Manzanares M, De Adres EF, Tenorio JL, Ayerbe I. 1998. Turgor maintenance osmotic adjustment and soluble sugar and proline accumulation in 49 pea cultivars response to water stress. Field Crop Res. 59: 225-235.

Saneoka H, Nagasaka C, Hahn DT, Yang WJ, Premachandra GS, Joly RJ, Rhodes D. 1995. Salt tolerance of glycinebetaine-deficient and -containing maize lines. Plant Physiol. 107: 631-638.

Sasse JM. 1997. Recent progress in brassinosteroid research. Plant Physiol. 100: 696-701.

Sassikumar D, Sudhagar R, Gopalan A. 2003. Screening of sunflower inbreds for sodicity through component analysis. Agricultura-Tropica-et-Subtropica. 36:73-77.

Sassikumar D, Sudhagar R, Gopalan A. 2004. Isolation of salt endurant sunflower, Helianthus annus L. genotypes through in vitro screening techniques. J Oilseeds Res. 21(1): 69-72.

Scardaci SC, Eke AU, Hill JE, Shannon MC, Rhodes JD. 1996. Water soil salinity studies on California rice. Rice Pub No. Coop Ext University, California.

Schaeffer GW. 1983. Recovery of heritable variability in anther derived double haploid rice. Crop Sci. 22: 1160-1164.

Schilling G, Schiller C, Ottas S. 1991. Influence brassinosteroids on organ relation and enzyme activities of sugarbeet plants. In: Brassinosteroids– Chemistry, Bioactivity and Applicant, Cutter HG, Yokota T, Adam G (eds.), ACS Symp Ser AM Chem Soc, Washington DE. pp. 208-219.

Schum A, Preil W. 1998. Induced mutations in ornamental plants. In: Somaclonal Variation and Induced Mutation in Crop Improvement, Jain SM, Brar DS, Ahloowalia BS (eds.). pp. 333-366.

Senadhira D, Zapata Arias FJ, Gregoria GB, Alejar MS, de la Cruz HC, Padolina TF, Galvez AM. 2002. Development of the first salt-tolerant rice cultivars through indica/indica anther culture. Field Crops Res. 76: 103-110.

Shadi AI, Rashed MA, Abou-Deif MH, Atta AH, Sarwat MI, El-Din MAT. 2003. Effect of absciric acid on some salt stressed maize inbreds. Bulletin Natl Res Centre Cairo. 28(4): 431-451.

Shadi AI, Sarwat MI, El-Din MAT, Abou-Deif MH. 2001. Effect of Benzyladenine treatment on chemical composition and salt tolerance of some maize inbreeds under salt stress. Arab Universities J Agril Sci. 9(1): 95-108.

Shankadhar D, Shankadhar SC, Mani SC, Pant RC. 2000. In vitro selection for salt tolerance in rice. Biologia Plant. 43: 477-480.

Shannon MC, Rhodes JD, Draper JH, Scardaci SC, Spyres MD. 1998. Assessment of salt tolerance in rice cultivars in response to salinity problems in California. Crop Sci. 38: 394-398.

Shao GH, Chang RZ, Chen YW, Yan SR. 1994. Study on inheritance of salt tolerance in soybean. Acta Agronomica Sinica. 20(6): 721-726.

Sharma KK, Kumar S. 1985. Studies on salt resistance in wheat. I. Growth and development Curr Sci. 54(10): 482-485.

Shen B, Jensen RG, Bohnert HJ. 1997. Increased resistance to oxidative stress in transgenic plants by targeting manitol biosynthesis to chloroplasts. Plant Physiol. 113: 1177-1183.

Shi H, Ishitani M, Lium C, Zhu J-K. 2000. The Arabidopsis thaliana salt tolerance gene SOS1 encodes putative Na^+/H^+ antiporter. Proc Nalt Acad Sci, USA. 97: 6896-6901.

Sibi Ml, Fakiri M. 2000. Androgenese et gynogenese sources de vitrovariation et de tolerance la salinite chez I. Orge Hordeum valgare Secheresse. 11(2): 125-132.

Singh AK, Singh RA, Sharma SG. 2001a. Influence of salinity on activity of hydrolytic and oxidative enzymes in chickpea seedlings. Indian J Plant Physiol. PPP

Singh BD. 2003. Breeding for Resistant to Abiotic Stresses. II. Mineral Stresses. In: Plant Breeding, Kalyani Publisher, New Delhi. pp. 410-435.

Singh RA, Roy NK, Haque MS. 2001b. Changes in growth and metabolic activity in seedlings of lentil (Lens culinaris Melic) genotypes during salt stress. Indian J Plant Physiol. 6(4): 406-410.

Singh SP, Singh BB, Singh M. 1994. Effect of kinetin on nitrogen and proline on mung bean (Vigna radiata) under saline conditions. Indian J Plant Physiol. 1: 37-39.

Smironoff N. 1993. The active oxygen in the response of plant to water deficit and desiccation. New Phytol. 125: 27-58.

Smith SE, Johanson DW, Conta DM, Hotchkiss JR. 1994. Using climatological geographical, and information to identify sources of nature-plant salt tolerance in alfalfa. Crop Sci. 34(3): 690-694.

Snape JW. 1989. Double haploid breeding: Theoritical basis and practical applications. In: Review of Advances in Plant Biotechnology, 1985-1988, Mujeeb-Kazi A, Sitch LA (eds.), CIMMYT, Mexico, DF Mexico and IRRI, Los Banos, Philippines. pp. 19-30.

Spivakov NS. 1990. Genetic nature and inheritance of salt resistance in sorghum. Doklady Vsesoyuznoi Krasnogo Znameni Akademii Sel'skokhozyaistvennykh Nauk in V.I. Lenina. 4: 24-27.

Subhashini K, Reddy GM. 1989. Evaluation of the progeny under stress of regenerated salt tolerant rice. J Genet Breed. 43: 125-129.

Sugino M, Hibino T, Tanaka Y, Nii N, Takabe T, Takabe T. 1999. Overexpression of DnaK from a halotolerant cyanobacterium Aphanothece halophytica acquires resistant to salt stress transgenic tobacco plants. Plant Sci. 146: 81-88.

Sukarin W, Kunpatcharamurak W, Okabe T. 1993. Simple testing methods for salt tolerance. Japanese J Tropical Agri. 37(1): 42-45.

Sureena, Dhingra HR. 2003. Evaluation of relative salt tolerance in Brassica on the basis of pollen performance. Indian J Plant Physiol. 8(2): 133-137.

Tain CY, Feng G, Li XL, Zhang FS. 2004. Different effects of arbuscular mycorrhizal fungal isolates from saline or non-saline soil on salinity tolerance of plants. Appl Soil Ecol. 26(2): 143-148.

Tanaka Y, Hibino T, Hayasi Y, Tanaka A, Kishitani S, Takabe T, Yokota S, Takabe T. 1999. Salt tolerance of transgenic rice overexpressing yeast mitrochondrial Mn-SOD in chloroplasts. Plant Sci. 148: 131-138.

Tanvir MK, Rustam Ali, Saleem A, Sajid-Jr R. 2002. In invo evaluation of sugarcane somaclones for salt tolerance. Pakistan Sugar J. 17(6): 6-8.

Tarezynski MC, Tensen RG, Bohnert H.J. 1993. Stress protelion of transgenic tobacco by production of the osmolyte manitol. Science: 259. 508-510.

Thomas JC, Sepathi M, Arendall B, Bohnert HJ. 1995. Enhancement of seed germination in high salinity by engineering manitol expression in Arabidopsis thaliana. Plant Cell Environ. 18: 801-806.

Tian C, He X, Zhong Y, Chen J. 2002. Effects of VA mycorrhizae and Frankia dual inoculation on growth and nitrogen fixation of Hippophae tibetana. Forest Ecol Mangt. 170: 307-312.

Tisdale SL, Nelson WJ, Beaton JD. 1993. Soil fertility and fertilizers. McMillan Publishing Company, New York.

Toenniessen GH, Khush GS. 1991. Prospects for the future. In: Biotechnology in Agriculture, Vol. 6 Rice Biotechnology, Khush GS, Toenniessen GH (eds), CAB International / IRRI, Wallingford, UK/Manila, Philippines. pp. 309-313.

Tripathi JN, Zhang J, Robin S, Nguyen TN. 2000. QTLs for cell-membrane stability mapped in rice (Oryza sativa L.) under drought stress. Theor Appl Genet. 100: 1197-1202.

Trivedi HB, Ramana Rao, TV, Bagdi DL, Gururaja Rao G. 2004. Influence of silicon on growth and salt uptake in wheat under salinity. Indian J Plant Physiol. 9(4): 360-366.

Tsuchiya M, Bonilla P, Kumano S. 1995b. Physiology response to salinity in rice plant. V. Varietal difference in ion exclusion of root exposed to NaCl stress under different hydraulic pressures. Japanes J Crop Sci. 64(1): 102-108.

Tsuchiya M, Miyake M, Bonilla P, Kumano S. 1995a. Physiological response to salinity in rice. IV. Ion exclusion in the excised root-exposed to NaCl stress with hydraulic pressure. Japanese J Crop Sci. 64(1): 93-101.

Turhan H. 2005. Salinity response of transgenic potato genotypes expressing the oxalate oxidase gene. Turkish J Agri Forestry. 29(3): 187-195.

Uma S, Ravishankar KV, Prasad TG, Reid JL, Kumar MU. 1993. Abscisic acid responsive proteins induce salinity stress tolerance in finger millet (Eleusine coracana Gaertn.) seedlings. Curr Sci. 65(7): 549-554.

Van Sint Jan V, Skali-Senhaji N, Bouharmont J. 1990. Comparison de differences varieties de riz. (Oryza sativa L.) pour leur aptitude a la culture in vitro. Belg J Bot. 123 (1/2): 36-44.

Veena Reddy VS, Sopory SK. 1999. Glyoxalase I from Brassica juncea: molecular cloning, regulation and its over-expression confer tolerance in transgenic tobacco under stress. Plant J. 17: 385-395.

Verma BL, Sharma Y, Verma A, Singhania RA. 2003. Screening of elite genotypes of mustard for cultivation under irrigation with quality waters. Crop Res, Hissar. 26(2): 237-239.

Walton, M. 2005. Scientists: Sumatra quake longest ever recorded. CNN. May 20, 2005.

Wang QL, Chen RX, Wang YH. 1991. Effect of vitamine on salt resistance in cotton. China Cotton. No. 4(30): 35.

Wang WX, Vinocur B, Altman A. 2003. Plant respons to drought, salinity and extreme temperatures: Towards genetic engineering for stress tolerance. Plant. 218(1): 1-14.

Weretilnyk EA, Bendarek S, McCue KF, Rhodes D, Hanson AD. 1989. Comparative biochemical and immunological studies of the glycine betaine synthesis pathway in diverse families of dicotyledons. Planta. 178: 342-352.

West M, Sanches John J, McNutt SR. 2005. Periodically Triggered Seismicity at Mount Wrangell, Alaska, After the Sumatra Earthquake. Science. 308(5725): 1144–1146.

Wilen RW, Sacco M, Gusta LW, Krishna P. 1995. Effect of 24 epibrassinolide on growing and thermo tolerance of brome grass (Bromus inermis) cell culture. Physiol Plant. 95: 195-202.

Winicov I, Bastola DR. 1999. Transgenic over expression of the transcription factor Alfin1 enhances expression of the endogenous MSPRP2 gene alfalfa and improve salinity tolerance of the plant. Plant Physiol. 120: 473-480.

Won YJ, Hen MH, Koh HJ. 1992. Cation content of salt tolerant and salt susceptible cultivars and its inheritance in rice. Korean J Crop Sci. 37(1):1-8.

Wong CK, Woo SC, Ko SW. 1986. Production of rice plantlets on NaCl stressed medium and evaluation of their progenies. Bot Bull Acad Sinica. 24: 11-23.

Wu LQ, Fan ZM, Guo L, Li YQ, Chen ZL, Qu LJ. 2005. Over-expression of the baterial nhaA gene in rice-enhances salt and drought tolerance. Plant Sci. 168(2): 297-302.

Xu, D, Duan X, Wang B, Hong B, Ho TD, Wu R. 1996. Expression of a late embryogenesis abundant protein gene, HVA1, from barley confers tolerance to water deficit and salt stress in transgenic rice. Plant Physiol. 110: 249-257.

Xue ZY, Zhi Dy, Xue GP, Zhang H, Zhao YX, Xia GM. 2004. Enhance salt tolerance of transgenic wheat (Triticum aestivuns L.) expressing a vacuolar Na^+/H^+ antiporter gone with improved grain yields in saline soils in the field and a reduced level of leaf Na^+. Plant Science. 167(4): 849-859.

Yadav JSP, Gupta IC. 1984. Usar Bhumi Ka Sundar (in Hindi), ICAR, New Delhi.

Yadav JSP. 1993. Salt affected soils and their management with special rererence to UP. J Indian Soc Soil Sci. 41: 623.

Yang YW, Newton RJ, Miller FR. 1990a. Salinity tolerance in sorghum. I. Whole parent response to sodium chloride in S. bicolor and S. halepense. Crop Sci. 30: 775-781.

Yang YW, Newton RJ, Miller FR. 1990b. Salinity tolerance in sorghum. II. Cell culture response to sodium chloride in S. bicolor and S. halepense. Crop Sci. 30: 781-785.

Yano-Melo AM, Saggin OJ, Maia LC. 2003. Tolerance of mycorrhized banana (Musa sp. Cv. Pacovan) plantlets to saline stress. Agric Ecosystems Environ. 95(1): 343-348.

Yasar F, Yzal O, Tufenkci S, Yildiz K. 2006. Ion accumulation in different organs of green bean genotypes grown under salt stress. Plant Soil and Envrion. 52(10): 476-480.

Yeo AR, Flower SA, Rao GG, Welfare K, Senanayake N, Flower TJ. 1998. Silicon reduces sodium uptake in rice in saline conditions and this is accounted for by a reduction in the transpirational by pass flow. Plant Cell Environ. 22: 559-565.

Yilmaz K, Akinci IE, Akinici S. 2004. Effect of salt stress on growth and Na, K contents of pepper (Capsicum annum L.) in germination and seedling stages. Pakistan J Bio Sci. 7(4): 606-610.

Yoshida S, Navsera SA, Ramirez EA. 1969. Effect of silica and nitrogen supply on some leaf characters of rice plant. Plant Soil. 31: 48-56.

Yoshida Y, Kiyouse T, Nakashima K, Yamaguchi-Shinozaki K, Shinozaki K. 1997. Regulation of levels proline as an osmolyte in plants under water stress. Plant Cell Physiol. 38: 1095-1102.

Zaidi PH, Singh BB. 1996. Modulation of adverse effects of salinity by growth regulators in soybean II. Growth, nutrient concentrations and balances. Plant Physiol Biochem, New Delhi. 23(2): 139-144.

Zandavalli RB, Dillenburg LR, Paula VD. 2004. Growth responses of Araucaria angustifolia (Araucariaceae) to inoculation with the mycorrhizal fungus Glomus clarum. Appl Soil Ecol. 25 (3): 245-255.

Zapata FJ, Alejar MS, Torrizo LB, Novero AY, Singh UP, Senadhira D. 1991. Field performance of anther culture-derived lines from F_1 cross of indica rice under saline and non-saline conditions. Theor Appl Genet. 83: 6-11.

Zeng L, Shannon MC, Lerch SM. 2001. Timing of salinity stress affects growth and yield components. Agril Water Management. 48: 191-206.

Zhang GY, Guo Y, Chen SL, Chen YS. 1995. RFLP tagging of a salt tolerance gene in rice. Plant Sci. 110: 227-234.

Zhang H-X, Blumwald E. 2001. Transgenic salt tolerant tomato plants accumulate salt in the foliage but not in the fruits. Nature Biotechnol. 19: 765-768.

Zhang H-X, Hadson J, Williams JP, Blumwald E. 2001. Engineering salt tolerant Brassica plants: characterization of yeild and seed oil quality in transgenic plants with increased vacuolar sodium accumulation. Proc Natl Acad Sci USA. 98: 12832-12836.

Zhang J, Kirkhan MB. 1994. Drought stress induced changes in activities of superoxide dismutase catalase and peroxidase in wheat species. Plant Cell Physiol. 35: 785-791.

Zhang LR, Lin GY, Foolad MR, 2003. QTL comparison of slat tolerance during seed germination and vegetative stage in a Lycopersicon esculentum L. p. mpinellifolium RIL population, In: Environmental stress and horticulture crops, a proceedings of the XXVI International Horticultural Congress on Tomato, Canada, 11-17 August, 2002. Acta-Horticulture, No. 618, 59-67.

Zhang SZ, Yang BP, Feng CL, Tang HL. 2005. Genetic transformation of tobacco with the trehalose synthase gene from Grifola forndosa Fr. Enhances the resistance to drought and salt in tobacco. J Intensive Plant Biol. 47(5): 579-587.

Zheng LH. 2004. Response and correlated response of yield parameters to selection for salt tolerance in rice. Cereal Res Comm. 32(4): 477-484.

Zhu B, Su J, Chang MC, Verma DPS, Fan YL, Wu R. 1998. Over expression of a pyrroline 5 carboxylate synthatase gene and analysis of tolerance to water and salt stress in transgenic rice. Plant Sci. 139: 41-48.

Unit-II
WATER STRESSES

3

Waterlogging Tolerance

3.1. INTRODUCTION

Transitory waterlogging occurs extensively in both irrigated and dry land agriculture on clay flats and duplex soils. Most of the humid tropical lowlands, with adequate, rain-fed but poor drainage, have the problems of waterlogging, which is further intensified by low water table and poor root aeration. In submerged condition, only root is exposed to excess water and the injury to the shoot is not a primary effect.

The term *'waterlogging'* is defined as a condition of the soil where excess water inhibits gas exchange of roots with the atmosphere. Waterlogging is different from 'flooding' because the latter results in addition factors of partial or complete submergence of the shoot. The waterlogging replaces a gaseous air by liquid water, leading to the gas stresses. Schematic representation of adverse effects of waterlogging on plant growth and survival has been given in Fig. 3.1.

Severe soil drainage constraints are estimated to adversely affect approximately 10% of the global land area (FAO, 2002). Water logging adversely affects bread wheat production in 4.7 mha in irrigated soils of the Indo-Gangetic Plains of Northern India (CSSRI, 1997). It includes 2.5 m ha of sodic soils (Sharma and Swarup, 1988) and 2.2 m ha affected by seepage from irrigation canals (CSSRI, 1997). Such problems become more acute when the soils are not leveled or irrigation is followed by excess rain (Gill *et al.*, 1992). Large areas of waterlogging occur in the irrigated rice-wheat rotation systems used throughout South and South-east Asia including Pakistan, India, Bangladesh, Nepal and China. Wheat is exposed to waterlogging in these system since the soil preparation used for rice cultivation specifically results in sub soil compactness to optimize flooding conditions for rice (Samad *et al.*, 2001). Another major cause of waterlogging in these countries is the use of fertilizer containing high carbonate and bicarbonate concentrations which induces sodicity in this typically fine texture soil, intermittent waterlogging to the heavy clay vertisols, which can be characterized by long duration of waterlogging.

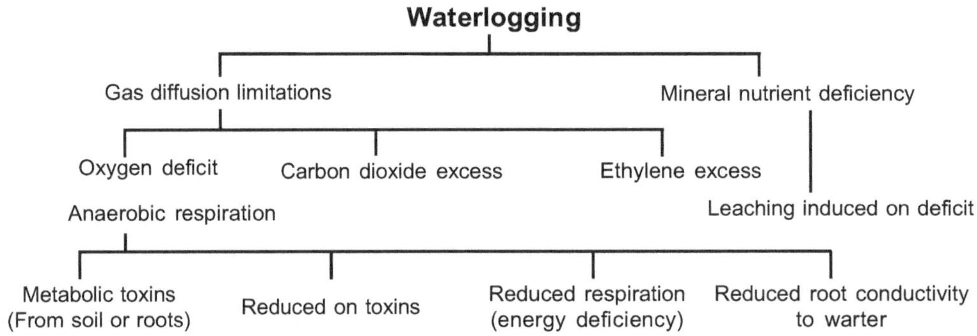

Fig. 3.1. Schematic representation of adverse effects of waterlogging on plant growth and survival (Setter and Waters, 2003)

The rapid fluctuations in waterlogging and in drainage on duplex soils can be explained by the soil profile. During rainfall, water rapidly penetrates through the sandy topsoil of the 'A' horizon and accumulates within and above the compacted gravel and 'B' horizon; this water accumulation occurred at about 60 cm depth in the soil, and resulted in a perched water table many meters above the ground water table. When water influx causes, the perched water table can rapidly fall as it drains to the ground water table. This rapid drainage occurs through cracks and root channels in the 'B' horizon.

3.2. CAUSES OF WATERLOGGING

1. Transient waterlogging occurs primarily in sandy duplex soils, where rainfall rapidly penetrates sandy topsoil and accumulates above compact clay subsoil with low hydraulic conductivity at 5-100 cm depth (Tennant *et al.*, 1992; Samad *et al.*, 2001).

2. Rising of groundwater and flooding in river basins are another cause of waterlogging (Grieve *et al.*, 1986; McDonald and Gardner, 1987; Meyer and Barrs, 1988).
3. Waterlogging timing; usually be concurrent with irrigation schedules, high rainfall or surface flooding events (Williamson and Kriz, 1970). Continuous rainfall coupled with inadequate drainage leads waterlogging.

3.3. CHARACTERISTICS OF WATERLOGGED SOILS

The major changes in the waterlogged soils are physical, biological and chemical and these changes have been briefly described below.

3.3.1. Physical Changes

Upon flooding, the pore spaces (filled with air) in the soil become saturated with water, as a result, the soil swells. Since the exchange of air between the atmosphere and the soil is impeded, and since the water particles are held by soil particles and prevent from percolating downward and escaping.

3.3.2. Biological Changes

The absence of soil air in waterlogged condition causes a change in the varieties of microbes, microscopic organisms which live in the soil. Anaerobic microbes tend to be much slower; less efficient decomposer of organic matter than the aerobic microorganisms. Consequently, the rate of decay of organic matter tends to be slow in flooded soils. Also the end products produced by anaerobic decomposition differ some are toxic to crop plants; particularly those released during the first two weeks after decomposition begins and the toxicity produced during decomposition may stunt the growth of crop plants.

3.3.3. Chemical Changes

Flooded soils develop two distinct chemical zones (Fig. 3.2); oxidized and reduced zones. The upper oxidized zone (1-10 mm) absorbs oxygen from the water, turns brown in colour, and reacts to nitrogen like an unfolded soil. The lower reduced zone, which extends down as far as the water, is extremely low in available oxygen, turns dark blue or gray in colour, and takes on chemical properties quite different from those of oxidized layer above.

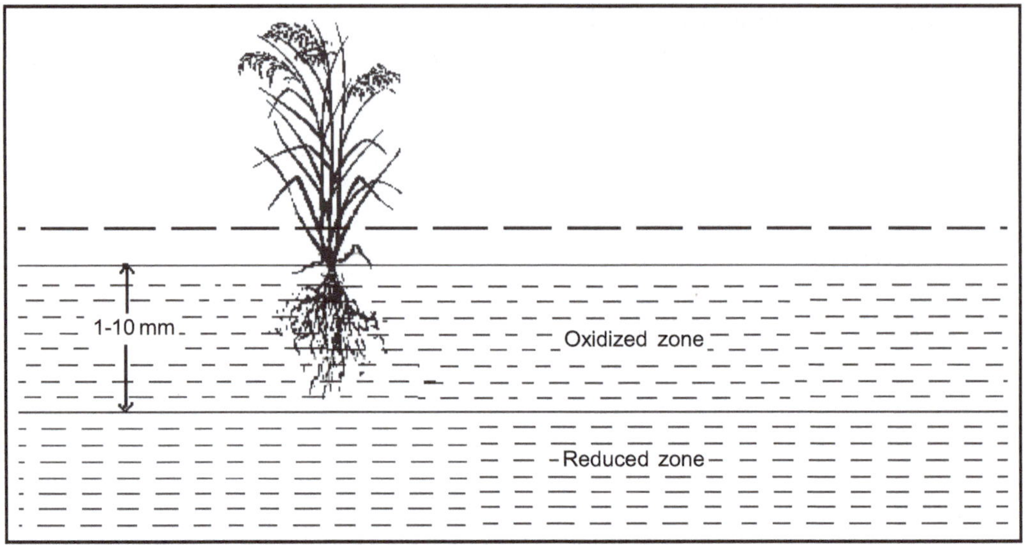

Fig. 3.2. Chemical zones of flooded soil

3.4. WATERLOGGING SYMPTOMS ON PLANT

1. Leaf rolling under excessive moisture is the primary leaf.
2. The visible effects of waterlogging on plants includes wilting, reduced plant growth, chlorosis, senescence, abscission of lower leaves, defoliation (Fig. 3.3), hypertrophy, epinasty (downward growth of petioles), lenticels formation, leaf curling, and ultimately reduced yield.

Fig. 3.3. Effect of waterlogging on potato plant followed by removal of water - stunted growth and defoliation of lower leaves. *(For colour version of this figure see page 535)*

3. Flooding suppresses the formation of root hair and branching of roots. On flooding, all terrestrial plants become chlorotic within 3-4 days.
4. Flooding barley plants for 7 days decreased chlorophyll *a* & *b* content of leaves and weakened the chlorophyll-protein bond (Mikhailova, 1977).
5. Brace root development from the above the ground nodes in maize (Zaidi and Singh, 2001; Zaidi *et al.*, 2003).
6. Field flooding causes premature senescence in soybean, which results in leaf chlorosis, necrosis, defoliation, cessation of growth and reduced seed yield (Oosterhuis *et al.*, 1990; VanToai *et al.*, 1994; Linkemer *et al.*, 1998). On-farm research indicated that flooding for as little 3 days at the early vegetative growth stages can kill soybean (Sullivan *et al.*, 2001).

3.5. INJURY MECHANISMS

Both root and shoot growth negatively affect by waterlogging. It significantly decreases with the chlorophyll content, CO_2 assimilation rate, and mineral quantum efficiency of photosystem II (Pang *et al.*, 2004) etc.

3.5.1. Reduce Gas Exchange

The environmental factors associated with waterlogging act independently, when it is most likely that several of these factors act simultaneously (Jackson and Drew, 1984). Such as, interactions of effects of low O_2, high CO_2 and high ethylene on many crops exposed to either waterlogging or flooding. In waterlogging condition, gaseous air space is displaced with liquid water. Root systems are this suddenly plunged into an anaerobic condition by waterlogging. The gas exchange between the roots and atmosphere is reduced drastically, because gases diffuse 1000 times more slowly in water than in air (Armstrong, 1979). Restricted soil aeration also impairs root respiration by retarding nutrient and water uptake; limits growth of root systems; stimulate toxic metabolism, disturb hormone metabolism and prevent the ordinary functioning of essential biological, chemical and physical processes associated with fertile and productive soils. With high temperature in tropics, the problem is further aggravated, as more O_2 is required to produce a given growth response than at lower temperature. Microorganisms affect the distribution of O_2 influence the supply of nutrients and under anaerobic conditions produce extremely toxic compounds, like sulphide and butyric acid.

Oxygen is vital in the central energy providing pathway of the cell, and the presence or absence of oxygen determines metabolic activity and energy production. Oxygen serves as an electron acceptor in the oxidative phosphorelation pathway, which generates ATP, the prime source of energy for cellular metabolism. Other biochemical pathways that involve cytochromes, oxidases and desaturases, for example, in sterol and fatty acid biosynthesis are oxygen-dependent. With time of waterlogging the soil gradually loses much or all of its O_2, concentrations of other gasses increase, certain microelements are reduced and increase in concentration in the soil solution and phytotoxins accumulate in waterlogging environment. Oxygen will rapidly accumulate and leads to *hypoxic* (low oxygen) or *anoxic* (no oxygen) conditions around the roots, which are major determinants of the adverse effects of flooding. When a soil becomes waterlogged, the rate of O_2 depletion is dependent on several factors, including the respiration rate of plant roots and microorganisms, the solubility of O_2 in water, and the rate of O_2 diffusion through the soil (Trought and Drew, 1982). Damage and death of flooded plants have been attributed to the lack of oxygen to support root respiration (Setter and Belford, 1990; Crowford, 1992). Permeability of the root cells decrease due to the shortage of oxygen. Permeability to water decreases by a factor of over three in few hours, but decrease in the permeability to nutrients is not so immediate (Russell, 1977). The xylem vessels get plugged and then blocked due to the slimy substance produced by bacteria. Thus root system in such plants becomes dead and stops its function of water and nutrient uptake. Translocation of assimilates from leaves to the roots is also stopped (Kosmakova and Zvereva, 1973). Root growth and penetration in the soil is reduced in cereals like wheat, barley and rye (Watson *et al.*, 1976).

CO_2 causes cellular toxicity in plants. CO_2 is known to penetrate the cytoplasmic membrane readily, resulting in toxic cellular acidification. Anoxic condition results in accumulation of CO_2, ethylene and other gases such as, hydrogen sulphide, ammonia and methane. Increased concentrations of these gases lead in various stress injuries. Being a growth inhibitor ethylene further complicates the effect of O_2 deficit stress on plant growth and development. Excess CO_2 concentration in the rhizosphere may lead to accumulation of toxic quantities of acetaldehyde and ethanol from pyruvate, probably due to strong controlling effects of CO_2 on microbial activities. High concentrations of CO_2 also suppress the activity of catalase, glycolate oxidase and nitrate reductase.

Sea water saturated with CO_2 halts protoplasmic streaming and increase its viscosity in *Nitella clavata* (Fox, 1933). Kramer and Jackson (1954) reported that field soil saturated with CO_2 inhibited water uptake by reducing the permeability of tobacco cell membranes. In addition to the reduction of water absorption and transport, CO_2 reduces absorption of K, N, P, Ca and Mg by roots of wheat, maize and rice and also lead to P and Fe deficiency in sorghum plants (Chang and Loomis, 1945; Matocha and Mostaghimi, 1988).

Soybean plants are very tolerant of excess water and anaerobiosis, but are injured under anaerobic conditions when root zone, CO_2 concentration increase to those found in flooded soybean fields (30%). Rice, a flooding-tolerant species, is much more tolerant of elevated root-zone. CO_2 levels than in soybean. Boru *et al.* (2003) suggested that CO_2 toxicity is a factor affecting soybean tolerance to flooded soils.

3.5.2. Water Deficit in Aerial Parts

Standing water or water-saturated soils face water deficit in the aerial parts, and they develop symptoms of drought stress. Excess moisture-induced injuries on cell membrane of root tissues affect the effectiveness of roots for water and nutrient uptake. It affects the membrane integrity, and nutrient uptake. The stress condition affects membrane integrity, and therefore, membrane become leaky. This disturbs osmotic gradient of the root cortex, and cause poor water uptake by roots. Thus, root can not fulfill the demand of water requirement by the aerial parts leading to water stress in the upper parts of the plant.

3.5.3. Nutrients Imbalance

Nutrient deficiency appears to be the major cause for the poor plant growth in waterlogged soils. In general, excess moisture stress causes nutrient imbalance by affecting nutrient ion homeostasis. Due to O_2 deficiency in the root zone, synthesis of ATP may be inhibited leading thus to a decrease in nutrient uptake. In some cases, it also cause nutrient toxicity, such as Fe and Mn. In contrary, water culture experiment conducted by Steffens *et al.* (2005) with anaerobic N_2 and aerobic aeration confirmed that O_2 deficiency did not induce nutrient toxicity (Fe and Mn), but caused suboptimum nutrient supply (N, P, K, Mn, Cu and Zn) for wheat and barley plants. In common, nutrient stress due to waterlogging may be due to following reasons:

1. O_2 deficit in root environment may produce mineral nutritional deficiency due to decreased uptake of ions, and poor distribution within the plant parts.

2. Low redox potential due to low O_2 partial pressure in submerged soil may reduce some elements in more soluble and toxic forms. For example, Fe^{3+} reduces to Fe^{2+}.
3. Nutrient stress may be due to leaching of essential nutrients from rhizosphere and intermediate metabolites from roots.

3.5.4. Change in Phytohormones Concentration and Activity

Concentrations of root-zone CO_2 more than 2% were shown to inhibit ethylene binding and action in sunflower and photo plants (Govindarajan and Pooraiah, 1982; Finlayson and Reid, 1996). Ethylene is known to be necessary for the formation of adventitious roots and aerenchyma in the acclimation responses to flooding (Jackson *et al.*, 1985; Justin and Armstrong, 1991). Soybean plants treated with elevated CO_2 produced fewer adventitious roots than did plants in the N_2 bubbling and non-aerated treatments (Boru *et al.*, 2003). Thus, some of the injurious effects of high root-zone CO_2 in soybean might be caused indirectly by the lack of ethylene action. In rice, treatment with 30% CO_2 in N_2 had no adverse effect on the number of adventitious roots. Actual concentrations of CO_2 depend upon many factors including soil water content, soil type, organic matter content and microbial activity (Duenas *et al.*, 1995; Bouma *et al.*, 1997). In many plant species, CO_2 absorbed by the root is transported to the shoot through the transpiration stream at a very low concentration (Enoch and Olesen, 1993). A second transport is present in rice plants: in this species, CO_2 can be transported in the gaseous phase via an extensive network of lysigenous gas spaces at concentrations of 3 to 4 fold higher than in other species (Higuchi *et al.*, 1984).

Ethylene concentration increases in the tissues of various species. Premature leaf senescence and chlorosis, and adventitious root production could be affected by anoxia-induced ethylene production (Jackson and Campell, 1976). They also showed relationship between waterlogging, O_2 deficiency, ethylene production and leaf epinasty in tomato. Ethylene concentration in sunflower cuttings increased 5-fold within 6 h after submergence in distilled water and then declined (Kawase, 1976).

Flooding markedly reduce gibberellins levels in xylem sap of tomato plants (Reid *et al.*, 1969). Gibberellin level falls after 1 or 2 days of flooding, and the exogenous application of gibberellic acid stimulates stem growth of the waterlogged plants (Reid and Crozier, 1971). Upward transport of cytokinins is also inhibited in waterlogged condition which leads reduction in plant growth, enhance chlorosis and abscission. Reduced cytokinin production leads to symptoms of early senescence (Livine and Yaadia, 1972). Wright and Hiron (1970) suggested that ABA concentration increase in flooded bean seedlings and their increase may be the results from an incipient wilting of leaves. Increased levels of ABA influence GA-biosynthesis, stem elongation, senescence and abscission.

3.5.5. Disturbance in Root Metabolism

Oxygen deficiency in the root system disrupts root metabolism, which forcing the plant to switch from aerobic to anaerobic respiration resulting in a decrease of ATP production, accumulation of toxic end-products of anaerobic respiration, and rapid depletion of organic compounds. Continuous supply of fermentable sugar to root tissue is critical for long-term survival of plants under excessive moisture condition. Accumulated starch in leaves and stems has low translocation ability to roots, which results in lower level of root metabolism and retarded root growth. The limited supply of available energy reduces the absorption and translocation of water and nutrients. The disruption of root metabolism caused by inadequate oxygen supply also adversely affects the hormonal balance of the shoot and suppresses the synthesis and translocation of hormones in the root.

3.5.6. Decreased Leaf Epidermal Conductance

Leaf epidermal conductance to water vapour decreased by 47% after 24 h flooding of tomato plants accompanied by an increased in the sensitivity of stomata to change in CO_2 concentration. The assimilation rate fell by 27% and the inhibition could not be overcome by elevated CO_2 partial pressure (Kent, 1983). Waterlogging reduced the total photosynthetic area and the rate of photosynthesis of sorghum (Bhagwat *et al.*, 1986).

3.5.7. Morpho-physiological Changes

Morphological characters, namely culm height, number of tiller per plant, number of functionally active leaves, number of panicles per plant, percentage of filled grains per panicle, grain yield and harvest index significantly reduced in rice under waterlogging condition (Neog *et al.*, 2002). Higher tissue iron (Fe^{2+}) content under waterlogging condition caused the bronzing appearance symptoms, which reduced grain yield. During waterlogging, the growth rate roots of wheat decreased more than that of shoot and plant growth and plant growth was reduced proportionately as the water level was increased (Malik *et al.*, 2001). The major physiological changes take place under waterlogging condition has been briefed here.

3.5.7.1. *Photosynthesis*

The rate of photosynthesis reported to be reduced photosynthesis under waterlogged condition. Waterlogging increased malondialdehyde content in flag leaf and decreased root vigour, chlorophyll content, net photosysnthesis rate, nitrate reductase activity, relative water content of flag leaf and crop yield in wheat (Zhou and Zhu, 2002). Neog *et al.* (2002) also found reduced in total chlorophyll content under waterlogged environment. The causes of reduction in photosynthesis may be as follows:

1. Reduced stomatal conductance, which interferes with CO_2 diffusion in leaf.
2. Reduced chlorophyll content of leaf is due to enhanced senescence.
3. Reduced in the activities of key enzymes, such as ribulose-bisphosphate carboxylase.
4. Inhibition in translocation of photosynthates due to interference in phloem translocation and stressed-induced reduction in sink area.

3.5.7.2. *Respiration*

A drastic reduction in aerobic respiration in root has been reported under excess water in the soil. Due to anoxic condition, there is switch over from aerobic to anaerobic respiration. To generate energy plants require a large amount of carbohydrate because of inefficiency of anaerobic respiration compared to aerobic respiration. Switchover from aerobic to anaerobic respiration severely curtails energy availability by 88%. It produces only two ATP per mole of glucose against 36 ATP during aerobic respiration. The major source of ATP production cycle, the Krebs cycle is absent under anaerobic environment, and ADH is responsible for recycling NAD^+ glycolysis pathway to continue (Saglio *et al.*, 1980) and the end-product is ethanol instead of CO_2 + H_2O (Fig. 3.4). Thus, increase in anaerobic respiration because rapid depletion of carbohydrate in roots leading to *carbohydrate starvation* during excessive water condition of soil (Setter *et al.*, 1987). In addition, the end-products of anaerobic respiration are toxic to plants, for example, ethanol, lactate, malate, and alanine.

Saches (1993) found that excess water tolerance was related to ADH-activity in plant. Liao and Lin (1995) also suggested that ADH-activity has direct correlation with the magnitude of excess moisture, and species with higher ethanol production were less tolerant. Hoffman *et al.* (1986) reported that activity of ADH increased by 20-fold during prolonged anoxia condition, which may be an important factor contributing to long-term adaptation to excess moisture condition. Thus, ethanol production and accumulation in plants may have a *self-poisoning* role.

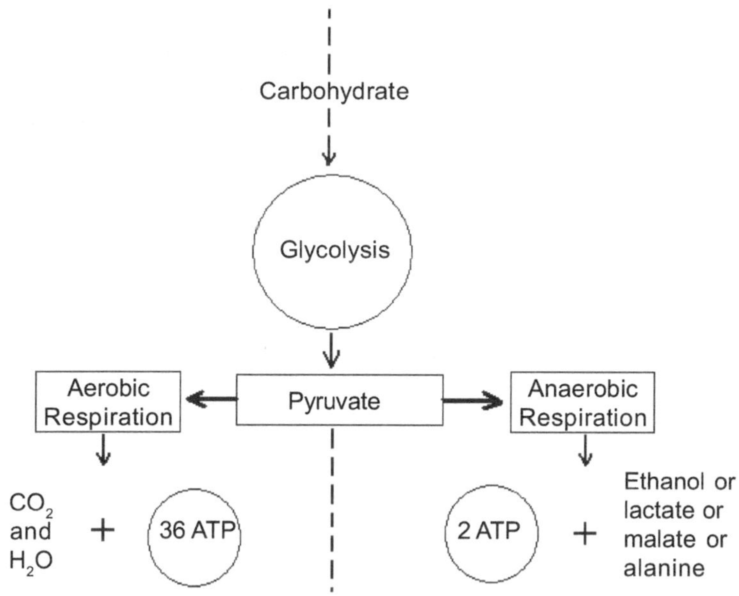

Fig. 3.4. Aerobic and anaerobic respiration in plants

3.6. TOLERANCE MECHANISMS

Mechanisms of survival or maintenance of high biomass production and grain yields during waterlogging may be important at the germination and emergence stages, during vegetative and reproductive stages, or both. Much research has supported the benefits of additive traits for waterlogging including increases in aerenchyma and root porosity, ethanolic fermentation, carbohydrate reserves, tolerance to post anoxic shock, and recovery mechanisms. The physiological mechanisms for waterlogging tolerance are diverse and can be grouped into adaptive traits relating to 1) phenology, 2) morphology and anatomy, 3) nutrition, 4) metabolism including anaerobic catabolism and anoxia tolerance and 5) post anoxic damage and recovery (Table 3.1).

Table 3.1. Adaptive traits for waterlogging tolerance of wheat, barley and oats identified from experiment in waterlogged soil and anaerobic solution cultures (from Setter and Waters, 2003)

Type	Traits	References
Phenology	1. Seedling vigour	Gardner and Flood (1993)
	2. Long season	McDonald and Gardner (1987), Gardner and Flood (1993)
	3. Dormancy (seed or whole plant tissues)	Setter (2000)
	4. Slow growth	McDonald *et al.* (2001a)
Morphology and anatomy	1. Nodal/adventitious root development	Belford (1981), Trought and Drew (1982), Barrett-Lennard *et al.* (1988), Thomson *et al.* (1992), Huang *et al.* (1994a, 1997), Malik *et al.* (2002)
	2. Survival of seminal roots	Armstrong (1979), Trought and Drew (1982), Barrett-Lennard *et al.* (1988), Thomson *et al.* (1990), Watkin *et al.* (1998)
	3. Aerenchyma	Armstrong (1979), Benjamin and Greenway (1979), Belford (1981), Erdmann and Wiedenroth (1986, 1988), Barrett-Lennard *et al.* (1988), Thomson *et al.* (1990, 1992), Huang *et al.* (1994a, b), Varade *et al.* (1970), Trought and Drew (1980b), Drew and Sisworo (1979), Watkin *et al.* (1998), McDonald *et al.* (2001a, b).
	4. Increased root porosity/intercellular spacer	Yu *et al.* (1969), Erdman *et al.* (1986), Thomson *et al.* (1992), Huang *et al.* (1997), Malik *et al.* (2001), McDonald *et al.* (2001a, b)
	5. Increased suberin/lignin; barriers to radial O_2 loss	Arikado (1959), Jackson and Drew (1984), Watkin *et al.* (1998), McDonald *et al.* (2001 b)
	6. Root membrane integrity	Hiatt and Lowe (1967), Greenway *et al.* (1992), Sangen *et al.* (1996).
Nutrition and nutrient toxicities	1. Root length and depth	Erdmann and Widedenroth (1986), Huang *et al.* (1997), McDonald *et al.* (2001a, b).
	2. Cell function for nutrient uptake, including K^+/Na^+ sensitivity	Leyshon and Sheard (1974), Drew and Siswarow (1977, 1979), Trought and Drew (1980a, b), Buwalda *et al.* (1988), Thomson *et al.* (1989), Greenway *et al.* (1992), Akhtar *et al.* (1994), Huang *et al.* (1995).
	3. Leaf chlorosis	Drew and Sisworo (1977, 1979).
	4. Microelement tolerance - Fe^{2+}	Stieger and Feller (1994), Ding and Musgrave (1995), Huang *et al.* (1995), Yuanmin *et al.* (1997).
	5. Microelement tolerance - Mn^{2+}	Sparrow and Uren (1987), Stieger and Feller (1994), Huang *et al.* (1995), Ding and Musgrave (1995), Yuanmin *et al.* (1997).

Contd...

Table 3.1. Contd...

Root metabolism: Respiration, anaerobic catabolism and anoxia tolerance	1. Reduced respiration	Huang and Johnson (1995)
	2. Anaerobic catabolism	Harberd and Edwards (1982), Thomson *et al.* (1989), Waters *et al.* (1991b)
	3. High carbohydrate concentrations	Limpinuntana and Greenway (1979), Benjamin and Greenway (1979), Barrett-Lennard *et al.* (1988), Albrecht *et al.* (1993), Huang and Johnson (1995).
	4. Anoxia tolerance	Waters *et al.* (1991 a, b), Greenway *et al.* (1992)
	5. Phytotoxin tolerance	Drew and Lynch (1980).
Recovery and prevention of post anoxia damage	1. Recovery ability	Barrett-Lennard *et al.* (1988), Buwalda *et al.* (1988), Albrecht *et al.* (1993), Malik *et al.* (2002)
	2. Antioxidants and antioxidative enzymes	Albrecht and Wiedenroth (1994), Wang *et al.* (1996), Biemett *et al.* (1998).

Mechanisms of survival during waterlogging may be important at the germination stage or during vegetative growth. This mechanisms may involve avoidance of waterlogging effects, i.e. by phenological development; adoption to waterlogging, e.g. by aerenchyma or metabolic change and recovery mechanisms following waterlogging (Greenway *et al.*, 1994). One or all of these mechanisms may be used by plants for waterlogging avoidance or tolerance. The particular mechanism use in crops may vary depending on the crops, the growth habit of the cultivar, and the duration of waterlogging.

3.6.1. Morphological and Anatomical Mechanisms of Tolerance

3.6.1.1. Root porosity

Adventitious roots develop, which are not physiologically adapted to excessive moisture conditions but help the plant to recover when the excess water is drained. Higher root porosities were found in the tolerant plants of the wheat (cv. Pato), corn and sunflower (Yu *et al.*, 1969). Many other plant species develop large continuous intercellular spaces. Oxygen moves from the shoot through such space to the root (Hook *et al.*, 1971). This oxygen may even go out of the roots into the soil and cause oxidation of substances in the soil. The enhanced root porosity provides a low resistance pathway for CO_2 movement to the root tip, and the barrier to the radial O_2 loss in basal zones, which further enhance longitudinal O_2 diffusion towards the apex, by diminishing losses to the rhizosphere (Colmer, 2003). Adventitious root porosity was enhanced up to 10-fold for wheat plants grown in waterlogged soil, depending on water level and position along the roots (Malik *et al.*,

2001). Porosity also increased in basal zones of roots above the water level when the younger tissues had penetrated the waterlogged zone. Plants which have either an abundantly branched surface root system, as rice, or the capacity to regenerate adventitious roots after flooding, as maize, are able to survive (Joshi and Dastane, 1965; Grineva *et al.*, 1970). Adventitious roots generated have higher rate of respiration (Grineva *et al.*, 1970) and synthesize cytokinin and gibberellins, and transport them to the shoot (Michniewicz *et al.*, 1970; Reid and Crozier, 1971).

Rice is well adapted to flooding of the roots because of its ability to transport oxygen efficiency from the arial parts of the plant to the roots. Formation of increased intercellular air spaces (aerenchyma) in the cortex, those provide/canals parallel to the axis of the root through which gases can diffuse longitudinally.

3.6.1.2. Adventitious root

Another mechanism of waterlogging tolerance is the ability to produce new adventitious roots (Kahn *et al.*, 1985). When plant grows for sometime under waterlogged condition, it adapts to a certain extent to the new environment, morphologically. Plants which are able to adjust the new environment, tolerate, and others which fail to adjust, die. Formation of adventitious roots close to the soil surface where oxygen tension is more usually higher or more quickly restored after transient waterlogging for instance in tomato (Jackson, 1979; Wample and Reid, 1978). Boru *et al.* (2003) also found that plants adapted to the prolonged low oxygen and no oxygen conditions by producing adventitious roots, undergoing stem hypertrophy and developing aerenchyma for transportation of oxygen to the roots. Similar results were reported by Bacanamwo and Purcell (1999) in soybean. These adaptive mechanisms allow the plants to produce functional nodules and probably explain the absence of leaf chlorosis.

Soybean is very sensitive to waterlogging compared to other crops. Flood inundation of more than 48 hours resulted in yield decrease and the degree of decrease was more severe at reproductive growth stage than at the vegetative stage. The response to excess water in soybeans is associated with a number of biochemical, morphological and anatomical changes in both the root (Richard *et al.*, 1994; Bacanamwo and Harper, 1997). Bacanamwo and Purcell (1999) reported that morphological mechanisms of acclimation to flooding stress in soybean appear to involve an avoidance of water loss by transpiration and a facilitated transport of atmospheric O_2 to the submerged roots through the flood-induced formation of adventitious roots and aerenchyma.

Singh (1988) found that the stems of tolerant soybean cultivars exposed to continuous waterlogging for 28 days developed slits in cortical, tissues near the water surface and produced adventitious roots in the slits, which contribute to increasing root respiration. Colmer (2003) found increased number of adventitious roots per stem of rice grown under waterlogged condition than drained soil. The number of adventitious roots formed per stem of wheat plant grown in waterlogged soil increased up to 1.5 times (Malik *et al.*, 2001).

3.6.1.3. Formation of aerenchyma tissue

Roots which survive decay become spongy in some plants and air cavities start forming in them. Under waterlogged condition, roots develop large air space called *aerenchyma*. There are two main types of aerenchyma which are usually associated with different plant types. *Schizogenous* aerenchyma arises by the separation of cells and is involved in the increases in root porosity in several wetland plants (Justin and Armstrong, 1987). *Lysigenous* aerenchyma arises by the partial breakdown of the root cortex and this is the type formed in roots of the Gramineae (Armstrong *et al.*, 1994; Huang *et al.*, 1994a, b; McDonald *et al.*, 2001 a, b; Watkin *et al.*, 1998; Jackson and Armstrong, 1999). The arenchyma system in root cortex greatly facilitates gas transport under excessive moisture condition in the soil.

The deficiency of oxygen triggers anaerobic formation of ethylene, causing an increase in cellulose activity which then leads aerenchyma formation. Jackson (1990) suggested that the ethylene is found to be the principal mediator for promoting the development of arenchyma in plants (Fig. 3.5). At the same time, auxin and gibberellins are prerequisite for initiate the action of ethylene rather than regulatory functions. ABA concentration was found to increase in roots of pea plants under waterlogging condition causing stomata closure and enriching the leaves with hormones (Zhang and Davies, 1987). Low oxygen partial pressure, under excessive soil moisture environment stimulates biosynthesis of ethylene by increasing 1-aminocyclopropane-1-carboxilic acid (ACC) synthase activity. Ethylene-induced cell lyses leading to aerenchyma formation is a process of progressive cell degradation, collapse of cells in the mid-cortex. Dissolution of protoplasm and much of the cell wall leaving gas-filled spaces (lacunae) that inter connect to gas space system in the shoot.

Aerenchyma provides a diffusion path of low resistance for the transport of O_2 from aerial parts to roots under anoxic conditions. It also provides a path for diffusion of volatile compounds such as ethylene, methane, CO_2, ethanol and acetaldehyde.

Dun and Zhu (2000) studied the effects of aerenchyma formation pattern of the secondary root cortex on the waterlogging tolerance in wheat. They found two patterns of aerenchyma formation in secondary roots cortex. The aerenchyma in the secondary root cortex formed before waterlogging-injury was noted as pattern I. This aerenchyma is protected by the disappearance of the protoplasm from the cell, and the folding and connecting of cell wall. The shape of this kind of aerenchyma was rectangular and it was not easy for the wall to collapse. The aerenchyma formed by the pattern II was produced by multicell fusion, collapse, which lead to the epidermis and the cortex separated from the stele. Aerenchyma formed in the pattern I had higher waterlogging tolerance than that in pattern II. Sunflower and bean plants adapt to waterlogged conditions by the formation of spongy aerenchyma tissue, which facilitates transport of oxygen from the leaves to the roots and rhizosphere (Kawase, 1978; Bidwell *et al.*, 1968). Pang *et al.* (2004) suggested that difference in tolerance ability among the barley genotypes might be partially due to a significant difference in the pattern of areanchyma formation in roots.

Excessive soil moisture

↓

Low O_2 partial pressure

↓

Increased ACC-synthase formation

↓

Enhanced ethylene formation

↓

Increased cellulose activity

↓

Lyses of cortical cell wall

↓

Aerenchyma development

Fig. 3.5. Schematic representation of aerenchyma development in root cortex under excessive soil moisture environment.

3.6.1.4. Changes in root geotropism

Under waterlogged condition, roots grow toward ground surface. Such change in geotropism may have adaptive role to excessive waterlogging environments because the visible root tips and shallow placed roots are under hypoxic rather than complete anoxic state, thus plants can retain partial aerobic respiration. The roots with fine diameter have less resistance to radial oxygen diffusion, and thus do not develop an anerobic core as would occur in thicker roots. Furthermore, a greater surface area of root volume may enhance the exploitation of the limited rhizosphere volume.

3.6.2. Metabolic Rescue Mechanisms of Tolerance

In plants, research has mainly focused on the presence and function of fermentation pathways as a metabolic rescue mechanism when respiration is arrested. Three main fermentation pathways are active in plants during flooding: ethanol, lactic acid, and a plant specific pathway which produces alanine from glutamate and pyruvate, involving alanine amino transferase.

In waterlogging-tolerant species/genotypes tri-carboxilic acid (TCA) cycle switch off, leading in increased photofuctokinase (PFK) activity, which resulted in enhanced rate of glycolysis. The phenomenon is referred as *Pasture effect*. The high rate of pasture effect under anaerobiosis helps the cells in maintaining sufficiently high energy supply, provided carbohydrate supply is sustained in the tissues under anoxia.

Another important metabolic adaptation of respiration is the regulatory induction of alcohol dehydrogenase (ADH) under anoxic environment. Its content in roots is related to the tolerance of waterlogging. Alcoholic fermentation is the key catalytic pathway for recycling NAD to maintain glycolysis and substrate level phosphorylation in absence of oxygen (Ap Rees *et al.*, 1987). Increased alcoholic fermentation is thus one way to alleviate the adverse effect of anoxia on energy production. ADHs play an important role in waterlogging tolerance in maize by generating oxidizing power in the absence of oxygen. Rapid increase in ADHs in maize primary roots during anaerobiosis is an adaptive reaction to waterlogging (Chow, 1984). Plants which are more flooding tolerant have a more active alcohol fermentation pathway and ADH null mutants are more flood sensitive (Kennedy *et al.*, 1992; Roberts *et al.*, 1984). After 6.5 days of flooding the level of alcoholdehydrogenase in the roots of tolerant barley variety Atlas 57 was greater than in the flooding sensitive variety, Psaknon (Wignarajah *et al.*, 1976). Lai (1979) reported that the rice cultivar, Tainan 5 posses higher oxidizing ability than that of Taichung Native 1. He also recommended that this kind of difference in adaptation to grow in submerged soil should be included among the selection criteria for this character.

In general, the petiole ureide concentration is increased by drought in soybean. By contrast, Paiatti and Sodek (1999) reported that ureide concentration in xylem decreased under flooded conditions.

A number of synthesized anaerobic polypeptides (ANP) have been identified (Table 3.2). Fermentation of carbohydrate enables the plant to maintain ATP production in the absence of oxygen, albeit with a reduced energy yield.

Table 3.2. The anaerobic polypeptides identified via cDNA cloning which are induced under low oxygen conditions under flood environment

Anaerobic polypeptide	Detection level	Function	Reference
Sucrose synthase	RNA, protein	Sucrose breakdown	Springer *et al.*, 1986 Martin *et al.*, 1993
-amylase	RNA, protein	Sucrose breakdown	Perata *et al.*, 1993
Glucose-6-phosphate isomerase	RNA, protein	Glycolysis	Kelley and Freeling, 1984a Sachs *et al.*, 1996
Pyrophosphate-dependent phosphofructokinase	Protein	Glycolysis	Botha and Botha, 1991 Mertens, 1991
Hexokinase	Proten	Glycolysis	Bouny and Saglio, 1996 Fox *et al.*, 1998
Fructose-1, 6-bisphosphate aldolase	RNA, protein	Glycolysis	Kelly and Freeloing, 1984b Dennis *et al.*, 1988
Glyceraldehyde-3-phosphate dehydrogenase	RNA, protein	Glycolysis	Sachr *et al.*, 1996
Alcoholdehydrogenase	RNA, protein	Alcohol formation	Freeling and Benett, 1985
Pyruvate decarboxylase	RNA, protein	Alcohol formation	Kelly, 1989
Lactate dehydrogenase	RNA, protein	Alcohol formation	Hoffman *et al.*, 1986 Germain *et al.*, 1997
Alanine aminotransferase	RNA, protein	Alanine fermentation	Good and Crossby, 1989
Glutamine synthase	Protein	Nitrogen metabolism	Mattana *et al.*, 1994a
Nitrate reductase	Protein	Nitrogen metabolism	Mattana *et al.*, 1994b
Formate dehydrogenase	RNA, protein	C1 metabolism	Hourton-Cabassa *et al.*, 1998
Xylocan	RNA	Cell wall loosening	Sachs *et al.*, 1996
AtmYB2	RNA	Transcription factor	Hoeren *et al.*, 1998

(From Dennis *et al.*, 2000)

The response of ANPs in oxygen deficit environment has been divided into three stages. The first stage (0-4 h) consists of the rapid induction or activation of signal transduction components, which activities the second stage (4-24 h), including the induction of glycolytic and fermentation pathway genes which are necessary to safeguard a continued energy production. Aminocyclopropane carboxylic acid (ACC) synthase, a critical enzyme in ethylene biosynthesis is also induced in this stage (Olson *et al.*, 1995), and responsible for accelerating ethylene, production of the various plant hormones studied one of the most striking hormonal change found in waterlogged plants is the dramatic increase in ethylene concentration. Naturally one would believe that the waterlogging resistant cultivars should have a mechanism to restrict the rise in ethylene concentration. Dodds *et al.*, (1984) found that *Vicia faba* cultivars resistant to waterlogging exhibited rapid conversion of ethylene to ethylene oxide. The third stage (24-48 h), which is important for survival of prolonged exposures to low oxygen tension, involves the formation of aerenchyma in the roots. Formation of aerenchyma is not a direct consequence of oxygen deficiency, but is presumably triggered, accumulation of ethylene (Drew, 1997).

3.6.4. Nutrition

Under low O_2 and high CO_2 in waterlogged soils decrease permeability of cell membrane, and reduce ion uptake. Flooding also causes decay of the fine roots and root hair, reducing surface area for nutrients absorption. The products of anaerobic decomposition of organic matter have a solubilizing effect on many heavy metals, which often account for changes in their availability and interaction with other elements. Uptake of N, P, K, Ca, Mg and Zn reduced by wheat under waterlogged condition, while the absorption of Na, Fe and Mn increased (Sharma and Swarup, 1988). Maranville *et al.*, (1986) also observed decreased concentrations of N, P, K, Ca, Mg and S in leaves of sorghum in waterlogged condition.

3.6.5. Others

It is possible that intermittent nature of the waterlogging in many parts of the world may influence different strategies for waterlogging tolerance of plants. Short term or intermitted waterlogging primarily requires plants to maintain processes associated with survival, while growth is a secondary priority. Strategies that could be used include diverse traits such as high rates of alcoholic fermentation to overcome anoxia, high carbohydrate concentrations to sustain alcoholic fermentation, reduced metabolic leakage, efficiency of nutrient uptake, and damage due to oxygen free radicals associated with return to aerobic conditions following waterlogging events. Tolerance to long term, waterlogging requires plants not only to survive but also to grow during the waterlogging event(s). The key strategy to long term waterlogging is to development of aerenchyma in roots to facilitate gas diffusion (Blom, 1999; Jackson and Armstrong, 1999). Other important traits in long term adaptation include suberisation of nodal roots, which contributes to effective aerenchyma development.

3.7. GENETICS OF TOLERANCE

Understanding of genetic behavior of waterlogging tolerance is important for breeding varieties with waterlogging tolerance. Tolerance of plants to waterlogging is related to many morphological and physiological traits that are under a strong environmental influence. Often their genetic control is confounded by environmental stress. Inheritance of submergence tolerance in rice was studied by Haque *et al.* (1989). A strong prepotency of parents was found in transmitting the character to their offspring. Additive and non-additive gene effects were highly significant. The parents, which were highly tolerant to submergence also, had high gca effects, and F_1S between two tolerant parents were found to be the most tolerant of the diallel combinations. The additive dominance model was found valid. Submergence tolerance was partially dominant over susceptibility and recessive alleles were more concentrated in the susceptible parents.

Collaku and Harrison (2005) studied the heritability and correlation of wheat for waterlogging tolerance. They found that grain yield had the lowest heritability ($h^2 = 0.25$). The highest heritability estimates were found for kernel weight (0.47), chlorophyll content (0.37), and tiller number (0.31). Strong genetic correlations were observed between grain yield and kernel weight ($r = 0.56$), and between grain yield and tiller number ($r = 1$). Selection for relatively highly heritable traits, such as kernel weight, would be an effective way to improve waterlogging tolerance in early generations, as grain yield has low heritability. Genotypic and phenotypic information about traits were used to constrict selection indices. A yield improvement of 17% is expected by selection on the basis of the index: grain yield-kernel weight-tiller number. Hamachi *et al.* (1989) found the frequency distributions of damage in F_2S showed continuous variation. The realized heritability for flooding tolerance in 4 F_4 to F_6 populations ranged from 0.12 to 0.48; base on percentage of dead leaf. Realized heritability estimates for 3 of the crosses ranged from -0.02 to 1.06

and from 0-0.12 to 0.32 on the basis of the tolerance index of culm length and grain yield, respectively (Hamachi *et al.,* 1990).

Heterosis for waterlogging endurance expressed as a reduction in damage was noted in F_1 or barley populations and the frequency distribution of damage in F_2 showed continuous variation (Hamachi *et al.,* 1989). A complete diallel cross was made among six barley varieties (Zhou *et al.,* 2004). They observed that the inheritance of waterlogging tolerance followed an additive-dominant genetic model. Even though the study showed more recessive tolerant genes in one of the tolerant varieties and more dominant susceptible genes in the Japanese variety, the statistical test found no evidence of a dominance effect. Selection in early generations could be very efficient in discarding the plants with severe leaf chlorosis.

Heritability of submergence tolerance in sugarcane (*Saccharum* sp.) calculated by parent-offspring regression ranged from 0.29 to 0.51 (Deren *et al.,* 1991). The diploid form of winter rye is more tolerant of excess soil moisture than are tetraploids (Trompel, 1978).

F_1 progenies of Nonglmi 46/Ningmai 3 and Nonlin/Zhen 7853 crosses survived waterlogging with a level of tolerance similar to Nonglin 46, this indicated that waterlogging tolerance in Nonglin 46 is dominant (Cao *et al.,* 1992). Segregation occurred in the waterlogging tolerance of F_2 plants. Chi squared tests showed that segregation ratios were consistent with a 3:1 ratio indicating that the waterlogging tolerance of Nonglin 46 in genetically controlled by a single dominant gene. Backcrossing experiments also confirmed that waterlogging tolerance of Nonglin 46 is controlled by one dominant gene. The heritability of grain weight per plant was 75% (Cao *et al.,* 1992, 1995).

F_1 and F_2 hybrids of barley were grown under waterlogged conditions at the internode elongation stage, selecting for a reduction in numbers of dead leaves as the waterlogging tolerance indicator (Hamachi *et al.,* 1989). The results indicated that screening for waterlogging tolerance by the amount of dead leaves was a useful criterion and that endurance was under polygenic control.

Boru (1996) proposed that different gene could relate to different mechanisms of tolerance to waterlogging, therefore, water logging tolerance could be substantially improved by combining all tolerance genes into one genotype. Cao *et al.* (1994) also indicate that additive gene action is the major determinant of inheritance of waterlogging tolerance.

3.8. BREEDING FOR WATERLOGGING TOLERANCE

Plant breeding is the genetic improvement of the plants ability to produce economically soluble products. Stability of crop production can be achieved through avoidance, tolerance, and/or resistance to important biotic and abiotic stresses. The major criteria required for germplasm improvement (Lagudah and Appels, 1994; Reynolds *et al.,* 2001) are:

1. Genetic diversity for tolerance
2. Accurate phenotyping including elucidation of mechanisms of tolerance and reselection in a breeding programme and
3. Heritability of traits.

3.8.1. Germplasm Diversity for Waterlogging Tolerance

Screening of crop germplasm for waterlogging tolerance is an important step in plant breeding, which depends on existing biodiversity of the particular crops. The tolerant ability differ species to species and among the genotypes within the species (Table 3.3 and 3.4). Huang *et al.* (1994a) showed that there is good genetic diversity for tolerance of wheat to hypoxic solution cultures. Genetic variability has been observed for seed germination and early seedling growth in maize (Porto, 1997). Such variability may be related to the ability of different maize genotypes to utilize the stored assimilate in the endosperm through anaerobic metabolism for germination of radicle/coleoptile development (Xia and Saglio, 1992).

3.8.2. Critical Stages of Crops

Waterlogging may appear at any stage of development due to excess rainfall. Evaluation of genetic diversity for waterlogging tolerance during different stages of development is therefore essential. Significant variations have been observed at different stages of crops.

Large reductions in grain yield for wheat, barley and oats were caused by 6 weeks of continuous waterlogging starting at 2 weeks after sowing, in comparison to starting at 6 weeks or 10-14 weeks (ear emergence) after sowing (Watson *et al.,* 1976). Similar results were found in winter wheat grown in lysimeter (Cannell *et al.,* 1980), where immediately after germination, waterlogging for 16 days at 12 ^{0}C killed all seedlings and waterlogging for 6 days reduced population to 12-28% of the non-waterlogging plants.

Table 3.3. Waterlogging tolerant genotypes of different crops

Crop	Tolerant variety	Reference
Wheat *Triticum* spp.	Savannah, Gore, FL302, C9766, Bayles, BR34	Hung *et al.*, 1994a
	96W639-D1-17, 96W639-D6-5, 96W639-D6-32	Setter *et al.*, 1999
	SARC1, Ducula-4, Chara	Setter and Water, 2003
	Nong 8675, Pato, Shuilaomai, Schuilizhan, Nonglin 46, Yang 85-85	Cao and Cai, 1991
	Champtel, Currawong, Carnamah	Setter *et al.*, 2001
	Zhemani No. 2, Zhengzhou 761, Nonglin No. 46	Lin *et al.*, 1994
Rice *Oryza sativa*	Panidhan-1 (a selection from IR 442)	Sahai *et al.*, 1979
	CR1002	Mukherjee and Biswas, 1979
	NC492	Basuchoudhury and Gupta, 1985
	NC491	Sinha and Bandhypodhyay, 1986
	IR39558-147-1-3, IR10198-66-2-1	Gupta, 1988
	IR4595-4-1-13-2	
	Amaji, Seijo	Hamachi *et al.*, 1988
	Deepwater rice: PLA-2, EB-14, Jaisuria, Jalmgna, AR 614-25B, Pt6-15, Ptb-16, TNR-1, TNR-2, Jaladhi-I, Jaladhi-II, Intermittent flooding: Jalaplaban-I, FR13A, FR 43A, PLA4, Ptb-15, Ptb-16, Co-14, Madhukar, Chakia-59	Gangadharan, 1985
	CR1030, CN540, IET6207, RAU 21, 168-1-2, NC492	Bhattacharjee, 1984
	CR1009	Sundaram *et al.*, 1988
	IET10016, IET10021, Bogaborodhan, Kushal, Monrian, IET13119	Neog *et al.*, 2002
Oat *Avena sativa*	Dalyup, Toodyay	Setter and Water, 2003

Contd...

Table 3.3. Contd...

Barley *Hordium valgare*	Kara Marumi 1, Ou2, Bizen Wase 53, Kaikei 39, Benkei 3, Sazanshu, Koshu Rokkaku, Yatomi Mochi, Zenkoji, Hachikoku, Raiden, Tankiaze 105, Yukiwarimugi, Koike. Rokkaku 2, Konosu 60, Rokujo, Omungi 5, Genpachi, Shirogoro, Warejiro, Takayama, Tokushina, Mochimugi 1, Monosu 50, Mochi Hadaka, Boseong Baitori 1, Suweon Shin 1, Haman Waedong 2, Tayeh 9, Hsinqntica 3, Chengchou 5, Trisuli Bazar 8, Keronja 3	K Takeda, Personal communication
	Atlas 57	Wingnarajah *et al.*, 1976
	Striting, Fitzgerald	Setter *et al.*, 2001
Soybean *Glycine max*	Pungsannamulkong, Mukhankong, GC30182-2-6, Shih Shih	Lee *et al.*, 2004 AVRDC, 1979
Tomato *Solanum esculentum*	L123	Kuo and Chen, 1979
Maize *Zea myase*	CH106-4, CH108-2, CH0348, CH0314	Chow, 1984
	Kisan	Goswami *et al.*, 1975
Mung bean *Vigna radiata*	VC1006-40-1, V1968, V2984, V3092, V3372	AVRDC, 1979
Sugarcane *Saccharum officiarum*	CoH12, Co7717, CoH14,	Sethiya *et al.*, 1986
	CB4013, H59-3775, H49-3533, PT4352	Somrajan and Rangarajan, 1985
Lentil *Lens culinaris*	ILL-6499, ILL-6778, ILL-6793	Asraf and Chishti, 1993
Sweet potato *Ipomya batatus*	LO-360, LO-323	Ton, 1978
Potato *Solanum tuberosum*	Gem, Jaspar	Martin, 1984
Vicia faba	Maries Blaze	Alvino *et al.*, 1983

Table 3.4. Flood tolerance ability of fruit trees

Tolerance level	Common name	Botanical name
Tolerant	Guava	*Psidium guajava* L.
Moderate tolerant	Banana	*Musa* sp.
	Black sapota	*Dispyros digyna*
	Cainito	*Chrysophyllum cainito*
	Canistel	*Pouteria campechiana*
	Carambola	*Averrhoa carambola* L.

Contd...

Table 3.4. Contd...

	Cashew	*Anacardium occidentale* L.
	Cherry	*Rubus moleccanus* L.
	Citrus	*Citrus* sp.
	Coconut	*Cocos nucefera* L.
	Indian jujabe	*Ziziphus jujube* (L.) Lamk
	Java plum	*Syzygium cuminii*
	Lychee	*Litchi chinensis*
	Macadamia Nut	*Macadamia integrifolia*
	Mango	*Mangifera indica* L.
	Oriental Persimmon	*Diospygros* sp.
	Rose apple	*Syzygium jambos* (L.) Alston
	Sapodilla	*Achas* sp.
	Surinam Cherry	*Eugenia uniflora* L.
	Tangerine/ Mandarin	*Citrus reticulate* Blanco
	Wax Jambu	*Syzygium samarangense* (bl.) Merr & Perry
Susceptible	Akee	*Blighia sapida*
	Ambarella	*Spondias pinnata*
	Apple	*Malus pumila* Mill.
	Barbados cherry	*Malpighia glabra* L.
	Carissa	*Carrisa carandas*
	Cattley guava	*Psidium callcianum*
	Cherimoya	*Annona chemola*
	Custard apple	*Annona squamosa*
	Fig	*Ficus carica*
	Grumichama	*Engenia brasiliensis*
	Ilama	*Annona diversifolia*
	Jaboticaba	*Myrciaria cauliflower*
	Jackfruit	*Arocarpus heterophyllus* Lamk.
	Jelly palm	*Butea eriospatha*
	Kei apple	*Dovyalis caffra*
	Kwai muk	*Artocarpus hypargyracea*
	Longan	*Euphoria longam* (Lour.) Steud.
	Loquat	*Eriobotrya japonica* (Thunb.) Lindl.
	Miracle fruit	*Synsepalum dulcificum*
	Muscadine grapes	*Vitis rotundifolia*
	Peach	*Prunus persica* (L.) Batsch.
	Pineapple	*Ananas comosus* (L.) Merr.
	Pineapple guava	*Feijoa sellowiana*
	Piloma	*Eugenia luchsnathiana*
	Pomegranate	*Punica granatum* L.

In wheat, waterlogging after germination but before emergence is most critical with very few plants surviving more than 6 days (Beldford *et al.*, 1984). After emergence, plant population was not effected by waterlogging although tiller was restricted and yield loss was associated with fewer fertile years. Grain yield in wheat is associated with severity of waterlogging before anthesis.

Gardner and Flood (1993) suggested that the early reproductive phases are more adversely affected by water logging than tillering stages, because the earlier maturing genotypes yield much less than late maturing genotypes on non-drained plots relative to drained plots, and the yield reductions are a consequence of reductions in grain per ear. An additional explanation could be a longer recovery period for the late maturing varieties which might enable a reduction in spikelet sterility. Great genotypes differences in lentil for waterlogging tolerance were evident with early maturing lines apparently more sensitive than late maturing lines (Bejiga and Anbessa, 1995). Bao (1997) arranged different stages of wheat according to the waterlogging tolerant ability:

Booting stage > tillering stage > grain filling stage.

Waterlogging reduces tillering and leaf area expansion of sorghum. Grain yield was reduced when waterlogged at panicle initiation, largely as a result of decrease in grain number (Orchard and Jessop, 1984). In sunflower, leaf expansion and stem extension were inhibited by waterlogging at the 6-leaf and visible bud stages; and leaf desiccation occurred at anthesis (Orchard and Jessop, 1984).

Wheat, barley, oat and triticale varieties that had previously identified as having waterlogging tolerance at the whole plant stage (Sayer *et al.*, 1994, Van Ginkel *et al.*, 1992). Seeds were sown beneath 30 mm of soil in beakers at 15 ^{0}C, and exposed to waterlogging for 4 days; there were 4 replicates of 50 seeds. After treatment, seeds were removed from the soil and sown on filter paper. Survival was assessed by germination tests relative to non-waterlogging seed according to the International Seed Testing Association (ISTA, 1984) guidelines. The mean survival of all varieties waterlogged in soil was 86, 75, 68 and 41% for oats, triticale, Australian wheat and Australian barley, respectively.

Gardner and Flood (1993) suggested that late season wheat appeared to have a clear yield advantage over early season wheat. McDonald and Gardner (1987) supported the use of long season wheat, because, they will enable early sowing so as to avoid waterlogging damage at the tolerant stage of germination and emergence, and this will allow anthesis to occur late enough to avoid waterlogging damage in spring.

3.9. SELECTION CRITERIA

Physiological traits evaluated for waterlogging tolerance include root traits (aerenchyma, suberisation, nodal/seminal roots), shoot/root carbohydrates and phenology. For spring wheat cultivars, yield is positively correlated to percent nodal root aerenchyma ($r^2 = 0.50$); while for early there is little or no correlation to root arenchyma (Setter *et al.,* 2001). Collaborative international research is aimed at improving tolerance of barley and wheat to waterlogging through further evaluations and selection, and a mechanistic approach to germplasm improvement.

Christiane *et al.* (2003) used leaf photosynthesis, pigment composition, PS II photochemistry, and plant growth characteristics as screening criteria for waterlogging in lucern (*Midicago sativa*).

Dead leaf (%) under excess soil moisture was thought to be the best criterion for selection for flooding tolerance in early generations because its heritability values are relatively constant and it is easy to measure (Hamachi *et al.,* 1990) and was correlated with reduction of grain yield per plant and culm length (Hamachi *et al.,* 1989). Based on leaf chlorosis, single plant selection for waterlogging tolerance could start as early as F_2 since no significant dominance and non-allelic effects existed (Zhou *et al.,* 2004).

Nodule activity, photosynthetic activity rate, stomatal conductance, water use efficiency, ureide content in petiole and the root distribution are promising indications for screening soybeans tolerant of excess water (Lee *et al.,* 2004).

Lin *et al.* (1994) used a '*waterlogging tolerance index*', i.e. response of waterlogged plants relative to non-waterlogged plants and they calculated an indices sum based on the four key traits- number of green leaves per main stem, grain per ear, 1000 grain weight and seed setting rate per ear.

Leaf chlorosis has been used successfully at CIMMYT and in China for many years, results are strongly correlated to grain yield in specific locations, and this is highly heritable traits (Samad *et al.,* 2001).

3.10. SCREENING METHODS

The highly variable nature of waterlogging in both space and time emphasis the complexity of the problems of screening germplasm in the field. This necessitates a detailed characterization of the field environment including use of specific approaches, e.g. specific site sampling or use of characterized waterlogging gradients, to relate yields to localized conditions for obtaining accurate assessment of genetic diversity. These conditions, and the transient nature of waterlogging in the field, make it obvious that simulation of field conditions in pot as controlled experiments requires more than a simple flooding of a pot.

Screening without non-waterlogged 'controls' obviously has advantages since twice number of genotypes can be evaluated. The positive impact of basing varietal selections are on such screening strategies is that yields may be high in this germplasm when grown in waterlogged environments, however, this may have nothing to do with waterlogging tolerance. In breeding programmes for waterlogging tolerance, it is important to first accurately select for the key trait, and once found, add it in additional traits required for combine waterlogging tolerance genes with high grain yield genes.

It may be useful for plant breeders and growers simply to know that some genotypes yield well under waterlogging even if they are not truly tolerant; this approach has been used by Callaku and Harrison (2002) to characterize high grain yield of wheat during waterlogging.

Christane *et al.* (2003) made a comparative analysis of waterlogging effects on leaf photosynthesis, pigment composition, PS-II photochemistry, and plant growth characteristics of lucern (*Medicago sativa*). Two-month-old plants, grown in half-strength Hoagland nutrient solution were waterlogged for 16 days, and plant physiological characteristics were monitored at regular intervals. All cultivars had significantly reduced fresh and dry weight for both shoots and roots after 16 days of waterlogging. Root biomass showed a greater percentage of reduction than did shoot biomass. As waterlogging stress developed, chlorophyll content, CO_2 assimilation rate, transpiration rate, stomatal conductance and maximum quantum efficiency of PS-II decreased significantly. Chlorophyll *a* and *b* content gradually decreased over the time of the experiment in the stressed cultivars, and leaf chlorosis became increasingly evident. It was concluded that for practical screening purpose maximal quantum efficiency ratio is the most appropriate.

Takeda and Fukuyama (1986) tested 3457 varieties barely in the world collection by submerging 50 sterilized grains of each in deionized water in a test tube for 4 days at 25 ^{0}C and subsequently determining their germination (%) after 4 days on moistened filter paper at 25 ^{0}C. The germination (%) ranged from 0-100%. The collection from China, Japan and Korea contained many tolerant varieties, while those from North Africa, Ethiopia and South-west Asia showed few tolerant varieties. The most tolerant varieties retained complete germination-ability after 8 days soaking at 25 ^{0}C. Qiu and Ho (1991), after testing 4512 barley varieties, reported that some showed very high level of waterlogging tolerance germplasm. Waterlogging caused significant chlorosis of the older leaves of all the varieties of barley under study (Zhou *et al.*, 2004). Varieties showed significantly different tolerance to waterlogging. They found that three Chinese varieties, TX 9425, DYSYH, and YYXT showed much less yellow leaf percentage.

Lee *et al.*, (2004) screened five soybean varieties based on tolerance ability to nodule activity, photosynthetic rate, stomatal conductance, water use efficiency, ureide content in petiole and root distribution. The specific nodule weight was reduced in all test varieties in response to waterlogging, but the degree of decrease was much less in Pungsannamulkong and Muhankong than Jangyuupkong and Myungjunamulkong. The similar tendencies were showed in photosynthetic rate, stomatal conductance and water use efficiency. In general, the leaf water potential decreases under excess water in all plants. So, stomatal conductance decreases to minimize the water content loss in plant tissue. But the water use efficiency also decreased here, indicating damage to the photosynthetic machinery from waterlogging.

Cao and Cai (1991) screened over 1000 varieties and breeding lines of wheat they defined as waterlogging tolerance i.e. low percentage of leaf damage, maintenance of 1000 grain weight, or grain per mainstream. They found that the cultivars Ning 8675, Nonglin 46, Yang 85-85, Pato and *Triticum macha* were waterlogging tolerant also having good agronomic traits. Lin *et al.* (1994) have screened Chinese wheat based on plant and grain characters. They identified Zhemani No. 2, Zhengzhou 761 and Nonglin No. 46 as tolerant varieties.

Oats appears to have one of the greatest ability for recovery from waterlogging; this is supported by observations of Cannell *et al.* (1985). During waterlogging of oats for 90 days in a sandy loam soil, the shoot dry weights were only 60%, and total root weight was reduced more than 50%, relative to plants in freely drained treatment. However, by maturity, shoot weights and grain yield were both about 93% of freely drained plants. Oats and triticale are more waterlogging tolerant than wheat and barley, and they tend to maintain green leaves during waterlogging, while other crops like wheat and barley often become chlorotic. Varieties of Australian oats indicated some diversity for tolerance, while overall mean (%) grain yield for these varieties (52%) was greater than that measured for Australian wheat (31%) and barley (31%) grown in same environments and under identical waterlogging treatments (Setter *et al.*, 1999; Setter, 2000).

Development of wide scale screening protocols for root traits in a large number of breeding lines using deoxygenated nutrient solutions to simulate waterlogging can be difficult and expensive. Wiengweera *et al.* (1997) used stagnant nutrient solution containing agar on wheat to simulate the changes in gas composition associated with waterlogging. They found that wheat grown in stagnant agar solutions developed up 10-fold greater aerenchyma in adventitious roots than roots in N_2 flushed solutions or in non-flushed solution without agar. Measurement of root lengths of seedlings grown in stagnant agar solutions might therefore be used be as a rapid screening protocol to select for the combined effects of root aerenchyma and low radial O_2 loss for donor parents (Watkin *et al.*, 1998).

3.11. BIOTECHNOLOGY IN WATERLOGGING TOLERANCE

Gene of waterlogging tolerance can be incorporated from the wild relatives. The presence of combination of traits such as high root porosities and low radial O_2 loss for waterlogging tolerance in species within the tribe Triticeae such as *Hordeum marinum* (McDonald *et al.*, 2001a) offer good prospects for enhancing these traits in cereal crops because wide-hybridizations are possible between *Hordeum* and *Triticum* (Jiang and Dajum, 1987).

3.11.1. Marker Assisted Selection

Submergence tolerance is the most useful mechanism for tolerance to waterlogging. The most widely source used for submergence tolerance is the indica rice cultivar 'FR13A'. A major QTL was shown to control submergence tolerance in this trait (Xu and Mackill, 1996). FR13A also has additional QTLs that contribute to its tolerance (Nandi *et al.*, 1997; Toojinda *et al.*, 2003). The major QTL from FR13A, designated *Sub1*, has been fine-mapped to an interval of less than 0.5 cM (Xu *et al.*, 2000). Simple sequence repeat markers closely linked to this locus have been used to refer it into different genotypes of rice (Siangliw *et al.*, 2003; Xu *et al.*, 2004).

Van Toai *et al.* (1999) identified a putative QTL associated with soybean growth and productivity. One hundred twenty-two recombinant inbred lines (RIL) of the Archer ´ Minsoy and 86 RIL of Archer ´ Noir populations were grown and were subjected to soil waterlogging for two weeks at the R1 stage. One single QTL, the SAT-064 from Archer parent, was associated with 50% of the variation in plant growth under flooded conditions. The same QTL was also associated with more than 50% of the variation in seed yields under flooding. This highly significant QTL is uniquely associated with flooding tolerance.

QTLs for internode elongation ability were mapped in a RIL population from IR74 (lowland) crossed with deepwater variety Jalmagna (Sripongpangkul *et al.*, 2000). A major locus near or at the semi-dwarf locus *sd1* was responsible for plant height and increase in plant height and internode length in response to raising water level, with less important loci on other chromosomes.

Two mapping population with 103 and 67 F_2; six lines from crosses ASG A5403 × Archer and PIO 964 × Archer, respectively were used by Cornelious *et al.* (2004) to map QTLs condition flooding tolerance in rice. The result showed that eight out of 18 polymorphic markers in the population from the other cross of ASG A5403 × Archer and 19 markers from the other population were significantly associated with the trait. These significant markers are located in linkage group A1, C1, D1a, I, K, F, N and J. Three markers, Satt252, Satt159 and Satt412, located in linkage groups F, N and J were significant in both the populations.

3.11.2. Transgenic Development

Genetic engineering provides opportunities for germplasm improvement as well as evaluating the impact of different mechanisms of tolerance to waterlogging without confounding effects of complete changes in the background. Two different approaches have been used to try and identifying limiting factors in the response to waterlogging. First is the under-expression of single candidate genes, e.g. for ethanol synthesis, using sense and anti-sense constructs. Second is the over-expression of transcription factors (Dennis *et al.*, 2000). It was anticipated that both approaches may have a beneficial effects in switching on the longer term adaptation response to low oxygen stress. Few achievements in transgenic development have been presented in Table 3.5.

Preliminary results of rice transgenic containing the cotton *adh* cDNA suggested that the over expressing *pdc* do not show increased tolerance to submergence (Ellis and Setter, 1999). Quimio *et al.* (2000) found that Taipei 309 transferred with *pdc1* linked to a constitutive 35S promoter had up to 3-fold higher *PDC* activities and ethanol synthesis rates when exposed to anoxia compared to non-transformed controls. They also reported that increasing ethanol production up to 6-fold in a range of transgenic lines exposed to anoxia was correlated with an 8-fold increase in percentage survival of lines during submergence under hypoxic conditions. In contrast, Rahaman *et al.* (2001) studied Taipei-309 transformed with *pdc1* and found that two transgenic lines had over 2-fold greater *PDC* activity and they had up to 43%

greater rate of ethanol synthesis, however, survival of seed lines exposed to anoxia was even less than that of non-transformed plants. Therefore, the results of rice transgenic need to be repeated for its potential extrapolation.

Table 3.5. Transgenic developed in different crop plants against waterlogging tolerance.

Gene	Gene action	Species	Phenotype	Reference
pdc1	Pyruvate decarboxylase	rice	Increased submergence toler ance	Quimio *et al.*, 2000
ADH		cotton	Increased ethanol fermentation	Dennis *et al.*, 2000
ACC	ACC deamina se	tobacco	Increased flooding (waterlogging) tolerance	Grickko and Glick, 2001
pdc1	Pyruvate decarboxylase	rice	Rice over 2-fold greater PDC activity and they had up to 43% greater rate of ethanol synthesis	Rahaman *et al.*, 2001
ipt	Cytokinin biosynthesis	*Arabidopsis*	Increased complete submergence and waterlogging	VanToai, 2001

Transgenic cotton plants containing the *ADH* cDNA driven by a constitutive 35S promoter showed 10-30 fold increased in ADH activity and a significant increase in the rate of ethanol fermentation (Dennis *et al.*, 2000). Cotton plants with rice *pdc1* cDNA driven by a constitutive 35S promoter produce more *pdc* protein but had only marginal more *PDC* activity. They also observed that neither *pdc* or *adh* transgenic cotton plants, nor plants containing both constructs showed increased tolerance of hypoxia stress.

To determine the effects of auto-regulated cytokinin production on flooding tolerance, a chimeric gene containing the senescence specific *SAG12* promoter and the *ipt* gene coding for cytokinin biosynthesis was constructed and introduced into *Arabidopsis* plants (Van Toai, 2001). The transgenic were consistently more tolerant to complete submergence, waterlogging than wild-type plants and after one week of stress, he observed that the SAG12: *ipt* plants contained 3 to 10-fold more cytokinin.

3.12. REFERENCES

Akhtar J, Gorham J, Qureshi RH. 1994. Combined effect of salinity and hypoxia in wheat (*Triticum aestivum* L.) and wheat *Thinophyrum* amphiploids. Plant Soil. 166: 47-54.

Albrecht G, Kammerer S, Pranznik W, Wiedenroth EM. 1993. Fructan content of wheat seedling (*Triticum aestivum* L.) under hypoxia and following reaeration. New Phytol. 123: 471-476.

Albrecht G, Wiedenroth EM. 1994. Protection against activated oxygen following re-aeration of hypoxically pretreated wheat roots. The response of the glutathione system. J Exp Bot. 45: 449-455.

Alvino A, Zeobi G, Frusciante L, Monte LM. 1983. Evaluation of field bean lines grown with shallow water table maintenance at different levels. Field Crop Res. 6: 176-188.

Ap Rees T, Jenkin LET, Smith AM, Wilson PM. 1987. The metabolism of flood tolerant plant. In: Plant Life in Aquatic and Amphibious Habitat, Crawford RMM (ed.). Blackwell Scientific Publishers, Oxford. pp. 227-238.

Arikado H. 1959. Supplementary studies on the development of the ventilating systems in various plants growing on lowland and on upland. Bull Fac Agric Mie Univ. 20: 1-24.

Armstrong W, Brandle R, Jackson MB. 1994. Mechanisms of flood tolerance in plants. Acta Bot Neerl. 43: 307-358.

Armstrong W. 1979. Aeration in higher plants. In: Advances in Botanical Research, Vol. 7, Woolhouse HW (ed.). Academic Press, New York. pp. 225-232.

Ashraf M, Chishti SN. 1993. Water logging tolerance of some accessions of lentil (*Lens culinaris* Medic.). Trop Agri (Trinidad). 70: 60-70.

AVRDC, Taiwan. 1979. Progress Report for 1978. Shanhua, Taiwan. pp. 173.

Bacanamwo M, Harper JE. 1997. The feedback mechanisms of nitrate inhibition of nitrogenase activity in soybean may involve aspargine and/or products of its metabolism. Physiol Plant. 100: 371-377.

Bacanamwo M, Purcell LC. 1999. Soybean root morphological and anatomical traits associated with acclimation to flooding. Crop Sci. 39: 143-149.

Bao X. 1997. Study on identification stage and index of water logging tolerance in various wheat genotypes (*Triticum aestivum* L.). Acta Agriculturae Shanghai. 13 (2): 32-38.

Barrett-Lennard EG, Leighton P, Buwalda F, Gibbs J, Armstrong W, Thomson CJ, Greenway H. 1988. Effects of growing wheat in hypoxia nutrient solutions and of subsequent transfer to aerated solutions I. Growth and carbohydrate status of shoots and roots. Aust J Plant Physiol. 15: 585-598.

Basuchoudhury P, Gupta DKD. 1985. NC492 promising rice for water logged fields in West Bengal. Int Rice Newsl. 10 (2): 8-9.

Bejiga G, Anbessa Y. 1995. Water logging tolerance in lentil. LENS Newsl. 22: 8-10.

Beldford RK, Christian DG, Cannell RQ, Cross MJ. 1984. Drainage, cultivation and the growth of winter cereals on clay soils. In: Fonctionement hydrique et comportement des sols. Plaisir, France, Association Francaise Pour I'tlude due sols (AFES). pp. 167-174.

Belford RK. 1981. Response of winter wheat to prolonged water logging under outdoor conditions. J Agric Sci Camb. 97: 557-568.

Benjamin LR and Greenway H. 1979. Effects of a range of O_2 concentrations on porosity of barley roots and on their sugar and protein concentrations. Ann Bot. 43: 383-391.

Bhagawat KA, Gore SR, Banerjee G. 1986. Water logging injury in sorghum seedling and their kinetin effected rapid recovery. Plant Growth Regulation. 5: 23-31.

Bhattacharjee DP. 1984. Varietal improvement for rainfed lowland rice. Oryza. 21: 106.

Bidwell OW, Gier DA, Cipra JE. 1968. Ferromanganese pedotubules on roots of *Bromus inermis* and *Andropogon gerardii*. Trans 9th Int Congr Soil Sci. 4: 683-692.

Biemett S, Keetman U, Albrecht G. 1998. Re-aeration following hypoxia or anoxia leads to activation of the antioxidative defence system in roots of wheat seedlings. Plant Physiol. 116: 651-658.

Blom CWPM. 1999. Adaptations of flooding stress: From plant community to molecule. Plant Biol. 1: 261-273.

Boru G, VanToai T, Alves J, Hua D, Knee M. 2003. Response of soybean to oxygen deficiency and elevated root zone carbon dioxide concentration. Ann Bot. 91: 447-453.

Boru G. 1996. Expression and inheritance of tolerance stresses in wheat (*Triticum aestivum* L.). Ph.D. Thesis, Oregon State University. pp. 88.

Botha AM, Botha EC. 1991. Effect of anoxia on the expression and molecular form of the pyrophosphate dependent phosphofructokinase. Plant Cell Physiol. 32: 1299-1302.

Bouma TJ, Nielson KL, Eissenstat DM, Lynch JP. 1997. Estimating respiration of roots in soil: interaction with soil CO_2, soil temperature and soil water. Plant Soil. 195: 221-232.

Bouny JM, Saglio PH. 1996. Glycolytic flux and hexokinase activities in anoxic maize root tips acclimated by hypoxic pretreatment. Plant Physiol. 111: 187-194.

Buwalda F, Barrett-Lennard EG, Greenway H, Davies BA. 1988. Effects of growing wheat in hypoxic nutrients solutions and of subsequent transfer to aerated solutions. II. Concentrations and uptake of nutrients and sodium in shoots and roots. Aust J Plant Physiol. 15: 599-612.

Callaku A, Harrison SA. 2002. Losses in wheat due to water logging. Crop Sci. 42 (2): 444-450.

Cannell RQ, Belford RK, Blackwell PS, Govi G, Thomson RJ. 1985. Effects of water logging on soil aeration and on root and shoot growth and yield of winter oats (*Avena sativa* L.). Plant Soil. 85: 361-373.

Cannell RQ, Belford, RK, Gales K, Dennis CW and Prew RD. 1980. Effects of water logging at different stages of development on the growth and yield of winter wheat. J Sci Food Agric. 31: 117-132.

Cao Y, Cai S, Wu Z, Zhu W, Fang X, Xiong E. 1995. Studies on genetic features of water logging tolerance in wheat. Jiangsu J Agril Sci. 11 (2): 11-15.

Cao Y, Cai SB, Zhu W, Fang XU. 1992. Genetic evaluation of water logging tolerance in the wheat variety Nonglin 46. Report of the Jiangshu Academy of Agricultural Sciences. Crop Genetic Resources. 4: 31-32.

Cao Y, Cai SB, Zhu W, Xiong E, Fang XU. 1994. Combining ability analysis of water logging tolerance and main agronomic traits in wheat. Sci Agric Sinica. 27 (6): 40-55.

Cao Y, Cai SB. 1991. Some water logging tolerant wheat varieties. Crop Genetic Resources. No. 2: 25-26.

Chang HT, Loomis WE. 1945. Effects of carbon dioxide on absorption of water and nutrients by root. Plant Physiol. 20: 221-232.

Chow KH. 1984. Alcohol dehydrogenase synthesis and water logging tolerance in maize. Trop Agri. (Trinidad). 61: 302-304.

Christiane E, Smethurst, Shabala S. 2003. Screening methods for water logging tolerance in Lucerne: comparative analysis of water logging effects on chlorophyll fluorescence, photosynthesis biomass and chlorophyll content. Flmetional Plant Biol. 30 (3): 335-343.

Collaku A, Harrison SA. 2005. Heritability of water logging tolerance in wheat. Crop Sci. 45: 722-727.

Colmer TD. 2003. Aerenchyma and an inducible barrier to radial oxygen loss facilitate root aeration in upland paddy and deep-water rice (*Oryza sativa* L.). 91(Special Issue): 301-309.

Cornelious B, Chen Y, Wang D, de Leon N, Chen P. 2004. Genetic mapping of QTLs underlying flood tolerance in soybean. In: Palnt & Animal Genomes XII Conference, January 10-14, 2004, Town & County Convention Center, San Diego, CA. P547.

Crowford RMM. 1992. Oxygen availability as an ecological limit to plant distribution. Adv Eco Res. 23: 93-185.

CSSRI. 1997. *Vision*-2002. CSSRI Perspective Plan. Indian Council of Agricultural Research, Director, Central Soil Salinity Research Institute, Karnal, India. p. 95.

Dennis ES, Dolferus R, Ellis M, Rahaman M, Wu Y, Hoeren FU, Grover A, Ismod KP, Good AG, Peacock WJ 2000. Molecular strategies for improving water logging tolerance in plants. J Exp Bot. 51: 89-97.

Dennis ES, Gerlach WL, Walker JC, Lavin M, Peacock WJ. 1988. Anaerobically regulated aldolase gene of maize. A chimeric origin? J Biol Chem. 202: 759-767.

Deren CW, Snjder GH, Miller JD, Porter PS. 1991. Screening and heritability of flood tolerance in Florida (CP) sugarcane breeding population. Euphytica. 56: 155-160.

Ding N, Musgrave ME. 1995. Relationship between mineral coating on root and yield performance of wheat under water logging stress. J Exp Bot. 46: 939-945.

Dodds JH, Smith AR, Hall MA. 1984. The metabolism of ethylene to ethylene oxide in cultivars of *Vicia faba*, relationship with water logging resistance. Plant Growth Regulation. 3: 203-207.

Drew MC, Lynch JM. 1980. Soil anaerobiosis, microorganisms and root function. Ann Rev Phytopath. 18: 37-66.

Drew MC, Sisworo EJ. 1977. Early effects of flooding on nitrogen deficiency and leaf chlorosis in barley. New Phytol. 79: 567-571.

Drew MC, Sisworo EJ. 1979. The development of water logging damage in young barley plants in relation to plant nutrient status and changes in soil properties. New Phytol. 82: 301-314.

Drew MC. 1997. Oxygen deficiency and root metabolism: injury and acclimation under hypomia and anomia. Annual Rev Plant Physiol Plant Mol Biol. 48: 223-250.

Duenas C, Fernandez MC, Carreter T, Liger E, Perez M. 1995. Emission of CO_2 from soils. Chemosphere. 30: 1875-1889.

Dun XP, Zhu XT. 2000. Effects of aerenchyma formation patter of secondary roots cortex on the water logging tolerance in wheat. J Huazhong Agril University. 19(4): 307-309.

Ellis MH, Setter TL. 1999. Hypoxia induces anoxia tolerance in completely submerged rice seedlings. J Plant Physiol.

Enoch HZ, Olesen JM. 1993. Plant response to irrigation with water enriched with carbon dioxide. New Physiologist. 125: 249-258.

Erdmann B, Hoffmann P, Wiedenroth EM. 1986. Changes in the root system of wheat seedlings following root anaerobiosis. I. Anatomy and respiration in *Triticum aestivum* L. Ann Bot. 58: 597-605.

Erdmann B, Wiedenroth EM. 1988. Changes in the root system of wheat seedlings following root anaerobiosis. III. Oxygen concentration of the root. Ann Bot. 62: 277-286.

Erdmann B, Wiednroth EM. 1986. Change in the root system of wheat seedling following anaerobiosis. II. Morphology and anatomy of evaluation forms. Ann Bot. 58: 607-616.

FAO, 2002. http://www.fao.org/waient/FAOINFO/AGRICULT/agl/agll/nav.html on March 18, 2002.

Finlayson SA, Reid DM. 1996. The effect of CO_2 on ethylene evaluation and elongation rate in roots of sun flower (*Helianthus annus*) seedlings. Physiologia Plantarum. 98: 875-881.

Fox CF, Green BJ, Kennedy RA, Rumpho ME. 1998. Changes in hexokinase activity in *Echinochloa phyllopogon* and *Echinochloa crub-pavonis* in response to abiotic stress. Plant Physiol. 118: 1403-1409.

Fox DL. 1933. Carbon dioxide nacrosis. J Cell Physiol. 3: 75-100.

Freeling M, Bennett DC. 1985. Maize *Adh1*. Annual Re Genet. 19: 297-323.

Gambrell RP, Delaune RD, Patrick WH Jr. 1991. Redox processes in soils following oxygen depletion. In: Plant Life Under Oxygen Deprivation, Jackson M, Davies DD, Lambers H (eds.), SPB Academic, the Hauge, the Netherlands. pp. 101-107.

Gangadharan C. 1985. Breeding. In: Rice Research in India, Jaiswal *et al.* (eds.), Indian Council of Agricultural Research, New Delhi, India. pp. 73-109.

Gardner WK, Flood RG. 1993. Less water logging with long season wheat. Cereal Res Comm. 21: 337-343.

Germain V, Raymond P, Richard B. 1997. Differential expression of two lactate dehydrogenase genes in response to oxygen deficit. Plant Mol Biol. 35: 711-721.

Gill KS, Qadar A, Singh KN. 1992. Response of wheat (*Triticum aestivum*) genotypes to sodicity in association with water logging at different stages of growth. Indian J Agri Sci. 62: 124-128.

Good AG, Crosby WL. 1989. Anaerobic induction of alanine aminotransferase in barley root tissue. Plant Physiol. 90: 1305-1309.

Goswami NN, David MS, Goel SK. 1975. Preliminaries studies on soil plant adjustment under water logging conditions with reference to maize (*Zea mays* L.). Sci Cult. 41: 562-564.

Govindarajan AG, Pooraiah BW. 1982. Effect of root-zone carbon dioxide enrichment on ethylene inhibition of carbon assimilation in potato plants. Physiologia Plantarum. 55: 465-469.

Greenway H, Gibbs J, Setter T. 1994. Mechanism of tolerance to water logging and submergence. IRRI, The Philippines. UWA and IRRI, Publ. pp. 136.

Greenway H, Waters I, Newsone J. 1992. Effects of anoxia on uptake and loss of solutes in roots of wheat. Aust J Plant Physiol. 19: 233-247.

Grichko VP, Glick BR. 2001. Flooding tolerance of transgenic tomato plants expressing the bacterial enzyme ACC deaminase controlled by the 35S ro1D or PRB-1b promoter. Plant Cell Physiol. 42: 245-249.

Grieve AM, Dunford E, Marston D, Martin RE and Slavich P. 1986. Effects of water logging and soil salinity on irrigated agriculture in the Murray Valley: A review. Aust J Exp Agric. 26: 761-777.

Grineva GM, Andreeva IN, Stupishina EA. 1970. Effect of flooding on growth, respiration and oxygen concentration in tissues of embryonic and stem roots of maize. Soviet Pl. Physiol. 17: 544-550.

Gupta S. 1988. Screening rice entries for coastal salinity and tidal swamp condition. Int Rice Res Newsl. 13 (4): 16-17.

Hamachi Y, Furusho M, Yoshida T, Ito M. 1988. Selection of high wet endurance cross parents for matting barley breeding programme. Jap J Crop Sci. 57: 715-721.

Hamachi Y, Furusho M, Yoshida T. 1989. Heritability of wet endurance in malting barley. Jap J Breed. 39: 195-202.

Hamachi Y, Yoshino M, Furusho M, Yoshida T. 1990. Index of screening for wet endurance in malting barley. Jap J Breed. 40: 361-366.

Haque QA, Hille Ris Lambors D, Tepora NM, Dela Cruz QD. 1989. Inheritance of submergence tolerance in rice. Euphytica. 41: 247-251.

Harberd NP, Edwards KJR. 1982. The effect of a mutation causing alcohol dehydrogenase deficiency on flooding tolerance of barley. New Phyto. 90: 631-644.

Hiatt AJ, Lowe RH. 1967. Loss of organic acids, amino acids, K and Cl from barley roots treated anaerobically and with metabolic inhibitors. Plant Physiol. 42: 1731-1736.

Higuchi T, Yoda K, Tensho K. 1984. Further evidence for gaseous CO_2 in rice plants. Soil Science Plant Nutr. 30: 125-136.

Hoeren FU, Dolfeous R, Wu Y, Peacock WJ, Dennis ES. 1998. Evidence for a role of AtMYB2 in the induction of the *Arabidopsis* alcohol dehydrogenase gene (ADH1) by low oxygen. Genetics. 149 ; 479-490.

Hoffman NE, Bent AF, Hanson AD. 1986. Induction of lactate dehydrogenase isozymes by oxygen deficit in barley root tissue. Plant Physiol. 82: 658-663.

Hook DD, Brown CL, Korwaink PP. 1971. Inductive flood tolerance Swamp Tupelo (*Nyrsa sylvatica* var. *biflora*). J Exp. Bot. 22: 78-89.

Hourton-Cabassa C, Ambard-Bretteville F, Moreau F, Davy de Virville J, Remy R, Colas des Frans-Small C. 1998. Stress induction of mitochondrial formate dehydrogenase in potato leaves. Plant Physiol. 116: 627-635.

Huang B, Johnson JW, Box JE, Nesmith S. 1997. Root characteristics and physiological activities of wheat in response to hypoxia and ethylene. Crop Sci. 27: 812-818.

Huang B, Johnson JW, Nesmith DS, Bridges DC. 1995. Nutritional accumulation and distribution of wheat genotypes in response to water logging and nutrient supply. Plant Soil. 173: 47-54.

Huang B, Johnson JW, Nesmith S, Bridges DC. 1994b. Growth, physiological and anatomical response of two wheat genotypes to water logging and nutrient supply. J Exp Bot. 45: 193-202.

Huang B, Johnson JW. 1995. Root respiration and carbohydrate status of two wheat genotypes in response to hypoxia. Ann Bot. 75: 427-432.

Huang BR, Johnson JW, Nesmith DS, Bridges, DC. 1994a. Root and shoot growth of wheat genotypes in response to hypoxia and subsequent resumption of aeration. Crop Sci. 34: 1538-1544.

ISTA. 1984. Seed Science and Technology. International Rules for Seed Technology. 1985. Proceedings of the International Seed Testing Association. Vol. 13, Number 2.

Jackson MB, Armstrong W.1999. Formation of aerenchyma and processes of plant ventilation in relation to soil flooding and submergence. Plant Biol. 1: 274-287.

Jackson MB, Campell DJ. 1976. Water logging and petiole epinasty in tomato: the role of ethylene and low oxygen. New Phytol. 76: 21-29.

Jackson MB, Drew MC. 1984. Effects of flooding on growth and metabolism of herbaceous plants. In: Flooding and Plant Growth, Kozlowski TT (ed.). Academic Press, New York. pp. 47-128.

Jackson MB, Fenning TM, Drew MC, Saker LR. 1985. Stimulation of ethylene production and gas-space (aerenchyma) formation in adventitious roots of *Zea mays* L. by small partial pressure of oxygen. Planta. 165: 482-492.

Jackson MB. 1979. Is the diageotropic tomato ethylene dieficient? Physiol Plant. 46: 347-351.

Jackson MB. 1990. Hormones and developmental changes in plants subjected to submergence and soil waterlogging. Aquatic Bot. 38: 49-72.

Jiang J, Dajum L. 1987. New *Hordium triticum* hybrids. Cereal Res Comm. 15: 95-99.

Joshi MS, Dastane NG. 1965. Excess water tolerance summer cereals. Indian J Agron. 10: 289-298.

Justin SHF, Armstrong W. 1987. The anatomical characteristics of roots and plant response to soil flooding. New Phytol. 106: 465-495.

Justin SHF, Armstrong W. 1991. Evidence of the involvement of ethane in aerenchyma formation in adventitious roots of rice (*Oryza sativa* L.). New Physiologist. 118: 49-62.

Kahn BA, Stoffella PJ, Sandsted RE, Zobel RW. 1985. Influence of flooding on root morphological components of young black beans. J Am Soc Hortic Sci. 110: 623-627.

Kawase M. 1976. Ethylene accumulation in flooded plants. Physiol Plant. 36: 236-241.

Kawase M. 1978. Aerenchyma formation: how plants adapt to water logging. Ohio Report on Research and Development. 63: 14-15.

Kelley PM, Freeling M. 1984a. Anaerobic induction of maize glucose phosphate isomerase I. J Biol Chem. 259: 673-677.

Kelley PM, Freeling M. 1984b. Anaerobic expression of maize fructose-16-diphosphate adholase. J Biol Chem. 259: 14180-14183.

Kelley PM. 1989. Maize pyruvate decarboxylase mRNA in induced anaerobically. Plant Mol Biol. 13: 213-222.

Kennedy RA, Rumpho ME, Fox TC. 1992. Anaeobic metabolism in plants. Plant Physiol. 84: 1204-1209.

Kent JB. 1983. Effect of soil flooding on leaf gas exchange of tomato plants. Plant Physiol. 73: 475-479.

Kosmakova VE, Zvereva EG. 1973. Distribution of assimilates in soybean plants grown under flooded conditions. Trudy Biologo-Pochvennogo Institute, Dalevostochnyi Nauchnyi Tsentr AN SSSR. 20: 204-208.

Kramer PJ, Jackson WT. 1954. Causes of injury to flooded tobacco plants. Plant Physiol. 29: 241-245.

Kuo CG, Chen BW. 1979. Evaluation of tomato varieties for resistance to flooding (Abstr.). Hort Sci. 14 (3, III): 413.

Lagudah ES, Appels R. 1994. Plant genetic enhancement. In: Encyclopedia of Agricultural Science. Arntzen CJ, Ritter M (eds.), Vol. 3. Academic Press, Inc., New York. pp. 283-293.

Lai KL. 1979. Studies on the formation and activity of the root system of paddy rice plants (*Oryza sativa* L.). In: The Causes of Low Yield of the Second Crop Rice in Taiwan and the Measures for Improvement, Hisich SC, Liu DJ (eds.). Proc Symp held at Taiwan, Agri Res Inst, June 7-8, 1978, National Science Council, Taipei, Taiwan. pp. 271.

Lee JE, Kim HS, Kim WH, Park SJ, Kwon YU, Kim JK. 2004. Identification of physiological indicator for establishing screening techniques related to tolerance of excess water in soybean [*Glycine max* (L.) Merrill]. In: Proceeding of the 4th International Crop Science Congress Brisbance, Australia, 26 Sep-1 Oct. 2004. pp.

Leyshon AJ, Shread RW. 1974. Influence of short term flooding on the growth and plant nutrient composition of barley. Can J Soct Sci. 54: 463-473.

Liao CT, Lin CH. 1995. Effect of flood stress on morphology and aerobic metabolism of *Memordica charantia*. Environ Exp Bot. 35: 105-113.

Limpinuntana V, Greeneay H. 1979. Sugar accumulation in barley and rice grown in solution with low concentrations of oxygen. Ann Bot. 43: 373-381.

Lin Y, Yang X, Liu F. 1994. A study on evaluation of water logging tolerance in wheat varieties (*Triticum aestivum* L.). Acta Agric Shanghai. 10 (2): 79-84.

Linkemer G, Board JE, Musgrave ME. 1998. Water logging effect on growth and yield components of late-planted soybean. Crop Sci. 38: 1576-1584.

Livine A, Yaadia Y. 1972. Water deficits and hormone relations. In: Water Deficits and Plant Growth, Vol III, Kozlowski (ed.). Academic Press, New York.

Malik AI, Colmer TD, Lambers H, Setter TL, Schortemeyer M. 2002. Short term water logging has long term effects on the growth and physiology of wheat. New Phytol. 153: 225-236.

Malik AI, Colmer TD, Lambers H, Schortemeyer M. 2001. Changes in physiological and morphological traits of roots and shoots of wheat in response to different depths of waterlogging. Australian J Plant Physiol. 28(11): 1121-1131.

Maranville JW, Rosario DA, Dalmacio SA, Clak RB. 1986. Variability in growth and nutrients accumulation in sorghum grown in water logged soils. Commu Soil Sci Plant Analysis. 17: 1089-1108.

Martin FW. 1984. Differences in stress resistance of rooted sweet potato leaves. J Agri Univ of Pureto Rice. 68: 235-242.

Martin T, Frommer WB, Salanoubat M, Willmitzer L. 1993. Expression of an *Arabidopsis* sucrose synthase gene indicates a role in metabolization of sucrose both during phloem loading and in sink organs. The Plant J. 4: 367-377.

Matocha JE, Mostaghimi S. 1988. Effects of carbon dioxide and iron enrichment of a calcareous soil on Fe-chlorosis, root and shoot development of grain sorghum. J Plant Nutr. 11: 1503-1515.

Mattana M, Bertani A, Reggiani R. 1994a. Expression of glutamine synthetase during the anaerobic germination of *Oryza sativa* L. Planta. 195: 147-149.

Mattana M, Coraggio I, Bertani A, Reggiani R. 1994b. Expression of enzymes of nitrate reduction during the anaerobic germination of rice. Plant Physiol. 106: 1605-1608.

McDonald GK, Garder WK. 1987. Effect of water logging on the grain yield response of wheat to sowing date in south-western Victoria. Aust J Exp Agric. 27: 661-670.

McDonald MP, Galwey NW, Colmer TD. 2001a. Water logging tolerance in the tribe *Triticeae*: The adventitious roots of *Critesion marinum* have a relatively high porosity and a barrier to radial oxygen loss. Plant Cell Environ. 24: 585-596.

McDonald MP, Galwey NW, Ellneskog Staam P, Calmer TD. 2001b. Evaluation of *Lophopyrum elongatum* as a source of genetic diversity to increase water logging tolerance of hexaploid wheat (*Triticum aestivum*). New Phytol. 151: 369-380.

Mertens E. 1991. Pyrophosphate-dependent phosphofructokinase, an anaerobic glycolytic enzyme? FEBS Letters. 285: 1-5.

Meyer WS, Barrs HD. 1988. Response of wheat of single, short-term water logging during and after stem elongation. Aust J Agric Res. 39: 11-20.

Michniewicz M, Kriesel K, Krassowska J. 1970. Effect of adventitious roots on level of endogenous gibberellin-like substances in buds and nearly formed shoots of willow cuttings (*Salix viminalis* L.). Bull Acad Polon Sci Ser Sci Biol. 18: 355-359.

Mikhailova AV. 1977. Some characteristic apparatus of barley leaves affected by flooding and vitamin PP application. Nauchnye Doklady Vysshei Shokoly, Biologicheskie Nauki. No. 9: 104-108.

Mukherjee D, Biswas S. 1979. CR1002: A promising rice cultivar for rainfed water logged conditions. Int Rice Res Newl. 4 (6): 9.

Nandi S, Subudhi PK, Senadhira D, Manigbas NL, Sen-Mandi S, Huang N. 1997. Mapping QTLs for submergence tolerance in rice by AFLP analysis and selective genotyping. Mol Gen Gent. 255: 1-8.

Neog B, Gogoi N, Baruah KK, Gogoi N. 2002. Morphological changes associated with waterlogging in rice (*Oryza sativa*). Indian J Agril Sci. 72(7): 404-407.

Olson DC, Oetiker JH, Yang SF. 1995. Analysis of LE-ACS3, a 1-aminocyclopropane-1-carboxlic acid synthase gene expressed during flooding in the roots of tomato plants. J Biological Chem. 270: 14056-14061.

Oosterhuis DM, Scott HD, Hampton RE, Wullscheger SD. 1990. Physiological response of two soybean [*Glycine max* L. Merr] cultivars to short-term flooding. Envi Expt Bot. 30: 85-92.

Orchard PW, Jessop RS. 1984. The response of sorghum and sunflower to short term water logging. I: Effect of stage of development and duration of water logging on growth and yield. Plant Soil. 81: 119-132.

Paiatti M, Sodek KL. 1999. Waterlogging affects nitrogen transport in xylem of soybean. Plant Physiol. 37(10): 767-773.

Pang JY, Zhou MX, Mendhan N, Shabala S. 2004. Growth and physiological responses of six barley genotypes to waterlogging and subsequent recovery. Australian J Agril Res. 55(8): 895-906.

Perata P, Geshi N, Yamaguchi J, Akazawa T. 1993. Effect of anoxia on the induction of a-amylase in cereal seeds. Planta. 191: 402-408.

Porto MP. 1997. Method of selection for tolerance to soil waterlogging in maize plants. Pesq Agrop Gaucha. 3: 187-190.

Qiu JD, Ho YA. 1991. Study of determination of wet tolerance of 4572 barley germplasm resources. Acta Agriculturae Shanghai. 7: 27-32.

Qunimio CA, Torrizo LB, Setter TL, Elllis M, Grover A, Abrigo EM, Oliva NP, Ella ES, Carpena AL, Ito O, Peacock WJ, Dennis E, Datta SK. 2000. Enhancement of submergence tolerance in transgenic rice overproducing pyruvate decarboxylase. J Plant Physiol. 156: 516-521.

Rahaman M, Grover A, Peacock WJ, Dennis ES, Ellis MH. 2001. Effects of manipulation of pyruvate decarboxylase levels on the submergence tolerance of rice. Aust J Plant Physiol. 28: 1231-1241.

Reid DM, Crozier A, Harvey BMR. 1969. The effect of flooding on the export of gibberellins from the root to the shoot. Planta. 89: 376-379.

Reid DM, Crozier A. 1971. Effects of water logging on the gibberellin content and growth of tomato plants. J Exp Bot. 22: 39-48.

Reynolds MP, Trethowan RM, Van Ginkel M, Rajaram S. 2001. Application of physiology in wheat breeding. In: Application of Physiology in Wheat Breeding, Reynolds MP, Ortiz, Monsasterio, JI, McNab, A (eds.). CIMMYT, Mexico. pp. 2-10.

Richard B, Couee I, Raymond P, Saglio PH, Saint Ges V, Pradet A. 1994. Plant metabolism under hypoxia and anoxia. Plant Physiol Biochem. 32: 1-10.

Roberts JKM, Callis J, Wemmer D, Jardetzky O, Walbot V. 1984. Cytoplasmic acidosis as a determinant of flooding intolerance in plants. Proceedings of the National Academy of Sciences, USA. 81: 6029-6033.

Russell RC. 1977. Plant Root Systems: Their Function and Interaction with the Soil. McGrow Hill Book Co. (UK) Ltd.

Sachs MM, Subbaiah CC, Saab IN. 1996. Anaerobic gene expression and flooding tolerance in maize. J Exp Bot. 47: 1-15.

Sachs MM. 1993. Molecular genetic basis of metabolic adaptation to anoxia in maize and its possible utility for improving tolerance of crops to soil waterlogging. In: Interacting Stress on Plants in a Changing Environment, Jackson MB, Black CR (eds.). Spriger-Verlag, Berlin. pp. 375-393.

Saglio PH, Raymond P, Pradet A. 1980. Metabolic activity and energy change of excised maize root tips under anoxia. Plant Physiol. 66: 1053-1057.

Sahai VN, Ghosh S, Suri HK, Saran S, Haque MS. 1979. Flowering and yield response of late maturing varieties under normal and water logged conditions. Int Rice Res Newsl. 4 (3): 11-12.

Samad A, Meisner CA, Saifuzzaman M, Van Ginkel M. 2001. Water logging tolerance. In: Application of Physiology in Wheat Breeding. Reynolds MP, Ortiz Monasterio JI, McNab A (eds.). CIMMYT, Mexico. pp. 136-144.

Sangen W, He L, Li Z, Zeng J, Chai Y, Hou L. 1996. A comparative study of the resistance of barley and wheat to water logging. Acta Agron Sinica. 22(2): 228-232.

Sayyre KD, Van Ginkel M, Rajaram S, Ortiz-Monasterio I. 1994. Tolerance to water logging losses in spring bread wheat: effect of time of onset on expression. In: Annual Wheat Newsletter, Colorado State University. 40: 165-171.

Sethiya HL, Verma SS, Aneja DR. 1986. Nitrate reductase activity, a selection tool for water logging tolerant genotypes of sugarcane. Indian Sugar. 35: 649-653.

Setter TL, Belford B. 1990. Water logging: how it reduces plant growth and how plants overcome its effects. Western Aust J Agric. 31: 51-57.

Setter TL, Burgess P, Waters I, Kuo J. 2001. Genetic diversity of barley and wheat for water logging tolerance in Western Australia. In: Proceedings of the 10th Australian Barley Technical Symposium, Canberra, ACT, Australia. 16 -20 Sep 2001.

Setter TL, Burgess P, Waters I, Kuo J. 1999. Genetic diversity of barley and wheat for water logging tolerance in Western Australia. In: 9th Barley Technical Symposium, Melbourne, Australia, Sept. 1999. pp. 2.17.1-2.17.7.

Setter TL, Kupkanchanakul T, Kupkanchanakul K, Bhekasut P, Weingweera A, Greenway H. 1987. Concentrations of CO_2 and O_2 in floodwater and in internodal lacunae of floating rice growing at 1-2 meter water depths. Plant, Cell and Environ. 10: 767-776.

Setter TL, Waters I. 2003. Review of prospects for germplasm improvement for water logging tolerance in wheat, barley and oats. Plant and Soil. 253: 1-34.

Setter TL. 2000. Farming systems for water logging prone sandplain soils of the south coast. Final Report of GRDC Project No. DAW292, Department Agriculture, Western Australia. pp. 68.

Sharma DP, Swarup A. 1988. Effects of short term flooding on growth yield and mineral composition of wheat on sodic soil under field conditions. Plant Soil. 107: 137-143.

Siangliw M, Toojinda T, Tragoonrung S, Vanavichit A. 2003. Thai jasmine rice carrying QTLch9 (SubQTL) is submergence tolerant. Ann Bot. 91: 255-261.

Singh BB. 1988. Tolerance of soybean varieties to water logging. Indian J Plant Physiol. 31: 410-412.

Sinha TS, Bandyopadhyay AK. 1986. Varietal evaluation of rice in water logged saline soil. Int Rice Res. Newsl. 11 (3): 14-15.

Somrajan KH, Rangarajan R. 1985. Performance of certain foreign commercial cane varieties under water logged condition. Indian Sugar Crops J. 11: 46-48.

Sparrow LA, Uren NC. 1987. The role of manganese toxicity in crop yellowing on seasonally water logging and strongly acidic soils in north-eastern Victoria. Aust J Agric. 27: 303-307.

Springer B, Werr W, Starlinger P, Bennett DC, Zokolica M, Freeling M. 1986. The *shrunken* gene on chromosome nine of *Zea mays* L. is expressed in various plant tissues and encodes an anaerobic protein. Mol General Genet. 205: 461-468.

Sripongpangkul K, Posa GBT, Senadhira DW, Brar D, Huang N, Khush GS, Li ZK. 2000. Genes/QTLs affecting flood tolerance in rice. Theor Appl Genet. 101: 1074-1081.

Steffens D, Hutsch BW, Eschholz T, Losak T, Schubert S. 2005. Waterlogging may inhibit plant growth primarily by nutrient deficiency rather than nutrient toxicity. Plant Soil Environ. 51(12): 545-552.

Stieger PA, Feller U. 1994. Nutrient accumulation and translocation in manuring wheat plants grown water logged soil. Plant Soil. 160: 87-95.

Sullivan M, VanToai TT, Fausey N, Beuerlin J, Parkinson R, Soboyejo A. 2001. Evaluating on-farm flooding impacts on soybean. Crop Sci. 41: 93-100.

Sundaram T, Pillai OR, Seougaperumal S, Robinson JG, Mathar AS. 1988. CR 1009: Suitable variety for water logged conditions. Int Rice Res Newsl. 13 (3): 17.

Takeda K, Fukuyama T. 1986. Variation and geographical distribution of varieties for flooding tolerance in barley seeds. Barley Genetics Newsl. 16: 28-29.

Tennant D, Scholz G, Dixon J, Purdie B. 1992. Physical and chemical characteristics of duplex soils and their distribution in south-west of Western Australia. Aust J Exp Agric. 32: 827-843.

Thomson CJ, Armstrong W, Waters I, Greenway H. 1990. Aerenchyma formation and associated oxygen movement in seminal and nodal roots of wheat. Plant Cell Environ. 13: 395-403.

Thomson CJ, Atwell BJ, Greenway H. 1989. Response of wheat seedling to low O_2 concentrations in nutrient solution. I. Growth, O_2 uptake and synthesis of fermentative end-products by root segments. J Exp Bot. 40: 985-991.

Thomson CJ, Colmer TD, Watkin ELJ and Greenway H. 1992. Tolerance of wheat (*Triticum aestivum* cv. Gamenya and Kite) and triticale (*Triticosecali* cv. Muir) to water logging. New Physiol. 120: 335-344.

Ton CS. 1978. A study of high soil moisture levels and water submergence on the sweet potato (*Ipomoea batatas* (L.) Lam.). Diss Abstr Int. B. 38: 2991B.

Toojinda T, Siagliw M, Tragoonrung S, Vanavichit A. 2003. Molecular genetics of submergence tolerance in rice: QTL analysis of key traits. Ann Bot. 91: 243-253.

Trompel AF. 1978. Yield of drought and excessive soil moisture on yield in winter rye. In: Ustoichivost Zern Kultur K faktoram sredy Minsk, Bolorussian, SSR. pp. 30-40.

Trought MCT, Drew MC. 1980a. The development of water logging damage in wheat seedlings (*Triticum aestivum* L.) III. Accumulation and redistribution of nutrients by the shoot. Plant Soil. 56: 187-199.

Trought MCT, Drew MC. 1980b. The development of water logging damage in young wheat plants in anaerobic solution culture. J Exp Bot. 31: 1573-1585.

Trought MCT, Drew MC. 1982. Effects of water logging on young wheat plants (*Triticum aestivum* L.) and on soil solutes at different temperature. Plant Sci. 69: 311-326.

Van Ginkel M, Rajaram S, Thijssen M. 1992. Water logging in wheat: Germplasm evaluation and methodology development. In: the Seventh Wheat Workshop for Eastern, Central and Southern Africa, Tanner DG, Mwangi W (eds.), Nakuru, Kenya, Sept. 16-19, 1991. pp. 115-124.

VanToai T, Sneller C, Lark KG, Mrtin SKS, Dorrance A, Boru B, Bower G. 1999. Mapping of QTL for tolerance to soil water logging in soybean. In: Plant and Animal Genome VII Conference, January 17-21, 1999. Town and Country Hotel, San Diego, C.A. pp. 251.

VanToai TT, Beuerlein JE, Schmitthenner AF, St. Martin SK. 1994. Genetic variability for flooding tolerance in soybean. Crop Sci. 34: 1112-1115.

VanToai TT. 2001. Improving plants tolerance to water stress by molecular breeding and genetic transformation. In: Conference of the International Society for Plant Anaerobiosis, June 12, 2001.

Varade SB, Stalzoy LH, Letey J. 1970. Influence of temperature, light intensity and aeration on growth and root porosity of wheat, *Triticum aestivum*. Agron J. 62: 505-507.

Wample RL, Reid DM. 1978. Control of adventitious root production and hypocotyls hypertrophy of sunflower (*Helianthus annuus*) in response to flooding. Physiol Plant. 44: 351-358.

Wang S, Yuan Xi M, Sumei Z, Yum JZ, Jiu Xing L, HuaCen W, Wang C, Ma YX, Zhou SM, Zhu YJ, Li JX, Wang HC. 1996. Effects of water logging on the metabolism of active oxygen and the physiological activities of wheat root systems. Acta Agron Sinica. 22: 712-719.

Water I, Kuiper PJC, Watkin E, Greenway H. 1991a. Effects of anoxia on wheat seedlings. I. Interaction between anoxia and other environmental factors. J Exp Bot. 42: 1427-1435.

Water I, Morrell S, Greenway H, Colmer TD. 1991b. Effects of anoxia on wheat seedlings. II. Effects of O_2 supply prior to anoxia on tolerance to anoxia, alcoholic fermentation and sugars. J Exp Bot. 42: 1437-1447.

Watkin ELJ, Thomson CJ, Greenway H. 1998. Root development in two wheat cultivars and one triticale cultivar grown in stagnant agar and aerated nutrient solution. Ann Bot. 81: 349-354.

Watson ER, Lapins P, Barron RJW. 1976. Effect of water logging on the growth, grain and straw yield of wheat, barley and oats. Aust J Exp Agric An Husb. 16: 114-122.

Wiengweera A, Greenway H, Thomson CJ. 1997. The use of agar nutrient solution to stimulate lack of convection in water logged soils. Ann Bot. 80: 115-123.

Wignarajah K, Greenway H, John CD. 1976. Effect of water logging on the growth activity of alcohol-dehydrogenease in barley and rice. New Phytol. 77: 585-592.

Willianson RE, Kriz GJ. 1970. Response of agricultural crops to flooding, depth of water-table and soil gaseous composition. Trans Am Soc Agric Eng. 13: 216-220.

Wright STC, Hrion RWP. 1970. The accumulation of abscisic acid in plants during wilting and under other stress conditions. Proc. 7th Int. Conf. Plant Growth Substances, Carr DJ (ed.), Berlin, Springer-Verlag. pp. 291-298.

Xia JH, Saglio PH. 1992. Lactic acid efflux as a mechanism of hypoxia acclimation of maize root tips to anoxia. Plant Physiol. 93: 453-459.

Xu K, Deo R, Mackill DJ. 2004. A microsatellite marker and a co-dominant PCR based marker of Marker-Assitated-Selection of submergence tolerance in rice. Crop Science. 44: 248-253.

Xu K, Mackill DJ. 1996. A major locus for submergence tolerance mapped on rice chromosome 9. Mol Breed. 2: 21-224.

Xu K, Xu X, Ronald PC, Mackill DJ. 2000. A high-resolution linkage map in the vicinity of the rice submergence tolerant locus *sub1*. Mol Gen Genet. 263: 681-689.

Yu PT, Stolzy LH, Letey J. 1969. Survival of plants under prolonged flooded conditions. Agron J. 61: 844-847.

Yuanmin S, Guo S, Zhang J, Huang Y, Tang Y, Xhang H. 1997. Relationship between wheat water logging injury and accumulation of reducing matter in rhizosphere. Jiangsu J Agric Sci. 13 (4): 202-205.

Zaidi PH, Rafique S, Singh NN. 2003. Response of maize (*Zea mays* L.) genotypes to excess moisture stress: morpho-physiological effects and basis of tolerance. Eur J Agro. 19: 383-399.

Zaidi PH, Singh NN. 2001. Effect of waterlogging on growth, biochemical compositions and reproduction in maize. J Plant Biol. 28: 61-69.

Zhang J, Davies WJ. 1987. ABA in roots and leaves of flooded pea plants. J Exp Bot. 38: 649-659.

Zhou GS, Zhu XT. 2002. Changes of physiological characters of wheat after waterlogging at booting and relations between physiological characters and waterlogging tolerance. Agril Sci China. 1(8): 87-884.

Zhou MX, Li H, Mendham N, Salter S. 2004. Inheritance of water logging tolerance of barley (*Hordium vulgare* L.). In: Proceedings of the 4th International Crop Science Congress, Brisbane, Australia, 26 Sept-1 Oct., 2004.

4

Drought Tolerance

4.1. INTRODUCTION

Water stress is the single most severe limitation to the productivity of rice in the rainfed ecosystem (Widawsky and O'Toole, 1990). Drought has been defined as the inadequacy of water availability, including precipitation and soil moisture storage capacity, in quantity and distribution during the life cycle of the crop to restrict expression of its full genetic yield potential. Drought is actually a meteorological event which implies the absence of rainfall for a period of time, long enough to cause moisture-depletion in soil and water deficit with a decrease of water potential in plant tissues. Under drought conditions, water stress develops in the plants as the demand exceeds supply of water; this may occur either/or both due to atmospheric or soil conditions, and is reflected in a gradient of water potentials developed between the soil or soil-root interface and leaf, the transpiring organ. Moisture stress is likely to develop to a different rate in different plant organs along with the gradient. Drought like many other environmental stresses has adverse effects on plant growth and yield. The irrigated ecosystem accounts for 55% of the total rice area and contributes 75% of global rice production. The remaining 25% of rice comes from unfavourable rice-growing areas (45%): rainfed land (25%), upland (12%), and flood-prone areas (8%) (Datta *et al.*, 2002). Many countries face periodical serious drought periods and a continuous deficit good quality fresh water resources and many countries face it within 20-25 years. Drought stress constitutes a permanent and increasing agronomic problem in arid and semi-arid regions. It acts as a serious limiting factor in agricultural production by preventing a crop from reaching the genetically determined theoretical maximum yield (Begg and Turner, 1976).

4.2. SYMPTOMS OF DROUGHTS

1. As water stress increases, older leaves senescence to various degrees; this reduces the leaf area of plants. Young *et al.* (2004) also reported loss of leaf function and premature onset of senescence of older leaves. They suggested that ethylene may serve to regulate leaf performance throughout its lifespan as well as to determine the onset of natural senescence and mediate drought-induced senescence.
2. Stress could come at the time of different stages plant growth, which may influence flowering or emergence of spike in cereals, or cause abscission of flowers.
3. Water deficit at anthesis and during advance stages delay in flowering and grain development resulting high spikelet sterility in rice, which leads lower grain yield (Kumar and Kujur, 2003). The loss in grain yield in rice has been estimated as much as 11-58% (Ouk *et al.*, 2004).

4.3. INJURY MECHANISMS

Water stress directly affects cellular processes, plant growth and development, and finally economic yield. At cellular level, drought affects membrane structures, and structure of macromolecules like protein, nucleic acids and enzymes, create pressure differential across the membrane-cell wall complex.

At low water potential, stomata close due to the loss in turgidity of the guard cells leading cease in transpiration. In absence of transpiration, temperature of the leaves increase to lethal level and the leaves die. Such, low water potentials (high tension on xylem, water) are known to cause severe embolism formation in the xylem vessels of woody plants blocking water transport and potentiality causing shoot dieback.

4.4. TOLERANCE MECHANISMS

Drought tolerance is generally defined as the property of a given cultivar to show a relatively small yield reduction upon exposure to drought. The various mechanisms by which a crop can minimize the loss in yield due to drought are grouped into three categories- drought escape, dehydration avoidance and dehydration tolerance (May and Mithoorpe, 1962; Turner, 1979; Levitt, 1980; Chang *et al.,* 1986; McWilliam, 1989). The characteristics or mechanisms that impose to minimize the effect of drought should be reversible and should perform its potential in drought free situation.

4.4.1. Drought Escape

The ability of a plant to escape periods of drought, in particular the most sensitive periods of its development are flower initiation and grain filling is termed as drought escape. Or it is the ability of the plant to complete its life cycle or most drought sensitive stage before on set of drought. One breeding strategy is to shorten the life cycle of a crop to enable it to mature safely during rainfall period. Early maturity is one of the most important attribute of drought escape (Lewin and Sparrow, 1975; Chang *et al.,* 1986), and is suitable for environments subjected to late-season drought stress. For example, in the Sahel in Africa toward the Sahara Desert very short season cowpeas avoid drought by maturing before any substantial stress develops, in less than 65 days (Hall *et al.,* 1978). The plant breeding programmes should therefore, aim at developing high-yielding genotypes with phenological patterns to match probable seasonal soil moisture availability of a given environment (Fischer *et al.,* 1982). Most of the studies showed significant positive correlation of crop duration with economic yield (Roy *et al.,* 2005). Development of early maturing genotypes thus may reduce biomass production and subsequently the crop yield. However, several short duration genotypes of legumes show higher and more stable yields than longer duration types (Hall and Patel, 1985; Rose *et al.,* 1992; Basu and Nautiyal, 2004). Early maturing genotypes generally have shallow root system (Arihara and Okada, 1991), lower leaf area index, lower total evapotranspiration and lower yield potential (Ferers *et al.,* 1986). Thus, it appears that maturity duration may be more deeply involved in the plant-water relations than its mere effects on total evapotranspiration. Early maturity and high yield in upland cotton (*Gossipium hirsutum*) are correlated when timely rainfall does not occur during the growing seasons, and the crop is forced to make its yield on stored soil moisture (Quisenberry and Roask, 1976).

Drought escape also involve with seed germination. Dormant seeds, quasi-dormant or quiescent state seed survive the dry spell (Thomas, 1997) and germinate only under favourable conditions.

4.4.2. Dehydration Avoidance

The ability to endure or withstand a dry period by maintaining a favourable internal water balance under drought condition is termed as dehydration avoidance. The dehydration avoidance mechanisms are results of various physiological, biochemical and metabolic processes of plants that are involved in growth and yield not being internally exposed to stress and, thereby they are protected from water stress (Blum, 1988). Associated high water use efficiency with plants that avoid drought and described two-types of drought avoiders: (1) *Water savers* avoid drought by closing their stomata during the day even through soil water supplies may be adequate and they achieve high water use efficiency through efficient control of the total water supply. (2) Whereas, *water spenders*, avoid drought by extracting large quantity of water from soil per unit time. As a result, the water spender keep their stomata more open throughout the day, assimilate more CO_2 and therefore have more rapid growth rate than water savers. The common measure of dehydration avoidance is the tissue water status as expressed by water or tugor potential under conditions of water stress. This can be achieved either by reducing transpiration, water saver; the plants that minimize dehydration through modified shoot attributes or increasing water uptake efficiency, water spenders; the plants that maximize water uptake through improved root characteristics (Levitt, 1980). Levitt (1972) has confirmed that how stomatal behaviour can affect plants water use efficiency under semi-arid conditions. Both water savers and water spenders maintain themselves in tiding high water potential state when they are exposed to external water stress. Wild species are readily classifiable as water savers and water spenders, but crop plants generally exhibit a combination of both features.

Dehydration avoidance is a complex trait, expression of which depends on action and interaction of different morphological (reduced leaf area, leaf rolling, cuticular wax, deficient rooting system, awn, stability in yield), physiological (reduced transpiration, high water-use efficiency, stomatal closure, osmotic adjustment) and bio-chemical (accumulation of proline, polyamine, trehalose, etc., increase nitrate reductase activity and increased storage of carbohydrate) characters.

4.4.2.1. Shoot characteristics in dehydration avoidance

The shoot attributes include stomatal control of transpiration, leaf orientation, leaf shape, leaf morphology etc.

Reduced transpiration: Stomatal resistance has been investigated as a tool in determining drought tolerance in sorghum (Blum, 1974; Henzell *et al.*, 1976). Stomatal aperture acts plant protective mechanisms by decreasing water loss through closure it's during periods of plant water deficits as well as influencing the rates of photosynthesis and respiration. *Water saving* species reduces transpiration mostly by closure of their stomata in response to water deficit well before wilting. Stomata may remain open during the early morning hours and close as solar radiation increases. There is considerable genetic diversity within the species and among the species in sensitivity of stomata to drought stress. Variation in stomatal density, aperture size, sensitivity to changes in internal and external water status has been reported by Saxena *et al.* (1996), Dhopte *et al.* (1987). Stomatal frequency has also been considered as one of the indices of drought resistance (Shirinkina, 1990). Low stomatal frequency may be associated with drought tolerance. Stomatal activity *per se* is an imperfect criterion of resistance, unless it is related to both carbon fixation and transpiration under stress. Higher rate of stomatal resistance and lower rate of transpiration can be used for screening germplasm for drought tolerance (Yadav *et al.*, 1991).

Cuticular wax: It consists of amorphous intracellular wax embedded in cutin polymer, and epicuticular wax crystalloids that coat the outer plant surface and impart a whitish appearance. Cuticular wax is mainly composed of long-chain aliphatic compounds derived from very long chain fatty acids. Plant cuticular waxes play an important role in protecting aerial organs from damage caused by multiple environmental stresses such as drought, cold, UV radiation, pathogen infection, and insect pest attack. The interface between plants and environment plays a duel role as a protective barrier and a medium for exchange of gasses, water and nutrients. The primary aerial plant surfaces are covered by a cuticle, acting as the essential permeability barrier towards the atmosphere.

Transpiration also occurs through cuticle during longer period of desiccation. The rate and amount of cuticular transpiration depends on wax deposition within and over the cuticle. It is largely controlled by the amount and physiological configuration of the wax deposition (Chambers *et al.*, 1976). Cuticular wax increases with increase water stress (Hamission *et al.*, 1991). But, the effect of cuticular wax on transpiration is small, and for a given plant, increase in wax load beyond a given threshold would not reduce transpiration. The structure affects the spectral characteristics of leaves (Eller, 1979). Genotypic variation in amount and composition of cuticular wax was observed in oat, barley and rice (Bengston *et al.*, 1978; O' Toole *et al.*, 1979).

4.4.2.2. Leaf morphology and orientation

Size, shape, surface characteristics, angle and arrangement of leaves play important role in determining drought tolerance in crop plants. Leaf pubescence generally increases leaf reflectance and reduces net radiation resulting in lower leaf temperature under high radiation. Leaf thickness, hairiness, increase in size of protective and mechanical tissues and mesophyll integrity seem to be good selection criteria for drought tolerance in wheat (Hameed *et al.*, 2002). Net radiation can also be reduced by altering the leaves from horizontal to angular position, which receives the maximum radiation. Narrow, vertically oriented leaves result in higher leaf area index and are considered to be adaptive character to stress conditions as they are less stressed. Thus, in cereals, erect leaf genotype perform better than lax leaf genotypes under drought. In some crop plants, leaflets oriented perpendicularly to incident light in the absence of water deficit but parallel to it during water deficits (Squire, 1990). The paraheliotropic movement of leaves, particularly legumes during high radiation reduces the transpiration loss of water in drought prone areas. Ability to change leaf angle or orientation depends on the water content. Genotypic differences in paraheliotropism have been reported in ground nut, beans and legumes (Subbarao *et al.*, 1995). Dhopte *et al.* (1995) found positive correlation between yield and flag leaf area in sorghum. Green leaf area at physiological maturity is a good indicator of drought tolerance (Borrell *et al.*, 2000). Thickness of leaf cuticle, epidermis, hypodermis and number of stomata generally increased under water stress, while the number of hairs and stomatal length decreased (Hameed *et al.*, 2002).

Leaf rolling is the most familiar response toward water stress, particularly in cereals, which avoids drought. It is a kind of defense mechanism under stress in tolerant genotypes. Reduction in transpiration rate of 46 to 83% resulting from leaf rolling has been reported by Hurd (1969). Rolling of leaves indicates a decrease in turgor pressure, indicating that the plant is no longer completely avoiding water stress. Leaf rolling is delayed by osmotic adjustment. Hence, delayed leaf rolling is an indication of turgor maintenance as a component of dehydration avoidance; it is used as an important selection criterion for dehydration avoidance in rice and wheat (Chang *et al.*, 1974; Loresto *et al.*, 1976; O'Toole and Moya, 1978).

4.4.2.3. Root characteristics in dehydration avoidance

The majority drought research investigations have dealt with aerial portions of plants because of their ease of study. The great limitation in the study of root resides in assessing the root system using seedling and, by contrast, assessing production field grown adult plants. The difficulties in evaluating root system, the large environmental influences and the complex inheritance of root characteristics hinder the use of these traits in selection programmes in spite of the obvious positive relationship between root depth, root growth and yield under drought conditions (Medrano *et al.*, 1998).

Root systems obviously play an important role in water acquisition for plants and are a significant component of tolerance to water-deficit stress (Baker and Varughese, 1992; Weerathaworn *et al.*, 1992; McCully, 1999). Root morphology characters and stressed-induced response form important component of drought. Enhancing yield remains the principal objective of the most breeding programmes. Interaction between primary characters posses a formidable challenge, while dealing with grain yield under stress. A crop which is able to increase the amount of water transpired relative to another crop grown in the same water-limiting yield condition, accumulates greater plant biomass. Increased water transpiration can result from increased soil water extraction by a deeper and/or denser root system. Therefore, root length has been proven to be one of the most important parameter closely related to the ability of plant in acquiring nutrient and water, lodging and drought tolerance, ultimately the productivity. Thus, to uncover the mechanisms controlling root elongation will be the outstanding agronomic importance, especially in cereal plants. Root elongation is determined by two coordinating processes of cell division and cell elongation. Suardi (2002) reported that the rice genotypes with thick roots and deep root systems were more tolerant to drought. Water uptake depends mainly on the root depth, size, root-length density, root axil resistance, root radial resistance, morphology, hydraulic conductance and physiology.

Plants of many species respond to drought by increasing the proportion of assimilate diverted with root growth, thus increasing root/shoot ratio and volume of soil water available to the plant (O' Toole and Bland, 1987). Sponchiado *et al.* (1989) found positive correlation between deep root system and seed yield in bean. Drought tolerant trees are often characterized by deep and vigorous root system. In *Robusta* coffee, by growing four contrasting clones in large containers, it was observed that a deeper root system could be associated with better avoidance to soil water limitation in drought-tolerant clones than in drought-sensitive ones. Another study has associated drought tolerance with a larger root dry mass, shown by Ramos and Carvalho (1997) working with 29 coffee genotypes.

Some rice groups have shown significance root attributes which contribute to higher yield under water stress environments and their corresponding QTL have been identified (Hemamalini *et al.*, 2000; Zheng *et al.*, 2000). Others also have made progress in using markers in breeding modifications of root attributes in rice that correlate well with improved tolerance to dry environments (MacKill *et al.*, 1999). It is apparent from such studies that shifting root biomass from the upper soil zone to deeper regions would enhance the changes of encountering new water sources.

As the depth, width and branching of root systems increase, plant water decreases. It is often been speculated that plant tolerant to poor soil fertility should have greater root growth, and thus perform well under drought conditions in similar sites (White and Izaquierdo, 1991). In an attempt to show a relationship between early seedling root development and effects of drought on yield, study was conducted in Chile using recombinant inbred lines (RILs) differing in seedling root traits (Burker TC, Saab I, unpublished data). Sixteen RILs, classified as having poor early root development lines showed reduced adventitious and lateral root development. Ten vigorous early root development lines exhibited extensive adventitious and lateral root development. The class of RILs with poor early root development yielded better than the class with more vigorous early root development. These data were reminiscent of the observation of improve grain yield under drought stress that was accompanied by reduced root biomass (Bolanos *et al.*, 1993). It is likely that vigorous root growth may be at the cost of grain production despite the improved advantage of water acquisition in dry soil.

4.4.2.4. Critical stages

Tolerance ability of crop plants under abiotic stresses also influenced by the critical stages. Findings of Meena-Kumari *et al.* (2004) reveled that the flowering and grain filling were the critical growth stages in maize, during which they observed maximum adverse effects. Under severe stress anthesis-silking-interval range for resistant and susceptible genotypes were 3-5 and 9-17 days respectively. Chlorophyll stability index decreased with increased water stress in most of the genotypes. Tolerant lines showed lower leaf temperature under stress than the susceptible ones, but transpiration rate was comparatively high in tolerant ones.

Passioura (1994) stated that a crop which takes a conservative approach to vegetative growth, in the face of terminal drought stress, will normally have a higher harvest index. A crop which accumulates too much dry matter prior to anthesis will be subjected to 'having off' as water resources become increasingly limited to growth, resulting in a reduced harvest index. The contrasting ratios of dry matter accumulation between vegetative and reproductive growth stages, indicates strikingly different plant growth patterns between wheat and pea. Wheat invests heavily in vegetative growth, growing a large root system for vigourous soil exploration and a dense grass canopy for maximum photosynthesis; it is an 'optimistic' growth strategy because it is dependent on continued favourable climatic conditions during the reproductive growth phase. Conversely, pea appears to have a more 'pessimistic outlook' choosing to initiate reproductive growth earlier and devote photosynthate to the seed, resulting in a high water use efficiency for seed yield.

4.4.2.5. Osmotic adjustment

There are some traits that are responsible for tolerance of plants to drought; one of such physiological traits is osmotic adjustment (OA). Tissue OA to condition of gradual dehydration is an important mechanism to plant drought tolerance. Capacity for OA allows maintenance of water content and thereby cell turgor. OA can be defined as the active accumulation of solutes within the plant tissue in response to a lowering of soil water potential.

Osmotic adjustment is referred as a net increased in solute concentration, may be perceived as an important survival mechanism to drought stress (Turner, 1997) in addition to being frequently correlated with yield stability in dry environments (Blum, 1997). As water is being removed from the plant cells, its osmotic potential is reduced due to the simple effect of solute concentration. However, during the course of cellular water loss, solutes are actively accumulated and lead reduction in osmotic potential. This reduces the out flow of water from cell, thereby reducing loss of turgor and allows stomatal opening and expansion growth to continue progressively at lower water potentials. It has no adverse effect on water use efficiency, but contributes to grain yield by increasing the water use from sub-soil during reproductive phase of growth in several crops (Singh *et al.*, 1990; Ludlow and Muchow, 1990).

It has been well established that plant accumulate a variety of osmoprotectant solutes as an adaptive mechanism to environmental stresses in addition to changes in protein levels (Bartels and Nelson, 1994; Ingram and Bartels, 1996; Verslues and Sharp, 1999; Zinselmeier *et al.*, 1999). Osmolytes also protect macromolecules such as enzymes and proteins from electrolytes and temperature. These changes are associated with OA and the protection of membranes from damage as well as cell contents desiccate. Osmoprotectant solutes include sugar and sugar alcohols (polyols) manitol, proline, cholinge, glycine, a number of quaternary ammonium compounds (betaines) and tertiary sulphonium compounds. Under drought, the cells and tissues of some plants can adjust osmotically. That is, they increase their solute concentration before there is much water loss. This maintains turgor because water is drawn back into the cell, rather than being lost. This character might enhance the capacity of the crop to cope better with declining soil water reserve during the later stages of seed filling, as they clearly do with wheat (Morgan, 1983, Morgan *et al.*, 1986). The turgor maintenance through OA is positively associated with yield (Lee and Asahira, 1983). In oil seed *Brassica*, higher yield is closely associated with greater post-anthesis growth which, in turn, is correlated with a capacity for osmotic adjustment to drought (Lewis and Thurling, 1994). Studying two grain sorghum lines exhibiting contrasting drought tolerance. Premachandra *et al.* (1995) found that the tolerant line maintained significantly higher OA than the drought susceptible line. Kumar *et al.* (1984) reported positive relationship between OA and grain yield in *Brassica* sp. In a study on chickpea breeding lines, Morgan *et al.* (1991) found a positive association between controlled environment of OA and grain yield of chickpea in field experiments. High leaf water potential minimized the effects of water deficit on spikelet sterility and consequently grain yield (Jongdee *et al.*, 2002). OA has a positive influence on leaf rolling, tissue death, and retention of green leaf area in rice (Nguyen *et al.*, 1997). Group of high osmotic families extract more water from the profile during the stress period, had greater shoot biomass and harvest index at the physiological maturity and greater grain yield (Chimenti *et al.*, 2002). Osmotic adjustment is time dependent. Thus, a slow water deficit may allow the rice plant to accumulate solutes and may provide protection (Lilley and Ludlow, 1996; Nguyen *et al.*, 1997).

Glycinebetaine: The benefits and stress-alleviating effects of glycinebetaine have been demonstrated under laboratory conditions, often on isolated enzymes. On the basis of the beneficial concentrations, the quantity of glycinebetaine required made it uneconomical to use on the field scale. However, recent work in controlled environment (Naidu, 1995) and field experiments (Campbell *et al.*, 1996). Short spell of drought resulted in high accumulation of glycinebetaine in the vegetative parts of tolerant cultivars of barley (Bergmann *et al.*, 2002).

Soil application of betaine increased germination and seedling vigour of cotton and wheat (Naidu, 1995). Under field conditions there is a possibility of increasing cotton productivity. Under rainfed conditions, cotton yield increased 22% in response to seed treatment over an untreated control. In this situation, the yield increase may have come from increased water use efficiency (Naidu, 1995) or photosynthetic efficiency. Cotton seed treatment with glycinebetaine resulted stronger stems and roots, improved branching, earlier flowering, and a greater number of squares or bolls. These responses suggested a hormone-like activity of glycinebetaine and similar effects have been noted in grapes (Naidu, BP, unpublished).

In contrary, OA is not a general trait observed in coffee genotypes grown under rainfed conditions (Meinzer *et al.* 1990; DaMatta *et al.*, 2003). Furthermore, development of increased leaf water deficits upon discontinuing irrigation may circumstantially be more rapid in genotypes having greater amplitude of OA (Meinzer *et al.*, 1990; DaMatta *et al.*, 1993). These observations imply that OA may not be an effective mechanism of drought tolerance in coffee, as has also been reported for several other woody species (Fan *et al.*, 1994).

Polyamine: Polyamines play a role in plant growth and development especially under stressful growing conditions (Ben-Hoyyim *et al.*, 1994). Polyamines are low molecular weight polycationic molecules, which are thought to play important roles in a number of physiological and developmental processes. In plants, polyamines accumulate under several abiotic stress stimuli, including salt and drought. It has been suggested that increase in polyamine concentration could be considered as an indicator of plant stress. Using differential display, a key possibly rate-limiting enzyme, S-adenosylmethionine decarboxylase (SAMDC) involved in polyamine biosynthesis was recently shown to be regulated by various stresses including drought and ABA treatments (Li and Chen, 2000). The polyamine pathway is ubiquitous in living organisms. In plants, the diamine putrescine (the precursor of the higher polyamines spermidine and spermine) is catalysed by arginine decarboxylase (ADC). Additional reactions convert putrescine into spermidine and spermine. These steps are catalyzed by spermidine and spermine synthase, which add propylamino groups generated from S-adenosylmethionine by SAMDC.

Fructan: Fructans are polyfructose molecules that are produced in only 15% flowering plants species, including wheat and barley. It functions mainly as a storage carbohydrate (Hendry, 1993; Pilon-Smits *et al.*, 1995), but being soluble may help plants to survive periods of osmotic stress induced by drought or cold, by varying the degree of polymerization of the frutan pool. Fructans promote the process of root branching, thus increasing root surface and water uptake. The petal opening in daylily involves the conversion of fructans to low degree of polymerization products which help in osmotic adaptation.

Trehalose: Trehalose, an osmoprotectant non-reducing disaccharide, effectively stabilizes hydrated enzymes and lipid membranes in some plants.

It plays an important role in desiccation stress protection and its accumulation is effective way of increasing drought tolerance in both model and important crops using genes of bacterial origin. Increased trehalose accumulation in transgenic tobacco plants by over-expression of the yeast *TPS1*, resulted in the loss of apical dominance, stunted growth, lancet shaped leaves and some sterility (Lee *et al.*, personal communication). Altered phenotype was always correlated with drought tolerance; plants showing severe morphological alternations had the highest tolerance under stress conditions.

Proline: Proline accumulation has been demonstrated to be associated with abiotic stresses (Delauney and Verma, 1993). It is accumulated to very high concentration during stress; its concentration declines upon rehydration. Proline is a cell compatible solute. Its accumulation in large amount contributes to osmotic adjustment; it also serves as a cytoplasmic osmotic balance for potassium accumulation as the main osmoticum in the vacuole. It serves as protactant of various enzymes and biological membranes subjected to desiccation and heat stress.

Accumulation of proline, nitrate and ammonium ions may contribute to osmotic adjustment and reflect inhibition of protein synthesis (Stweart and Hanson, 1980; Frota and Tucker, 1978). Accumulation of proline in *in vitro* stressed seedling leaf segments and its association with number of physiological response induced by water deficit was studied in 14 wheat cultivars with known drought tolerance rankings (Tan and Halloran, 1982). The cultivars differed in their capacity to accumulate proteins which was also inter-correlated with increase in both total catabolic amino acids and sugar during stress.

Bhale *et al.* (1982) observed increase in proline accumulation due to stress. Further, they evaluated the magnitude of heterosis in 10 crosses for proline before and after stress at panicle initiation and panicle emergence stages. The range of heterosis was higher before stress treatment. Hybrids of 1201A × PD3-1-11, 1202A × 285 and 1202A × 168 accumulated significantly higher amount of proline than check M35-1 and revealed high heterosis. Proline content increased in leaves and roots of sweet potato under drought environment (Ravi and Saravanan, 1999). Quilambo (2004) reported high proline accumulation in the tolerant rice cultivar, Falcon. Proline content was effective criterion for detecting drought tolerance strategies taking into consideration of the growth stages and duration of the stress period.

4.4.2.6. Late embryogenesis abundant (LEA) proteins

The late embryogenesis abundant (LEA) proteins were the first to be identified as genes that are expressed during the maturation and desiccation phases of seed development and recognized that, these genes are also expressed in vegetative tissues during periods of water loss. Osmotic stress induces accumulation of a set of low-molecular weight proteins known as dress proteins in plant tissues such as LEAs and dehyrins. It is also accumulated during seed desiccation and in response to water stress (Dure, 1993). LEA proteins were first characterized in cotton as a set of proteins that are highly accumulated in the embryos at the late age of seed development. Subsequently many LEA proteins or their genes have been characterized from different plant species.

4.4.2.7. Phytohormones

Abscisic acid: Abscisic acid (ABA) is known as stress hormone. It is produced under drought conditions and plays important roles in tolerance against dehydration. Water deficit is sensed by root, which begin to synthesis of ABA and then it is transported to leaves. ABA appears to involve in stomata closure under drought, but since it is accumulated in other tissues, additional functions may be involved. ABA is produced to high levels in response to drought (Ingram and Bartels, 1996; Leung and Giraudat, 1998).

Larque-Saavedra and Wain (1976) reported that the leaves of maize and sorghum tolerant to drought contained substantially more free ABA levels, which would result in closure of stomata and thereby reduce transpiration. The ABA-insensitive genotypes showed high growth rate, which lead higher grain yield in wheat (Lu *et al.*, 1989). They suggested that selection for ABA-insensitivity might be an effective approach to improve drought resistance. Austin *et al.* (1982) found that the capacity for ABA accumulation is highly heritable. Further, Larque-Saavedra and Rodriquez (1989) suggested that ABA accumulation is inherited maternally.

In contrary, Durley *et al.* (1983) found that mean leaf ABA concentrations during flowering and early grain filling in drought stressed sorghum plants were significantly correlated with reduction in grain yield. The relationship between ABA accumulation and drought tolerance in Kentuky bluegrass (*Poa pratensis*) has been studied by Liu *et al.* (2003) and Wang *et al.* (2004). Leaf ABA content increased linearly with drought stress. They also reported that the rate of ABA accumulation during drought stress was positively correlated with the rate of decrease in turf quality, increase in electrolyte leakage and decrease in relative water content. They observed highly significant negative correlation between ABA content and leaf water potential, stomatal conductance, transpiration rate, and net photosynthetic rate, and a positive correlation between ABA content and electrolyte leakage. Thus, ABA accumulation in response to drought stress could be used as a metabolic factor to select drought tolerant plants.

Superoxide dismutase (SOD): SOD activity increased by water stress in tomato cultivars (Rahaman *et al.*, 2004) and the increase was much more rapid and pronounced in tolerant gentypes than the sensitive ones. The regression analysis of SOD suggested the possibility of using SOD activities as additional screening criteria for tomato drought tolerance improvement.

4.4.2.8. Water use efficiency

Water use efficiency (WUE) can be defined as the ratio of the rate of photosynthesis to the rate of transpiration (or unit yield/unit water used). Arid agriculture might have require drought-resistant plants with an ability to survive, endure, compensate for, or escape damage from wilting, and also need them to be efficient in water use. A crop which extracts the soil water in the surface quickly can be expected to minimize losses of this surface layer water. Rye is known to have higher photosynthetic rate than wheat, but similar transpiration rate, which suggests a better WUE (Waines *et al.*, 1979). The wheat-rye hybrid (triticle) has a transpiration ratio comparable to rye, thus has better performance in drought prone areas.

Earlier onset of reproductive growth in pea, coupled with its ability to be seeded very early, provides peas with an opportunity to escape the most severe drought stress. Pea has an ability to extract available water from the surface soil layer more quickly than wheat, minimizing losses to evaporation (Martin *et al.*, 1994).

In some species, accumulation of compatible solutes has been considered as an adaptation to drought stress. In coffee, however, such accumulation does not satisfactorily correlated with drought tolerance in various cultivars of coffee. It retains high leaf relative water content (RWC) under dehydration conditions, being considered a water-saving rather than a dehydrating tolerant species (DaMatta *et al.*, 1993). This may be attributed to an efficient stomatal control on transpiration (Josis *et al.*, 1983; Meguro and Magalhaes, 1983; Pinheiro, 2004), and/or low cell-wall elasticity (Meizer *et al.*, 1990; DaMatta *et al.*, 1993, 2003; DaMatta and Rena, 2002; Pinheiro, 2004). A small water loss should, therefore, cause a shift in turgor so that leaves tend to maintain a high RWC to retain a high symplast volume. Thus, it appears that under water deficit, the maintenance of a high RWC is more important than OA *per se* in conferring drought tolerance to the coffee plant (DaMatta *et al.*, 1993).

4.4.3. Dehydration Tolerance

The ability of plant cells to withstand water stress and continue metabolic functions is termed as dehydration tolerance. Cellular dehydration causes significant disorders in membrane structure, composition and function. Maintenance of membrane integrity and function under a given level of dehydration stress can be taken as a measure of tolerance. Dehydration tolerance of a genotype means that a significantly lower level of changes is induced in it than those in another genotype when both of them are subjected to the same level of dehydration.

4.4.4. Microorganisms in Drought Tolerance

Vascular arbicular mycorrhiza (VAM) influences leaf water potential, solute accumulation and oxidative stress in crop plants under drought environment. VAM also play a major role in the ability of the plant to take up phosphorus. It is now accepted that the contribution of VAM symbiosis to plant drought tolerance is the result of accumulative physical, nutritional, physiological and cellular effects. VAM colonization enhance growth of *Poncirus trifoliata*, subsequently soluble sugar content of the leaves and roots as well as water use efficiency increased in mycorrhizal than the non-mycorrhizal plants under water stress (Wu *et al.*, 2004a). The findings of Porcel and Ruiz-Lozano (2004) showed that VAM inoculated plants were protected against drought as shown by their significantly higher shoot biomass production. They reported higher leaf water potential and proline in VAM plants. Shivaputra *et al.* (2004) also found that VAM (*Glomus fasciculatum* and *Selerocystis dussii*) increase drought tolerance in papaya. VAM enhanced leaf relative water content, photosynthetic rate, stomatal conductance, and improved plant drought tolerance significantly in jujube (*Zizyphus spinosus*) seedlings grown in pots (Lu *et al.*, 2003).

4.5. GENETICS OF DROUGHT TOLERANCE

In formulating breeding programme, the study of genetic mechanisms controlling drought is the foremost important. It is a complex trait, expression of which depends on action, interaction of different morphological (earliness, reduced leaf area, leaf rolling, wax content, efficient rooting system, stability in yield and reduced tillering), physiological (reduced transpiration, high water use efficiency, stomatal closure, osmotic adjustment) and biochemical (accumulation of proline, polyamins, trehalose, etc. and several enzymes) characters. Very little is known about the genetic mechanisms that condition these characters (Mitra, 2001). The genetic control of the drought tolerant characters ranges from oligogenic to polygenic. A stress character controlled by a single gene with discrete quantitative effect may be readily transferred to be pertinent cultivars. In case of quantitative characters, the degree of heritability is important in producing the success in breeding and in developing the suitable breeding procedures.

Generally leaf characters likewaxy bloom, grossiness, glaucousness, glabourous leaves are under oligogenic control. Some other traits like ABA accumulation in wheat, constitutive proline accumulation in barley mutant and resistant to flower abscission and ability to support pod formation in rajma (*Phaseolus vulgaris*) seems to be determined by oligogenes (Singh, 2003). Ludlow and Muchow (1990) listed 16 genetic characters related to crop drought tolerance.

1. Matching phenology to water supply
2. Photoperiod sensitivity
3. Developmental plasticity
4. Mobilization of preanthesis day matter
5. Rooting depth and density
6. Low root hydraulic conductance
7. Early vigour (canopy)
8. Leaf area maintenance (stay-green)
9. Osmotic adjustment
10. Low lethal water status
11. Reduced stomatal conductance
12. Leaf movements
13. Leaf reflectance
14. Heat tolerance of seedlings
15. Low epidermal conductance
16. Transpiration efficiency

The identification of genes responsible for morphological and physiological traits, and their location on chromosome have not been possible, but their inheritance pattern and nature of gene action have been reported. Polygenic inheritance of root characters has been reported by Ekanayake *et al.* (1985). Association between root characters and plant height, tiller number and shoot were positive and significant. Five root characters were significantly correlated with the visual field drought tolerance scores and with leaf water potential, confirming the role of root characters in maintaining high leaf water potential under water stress. The long root and high root numbers are controlled by dominant alleles and thick tip by recessive alleles (Armento-Sato *et al.*, 1983). Leaf rolling (Singh and MacKill, 1991) and osmotic adjustment have sown monogenic inheritance. Tomar and Prasad (1996) reported a drought resistant gene (*Drt1*) in rice, which is linked with genes for the plant height, pigmentation, hull colour and awn, and has pleiotropic effect on root system. The relative water content of rice was noted as quantitative character controlled by polygene (Wu *et al.*, 2004b). Both the broad sense and narrow sense of heritability leaf relative water content were both high. They also concluded that the relative water content of rice was controlled by additive gene effects; though gene effects have also been reported by a few other workers.

Under the drought, seed yield per plant of canola showed high significant positive correlation with plant height, number of siliquae per plant, number of primary and secondary branches per plant (Sadaquat *et al.*, 2003). They also suggested that the number of primary and secondary branches per plant and number of siliqua may be used for indirect selection of high yielding rape cultivars.

Variability, heritability and genetic advance for physiological and economic characters were studied in rice landraces (Gomez and Kalamani, 2003). They found significant variation among the landraces for all the characters under study, especially biological yield per plant, root length and root weight. High heritability coupled with high genetic advance was recorded for biological yield per plant, plant height and number of panicles per plant, indicating that those characters may be considered during selection for drought tolerance in rice.

Drought tolerance quantitatively inherited with a complex physiological reaction; thus its genetic basis has received limited attention and the development of drought tolerant varieties was slow in most of the crops. Heyne and Bruson (1940) found that drought tolerance in maize was heritable and gene action was intermediate to drought. Williams (1966) reported partial dominance with some epistatic variability. Genetic control of drought resistance in maize is assigned to 3-5 gene pairs. Whereas, Williams *et al.* (1969) indicated that inheritance of drought resistance is controlled by partial or complete dominance. The association of different plant characters with drought tolerance capacity of the genotypes was studied in inbred lines open pollinated varieties, synthetics, composites and hybrids of maize (Ali and Naidu, 1982). There was a significant positive correlation between yield under stress and plant height, number of leaves, leaf area index, ear number, length and girth, and 1000 kernel weight. Significant negative correlation was observed between yield and stomata number. Reaction of seedlings to heat and stability of chlorophyll on heating, were found to be significantly associated with yield. Younis *et al.* (1988) tested three open pollinated populations of maize collected from dry areas, two commercial varieties, their selfed progenies and their 10 F_1 hybrids over 9 locations differing in degree of drought stress. The grain yield showed high heterosis.

Predominance of non-additive gene action was observed for all the traits in sorghum under water deficit environment (Khidse *et al.,* 1982).

4.6. MEASUREMENT OF DROUGHT TOLERANCE

Drought tolerance is measured by the degree of change in phenotype under exposure to water stress. These phenotypic changes include:

1. Change in growth patterns (i.e. retardation)
2. Changes in seed production
3. Electrolyte leakage from leaf segments
4. Leaf wilting
5. Relative leaf water content
6. Changes in the transcriptome

Drought tolerance is also measured in terms of survival. A particular plant is more drought tolerant than wilt-type, if a greater percentage of the samples survive after dehydration treatment as compared to wild-type (Kasuga *et al.,* 1999; Tripathi *et al.,* 2000).

4.7. SOURCES OF TOLERANCE

There are several informations on the sources of drought tolerance. Tolerance ability of crop plants varies among the species (Table 4.1) as well as among the genotypes within a species (Table 4.2). Tolerance may exist in land races, wild relatives, high yielding varieties, initial breeding materials and advance breeding materials. The first priority of a breeder should be to identify and utilize cultivated varieties as source of drought tolerance since this is the least problematic of all sources. The wheat species *Sphaerococcum, Vavilovii* of genus *Triticum; Variabilis,* Speltoides, *Umbellutata* and *Squarrosa* of genus *Aegilops* were reported to be drought tolerant (Bansal and Sinha, 1991a, b; Rekika *et al.,* 1998). Wild relatives of many crop plants are rich in drought tolerance and are regarded as valuable genetic resource.

Wild rice may serve as source of superior drought tolerance alleles for cultivated rice. Studies of Liu *et al.* (2004) indicated that *Oryza logistaminata* and *O. rufipogon* may serve as sources of novel alleles for maintenance of leaf elongation, stomatal conductance and membrane stability. *Oryza inandamanica,* a wild species found Andaman group of Islands poses drought tolerant potential, it roles leaves under drought situation.

Kumar *et al.* (2004) reported wide variability among the chickpea varieties under drought situation. BG365, BG364, Pusa 362, Pusa 3391, Pusa 256 and IPC 94-32 were identified as higher yielder under water stress, and they suggested that those genotypes may be used as a source of drought tolerance in chickpea. Pigeon pea is also an important drought tolerant food-legume grown in arid and semi-arid regions of the glob (Mathew and Saxena, 2005).

Table 4.1. Drought tolerant crop species

Scientific name	Common name	Tolerance level
I. Fruits		
Aegle marmelos	Bael	Tolerant
Anarcadium oxidantale	Cashew nut	Tolerant
Annona sqamosa	Custard apple	Tolerant
Feronia limonia	Wood apple	Tolerant
Carissa carandas	Karonda	Tolerant
Citrus lemon	Lemon	Moderately tolerant
Citrus paradisi	Grapefruit	Moderately tolerant
Diospyros kiki	Persimon	Tolerant
Ficus carica	Fig	Tolerant
Grewia subinaequalis	Phalsa	Tolerant
Mangifera indica	Mango	Moderately tolerant

Contd...

Table 4.2. Contd...

Psidium gujava	Guava	Moderately tolerant
Phonix ductylifera	Date palm	Moderately tolerant
Punica granatum	Pomegranate	Tolerant
Tamarindus indica	Tamarind	Moderately tolerant
Syzygium cumini	Jamun	Moderately tolerant
Zigipus jujube	Ber	Tolerant
II. Tree species		
Acer campestre	Hedge Maple	Tolerant
Albijia julibrissin	Mimosa Silk Tree	Tolerant
Celtis occidentalis	Common hackberry	Tolerant
Cercocarpus ledifolius	Curlleaf Mountain-mahygany	Tolerant
Corylus colurna	Turkish filbert	Tolerant
Cotinus coggygria	Common Smoke Tree	Tolerant
Cowania mexicana	Quinine bush	Tolerant
Crataegus phaenopyrum	Washington howthorn	Tolerant
Fraxinus velutina	Velvet Ash	Tolerant
Ginkgo biloba	Gingko	Tolerant
Gledistia triacanthos	Honey locust	Tolerant
Gymnocladus dioicus	Kentuty Coffee Tree	Tolerant
Pinus aristata	Bristle cone Pine	Tolerant
Pinus edulis	Pinyon Pine	Tolerant
Pinus mugo	Mugo Pine	Tolerant
Pinus thunbergiana	Japanese Black Pine	Tolerant
Pistacia chinensis	Chinese Pistache	Tolerant
Prosopis juliflora	Honey Mesquite	Tolerant
Quercus alba	White oak	Tolerant
Quercus macrocarpa	Bur oak	Tolerant
Robinia ambigua	Idaho flowering locust	Tolerant
Zelcova serrata	Japanese Zelkova	Tolerant

4.8. BREEDING APPROACHES

Three important elements of drought characterization for successful breeding of stress tolerance are timing, duration and intensity of the stress. Most period of stress encountered in the major grain-crop growing areas worldwide are transient, unpredictable and imprecisely measured, leading to difficulties in breeding efforts for drought tolerance (Zavale-Garcia *et al.*, 1992). A major factor that has prevented progress for improving yield in water-limited environments is the lack of knowledge of the critical traits that should be selected for achieving goal. There are four breeding approaches to select for drought tolerance.

Table 4.2: Drought tolerant genotypes of different crop species

Crop	Genotype	Reference
Rice *Oryza sativa*	ARC-10327, DJ-129, DV-110, DZ-41, DNJ-60, Lua Ngu, IR 36	Khush and Coffman, 1977
	MTU-17, Mettansannavari, CH-45, Sathi-34-36, PTB-29, B76, Lalnakanda-41, TKM-1, N-22, Sudha	Ram and Singh, 1994
	Salumpikit, OS4, Dular, Mi-48, IR442-2-58, Cisadane, Cipunegara, Kruseng, Aceh, Ayung	Suardi, 2002
Wheat *Triticum aestivum*	*Triticum sphaerococcum, T. vavilovii, T. aestivum* cv. C306	-
Maize *Zea mayse*	ZPBL-1304	Ristic *et al.*, 1998
	HI-209, HI-295, HI-536, HI-1040	Meena-Kumari *et al.*, 2004
Cowpea *Vigana radiata*	TVu 11979, TVu14914	Watanbe, 1998
	IT90K-59-2, Kanannado, Dan IIa	Singh *et al.*, 1999
Bean *Phaseolus vulgaris*	BG365, BG364, Pusa 362, Pusa 256, IPC 94-32	Kumar *et al.*, 2004
Mustard/rape *Brassica* spp.	Oscar, range, Tarnab-2	Sadaqat *et al.*, 20003

4.8.1. Breeding Under Optimum (water-stress free) Condition

The basic philosophy of this approach is a genotype superior under optimum level will also yield relatively well under drought condition. A high positive correlation exists between performance under optimum and stress conditions (Johnson and Frey, 1967). The yield potential of a cultivar under favourable moisture condition is important in determining yielding ability under moisture stress condition. Tolerance to drought may be present in such a variety which is expressed as an unidentified component of stability in performance over various locations. In the process of breeding yield and stability are tackled as one complex. The accumulation of environmentally stable yield genes equates with better performance under stress situations (Fischer *et al.*, 1982). The maximum genetic potential of yield is expected under optimum environmental conditions. However, the concept of expression of maximum genetic potential in optimum condition is debated (Blum, 1973) as genotype-environment interaction may restrict the high yielding genotype to perform well under drought.

4.8.2. Breeding Under Actual Drought Condition

This approach of breeding for drought tolerance has been suggested by Hurd (1971). Early domestication of crop plants concentrated their attention on species adapted to natural selection to their locality. The natural selection in heterogeneous population of crop plants should increase the frequency of genes for stress tolerance. It suffers from the problem that the intensity of drought is highly variable from season to season as well as within the experimental plots. Therefore, selection pressure on breeding material changes drastically from generation to generation. This problem is compounded with low heritability of yield (Roy and Murthy, 1970).

4.8.3. Breeding Under Artificially Created Environment

Germplasm may be screened by creating artificial drought conditions under controlled environment according to the requirements.

Polyethylene glycol (PEG) creates artificial moister stress in plant growing media. Rao and Singh (2004) screened maize F_1 in 0.5 Mpa PEG. They observed that non-additive variance was important for germination percentage, shoot length, and root: shoot ratio, while the additive gene action played a major role of most of the drought related traits in PEG. Dhanasekar *et al.* (2003) screened tolerant genotypes of cowpea using PEG6000. They used PEG solution (0.04M) in petri-plates lined with filter paper for germination of seeds at room temperature.

Wheat genotypes were screened at the seedling stage in wooden boxes in greenhouse conditions for drought tolerance (Tomar and Kumar, 2004). The boxes were filled with a mixture of soil: sand: FYM in a 50: 45: 5 ratio. They were irrigated with equal volume of water to ensure good germination. After sowing, no irrigation was given till most of the genotypes started wilting. Then, boxes were irrigated to study recovery response of the genotypes. Based on recovery response, the genotypes were grouped as either susceptible or tolerant.

4.8.4. Incorporation of Drought Tolerance

This approach involves in improvement of drought tolerance in pre-existing high-yielding genotypes through incorporation of morphological and physiological mechanisms of drought tolerance. The back cross method is a form of recurrent hybridization by which a gene for a superior characteristics may be added to desirable variety. A comprehensive scheme for breeding for drought resistance and high yield potential has been proposed, which takes advantage of recurrent selection procedure (Paroda, 1986). Photosynthetic rate could be improved by recurrent selection and thereby drought tolerance (Crosbic *et al.,* 1981). Eight cycles of recurrent full-sib selection for improved drought tolerance in a tropical maize population resulted in a yield gain of 500-800 kg/ha (Edmeades *et al.,* 1992). Recurrent selection for drought tolerance for three to eight cycles has increased grain yield under drought at flowering by 30-50% in three lowland tropical maize populations (Chapman and Edmeades *et al.,* 1999). They found that under drought condition, changes per cycle with recurrent selection were as grain yield (12.6%), fertile ears per plant (8.9%), grain per fertile ear (6.3%), grain number per square meter (12.2%), 1000-grain weight (no change), anthesis-silking interval (22.0%), days from sowing to 50% anthesis (0.7%), plant height (2.0%), primary tassel branch number (5.9%), and senesced leaf area (2.7%). Responses under well-watered conditions were smaller but generally of the same sign. Grain yield was strongly associated with grain number per square meter in both water-stressed and well-watered environments. Grain yield, ear per plant and grains per fertile ear were strongly correlated with anthesis-silking interval across entries under drought, though not when water was plentiful.

However, transferring drought tolerance in high-yielding genotypes is complicated due to lack of understanding of physiological and genetic basis of adaptation in drought condition.

The breeding methodology for drought tolerance remains the same as applied for other purpose. In general, pedigree and bulk method could be useful for self-pollinated crops (Ceccarelli, 1987). To transfer of traits relating to drought tolerance to a high-yielding cultivar, backcross method of breeding is appropriate (Edmeades *et al.,* 1992). Biparental mating provides scope to evolve the desired genotype of drought resistance.

4.9. SCREENING CRITERIA

For genetic improvement in drought tolerance, utilization of existing genetic variability requires an efficient screening technique, which should be rapid and capable of evaluating plant performance at the critical developmental stages and screening a large population using only a small sample of plant materials (Johnson, 1980). Selection for drought tolerance by any selection index requires a rigorous control over the stress environment, and needs to address moisture stress in terms of growth stage and stress intensity along with temperature gradients and radiation. Evidence of variations of moisture content within a field (Kitchen *et al.,* 1999) implies a need for a reasonable level of drought tolerance. Drought tolerance is an interactive result of different morphological, physiological and biochemical traits and thus, these different components could be used as selection criteria for screening appropriate plant type. A combination of different traits of direct relevance, rather than a single trait, should be used as selection criteria.

The selection criteria primarily based on morphological characters could be used by plant breeders for selection of parents as well as desirable segregants followed by hybridization. During selection characters with high heritabilities and high correlation with yield under stress across the environments may be preferred. Grain yield under stress conditions is usually the primary traits for selection. A suitable secondary trait should have (1) genetically associated with grain yield under drought, (2) highly heritable, (3) stable and feasible to measure, (4) not associated with yield loss under ideal growing conditions (Edmeades *et al.,* 2001). Very few secondary traits meet these criteria. Secondary traits of maize under drought are reduced bareness, anthesis-silking interval (ASI), stay green and to a lesser extent, epinasty or leaf rolling under drought (Banziger *et al.,* 2000). The various selection criteria used in breeding for drought tolerance in different crops are outlined below.

4.9.1. Yield Potential

As loss in yield is the main concern for the crop plant from agricultural point of view, plant breeder emphasize on yield performance under moisture stress condition. The yield potential of a cultivar under favourable moisture conditions is important in determining yielding ability under water stress. Drought index based on yield is an important criterion for selection for stress environment. Fischer and Wood (1979) suggested selection on drought tolerance in maize based on yield. They assessed the response to drought by a drought index based on yield under both fully irrigated and stress conditions. The *drought index* for a genotype is the ratio of its yield under stress to yield under no stress, relative to the ratio of the mean yield of all genotypes under stress to yield under no stress. A drought index > 1.0 suggests relative drought resistance and an index < 1.0 suggests relative drought susceptibility. Ramteiz and Kelly (1998) found that yield of bean under stress was correlated with yield under non-stress and negatively correlated with drought susceptibility index. In fact, in some studies, a cultivar that was high yielding under adequate moisture was lowest yielding under drought stress. A drought index, which provides a measure of drought based on loss of yield under drought condition in comparison to moist condition, has been used for screening drought tolerance genotype (Bruckner and Frohberg, 1987). Edmeades *et al.* (1992) also suggested that selection should be primarily for improved grain yield of maize under any array of water regimes, increased ears/plant, reduced anthesis-silking interval (ASI) under stress and reduced tassel size.

Molina-Gallan (1980) subjected to drought resistant maize Zacatecas 58, Cariollo de Mezquital and Cafime for 5 or 6 cycles of stratified mass selection based on a visual assessment of ear yield after cultivation under constant water stress. Genetic advance assessed in dry season trials ranged between 0.6 and 14.5% per cycle.

Grain yield in maize is normally highly correlated with the kernel number per unit area and per plant rather than with weight per kernel (Bolanos and Edmeades, 1996). Factors affecting grain set under drought, therefore, are of special interest in commercial breeding endeavours. Because, maize is an out-crossing species, pollen must move from the anther at the tops of the plant to the exposed silk of the same and surrounding plants. This process is risky because pollen and the delicate stigmatic tissues are exposed to a desiccating environment. One universal phenomenon observed when maize flowers are under drought is the delay of silking in relation to pollen shade, giving rise to ASI, whose duration is highly correlated with kernel set (Edmeades *et al.*, 2000). Under such conditions, pollen can arrive after it has desiccated, when silk have withered or senesced (Bassetti and Westgate, 1993a, b) or after ovaries have exhausted their starch reserves (Saini and Wastegate, 2000; Zinselmeier *et al.*, 2000). The flux of currently formed assimilates to the developing ear is a powerful determinant of ear growth, suggesting that stored reserves in the stem and ear shank have little to do with reproductive success (Schussler and Wastegate, 1995).

4.9.2. Leaf Characteristics

Visual scoring or measurement for leaf rolling, leaf wilting, leaf firing and leaf length have direct relevance to drought tolerance and should be taken into consideration during selection. Leaf wilting is the prominent observable phenomenon of water-stress in crop plants.

4.9.2.1. Leaf rolling

Leaf of any crop plant frequently rolls when stressed. This passive movement reduces the radiation effect, thereby counteracting the increase in leaf temperature arising from stomatal closure and preventing further development of leaf water deficit. Leaf conductance and transpiration rate decrease sharply with leaf rolling (Sobrado, 1987). During drought period, the resistant sorghum lines showed more leaf rolling than the susceptible lines, reducing the effective area of the uppermost leaves by 75% (Mathew *et al.*, 1990). Bogdanova *et al.* (1988) reported that the leaf rolling character in wheat is controlled by two dominant genes (designed R4 and R42), which were located on chromosome 6A and 41.

Leaf rolling is scored visually in rice either in the morning or at mid-day. It is a most familiar response for reducing the energy load on the leaf in species of low osmotic adjustment ability, viz., rice, and while in those with greater osmotic adjustment, viz., wheat and sorghum. Leaf rolling is a drought adaptive trait and a delayed leaf rolling is an indication of turgor maintenance as a component of dehydration avoidance. Lines with good drought tolerance and delayed leaf rolling recovered faster after the stress were removed in rice (Singh and MacKill, 1991). Delayed leaf rolling is used as an important selection criterion for drought tolerance in rice. However, care should be taken to avoid the confusion that may arise due to changes in leaf rolling symptom due to leaf age and plant growth stage. The leaf rolling character is associated with drought resistance, which could be improved by incorporating the gene(s) into better performing varieties.

4.9.2.2. Water retention by leaf

Leaf water retention may be useful in selection for drought tolerance. Water loss rates from excised leaves or plants have been used to assess genotype difference in water retention capacity. High water retention is the positive response. Quisenberry *et al.* (1985) had suggested that leaf turgidity of cotton under water deficit at field condition may be useful in selecting lines with enhanced drought tolerance. An association between greater leaf water retention and an improved yield performance under stress was found in wheat (Bushuk *et al.*, 1989). Blum (1988) described the procedure to measure leaf water retention. In this method, excised leaves are allowed to dehydrate for 6-24 hours under standard environment and the relative loss or retention of water in the dehydrated tissue is determined. A short time drying (2 hours) gives more weightage to the amount of water lost by stomatal transpiration. When excised leaves are allowed to dehydrate, stomatal closure occurs within 1-10 minutes after excision and this reduces the dehydration rate.

4.9.2.3. Leaf firing

Leaf firing is a result of insufficient transpirational cooling and subsequent heating of all or part of the leaf to lethal temperature. It is a symptom of water deficit in crops like maize, millet, sorghum and various legumes. Leaf rolling and leaf firing may be jointly used to contract a score for drought susceptibility. In maize, leaf senescence under stress was significantly negatively correlated with yield under stress.

4.9.2.4. Leaf stomata

Permanent damage to the plant processes caused by severe moisture stress may be directly or indirectly attributed to stomatal characters, such as, size and frequency per unit leaf area show high heritability in a range of species and breeding for these characters has proved successful. A variety that offers more resistance to water flow from the stomata into the atmosphere has a beneficial character towards drought tolerance. Lal and Moomaw (1977) reported that the rice variety GS 6 has higher stomatal diffusive resistance than IR 20, a relatively more drought sensitive variety. Roark and Quisenberry (1977) found that low conductance is a dominant character.

4.9.2.5. Delayed flag leaf senescence

Flag leaf in cereal crops plays an important role in grain filling and development (Evans *et al.*, 1975) particularly in wheat and rice. Therefore, the selection proceeded to identify genotypes which had almost erect flag leaf enabling photosynthesis for a longer duration (Yoshida *et al.*, 1976; Khush, 1995). As grain development commences, there is decline in flag leaf area coupled with leaf senescence.

Leaf attributes like dense pubescence, heavy glaucouness are scored visually. Epiculticular wax load can also be measured for drought tolerance. Leaf attributes can be incorporated in integrated selection index for improvement of drought tolerance in crop plants.

4.9.3. Root Characteristics

It is difficult to evaluate root mass and distribution directly in the field condition. In any case, the root attributes are generally reflected in the canopy response to drought stress. An effective banding herbicide metribuzin is introduced at a given depth and the genotype that are first to show symptoms of toxicity have faster root growth and penetration (Robertson *et al.*, 1985). Hydroponics culture under stress of 15 bar for screening root growth also can be used for drought tolerance assessment. Deep root system has been identified as the target for drought tolerance improvement (Boyer, 1996). One of the early selection criteria has been the number of embryonic roots, which has shown a fairly good correlation with survival and yield under moisture stress condition. The number and rate of appearance of embryogenic roots was correlated with drought tolerance (Vedrov, 1971a, b). Vedrov (1982) emphasized the importance of embryogenic roots as an important selection criterion in breeding for drought resistance. To improve drought resistance, the maternal parent should have more seminal roots (Darofeev and Tyselanko, 1982). Leont'ev and Petukhovaskii (1987) used the number and depth of penetration of seminal and crown roots as criteria in selecting parents for breeding drought-resistant wheat varieties. Narayan and Misra (1989) recorded deeper penetration of roots under moisture stress, and concluded that the depth of root penetration can provide a useful selection criterion for wheat breeding under drought stress. Richard and Passioura (1989) suggested selecting wheat plants with higher resistance for water flow in the seminal root. They have succeeded in developing a wheat variety which has greater root resistance to water flow.

A crop that is able to increase the amount of water transpired relative to another crop grown in the same water limiting yield conditions, accumulates greater plant biomass. Increased water transpiration can result from increased soil water extraction by a deeper and/or denser root system or by osmotic adjustment within its leaves to exert stronger tension.

The screening methods commonly uses have been detailed by many workers (Hurd and Spratt, 1975; Belford, 1988).

4.9.4. Seed Germination and Seedling Characteristics

Stress tolerance ability of crop plants vary with growth stages. Seed germination and seedling stages are most vulnerable. Adequate seed germination in moisture-stress field condition leads to a good crop stand. There was no consistency between germination rate in presence of osmoticum and field emergence and subsequently yield. Laboratory germination studies, in general, overestimate the field germination and emergence.

Many attempts have been made to relate the drought tolerance of young seedlings to that of adult plants. Seedling growth under PEG stress in *Hogland solution* (Yoshida *et al.,* 1976) is a useful selection criterion. Depending on the species, the extent of genetic variation and PEG concentration, a visual ranking of the response to PEG may be sufficient for selection. It is important that the seedling must not suffer root damage on transfer to the nutrient solution; a PEG may be applied stepwise to avoid osmotic shock. A study on the seedling growth of soybean in an aqueous medium suggested that the use of a medium with PEG6000 in a nutrient solution of -6 bars for a period of 14 days beginning at the first node growth stage could be a satisfactory and rapid technique for screening a large number of types for moisture stress-tolerance (Bouslama, 1983). Roy and Murty (1970) emphasized on early vigour, for selection in wheat and chickpea, respectively. Seed that emerged first and consequent early seedling establishment escaped moisture stress.

4.9.5. Canopy Temperature

Genetic improvement for drought resistance in wheat is possible using infrared thermometer (Blum, 1988). A sufficient level of water stress is a major prerequisite for this method. The crop must be fairly dense and free from skips. It is measured under clear sky at midday, windy moments should be avoided. Significant negative correlation was found between canopy temperature and grain yield in rice (Chang and Loresto, 1985) and maize (-0.56 to -0.73). Cultivars that were warmest under irrigation maintained highest relative yield when exposed to drought (Pinter *et al.,* 1990). Srivastava and Chaturvedi (1989) reported that differences in transpiration rates between controlled stressed plants were less in drought resistant variety (C306) than in relatively drought susceptible variety (Kalyan Sona) of wheat.

4.9.6. Plant Phenology

Drought stress delays or accelerates flowering depending upon the affected growth stage of the crop. Delayed heading under stress has been used as a selection index in rice. Similarly, the time interval between pollen shedding and, silking and anthesis silking interval (ASI) in maize under drought condition is shorter; this ASI has been used as selection index (DuPlessis and Dijkhuis, 1967; Bonalos and Edmeades, 1996; Edmeades *et al.*, 2000, 2001). Silim *et al.* (1993) suggested that early flowering of lentil can be used as selection criterion for severely drought prone areas. The spikelet sterility in rice and wheat is another important parameter for selection under drought.

4.9.7. Remobilization of Stem Reserve for Grain Filling

During drought stress, generally the photosynthetic rate decrease leading to reduction in energy flow from source to sink. Drought at reproductive stage, grain filling takes place by translocation of stem reserve. The mobilization of reserve from vegetative parts into the growing grains can be seen in the appreciable reduction in the stem dry weight in wheat (Blum, 1988; Turner, 1979). Remobilization of stem reserve, 20-30% in grain legumes, 40-70% in cereals and 20-24% in *Brassica*, has been reported under drought conditions (Blum, 1988; Ludlow and Muchow, 1990; Singh, 1991). In this method, the plants are defoliated by manually or chemically with magnesium per-chloride (4%) or potassium chloride (0.3%) at 14th day after anthesis. Grain filling now proceeds with the translocated plant reserve. The grain yield and 1000-grain weight is compared with that of corresponding non-treated plants. A smaller difference between the treated and non-treated plots indicates a more reserve translocation. This technique is effective in selection of wheat and rice genotypes for post flowering drought tolerance. Dry weight of pod walls in chickpea decreased with its development under rainfed conditions but not when irrigated, suggesting the possibility of assimilates remobilization from the pod to the seeds (Davies *et al.*, 1996).

4.9.8. Photosynthetic Rate

Net photosynthesis declines in drought situation. Selection with high photosynthetic rate and the ability of plants to continue relatively a high rate of photosynthesis under stress environment may contribute to increased production. Different methods are now available to measure the photosynthetic rate viz., respirometer infrared gas analyzer, $^{14}CO_2$ assimilation, etc. Diffusion porometry also can be used to measure leaf water conductance, which reflects on photosynthetic rate. The infrared analyzer may be kept in a laboratory; the method is nondestructive, simple and quick and therefore, it is suitable for large-scale screening of varieties or samples.

Photosynthesis in some genotypes is more susceptible to drought than others. A drought resistant line of sorghum was found to have a higher photosynthetic rate at low water potential than a less drought resistant line (Blum and Sullivan, 1972). The glossy lines of sorghum are able to maintain a higher photosynthetic capacity under moisture stress than the nonglossy lines (Maiti and Saucedo-Rodriguez, 1986). Cultivars showing similar rates of photosynthesis under normal soil moisture conditions may differ under moisture stress. Bunce (1988) reported that soybean varieties, Clark, Kent and Dorman, did not differ under normal moisture, but as the soil dried, rates of photosynthesis declined; Clark had the highest rate of photosynthesis and stomatal conductance.

4.9.9. Membrane Stability

Leakage electrolytes through cellular membrane indicate susceptibility under stress environment. Tissue water stress disrupts the membrane this can bc measured by solute leakage into deionized water conductometrically. In this process, leaf discs or segments are subjected to PRG-induced water stress and then measuring the solute leakage (Sullivan, 1972). The detail steps for measurement of solute leakage are being used as described in Chapter 2, section 2.8.8.

Plants of all materials must be stressed to the same relative water content (RWC) of about 70% (depending on crop species) before being sampled ('treatment' samples) into stoppered vials and brought to the lab. Samples should not desiccate between sampling and washing. Control samples are taken from similar leaves of well water plants at RWC close to 98%. Once washed, 20 cc of deionized water is added to all samples and they are directly placed for 24h incubation period (#4 above). *The remaining procedure is as described in Chapter 2, Section 2.8.8.*

CMS % = [1-(T1/T2)]/[1-(C1/C2)] x 100%

Injury = 100-CMS

4.9.10. Water Use Efficiency

Carbon isotope discrimination can be used for selecting increased water-use efficiency (Farquhar and Richards, 1984). Water use efficiency in groundnut can be increased either by the conventional breeding or trait based selection based on specific leaf area (Basu and Nautiyal, 2004). They also studied the physiology on low and high specific leaf area genotypes and SDS-PAGE of the wild *Arachis* species indicated that there is the possibility of identifying the molecular marker for high WUE and/or drought tolerance characteristics in groundnut.

4.9.11. Combination of Characters

A combination of different characters of direct relevance, rather than a single trait, could be considered into a single selection index. Roy and Murty (1970) suggested integration of early vigour, synchronous tillering, days to heading and ear length showed stability over different environments, and were important components of wide adaptation and productivity under moisture stress in maize. Morphological/phenological characters, viz., early maturity, early vigour, fast ground cover, larger seed, and physiological/ biochemical characters such as diffusive resistance of stomata, osmotic adjustment, leaf rolling, leaf water retention, and leaf senescence were associated with drought tolerance (Singh, 1993). Increased seed size, early maturity and reduced plant height at the drought prone location and early maturity at the drought free location were of prime importance in increasing seed yield (Singh *et al.*, 1995).

4.10. LIMITATIONS IN SELECTION

Although research on drought tolerance achieved many goals, certain constraints still remained unsolved. Drought tolerance has appeared to be difficult research aim in breeding programmes due to several technical limitations (Mitra, 2001).

1. Drought stress is highly heterogeneous in time (over the season and years) and space (between and within sites), and is extremely unpredictable. This makes it very difficult to identify a representative drought stress condition.
2. In any case, the environment is not defined in terms of rainfall, soil moisture, temperature and vapour pressure deficit. Consequently, the performance of the same genotypes varies year after year.
3. *Environmental influence:* Since the phenotype is the ultimate expression of the interaction between genotype and environment, assessment of the desired genotype is highly dependent on the proper environmental conditions. The unpredictable and variable forms in which drought stress will manifest itself, makes selection of promising individual plants and hence breeding for drought tolerance become extremely difficult.

4. Lack of repeatable and precise screening techniques. There is hardly any basis of selection in F_2 and F_3 segregating generations, but subsequently some lines may be identified which appear promising.

5. *Negative association of stress tolerant traits with yield:* Some drought tolerant characters are negatively associated with yield. For instance, high levels of tissue ABA may reduce productivity even though they may contribute to survival. Selection for high water-use efficiency often results in decrease in crop growth rate. Most of the plants evolve to maximize water-use efficiency through reduction in transpiration. Since, dry matter production is strongly associated with total transpiration, any reduction in transpiration results in reduced crop growth rate. Leaf rolling and stomata closure conserve water in plant, but reduce light interception and entry of carbon dioxide into leaf and in turn, reduce the yield.

6. *Multiple stresses:* Drought tolerance hardly is separated from other important abiotic stress such as temperature and salinity. Due to these interrelations, no single mechanism exists by which multiple stresses are alleviated. Drought reduces nutrient uptake and is associated with temperature stress, and at higher elevation with cold. This association makes the breeding programme more complicated.

7. So far breeding for drought tolerance mainly confined in cereal crops. The evaluation of rice genotypes in upland represents selection for drought tolerance. There is no such directed effort in pulses and oilseed.

8. Similarly, biotechnological approaches of breeding, viz., MAS and transgenic development, mainly centred on crops which are easy to handle, like *Arabidopsis, Tobacco, Oryza sativa, Triticum* and some extent in maize.

9. Knowledge is incomplete about reliable attributes as indices of drought tolerance, selection criteria and influence of environment or drought related traits (Paroda, 1986).

10. *Multiple genes involvement:* Drought tolerance has been noted to be a highly complex trait, influenced by many different genes. In fact, drought tolerance should be regarded as a unique heritable trait, but as complex of often fully unrelated plant properties.

11. *Multidisciplinary approach:* There is lack of multidisciplinary approach to understand the integrated plant response to drought and complex genetic control of different mechanisms of drought resistance.

4.11. BIOTECHNOLOGY IN BREEDING FOR DROUGHT TOLERANCE

4.11.1. Tissue Culture

In vitro screening for drought tolerance in sweet potato cv. Lizixiang was studied by Wang *et al.* (2003a). Embryogenic suspension cultures were subjected to stress medium (MS) containing 0-35% PEG6000. Their result indicated that 30% PEG6000 can be used for the potential selection for drought stress. Dobranski *et al.* (2003) studied the effects of manitol of different genotypes of potato in callus and plantlet culture. In callus-test, the relative increase of callus mass was a useful parameter for the determination of osmotic tolerance of genotypes at cellular level.

Sorghum varieties were screened for their ability to indicate callus, embryogenic callus and shoot formation on *in vitro* culture media supplemented with different concentration of PEG6000 (Makhouf *et al.*, 2002). They found that the callus induction, callus weight, and embryogenic callus and shoot formation in the media. Biswas *et al.* (2002) screened rice genotypes *in vitro* using PEG6000 in the culture medium. They obtained a number of putative drought-tolerant plants. Their findings revealed monogenic inheritance for drought tolerant characters in the majority of the progenies.

4.11.2. Genes for Drought Tolerance

Many genes have been identified and used for drought tolerance in different crop plants. Kang *et al.* (2004) reported that ABA biosynthesis and signaling can be regulated through *AtGLR1.1* gene to trigger pre- and post-germination arrest and changes in whole plant responses to water stress. They also suggested that *AtGLR1:1* integrates the different aspects of C, N, plant growth and development.

The molecular characteristics of *P5CS* (pyrroline-5-carbohydrate synthetase) gene in sugarcane is related to drought tolerance (Minarsih *et al.*, 2001). They analysed proline in water-stressed plants, and the amount of proline accumulated. The higher content of proline in stressed plants implied that the *P5CS* is inducible.

Dehydrins (*DHNs*, late embryogenesis abundant D-H) are a family of plant protein induced in response to environmental stress such as water stress, salinity and freezing or which occur during the late stage of embryogenesis. Dehydrin gene (*Dhn*) expression is associated with plant response to dehydration in barley. Suprunova *et al.* (2004) noted difference between resistant and sensitive genotypes and it was mainly in the expression of *Dhn1* and *Dhn6* genes, depending on the duration of dehydration. Their findings indicated that those genes had some functional role in dehydration tolerance in wild barley. Porat *et al.* (2004) also isolated dehydrin cDNA from orange and grapefruit.

Aharoni *et al.* (2004) identified an *Arabidopsis thaliana* activation tag gain-of-function mutant shine (*shn*) that displayed a brilliant, shiny green leaf surface with increased cuticular wax compared with the leaves of wild-type plants. The over-expression of *shn* clade genes, increased in *shn* plants, which ultimately lead in altered leaf and petal epidermal cell structure, number of trichomes, branching and stomatal index. *Shn* over-expression displayed significant drought tolerance and recovery, probably related to reduce stomatal density.

Messmer *et al.* (2004) constructed a genetic linkage map of maize. The anthesis-silking interval as well as the ear and silk dry weight at pollen shedding seem to be good secondary traits for drought tolerance, as indicated by the phenotypic correlations and the results of the QTL analysis.

4.11.3. Marker Assisted Selection (MAS)

Drought adaptation is the most complex aspect of crop improvement for rainfed dry-land production system. Despite over half a century of intensive genetic improvement efforts by plant breeders and physiologists for enhancing drought tolerance, successful impacts in farmers' fields continue to be rare and unpredictable. The common approach in breeding for drought tolerance is to select drought tolerance components, since several of these components are difficult to measure, indirect selection might be applied. The scientific community has struggled to develop effective field based phenotype evaluation and selection systems, which has constrained directed progress in conventionally. It is here that novel biotechnological tools can be of use. During the last two decades, plant molecular biologists have developed increasingly precise tools and methodologies for the genetic dissection of complex traits. Given genetic population of appropriate size and pedigree, plus phenotypic data of sufficient accuracy and precision, these approaches can now be used to identify a large range of genes contributing to the expression of a given quantitative trait such as the ones involved in drought tolerance. Breeder can select for molecular marker which have exhibited a strong genetic linkage with the component that is to be selected. This linkage may be purely physical; the molecular marker is a stretch of DNA which is located close to a gene influencing the drought tolerance component. Rapid advances in structural and functional genomics of model species (such as *Arabidopsis* and *Medicago*) and major crops (such as rice and maize) are opening many new avenues for increasing scientists understanding and ability to manipulate complex characters such as drought tolerance. The identification, isolation and manipulation of elements controlling gene expression hold particular promise for transgenic approaches. Whilst, large-scale sequencing of expressed sequence tags (EST) is likely to revolutionize marker-assisted selection through the application of candidate gene mapping for the development of gene-based markers for loci of known function.

The prospects of changing the phenotype through genetic manipulations or conventional breeding become greater if one or a few defined regions of chromosomes are of crucial importance. The identification of QTLs has, therefore, practical importance in attempts to enhance stress tolerance (Koyama *et al.,* 2001).

Plant physiologists are also making rapid progress in the functional dissection of complex traits, which will facilitate the genetic mapping and population enhancement of key component characters. Through a carefully balanced combination of conventional and new molecular based tools, a new paradigm in plant breeding is emerging that has the potential to develop economic adaptation to drought and thus real impact in farmers' fields. The advances in the model species and major crops are also likely to have a substantial impact on progress in lesser-studied crops.

The advantage of indirect selection for associated molecular markers, instead of the genetic trait itself, is that the presence of the marker in the off-spring of a crossing can be demonstrated, independently from interfering environmental stresses, at the plantlet level in the laboratory. Marker assisted selection becoming increasingly useful in the development of new germplasm with improved stress tolerance (Hoisington *et al.,* 1996; Quarrie *et al.,* 1999; Hemamalini *et al.,* 2000). Recent progress in crop plant genome analysis has made it possible to dissect complex characters into individual genetic factor by means of marker assisted approach. Marker assisted breeding depends on the availability of a large number of molecular markers. The availability of a genomic map, in which the genetic characters and the molecular markers of a species have been physically arranged, is also important.

The availability of genetic map saturated with molecular markers allows a breeder to do selection for certain characters much more efficiently and effectively. Quantitative trait loci (QTL) for root characteristics have been identified in rice (Ray *et al.,* 1996) and maize (Lebreton *et al.,* 1995). Some of best target QTLs identified in rice have been listed in Table 4.3.

Table 4.3. Target QTLs for marker-assisted backcrossing of drought tolerance in rice

Trait	Chr	Marker	LOD	R^2	Population*	Reference
CMS	3	RZ403	12.1	42.1	CT9993/**IR2266**	Tripathy *et al.*, 2000
CMS	9	RZ698-RM219	10.4	37.4	**CT9993**/IR2266	Tripathy *et al.*, 2000
CMS	8	RG598	07.5	29.4	**CT9993**/IR2266	Tripathy *et al.*, 2000

Chr: Chromosome; **LOD:** Logodds, which is likelihood score for the presence of the QTL; $\mathbf{R^2}$: which is the proportion (or percentage) of the phenotypic variation explained (PVE) by the QTL; **CMS:** Cell membrane stability; *: Parent in bold is considered source of the desirable allele.

Differentially expressed sequence tags (dESTs) and candidate genes for the drought response were mapped (Diab *et al.*, 2004) in barley (*Hordeum vulgare*). Composite interval mapping was used to identify QTL associated with drought tolerance including leaf relative water content, leaf osmotic potential at full turgor, water-soluble carbohydrate concentration, osmotic adjustment and carbon isotope dissolution. They have identified a total of 68 QTLs. The number of QTLs identified for each character varied from 1-12, indicating that the genome contains multiple genes affecting different characters. Two candidate genes and in differentially expressed sequence were associated with QTLs for drought tolerant characters.

Courtois *et al.* (2003) prepared linkage map for root characters in rice. They found that chromosomes 1, 4, 9, 11 and 12 carried QTLs for several characters. Hu and Yan (2004) identified seven QTLs related to chlorophyll content under drought, which were located on chromosome 3 and 7 in rice. The phenotypic variance explained by each QTL ranged from 5.50-8.57% by using the RIL populations and SSR marker linkage map. This may provide a good base for breeding rice with high photosynthetic efficiency and cloning relative genes. Nguyen *et al.* (2004) have developed 85 molecular markers (50 RFLPs, 5 SSRs, 12 DD cDNAs, 9 ESTs, 8 HSP- encoding cDNA and BSA-derived AFLPs markers) for saturation mapping of QTL region for drought tolerance in rice. QTLs for grain yield in rice were identified by Lanceras *et al.* (2004) using double haploid mapping population. They mapped seven QTLs for grain yield. QTLs flanked by marker RG104 to RM231, EMP2-2 to RM127 and G2132 to RZ598 on chromosomes 3, 4 and 8 were associated with grain yield under drought treatments. The aggregated effects of those QTLs will be useful for rainfed rice improvement, and will also contribute to yield under drought conditions at sensitive reproductive stage.

4.11.4. Transgenic Development

Plant genetic engineering techniques could be effectively utilized in exploiting untapped tolerance mechanisms to improve the harvestable crop yield. Although conventional breeding and marker assisted selection are being used to develop varieties more tolerant to water stress, these methods are time taking and resource consuming and germplasm dependent. Genetic engineering is alternative because of its potential to improve abiotic stress tolerance more rapidly. The techniques for gene transformation of crop plants have been applied for identification of genes responsible for drought resistance and their transfer (Grover *et al.,* 1998). Genes expressed during stress are anticipated to promote cellular tolerance to dehydration through protective functions in the cytoplasm, cell membrane, alternations of cellular water potential to promote water uptake, control of ion accumulation and further regulation of other gene(s). A comprehensive study on transgenic plants summarized and nature of tolerance has been listed in Table 4.4.

Bacterial trehalose-6-phosphate (*OtsB*) genes introduced in tobacco showed better growth under drought stress (Pilon-Smits *et al.,* 1995). Transgenic plants also showed a better capacity to retain water and preformed more efficient photosynthesis under stress. The gene *TPS1* found in yeast encodes for trehalose-6-phosphate synthetase and is involved in biosynthesis of trehalose. The transgenic tobacco plants containing the yeast *TPS1* gene exhibited multiple alteration and improved drought tolerance (Romero *et al.,* 1997). By determining the effect of withholding irrigation on the death and damage of leaf (Romero *et al.,* 1997), it has been shown that the transgenic plants have increased drought resistance. Further, Holmstrom (1998) transferred tobacco, resulting in the accumulation of trehalose and improved drought resistance. It is suggested that the improved drought tolerance is due to enhanced water retention capacity. Serrano *et al.* (1999) reported that yeast regulatory gene involved in stress tolerance, *TPS1,* have been expressed in transgenic plants. Almeida *et al.* (2004) transformed *Nicotiana tabaccum* with *Arabidopsis thaliana* gene (*AtTPS1*), which is involved in trehalose biosynthesis, by *Agrobacterium*-mediated genetic transformation. They observed high germination rates at higher levels of manitol than did wild type plants.

The trehalose synthase gene (*TSase*) gene from *Grifola frondosa* was transferred into sugarcane (Wang *et al.,* 2003b). The high frequency phosphinothricine-resistant plants were obtained from the transformation. Some transgenic plants showed multiple phenotypic alterations and some plants showed improved tolerance to osmotic stress. Another trehalose

Table 4.4. Transgenic plants over expressing enzymes that synthesize osmoprotectants

Gene	Gene action	Species	Phenotypic expression by transgenic plant	Reference
Sod	MnSod	*Solanum tobaccum*	Reduced cell damage, increased tolerance to water deficit	Bowler *et al.*, 1991
Sod	Cu/Z superoxide dismutase	*Solanum tobaccum*	Retained 90% photosynthesis under chilling and heat stress	Sengupta *et al.*, 1993
Sod	MnSod	*Medicago sativa*	Increased tolerance to water deficit	McKersie *et al.*, 1993, '99
tstB	Trehalose 6-phosphate synthetase	*Solanum tobaccum*	Increased weight and more efficient photosynthesis under drought stress	Pilon-Smits *et al.*, 1995
HVA1	Group 3 LEA protein gene	*Oryza sativa*	Increased tolerance of drought and salinity stress	Xu *et al.*, 1996
HVA1	Group 3 LEA protein gene	*Oryza sativa*	-	Jain *et al.*, 1996
otsA	trehalose-6-phosphate (Trehalose synthesis)	*Solanum tobaccum*	Increased dry weight and photosynthetic activity under drought	Pilon-Smits *et al.*, 1996
Sod	Mn Sod	*Medicago sativa*	Increased tolerance to water deficit	McKersie *et al.*, 1996
TPS1	Trehalose 6-phosphate synthetase (trehalose synthesis)	*Solanum tobaccum*	Increase drought tolerance	Romero *et al.*, 1997
imt 1	Myo-inositol o-methyl transferase (D-ononitol synthesis)	*Solanum tobaccum*	Improve performance under drought and salt as shown by better photosynthetic rate	Sheveleva *et al.*, 1997
IMT1	Myo-inositol-O-methyl transferase (D-ononitol synthesis)	*Solanum tobaccum*	Performed better under drought and salinity stress	Sheveleva *et al.*, 1997
Adc	Arginine decarboxylase	*Oryza sativa*	Reduced chlorophyll loss under stress	Capell *et al.*, 1998
Phes	Pyrroline carboxylate synthelase (proline synthesis)	*Oryza sativa*	Increased biomass production under drought and salinity stress	Zhu *et al.*, 1998
Adc	Arginine decarboxylase	*Oryza sativa*	Reduce chlorophyll loss under drought stress	Capell *et al.*, 1998
p5cs	Pyrroline carboxylate synthetase (Proline synthesis)		Increased biomass production under drought and salinity stress	Zhu *et al.*, 1998
DREB	Transcription factor	*Arabidopsis thaliana*	Increased tolerance of cold, drought and salinity	Kasuga *et al.*, 1999
sod	Mn superoxide dismutase	*Medicago sativa*	Increased tolerance to water deficit	McKersie *et al.*, 1999
OsCDPK7	Transcription factor	*Oryza sativa*	Increased tolerance of cold salinity and drought	Saijo *et al.*, 2000
HVA1	Group 3 LEA protein gene	*Triticum aestivum*	Increased biomass WUE under stress	Sivamani *et al.*, 2000

Contd....

Table 4.4 Contd....

Tsi1	Transcription factor	*Solanum tobaccum*	Increased osmotic stress tolerance	Park *et al.*, 2001
AhCMO	Chotine monoxygenase	*Solanum tobaccum*	Improved drought tolerance	Shen *et al.*, 2002
HVA1	Group 3 LEA protein gene	Pusa Basmati 1	Increased stress tolerance in terms of cell integra tion and growth	Rohlia *et al.*, 2002
TSase	Trehalose synthase	Sugarcane	Tolerant to osmotic stress	Wang *et al.*, 2003 a,b
TPS1	Trehalose synthase	*Solanum tobaccum*	High intergrity of chloroplast thylakoid membrane	Lee *et al.*, 2003
DREB1A	Transcription factor	*Triticum aestivum*	Increased drought tolerance	Pellegrneschi *et al.*, 2003, 2004
AtTPS1	Trehalose biosynthesis	*Solanum tobaccum*	High germination rates at high levels of manitol	Almeida *et al.*, 2004
NPK1	Experssion of nitrogen-activated protein kinase	*Solanum tobaccum*	Enhanced drought tolerance	Shou *et al.*, 2004

phosphate synthase (*TPS1*) gene from yeast was introduced into tobacco chloroplast (Lee *et al.*, 2003). Stable integration of *TPS1* into chloroplast genome of tobacco was confirmed by PCR and Southern blots analysis. Transgenic chloroplast thylakoid membranes showed high integrity under osmotic stress as evidenced by the retention of chlorophyll even when grown in 6% PEG6000, whereas chloroplasts of untransformed plants were bleached. Thus, they suggested that trehalose functions by protecting biological membranes rather than regulating water potential.

Transgenic rice carrying barley *HVA1* gene had shown drought resistance (Xu *et al.*, 1996). Gene *HVA1* encodes for a group of three LEA proteins which gets accumulated in vegetative organs during drought condition (Dure, 1993). Transgenic rice showed enhanced accumulation of the *HVA1* protein, which led to higher growth rates, delayed stress-related damaged systems, and improved recovery from the removal of stress conditions (Xu *et al.*, 1996). The transgenic rice plants exhibited constitutive high expression of *HVA1* protein ranging from 0.3-2.5% and 0.3-1.0% of the total soluble protein in leaf and root, respectively. Transgenic wheat plants containing the *HVA1* gene showed consecutive expression of the transgene resulting in improvement of growth characteristics under water deficient conditions, more biomass, and more efficient water use (Sivamani *et al.*, 2000). Rohila *et al.* (2002) transformed Pusa Basmati 1 with *HVA1* to increase tolerance against abiotic stresses. They developed transgenic lines showed increased stress tolerance in terms of cell integrity and growth after imposed salt- and water stresses. Their findings exhibited high levels of LEA3 accumulation in the leaves of transgenic Pusa Basmati 1 rice plants might have conferred the significant increase in tolerance against drought and salt stresses.

The bacterial gene *sacB* found in *Bacillus subtilis* encodes for levan sucrase, which takes part in fructan synthesis. Fructan promote the process of root branching, thus increasing root surface and water uptake. Pilon-Smits *et al.* (1995) have shown the that over-expression of *scaB* gene from *Bacillus subtilis* lead to high level of fructans in transgenic tobacco cells, leading in higher capacity for osmotic adjustment, and this is associated with increased drought tolerance. The additional carbohydrate gained may lead to deeper rooting and greater water-uptake (Pilon-Smits *et al.*, 1995; Sharp *et al.*, 1996; Schellenbaum *et al.*, 1999). The transgenic tobacco that produces bacterial fructans was produced and examined for growth of the transgenic plants was significantly higher both on fresh weight and dry weight basis under drought stress compared to the wild type tobacco. The transgenic tobacco exhibited significantly more biomass accumulation in roots under drought stress and higher non-structural carbohydrate content under all conditions.

Proline accumulation has been demonstrated to be associated with abiotic stress (Delauney and Verma, 1993). Some genes related to proline synthesis, transport and accumulation have been identified (Delauney and Verma, 1993). Transgenic tobacco overexpressing *P5CS* gene transferred from mothbean exhibited a high level of enzyme and produced 10-18 fold more proline than control plant (Kavi-Kishore *et al.*, 1995). The over-production of proline enhanced root biomass and flower development under drought condition. Zhu *et al.* (1998) also found that transgenic rice with *P5CS* gene enhanced root biomass and flower development under water stress. Transgenic soybean plants transferred with antisense *P5CR* (*L*-D'-pyrroline-5-carboxylate reductase) gene showed increased proline accumulation, leading to higher tolerance to water stress (De Ronde *et al.*, 2000).

In soil-bacterium *Arthrobacter globiformis*, *codA* gene (encoding for choline oxidase) is responsible for choline to glycinebetaine conversion. The genes *betA* encoding for choline dehydrogenase and *betB* encoding for betaine aldehyde dehydrogenase are involved in the glycinebetaine confers drought resistance.

In plants, polyamine, accumulate under several abiotic stress stimuli, including drought and salt. It has been suggested that this increase in polyamine concentration could be considered as an indicator of plant stress. With the availability of genes responsible for polyamine biosynthesis such as *ADC* (encodes for arginine decarboxylase), it is now possible to manipulate polyamine content using sense and antisense constructs of these genes in transgenic plants. Engineering of the plant polyamine biosynthesis pathway has concentrated mostly on two species, tobacco and rice (Kumar and Minocha, 1998; Capell and Christou, 2004). Further studies are required to understand the tolerance ability of these genes. The increase in putrescine levels in plants under stress might be the cause of stress-induced injury or alternatively a means of protection against stress. Roy and Wu (2001) expressing oat *adc* cDNA in rice under control of an ABA-inducible promoter resulted in transgenic rice plants with increased biomass when grown under salt stress. Capell and Christou (2004) have generated a diverse rice germplasm with altered polyamine content. Transgenic rice plant expressing the *SAMDC* DNA accumulated spermidine and spermine in seed at 2-3 fold higher levels compared to wild type. In another set of experiment, they have obtained ten fold putrescine accumulations in transgenic rice plants laboring oat *adc* cDNA compared to wild type. Spermidine and spermine de-novo synthesis in transgenic plants under drought stress is corroborated by the activation of the rice *SAMDC* gene. Transcript levels for rice *SAMDC* reach their maximum levels at 6 days after stress induction. Such increase in the endogenous spermidine and spermine pools of transgenic plants not only regulates the putrescine response, but also exerts an anti-senescence effect at the whole plant level, resulting in phenotypically normal plants. Wild type plants, however, are not able to raise their spermidine and spermine levels after 6 days of drought stress and consequently exhibit the classical drought stress response (Capell and Christou, 2004).

DREB1A, a transcription factor that recognizes dehydration response elements, has been shown in *Arabidopsis thaliana* to play a crutial role in promoting the expression of drought tolerant genes (Pellegrineschi *et al.*, 2003, 2004). They have transformed *DREB1A* gene into wheat. Plant expressing the gene demonstrated substantial resistant to water stress compared with the control under stress condition.

Sengupta *et al.* (1993) transferred *SOD* (superoxide dismutase) gene from pea to tobacco and the transgenic were found to be drought resistant. Transgenic tobacco plants containing oxidative stress related genes showed elevated levels of glutathione reductase, superoxide, dismutase and ascorbate peroxidase, resulting in enhanced drought tolerance (Van Rensburg and Kruger, 1994). In contrast, overexpressed Mn-SOD in chloroplast and mitochondria reduced cellular damage (Bowler *et al.*, 1991).

The gene designated *WXP1*, is able to activate wax production and confer drought tolerance in alfalfa (*Medicago sativa*). Overexpression of *WXP1* under the control of CaMV355 promoter led to a significant increase in cuticular wax loading on leaves of transgenic alfalfa (Zhang *et al.*, 2005). *WXP1* over expression induced a number of wax-related genes. Transgenic leaves showed reduced water loss and chlorophyll leaching – transgenic alfalfa plants with increased cuticular wax showed enhanced drought tolerance demonstrated by delayed wilting after watering has been ceased and quicker and better recovery when the dehydrated plants were re-watered.

A full length rDNA of dehydrin *BcDh2* from *Boea crassifolia* and its antisense nucleotide sequence was transferred into tobacco, var. NC89 under the control of CMV promoter (Shen *et al.*, 2004). Under progressive water stress, the photosynthetic rate, transpiration rate and stomatal conductance of sense and antisense plants decreased. However, those parameters increased after 24 hours of watering and the increase was higher in sense and antisense plants than the control.

Expression of nitrogen-activated protein kinase kinase kinase (*MAPKKK*) gene activates an oxidateive signal cascade and lead to the tolerance of freezing, heat, and salinity stressing transgenic tobacco. Shou *et al.* (2004)

transformed maize with a tobacco *MAPKKK* (*NPK1*). They found that the *NPK1* expression enhanced drought tolerance in transgenic maize. Under drought stress, it maintained significantly higher photosynthesis rates than did the non-transgenic control, suggesting that protected photosynthesis machinery from dehydration damage.

Choline monooxygenase (*CMO*) catalyzes the committed step of glycinebetaine (GlyBet) biosynthesis in many flowering plants. Over-expression of *AhCMO* improved drought tolerance in transgenic tobacco when cultured in medium containing PEG6000 (Shen *et al.*, 2002).

4.12. REFERENCES

Aharoni A, Dixit S, Jetter R, Thoenes F, Arkel G van, Pereira A. 2004. The SHINE clade of AP2 domain transcription factors activates wax biosynthesis, alters cuticle properties, and confers drought tolerance when overexpress in *Arabidopsis*. Plant Cell. 16(9): 2463-2480.

Ali SM, Naidu AP. 1982. Screening for drought tolerance in maize. Indian J Genet. 42: 381-388.

Almeida AM de, Araujo SS, Cardoso LA, Fevereiro MP, Tarne JM, Santos DM dos. 2004. Transformation of tobacco with an *Arabidosis thaliana* gene involved in trehalose biosynthesis as a model to increase drought resistance in crop plants. In: Genetic Variation for Plant Breeding, Proc of the 17th EUCARPIA General Congress, Tullen, Austria, 8-11 Sep 2004. pp. 285-289.

Arihara J, Ae N, Okada K. 1991. Root development of pigeonpea and chickpea and its significance in different cropping systems. In: Phosphorus Nutrition of Grain Legumes in the Semi-Arid Tropics, Johansen C, Lee KK, Sahrawat KL (eds.). ICRISAT, Patancheru, AP, India. pp. 183-194.

Armenta-Soto J, Chang TT, Loresto GC, O'Toole JC. 1983. Genetic analysis of root characters in rice. SABRAO J. 15: 103-116.

Austin RB, Henson IE, Quarrie SA. 1982. Abscisic acid and drought resistance in wheat, millet and rice. In: Drought Resistance in Crops with Emphasis on Rice. IRRI, Los Banos, Philippines. pp. 171-180.

Baker TC, Varughese G. 1992. Combining ability and heterosis among eight complete spring hexaploid triticle lines. Crop Science. 32: 340-344.

Bansal KC, Sinha SK. 1991a. Assessment of drought resistance in 20 accessions of *Triticum aestivum* and related species. I. Total dry matter and grain yield stability. Euphytica. 56: 7-14.

Bansal KC, Sinha SK. 1991b. Assessment of drought resistance in 20 accessions of *Triticum aestivum* L. and related species. II. Stability in yield components. Euphytica. 56: 15-26.

Banziger M, Edmeades GO, Beck D, Bellon M. 2000. Breeding for drought and nitrogen stress tolerance in maize. From Theory to Practive. Maxico D.F., CIMMYT.

Bartels D, Nelson D. 1994. Approaches to improve stress tolerance using molecular genetics. Plant Cell Envin. 17: 659-667.

Bassetti P, Westgate ME. 1993a. Senescence and receptivity of maize silks. Crop Sci. 33: 275-278.

Bassetti P, Westgate ME. 1993b. Water deficit affects reproductivity of maize silks. Crop Sci. 33: 279-282.

Basu MS, Nautiyal PC. 2004. Improving water use efficiency and drought tolerance in groundnut by trait based breeding programme in India. Proceedings of the 4th International Crop Science Congress, Brisbane, Australia, 26 Sep.-1 Oct. 2004.

Begg JE, Turner NC. 1976. Crop water deficits. Adv Agron. 28: 161.

Belford RD. 1988. *In situ* Measurement of crop root growth using video equipment. Crop Agronomy Branch, Divisional of Plant Industry, Western Australian Department of Agriculture, Final Report. pp. 1-8.

Bengston C, Larson S, Liljenberg C. 1978. Effect of water stress on cuticular transpiration rate and amount and composition of epicuticular wax in seedlings of six Oat varieties. Physiol Plant. 44: 319.

Ben-Hoyyim G, Damon JP, Martin-Tanguy J, Tepfer D. 1994. Changing root system architechture through inhibition of putrescine and feruloyl putrescine accumulation. FEBS Letters. 342: 145-148.

Bergmann H, Rost S, Machelett B. 2002. Improvement of drought tolerance and change of glycine betaine or proline accumulation in *Hordium vulgare* L. by choline and 2-aminoethanol treatment. J Appl Bot. 76(3-4): 87-95.

Bhale NL, Khidse SR, Borikar ST. 1982. Heterosis for proline accumulation in sorghum. Indian J Genet. 42: 101-104.

Biswas J, Chowdhury B, Bhattacharya A, Mandal AB. 2002. *In vitro* screening for increased drought tolerance in rice. *In Vitro* Cellular Dev Biol Plant. 38(5): 525-530.

Blum A, Sullivan CV. 1972. A laboratory method for monitoring net photosynthesis in leaf segments under controlled water stress experiments with sorghum. Phtotsynthesis. 6: 18-23.

Blum A. 1973. Components analysis of yield responses to drought of sorghum hybrids. Exp Agric. 9: 159-167.

Blum A. 1974. Genotypic Responses in Sorghum to Drought Stress. CRC Press, USA.

Blum A. 1988. Plant Breeding for Stress Environments. CRC Press, Inc., Boca Raton Florica.

Blum A. 1997. Crop responses to drought and the interpretation of adaption. In: Drought Tolerance in Higher Plants: Genetical Physiological and Molecular Biology Analysis, Belhassen E (ed.). Kluwer Academic Publishers, Dordrecht, The Netherland. pp. 57-70.

Bogdanova ED, Shulembaeva KK, Shegebaev O, Bondarenko OV. 1988. Monosomic analysis of the rolled leaves trait in winter bread wheat. Genetika, USSR. 24: 1710-1711.

Bolanos J, Edmeades GO, Martinez L. 1993. Eight cycles of selection for drought tolerance in lowland tropical maize III. Response in drought adaptive physiological and morphological traits. Field Crop Research. 31: 269-286.

Bolanos J, Edmeades GO. 1996. The importance of the anthesis-silking interval in breeding for drought tolerance in tropical maize. Field Crops Res. 48: 65-80.

Borrell AK, Hammer GL, Henzell RG. 2000. Does maintaining green leaf area in sorghum improve yield under drought. II. Dry matter production and yield. Crop Sci. 40: 1037-1043.

Bouslama M. 1983. Stress tolerance in soybeans (*Glycine max* (L.) Merr.) (Abstr), Diss Abstr Int. B. 43: 2006B.

Bowler C, Stoolen L, Vandenbranden S, Ryake RD, Botterman J, Sybesma C, Van MM, Inze D. 1991. Manganese superoxide dismutase can reduce cellular damage mediated by oxygen radicals in transgenic plants. EMBO J.10:1723-1732.

Boyer JS. 1996. Advance in drought tolerance in plants. Adv Agron. 56: 187-218.

Bruckner PL, Frohberg RC. 1987. Stress tolerance and adaptation in spring wheat. Crop Sci. 27: 31-36.

Bunce JA. 1988. Differential responses of photosynthesis to water stress in three soybean cultivars. Plant Physiol Biochem, France. 26: 415-420.

Bushuk W, Jana S, Tourley Smith TE. 1989. Canadian research on drought resistance in cereals. In: Drought resistance in cereals, Baker FWG (ed.). CAB International, UK. pp. 191-200.

Campbell JA, Naidu BP, Weaich K, Wilson JR. 1996. In: Sugarcane: Research Towards Efficient and Suitable Production, Wilson JR, Hogarthy DM, Campbell JA, Garside Al (eds.). CSIRO Tropical Crops and Pasture, Brisbane. pp. 181-182.

Capell T, Christou P. 2004. Enhanced drought tolerance in transgenic rice. Curr Opinion in Biotchnol. 15: 148-154.

Capell T, Escobar C, Liu H, Burtin D, Lepri O, Chistou P. 1998. Overxpression of oat arginine decaboxylase cDNA in transgenic rice (*Oryza sativa* L.) affects normal development pattern *in vitro* and result in putrecine accumulation in transgenic plants. Theor Appl Genet. 97: 246-254.

Ceccarelli S. 1987. Yield potential and drought tolerance of segregating populations of barley in contrasting environments. Euphytica. 36: 265-273.

Chambers TC, Ritchie LM, Booth MA. 1976. Chemical models for plant wax morphogenesis. New Phytol. 77: 43.

Chang TT, Armenta-Soto JL, Mao EX, Peiris R, Loresto GC. 1986. Genetic studies on the components of drought resistance in rice (*Oryza sativa* L.). In: Rice Genetics I, International Rice Research Institute, Manila, Philippines. pp. 387-398.

Chang TT, Loresto GC, Tagumpay O. 1974. Screening rice germplasm for drought tolerance. SABRAO J. 6: 9-16.

Chang TT, Loresto GC. 1985. Screening techniques for drought resistance in rice. In: Proc Natl Seminar on Breeding for Stress Resistance in Plants, Haryana Agricultural University, India.

Chapman SC, Edmeades GO. 1999. Selection improves drought tolerance in tropical maize population. Crop Sci. 39: 1315-1324.

Chimenti CA, Pearson J, Hall AJ. 2002. Osmotic adjustment and yield maintenance under drought in sugarcane. Field Crops Res. 75(2-3) 235-246.

Courtois B, Shen L, Petalcorin W, Carandang S, Mauleon R, Li Z. 2003. Locating QTLs controlling constitutive root traits in rice population IAC 165 × Co39. Euphytica. 134(3): 335-345.

Crosbic TM, Pearce RB, Mock JJ. 1981. Recurrent phenotypic selection for high and low photosynthesis in two maize populations. Crop Sci. 21: 736-740.

DaMatta FM, Chaves ARM, Pinheiro HA, Ducati C, Loureiro ME. 2003. Drought tolerance of two field grown clones of *Coffea canephora*. Plant Sci. 164: 111-117.

DaMatta FM, Maestri M, Barros RS, Regazzi AJ. 1993. Water relations of coffee leaves (*Coffea arabica* and *C. canephora*) in response to drought. J Hort Sci. 68: 741-746.

DaMatta FM, Rena AB. 2002. Relacoes Lidricas no cafeciro. In: Encarnacao RO, Agonso Jr PC, Rufino JLS (eds.), I Grimposia de Pesquisa dos cafes do Brasil: Palestras. Embrapa Café, Brasila. pp. 9-44.

Darofeev VF, Tyselanko AM. 1982. Number of seminal roots in spring wheat in the course of selecting pair for hybridization. Vestnik Selokokhozyist-vennoi Nauki. No. 8: 50-56.

Datta SK. 2002. Recent development in transgenic for aboitic stress tolerance in rice. JIRCAS Working Report (2002). pp. 43-53.

Davies S, Turner NC, Siddique KHM, Leport L, Plummer J. 1996. Influence of drought on seed growth in chickpea. Second International Crop Science Congress, Nov. 17-24, 1996. New Delhi, India. p. 234.

De Ronde JA. Spreeth MH, Cress WA. 2000. Effects of antisense Δ^{-1}-pyrroline-5-carbohydrate reductase transgenic soybean plants subjected to osmotic and drought stress. Plant Growth Reg. 32: 13-26.

Delauney AJ, Verma DPS. 1993. Proline biosynthesis and osmoregulation in plants. Plant J. 4: 212-223.

Dhanesekar P, Pandey RN, Henry A, Kumar D, Singh NB. 2003. *In vitro* screening for drought tolerance in cow pea. In: Proc in Natl Symp on Arid Legumes, for food nutrition security and promotion of trade, Hissar, India, 15-16th May 2002. Advances in Arid Legume Research. pp. 372-375.

Dhopte AM, Rahangdale SL, Jamadar SL. 1987. Physiological evaluation of forty exotic wild sorghum lines in relation to stomatal factors in drought resistance. Annal Plant Physiol. 1: 143-150.

Dhopte AM, Shekar VB, Pandrangi R, Wankhade SG, Rahangdale SL. 1995. Variation in leaf amino acid contents in drought tolerant and susceptible genotypes and relationship of physiological traits with yield stability in grain sorghum. Annals Plant Physiol. 9: 28-33.

Diab AA, Teulat-Merah B, This D, Qzturk NZ, Benscher D, Sorrells ME. 2004. Identification of drought-inducible genes and differentially expressed sequence tag in barley. Theor Appl Gene. 109(7): 1417-1425.

Dobranszki J, Magyar-Tabori K, Takacs-Hudak K. 2003. Growth and developmental responses of potato to osmotic stress under *in vitro* condition. Acta Biologica Hungarica. 54(3): 365-372.

DuPlessis DP, Dijkhuis FJ. 1967. The influence of time lag between pollen shedding and silking on the yield of maize. South Afr J Agril Sci. 10: 667-674.

Dure LIII. 1993. The LEA protein of higher plants. In: Control of Plant Gene Expression, Verma DPS (ed.). CRC Press, Boca Raton, FL. pp. 325-335.

Durley RC, Kannangara T, Seetharama N, Simpson GM. 1983. Drought resistance of *Sorghum bicolor*. Genotypic differences in the concentrations of free and conjugated abscisic, phaseic and indole-3-acetic acid in leaves of field-grown drought stressed plants. Canadian J Plant Sci. 63: 131-145.

Edmeades GO, Bolanos J, Chapman SC, Lafitte HR, Banziger M. 1999. Selection improves drought tolerance in tropical maize population. I. Gains in biomass, grain yield and harvest index. Crop Sci. 39: 1306-1315.

Edmeades GO, Bolanos J, Lafitte HR. 1992. Progress in breeding for drought tolerance in maize proceedings of the Fourthy-Seventh Annual Corn and Sorghum Industry Research Conference. pp. 93-111.

Edmeades GO, Bolanos J, Elings A, Ribaut JM, Banziger M, Westgate ME. 2000. The role and regulation of the anthesis-silking interval in maize. In: Westgate ME, Boote KJ (eds.). Physiology and modeling kernel set in maize. CSSA Special Publication No. 29. Madison, WI: CSSA. pp. 73-73.

Edmeades GO, Copper M, Lafitte R, Zinselmeier. C, Ribaut JM, Habben JE, Loffer C, Banziger M. 2001. Abiotic stress and staple crops. Proceedings of the Third International Crop Science Congress, Hamburg, Germany, August 18-23, 2000. CABI.

Ekanayake IJ, O' Toole JC, Garrity DP, Masajo TM. 1985. Inheritance of root characters and their relations to drought resistance in rice. Crop Sci. 25: 927-933.

Eller BM. 1979. The significance of layers of wax superimposed to the epidermis for leaf ecology and leaf radiation budget. Flora. 168-146.

Evans LT, Wardlaw IF, Fisher RA. 1975. In: Crop Physiology- some case stories. Evans LT (ed.), Cambridge University Press, London. pp. 101-149.

Fan S, Blake TJ, Blumwald E. 1994. The relative contribution of elastic and osmotic adjustment to turgor maintenance of woody plants. Physiol Plant. 90: 414-419.

Farquhar GD, Richards RA, 1984. Isotopic composition of plant carbon correlates with water use efficiency of wheat genotypes. Aust J Plant Physiol. 11: 539-552.

Fereres E, Gimenez C, Fernendez JM-1986. Genetics variability in sunflower cultivars under drought. I. Yield relationship. Aus J Agric Res. 37. 37: 573-582.

Fischer KS, Johnson EC, Edmeades, GO. 1982. Breeding and selection for drought resistance in tropical maize. In: Drought resistance in Crop Plants with Emphasis on Rice, IRRI Los Bonos, Philippines. pp. 377.

Fisher RA, Wood JT. 1979. Drought resistance in spring wheat cultivars. III. Yield associations with morpho-physiological traits. Australian J Agric Res. 30: 1001-1020.

Frota JNE, Tucker TC. 1978. Salt and water stress influences nitrogen metabolism in red kidney beans. Soil Sci Soc Am J. 42: 743-746.

Gomez SM, Kalamani A. 2003. Scope of landraces for future drought tolerance breeding programme in rice (*Oryza sativa* L.). Plant Archives. 3(1): 77-79.

Grover A, Pareek A, Singhala S-L, Minhar D, Katiyar S, Ghawana S, Dubey H, Agarwal M, Rao GU, Rathee J, Grover A. 1998. Engineering crops for tolerance against abiotic stress through gene manipulation. Curr Sci. 75(7): 689-296.

Hall AE, Foster KW, Waines JG. 1978. Crop adaptation to semi-arid environments. In: Agriculture in Semi-Arid Environments, Cannell GH, Hall AE, Lawton H (eds.), Springer-Verlag, New York. pp. 148-179.

Hall AE, Patel PN. 1985. Breeding for resistance to drought and heat. In: Singh SR, Rachie KO (eds.), Cowpea Research, Production and Utilization. Wiley, U.K. pp. 137-151.

Hameed M, Mansor U, Muhammad A, Rao AR. 2002. Variation in leaf anatomy in wheat germplasm from varying drought-hit habitats. Interl J Argil Sci. 4(1): 12-16.

Hamission M, Webster JA, Richardson PE, Todd GM. 1991. Ultra structural and physiological study of damage to *Sorghum bicolor* (L.) caused by drought and *Schizaphis granium* Randani. In: proceedings of Aphid Plant Interaction; population to molecule as asu centennial Events Still Water, Okeahama, USA. pp. 26-27.

Hemamalini GS, Shashidhar HE, Hittalmani S. 2000. Molecular marker assisted tagging of morphological physiological traits under two contrasting moisture regimes at peak vegetative stage in rice (*Oryza sativa* L.). Euphytica. 112: 69-78.

Hendry GAF. 1993. Evalutionary origins and natural functions of fructans: a climatological, biogeographic and mechanistic appraisal. New Phytol. 123 (1): 3-14.

Henzel RG, Mc Cree KJ, van Bavel CHM, Schertz KF. 1976. Sorghum genotypic variation in stomatal sensitivity to leaf water deficit. Crop Sci. 16: 660 -662.

Heyne EG, Bruson AM. 1940. Genetic studies of heat and drought tolerance in maize. J Am Soc Agron. 32: 803-814.

Hoisington D, Jiang C, Khairallah M, Ribuat JM, Bohn M, Melchinger A, Willcox M, Gonzalez-de-Leon D. 1996. QTL for insect resistance and drought tolerance in tropical maize: prospects for marker assisted selection. Symposium of the Society of Experimental Biology. 50: 39-44.

Holmstrom KO. 1998. Engineering plant adaptation to water stress. Acta Univ Agric Sueciae, Agraria No. 84-49.

Hu SP, Yan YM. 2004. Genetic analysis for chlorophyll content in rice by molecular markers. Agril Sci Technol, Hunan. 5(2): 8-11.

Hurd EA, Spratt ED. 1975. Root patterns in crop as related to water and nutrient uptake. In: Physiological Aspects of Dryland Farming, Gupta US (ed.). Oxford and IBH Pub-Co. Pvt. Ltd., New Delhi. pp. 167-235.

Hurd EA. 1969. A method of breeding for yield of wheat in semi-arid climates. Euphytica. 18: 217-226.

Hurd EA. 1971. Can we breed for drought resistance ? In: Drought Injury and Resistance in Crops Larso KL, Eastin JD (eds.) Crop Science Society of America, Madison, W.I. USA. pp. 77-78.

Ingram J, Bartels D. 1996. The molecular basis of dehydration tolerance in plants. Annual Rev Plant Physiol Mol Biol. 47: 377-403.

Jain RK, Jain S, Wu R. 1996. Stimulating effect of water stress on plant regeneration in aromatic indica rice varieties. Plant Cell Rep. 15: 449-454.

Johnson DA. 1980. In: Adaptation of Plants to Water Stress and High Temperature Stress, Turner NC, Kramer PJ (eds.). Wiley, New York. pp. 419-433.

Johnson GR, Frey KJ. 1967. Heritabilities of quantitative attributes of oats (*Avena* sp.) at varying levels of environmental stresses. Crop Sci. 7: 43-47.

Jongdee B, Fukai S, Cooper M. 2002. Leaf water potential and osmotic adjustment as physiological traits improved drought tolerance in rice. Field Crops Res. 76(2-3): 153-163.

Josis P, Ndayishimiye V, Renard C. 1983. Etude des relations hydriques chez *Coffea arabica* L. II. Evaluation de la resistance a la secheresse de divers cultivars a Ghisa (Burundi). Café Cacao The. 27: 275-282.

Kang JM, Mehta S, Turano FJ. 2004. The putative glutamate receptor 1.1 (AtGRL1.1) in *Arabidopsis thaliana* regulates abscisic acid biosynthesis and signaling to control development and water loss. Plant Cell Physiol. 45(10): 1380-1389.

Kasuga M, Liu Q, Miura S, Yamaguchi-Shinozaki K, Shinozaki K. 1999. Improving plant drought, salt and freezing tolerance by gene transfer of a single stress inducible transcription factor. Nature Biotechnol. 17: 287-291.

Kavi Kishor PB, Hong Z, Miao GH, Hu CAA, Verma DPS. 1995. Overexpression of Δ'-pyrroline-5-carboxylate synthase increases proine production and confers osmotolerance ion transgenic plants. Plant Physiol. 108: 1387-1394.

Khidse SR, Bhale NL, Borikar ST. 1982. Combining ability for leaf water deficit and grain yield in rabi sorghum. Indian J Genet. 42: 82-86.

Khush GS, Coffman WR. 1977. Genetic evaluation and utilization (GEU) program. Theor Appl Genet. 51: 97-110.

Khush GS. 1995. Breaking the yield frontier of rice. Geojournal 35: 329-332.

Kitchen NR, Sudduth KA, Drummond ST. 1999. Soil electrical conductivity as a crop productivity measures for claypan soils. J Production Agri. 12: 607-617.

Koyama ML, Levesley A, Koebner RMD, Flowers TJ, Yeo AR. 2001. Quantitative trait loci of component physiological determining salt tolerance in rice. Plant Physiol. 125: 406-422.

Kumar A, Minocha SC. 1998. Transgenic manipulation of polyamine metabolism. In: Lindsey K (ed.) Transgenic research in plants. Hardwood Academic Publishers, UK. pp. 187-199.

Kumar A, Singh P, Singh DP, Singh H, Sharma HC. 1984. Differences in osmoregulation in *Brassica* species. Annals Bot. 54: 537-541.

Kumar J, Jadav SS, Kumar SS. 2004. Influence of moisture stress on quantitative characters in chickpea (*Cicer arientium* L.). Indian J Genet. 64(2): 149-150.

Kumar R, Kurju R. 2003. Role of secondary traits in improving drought tolerance during flowering stage in rice. Indian J Plant Physiol. 8(3): 236-240.

Lal R, Moomaw JC. 1977. Technique for screening rice varieties for drought tolerance in rice. In: Rice in Africa, Workshop organized in cooperation with IRRI, IRPAT and WARDA, held at IITA, lbadan, March 7-11, 1977.

Lanceras JC, Pantuwan G, Jongdee B, Toojinda T. 2004. Quantitative trait loci associated with drought tolerance at putative stage in rice. Plant Physiol. 135(1): 384-399.

Larque-Saavedra A, Rodriquez MT. 1989. Evidences for maternal inheritance of abscisic acid in relation to drought tolerance in *Zea mays* L. Phyton, Argentina. 49: 145-150.

Larque-Saavedra A, Wain RL. 1976. Studies on plant growth regulating substances. XLII: Abscisic acid as a genetic character related drought tolerance. Ann Appl Biol. 83: 291-297.

Lebreton C, Lazic Jancic V, Steed A, Pekic S, Quarrie SA. 1995. Identification of QTL for drought responses in maize and their use in testing caused relationship between traits. J Exp Bot. 46: 853-865.

Lee SB, Kwon HB, Kwon SJ, Park SC, Jeong MJ, Han SE, Byum MO, Daniell H. 2003. Accumulation of trehalose within transgenic chloroplast confer drought tolerance. Mol Breed. 11(1): 1-13.

Lee WS, Asahira T. 1983. Studies on drought and salt resistance in vegetable crops. I. Varietal difference of drought resistance in peppers. J Korean Hortic Sci. 24: 107.

Leont'ev SI, Petukhovaskii SL. 1987. Use of root system traits in breeding spring wheat in the southern for rest steppe of Western Siberia. In: Genetica, Seletsiya I Semenovostva Zernovykh. V. Zapadnoi Sibiri OMSK USSR. pp. 4-10.

Leung J, Giraudat J. 1998. Abscisic acid signal transduction. Annu Rev Plant Physiol Plant Mol Priol. 49: 199-222.

Levitt J. 1972. Responses of Plants to Environmental Sress. Academic Press, New York.

Levitt J. 1980. Response of Pants to Evironmental Sresses Vol. II Water, Radiation Salt and other Stresses, 2nd edition, Academic Press, New York.

Lewin LG, Sparrow DH. 1975. The Genetic and Physiology of Resistance to stress. Proceedings of third International Barley Genetics Symposium, Garching. pp. 486-501.

Lewis GL, Turling N. 1994. Growth, development, and yield of three oilseed *Brassica* species in a water-limited environment. Aust J Exp Agri. 34: 93-103.

Li Z-Y, Chen S-Y. 2000. Differntial accumulation of the S-adenosylmethionine decarboxylase transcript in rice seedlings in response to salt and drought stresses. Theor Appl Genet. 100: 782-788.

Lilley JM, Ludlow MM. 1996. Expression of osmotic adjustment an desiccation tolerance in diverse rice lines. Field Crops Res. 48: 185-197.

Liu CL, Chen HP, Liu EE, Peng XX, Lu SY, Guo ZF. 2003. Multiple tolerance of rice to abiotic stresses and its relationship with ABA accumulation. Acta Agronomica Sinica. 29(5): 725-729.

Liu L, Lafitte R, Guan D. 2004. Wild *Oryza* species as potential source of drought-adaptive traits. Euphytica 138(2): 149-161.

Loresto GC, Chang TT, Tagumpay O. 1976. Field evaluation and breeding for drought resistance. Philippines J Crop Sci. 1: 36-39.

Lu DB, Sears RG, Paulsen GM. 1989. Increasing stress resistance by *in vitro* selection for abscisic acid insensitivity in wheat. Crop Sci. 29: 939-943.

Lu JY, Mao YM, Shen LY, Peng SQ, Li XL. 2003. Effects of VA mycorrhizal fungi inoculated on drought tolerance of wild jujube (*Zuzyphus spinosus* I Hu) seedlings. Acta Horticulturae Sinica. 30(1): 29-33.

Ludlow MM, Muchow RC. 1990. A critical evaluation of traits for improving crop yields in water limited environments. Adv Agron. 43: 107-153.

MacKill DJ, Nguyen HT, Zhang J. 1999. Use of molecular markers in plant improvement programs for rainfed low land rice. Field Crop Research. 64: 177-185.

Maiti RR, Saucedo-Rodriguez JM. 1986. Studies of some biochemical characters of 5 glossy and 5 nonglossy sorghum related to drought resistance. Sorghum Newsl. 29: 91-92.

Makhlouf A, Marbrouk Y, Et-Saied M, Mahdy M. 2002. *In vitro* selection for drought tolerance in sorghum [*Soghum bicolor* (L.) Moench: Callus] and regeneration of selected genotypes. Alexandria J Agril Res. 47(1): 77-88.

Martin I, Tenoria JL, Ayerbe L. 1994. Yield, growth and water use of conventional and semileafless pea in semiarid environments. Crop Sci. 34: 1576-1583.

Mathew C, Saxena KB. 2005. Prospects for pigeonpea cultivation in drought-prone areas of South Africa. Proc of the 1st Intern Edible Legume Conf in conjuction with the IVth World Cowpea Cong, Durban, South Africa, 17-21 April 2005. pp. 1-11.

Mathew RB, Azam-Ali, Peacock JM. 1990 Repsonse of four sorghum lines to mid-season drought II. Leaf characteristics. Field Crops Res. 5: 297-308.

May LH, Milthoorpe FL. 1962. Drought resistance of crop plants. Field Crop Abstr. 15: 171-179.

McCully ME. 1999. Roots in soil; unearthing the complexities of roots and their rhizospheres. Ann Rev Plant Physiol Plant Mol Biol. 50: 695-718.

McKersei BD, Chen Y, Debens M, Bowley SR, Bowler C, Inze D, D'Halluin K, Botterman J. 1993. Superoxide dismutase enhances tolerance for freezing stress in transgenic alfalfa (*Medicago sativa* L.). Plant Physiol. 103: 1155-1163.

McKersie BD, Bowely SR, Harjanta SR, Leprince O 1996. Water deficit tolerance and field performance of transgenic alfalfa over expressing superoxide dismutase. Plant Physiol. 111-1177-1181.

McKersie BD, Bowley SR, Jones KS. 1999. Winter survival of transgenic alfalfa overexpression superoxide dismutase. Plant Physiol. 199:839-849.

McWilliam JR. 1989. The dimensions of drought. In: Drought Resistance in Cereals, Baker EWG (ed.) Walling Ford Oxon, CAB International. pp. 1-12.

Medrano H, Chaves MM, Porqueddu C, Caredda S. 1998. Improving forage crops for semi-arid areas. Out Agric. 27: 89-94.

Meena-Kumari, Dass S, Vimala Y, Arora P. 2004. Physiological parameters governing drought tolerance in maize. Indian J Plant Physiol. 9(2): 203-204.

Meguro NE, Megalhaes AC. 1983. Water stress affecting nitrate reduction and leaf diffusive resistance in *Coffea arabica* L. cultivars. J Hort Sci. 58: 147-152.

Meinzer FC, Grantz DA, Goldstein G, Saliendra NZ. 1990. Leaf water relations and maintenance of gas exchange coffee cultivars grown in dry soil. Plant Physiol. 94: 1781-1787.

Messmer R, Fracheboud Y, Banziger M, Stamp P, Ribaut JM. 2004. Genetic analysis of drought tolerance in tropical maize. Genetic variation in plant breeding- Proc of the 17th EUCARPIA General Congress, Tulln, Austria, 8-11 Sep 2004. pp. 57-61.

Minarsih H, Santaso D, Fitranty N. 2001. Identification of *P5CS* gene on sugarcane by PCR heterologous primer. Menara Perkebunan. 69(1): 1-9.

Mitra J. 2001. Genetic and genetic improvement of drought resistance in crop plants. Curr Sci. 80: 758-763.

Molina-Galan JD. 1980. Mass selection for drought resistance in maize. Agrociencia No. 42: 69-76.

Morgan JM, Hare RA, Fleteher RJ. 1986. Genetic variation in osmoregulation in bread and durum wheat and its relationship to gain yield on a range of field environments. Aust J Agri Res. 37 (5): 449-457.

Morgan JM, Rodriguez Maribona B, Knights EJ. 1991. Adaptation to water deficit in chickpea breeding lines by osmoregulation: relationship to grain yield in the field. Field Crop Res. 27: 61-70.

Morgan JM. 1983. Osmoregulation as selection criterion for drought tolerance in wheat. Anst J Agric Res. 34: 607-614.

Naidu BP. 1995. Australian Patent No. 27071/95 (CSIRO Tropical Agriculture: Brisbane).

Narayan D, Misra RD. 1989. Free proline accumulation and water stress resistance in bread wheat (*Triticum aestivum*) and durum wheat (*T. durum*). Indian J Agri Sci. 59: 176-178.

Nguyen HT, Chandra Babu R, Blum A. 1997. Breeding for drought resistance in rice: physiology and molecular genetic considerations. Crop Sci. 37: 1426-1434.

Nguyen TTT, Klueva N, Chamareck V, Aarti A, Magpantay G, Millena ACM, Pathan MS, Nguyen HT. 2004. Saturation mapping of QTL regions and identification putative candidate genes for drought tolerance in rice. Mol Genet Genomics. 272(1): 35-46.

O' Toole JC, Cruz TT, Seiber JN. 1979. Epicuticular wax and cuticular resistance in rice. Physiol Plant. 47: 239-244.

O' Toole JC, Moya TB. 1978. Genotypic variation in maintenance of leaf water potential in rice. Crop Sci. 18: 873-876.

O'Toole. JC, Bland WL. 1987. Genotypic variation in crop plant root systems. Adv Agron. 41: 91-145.

Ouk M, Fukai S, Fischer KS, Basnayake J, Cooper M, Nesbitt H. 2004. Drought response index for identifying drought resistant genotypes for rainfed lowland rice in Cambodia. In: Water in Agriculture, Proc of a CARD-International Conference Research on Water in Agricultural Production in Asia for the 21st Century, Phnom Penh, Cambodia, 25-28 Nov 2003. pp. 203-214.

Park JM, Park CJ, Lee SB, Ham BK, Shin R, Paek KH. 2001. Overexpression of tobacco *Tsi1* gene encoding an EREBP/AP2-type transcription factor enhance resistance against pathogen attach and osmotic stress in tobacco. Plant Cell. 13: 1035-1046.

Paroda RS. 1986. Breeding approaches for drought resistance in crop plants. In: Approaches for Incorporating Drought and Salinity Resistance in Crop Plants, Chopra VL, Paroda RS (eds.). Oxford and IBH Publishing Co. Pvt. Ltd., New Delhi. pp. 87-107.

Passioura JB. 1994. The yield of crops in relation to drought. In: Physiology and Determination of Crop Yield. Boote *et al.* (eds.) ASA, CSSA and SSSA, Madison, WI. pp. 343-359.

Pellegrineschi A, Reynolds M, Pacheco M, Bitro RM, Almeraya R, Yamaguchi-Shinazaki K, Hoisington D. 2004. Stress-induced expression in wheat of the *Arabidopsis thaliana* DREB1A gene delays water stress symptoms under green house condition. Genome. 47(3): 493-500.

Pellegrineschi A, Ribaut JM, Tretowan R, Yamaguchi-Shinozaki K, Hoisington D, Vail IK. 2003. Preliminary characterization of DRFB genes in transgenic wheat. In: Plant Biotechnolgy 2002 and beyond. Proc of the 10th IAPTC and B Congress, Orlando, Florida, USA, 23-28 June 2002. pp. 1836-187.

Pilon-Smits EAH, Ebskamp MJM, Jeuken MJW, van der Meer IM, Weisbeck PJ, Visser RGF, Smeekens SCM. 1996. Microbial production in transgenic potato plants and tubers. Indust Crops Prod. 5: 35-56.

Pilon-Smits EAH, Ebskamp MJM, Paul MJ, Jeuken MJW, Weibeck PJ, Smeekens SCM. 1995. Improved performance of transgenic fructan-accumulating tobacco under drought stress. Plant Physiol. 107: 125-130.

Pinheiro HA. 2004. Physiological and morphological adaptations as associated with drought tolerance in robusta coffee (*Coffea canephora* Pierre var. *kouillou*). Vicosa, Universidade Federal de Vicosa, Ph. D. Thesis.

Pinter PJ Jr. Zipoli G, Reginato RJ, Jackson RD, Idso SB, Hohman JP. 1990. Canopy temperature as an indicator of different water use and yield performance wheat cultivars. Agr Water Management. 18: 35-48.

Porat R, Pasentsis K, Rozentzvieg D, Gerasopoulos D, Falara V, Samach A, Lurie S, Kanellis AK. 2004. Isolation of a dehydrin cDNA from orange and grapefruit citrus fruit that is specifically induced by the combination of heat followed by chilling temperatures. Physiologia Plantarum. 102(2): 256-264.

Porcel R, Ruiz-Lozano JM. 2004. Arbuscular mycorrhizal influence of leaf water potentioal, solute accumulation, and oxidative stress in soybean plants subjected to drought stress. J Expt Bot. 55(403): 1743-1750.

Premachandra GS, Hahn DT, Rhodes D, Joly RJ. 1995. Leaf water relations and solute accumulation in two grain sorghum lines exhibiting contrasting drought tolerance. J Exp Bot. 46: 1833-1841.

Quarrie SA, Lazic-Jancic V, Kovacevic D, Steed A, Pekic S. 1999. Bulk segregant analysis with molecular markers and its use for improving drought resistance in maize. J Exp Bot. 50: 1244-1306.

Quilambo OA. 2004. Proline content, water retention capacity and cell membrane integrity as parameters for drought tolerant in two peanut cultivars. South African J Bot. 70(2): 227-234.

Quisenberry JE, Roask B. 1976. Influence of indeterminate growth habit on yield and irrigation water-use efficiency in upland cotton. Crop. Sci. 16: 762-765.

Quisenberry JE, Wendt CW, Berlin JD, Mc Michael BL. 1985. Potential for using leaf turgidity to select drought tolerance in cotton. Crop Sci. 25: 294-299.

Rahman SML, Mackay WA, Nawata E, Sakuratani T, Uddin ASMM, Quebedeaux B. 2004. Super oxidedismutase and stress tolerance of four tomato cultivars. HortScience. 39(5): 983-986.

Ram HH, Singh HG. 1994. Rice. In: Crop Breeding and Genetics. Kalyani Publishers, New Delhi. pp. 58-103.

Ramos RLS. Carvalho A. 1997. Shoot and root evaluations on seedlings from Coffee genotypes. Bragantia. 56: 59-68.

Ramtirez VP, Kelly JD. 1998. Traits related to drought resistance in common bean. Euphytica. 99: 127-136.

Rao PS, Singh RD. 2004. Studies on combining ability for drought tolerance in maize (*Zea mays* L.) using polyethylene glycol (PEG) as an osmoticum. Annals Agri Bio Res. 9(2): 135-139.

Ravi V, Saravanan R. 1999. Proline metabolism and its regulation to drought tolerance in sweet potato. J Root Crops. 25(2): 135-142.

Ray JD, Yu L, Mc Couch SR, Champoux MC, Wang G, Nguyen HT. 1996. Mapping quantitative trait loci associated with root penetration ability in rice (*Oryza sativa* L.). Theor Appl Genet. 92: 627-636.

Rekika D, Havaux M, Araus JL, Monneveux P, Jaradat AA. 1998. Variation for physiological traits related to abiotic stress tolerance in *Aegeilops* species. Triticeae III. Proc Third Interl Triticeae Symp, Alleppo, Syria, 4-8 May, 1997. pp. 293-304.

Richard RA, Passioura JB. 1989. A breeding programme to reduce to diameter of major xylem vessel in the seminal roots of wheat and its effect on grain yield in rainfed environments. Aust J Agri Res. 40: 943-950.

Ristic Z, Yang GP, Martin B, Pullerton S. 1998. Evidence of association between specific heat shock protein (s) and the drought and heat tolerance phenotype in maize. J Pl Physiol. 153: 497-505.

Roark B, Quisenberry JE. 1977. Environmental and genetic components of stomatal behaviour in two genotypes of upland cotton. Plant Physiol. 59: 354-356.

Robertson BM, Hall AE, Foster KW. 1985. A field technique for screening genotypic differences in root growth. Crop Sci. 25: 1084-1090.

Rohila JS, Jain RK, Wu R. 2002. Genetic improvement of Basmati rice for salt and drought tolerance by regulated expression of a barley *Hva1* cDNA. Plant Sci. 163(3): 525-532.

Romero C, Belles JM, Vaiya JL, Serrano R, Culianez Macia FA. 1997. Expression of the yeast trechalose-6-phosphate synthetase gene in transgenic tobacco plants; pleiotropic phenotypes include drought tolerance. Planta. 201: 293-297.

Rose IA, McWhiter KS, Spurway RA. 1992. Identification of drought tolerance in early maturing indeterminate soybeans (*Glycine max* L. Merr.). Aus J Agric Res. 43: 645-357.

Roy B, Mandal AB, Basu AK. 2005. Genetic studies for rice improvement in humid tropics of Bay Islands. Tropical Agri (Trinidad. 82(2): 85-96.

Roy M, Wu R. 2001. Arginine decarboxylase transgene expressing and analysis of environmental stress tolerance in transgenic rice. Plant Sci. 160(5): 869-875.

Roy NN, Murty OR. 1970. A selection in wheat for stress environments. Euphytica. 19: 509-521.

Sadaquat HA, Tahir MHN, Hussain MT. 2003. Physiogenetic aspects of drought tolerance in canola (*Brassica napa*). Intr J Agri Biol. 5(4): 611-614.

Saijo Y, Hata S, Kyozuka J, Shimamoto K, Izui K. 2000. Over expression of a single Ca^{2+} dependent protein kinase confers both cold and salt/drought tolerance on rice plants. Plant J. 23: 319-327.

Saini HS, Westgate ME. 2000. Reproductive development in grain crops during drought. Adv Agro. 68: 59-96.

Saxena HK, Yadav RS, Mathur RK. 1996. Effect of moisture stress on metabolic activity and grain yield in wheat genotypes. Indian J Plant Physiol. 1: 303-306.

Schellenbaum L, Sprenger N, Schuepp. H, Wiemken A, Boller T. 1999. Effects of drought, transgenic expression of a fructan synthesizing enzyme and of mycorrhizal symbiosis on growth and soluble carbohydrate pools in tobacco plants. New Phytol. 142(1): 67-77.

Schussler JR, Westgate ME. 1995. Assimilate flux determines kernel set at water potential in maize. Crop Sci. 35: 1074-1080.

Sengupta A, Heinen JL, Haladay AS, Barke JJ, Allen RD, 1993. Increased resistance to oxidative stress in transgenic plants that overexpress chloroplastic Cu/Zn superoxide dismutase. Proc Natl Acad Sci, USA. 90: 1629-1633.

Serrano R, Culianz-Macia FA, Moreno V. 1999. Genetic engineering of salt and drought tolerance with yeast regulatory genes. Sci Horti. 78: 261-269.

Sharp RE, Boyer JS, Nguyen HT, Hsio TC. 1996. Genetically engineered plants resistant to soil drying and salt stress: How to interpret osmotic relations? Plant Physiol. 110: 1051-1052.

Shen Y, Jia WL, Zhan YQ, Hu YL, Wu LZ. 2004. Improvement of drought tolerance in transgenic tobacco plants by a dehydrin-like gene transfer. Agril Sci China. 3(8): 575-583.

Shen YG, Du BX, Zhang WK, Zhang JS, Chen SY. 2002. AhCMO, regulated by stress in transgenic tobacco. Theor Appl Genet. 105: 6-7.

Shevdeva E, Chmara W, Bohnett HJ, Jensen RG. 1997. Increased salt and drought tolerance by D-ononitol production in transgenic *Nicotiana tabacum* L. Plant Physiol. 115: 1211-1219.

Shirinkina EV. 1990. Leaf epidermis in sorghum in relation to drought resistance. Nouchno-Tekhnichekil Byulletes Vseyeznogo Ordena Lemia Ordema Druzhby Narodov-Issedovated, Skojo Instiuta Restenievodstwa Imen N.I. Vavilova. 197: 43-44.

Shivaputra SS, Patil CP, Swamy GSK, Patil PB. 2004. Effect of vescular-arbuscular mycorrhiza fungi and vermicompost on drought tolerance in papaya. Mycorrhiza News. 16(3): 12-13.

Shou HX, Bordallo P, Wang K. 2004. Expression of *Nicotiana* protein kinase (*NPK1*) enhance drought tolerance in transgenic maize. J Expt Bot. 55(399): 1013-1019.

Silim SN, Saxena MC, Erskine W. 1993. Adaptation of lentil to the Mediterranean environment. I. Factors affecting yield under drought conditions. Expl Agric. 29: 9-19.

Singh BD. 2003. Breeding for resistance to abiotic stresses. I. Drought resistance. In: Plant Breeding, Kalyani Publishers, New Delhi. pp. 381-409.

Singh BB, Kodomi YM, Terao T. 1999. A simple screening method for drought tolerance in cowpea. Indian J Genet. 59: 211-220.

Singh BN, MacKill DJ. 1991. Genetics of leaf rolling under drought stress. In: Rice Genetics. International Rice Research Institute, Los Banos, Philippines. pp. 159-160.

Singh DP, Choudhury BD. Singh P, Sharma HC, Karwasra SPS. 1990. Drought Tolerance in oil Seed *Brassica* and Chickpea. HAU, Hissar. pp. 10-55.

Singh DP. 1991. Evaluation of traits for improvement of drought tolerance in oilseeds and cereals. In: Proc Golden Jubilee Symp on Genetic Research and Education: Current Trends and Nest Fifty Years, Feb., 1991, New Delhi.

Singh KB, Bejiga G, Saxena MC, Singh M. 1995. Transferability of chickpea selection indices from normal to drought prone growing condition in Mediterranean environment. J Agron Crop Sci. 175: 57-63.

Singh KB. 1993 Problems and prospects of stress resistance breeding in chickpea. In: Breeding for Stress Tolerance in Cool-Season Food Legumes, Singh KB, Saxena MC (eds.), John Wiley, U.K. pp. 17-35.

Sivamani E, Bahieldin A, Waith JM, Al-Niemi T, Dyer WE, Ho THD, Qu RD. 2000. Improved biomass productivity and water use efficiency under water deficit conditions in transgenic wheat conductivity expressing the barley HVA1 gene. Plant Sci. 155: 1-9.

Sobrado MA. 1987. Leaf rolling: a visual indication of water deficit in corn (*Zea mays* L.). Maydica. 32: 9-18.

Sponchiado BN, White JW, Castillo JA, Jones PG. 1989. Root growth of four common cultivars in relation to drought tolerance in environments with contrasting soil types. Expl Agric. 25: 249-257.

Squire GR. 1990. The physiology of Tropical Crop Production. CAB International. U.K.

Srivastava JP. Chaturvedi SN. 1989. Influence of water deficit on transpirational and water relation parameters in wheat. Ann Arid Zone. 28: 257-266.

Stewart CR, Hanson AD. 1980. Proline accumulation as a metabolic response to water stress. In: Adaptation of Plants to Water and High Temperature Stress, Turner NC, Kramer PJ (eds.). Wiley, New York. pp. 173-189.

Suardi D. 2002. Root of rice in relation to drought tolerance and yield. J Penelitian dan Pengembangan Pertanian. 21(3): 100-108.

Subbarao GV, Johansen C, Slinkard AE, Nageswara Rao RC, Saxena NP, Chauhan YS. 1995. Strategies for improving drought resistance in grain legumes. Critical Rev Plant Sci. 14: 469-523.

Sullivan CV. 1972. Mechanisms of heat and drought resistance in grain sorghum and methods of measurements. In: Sorghum in the Seventies, Rao NGP, House LR (eds.). Oxford and IBH Publishing Co., New Delhi, India. pp. 247-263.

Suprunova T, Krugman T, Fahima T, Chen G, Shams I, Korol A, Nevo E. 2004. Differential expression of dehydrin genes in wild barley, *Hordium spontaneum*, associated with resistance to water deficit. Plant Cell Envin. 27(10): 1297-1308.

Tan BH, Halloran GM. 1982. Variation and correlation of proline accumulation in spring wheat cultivars. Crop Sci. 22: 459-463.

Thomas H. 1997. Drought resistance in plants. In: Mechanisms of environmental stress resistance in plants. Hardwood Academic Publishers. The Netherlands. pp. 1-42.

Tomar JB, Prasad SC. 1996. Indian J Agric Sci. 66: 459-465.

Tomar SMS, Kumar GT. 2004. Seedling survivability as a selection criterion for drought tolerance in wheat. Plant Breed. 123(4): 392-394.

Tripathy JN, Zhang J, Robin S, Nguyen TT, Nguyen HT. 2000. QTLs for cell-membrane stability mapped in rice (*Oryza sativa* L.) under drought stress. Theor Appl Genet. 100: 1197-1202.

Turner NC. 1979. Drought resistance and adaptation to water deficits in crop plants In: Stress Physiology in Crop Plants, Mussell. HW, Staples RC (eds.). Wiley Inter Science, New York.

Turner NC. 1997. Further progress in crop water relations. Adv Agron. 58: 293-338.

Van Rensburg L, Kruger GHJ. 1994. Applicability of abscisic acid and (or) proline accumulation as selection criteria for drought tolerance in *Nicotiana tabacum.* Can J Bot. 72: 1535-1540.

Vedrov NG. 1971a. Questions concerning the breeding of wheat for resistance to spring and summer drought. *Tr. Krasnoysrsk* S.-kh.in-ta. 22: 163-175.

Vedrov NG. 1971b. The use of the characters of the root system in breeding spring wheat. Sib. Vestn. S-kh., rauki, No. 6: 27-32.

Vedrov NG. 1982. Nature of root system development in spring wheat in the drought zone of eastern Siberia. *Selskhozyaistvennaya Biologiya.* 17: 196-198.

Verslues PE, Sharp RE. 1999. Proline accumulation in maize (*Zea mays* L.) primary roots water potentials. II. Matabolic source of increase proline deposition in the elongation zone. Plant Physiol. 119: 1349-1360.

Waines G, Ting IP, Lazzaro CA. 1979. Chromosomal location of genes controlling drought tolerance in rye, triticale and wheat/rye addition lines Genetics. 91 (4, II): 134.

Wang WX, Vinocur B, Altmann A, Wang WX. 2003b. Plant responses to drought, salinity and extreme temperatures: towards genetic engineering for stress tolerance. Planta. 2187(1): 1-14.

Wang YP, Liu QC, Li AX, Zhai H, Zhang SS, Liu BL. 2003a. *In vitro* selection and identification of drought-tolerant mutants in sweet potato. Agril Sci China. 2(12): 1314-1320.

Wang ZL, Huang BR, Bonos SA, Meyer WA. 2004. Abscisic acid accumulation in relation to drought tolerance in Kentucky bluegrass. HortScience. 39(5): 1133-1137.

Watanbe I. 1998. Drought tolerance of cowpea (*Vigna anguiculata* (L.) Walp.) I. Methods for evaluation of drought tolerance. JIRCASJ, 6: 21-28.

Weerathaworm P, Soldati A, Stamp P. 1992. Anatomy of seeding roots of tropical maize (*Zea mays* L.) cultivars at low water supply. J Expt Bot. 43: 1015-1021.

White JW, Izquierdo J. 1991. Physiology of yield potential and stress tolerance. In: Common Beans Research for Crop Improvement, van schoonhoven A, Voysest O (eds.). CAB International & CIAT. pp. 287-382.

Williams TV, Snell RS, Cress CE. 1969. Inheritance of drought tolerance in sweet corn. Crop Sci. 9: 19-23.

Williams TV. 1966. The inheritance in sweet corn. Diss Abstr Int B. 27. Order No. 66-11, 222: pp. 1688-1689.

Windawsky, DA, O'Toole JC. 1990. Prioritizing rice biotechnology research agenda for Eastern India. New York (USA); The Rockfeller Foundation.

Wu MG, Lin JR, Zhang GH. 2004b. Genetic analysis of leaf water content in paddy upland hybrid rice. Chinese J Rice Sci. 18(6): 570-5723.

Wu QS, Xia RX, Hu LM. 2004a. Effects of arbuscular mycorrhizal on growth and drought tolerance of trifoliate (*Poncirus trifoliata*) under non-sterilized soil conditions. J Fruit Sci. 21(4): 315-318.

Xu D., Duan X, Wang B, Hong B, Ho TD, Wu R. 1996. Expression of a embryogenesis abundant protein gene, HVA1, from barley confers tolerance to water deficit and salt stress in transgenic rice. Plant Physiol. 110: 249-257.

Yadava RBR, Ehatt RK, Katiyar DS. 1991. Phjysiological evaluation of bodder sorghum genotype for drought tolerance. Sorghum News Letter. 32: 59.

Yoshida S, Forno DA, Cock JH, Gomez KA. 1976. Routin procedure for growing rice plants in culture solution. In: Laboratory Manual for Physiological Studies of Rice. Los Banos, Philippines, IRRI. pp. 61-66.

Young TE, Meeley RB, Gallie DR. 2004. ACC synthase expression regulates leaf performance and drought tolerance in maize. Plant J. 40(5): 813-825.

Younis SEA, Omara MK, Saleh FM, Saba MF. 1988. Heterosis in dry matter and grain yield / plant under drought in varietal crosses among maize population. Aust J Agric Sci. 19: 239-256.

Zavale-Garcia F, Bramel Cox PJ, Eastin JD, Wilt MD, Andrews DJ. 1992. Increasing the efficiency of crop selection for unpredictable environments. Crop Sci. 32: 51-57.

Zhang J-Y, Broeckling C, Blancaflor E, Sledge M, Summer L. Wang Z-Y. 2005. Increased cuticular was accumulation and enhanced drought tolerance in transgenic alfalfa by over expression of a transcription factor gene. Plant & Annual Genomes XIII conference, January 15-19, 2005 Town & Country Convention Center, San Diego, CA, USA. p. WO99: Forage & Turf Plants.

Zheng HG, Babu RC, Pathan MS, Ali L, Huand N, Courtois B, Nguyen HT. 2000. Quantitative trait loci for root penetration ability and root thickness in rice: comparison of genetic backgrounds. Genome. 43: 53-61.

Zhu B, Su J, Chang MC, Verma DPS, Fan YL, Wu R. 1998. Over expression of a pyrroline-5-carboxylate synthetase gene and analysis of tolerance to water and salt stress in transgenic rice. Plant Sci. 139: 41-48.

Zinselmeier C, Habben JE, Westgate ME, Boyer JS. 2000. Carbohydrate metabolism in setting and aborting maize ovaries. In: Physiology and Modeling Kernel Set in Maize, Westgate ME, Boote KJ (eds.). CSSA special publication No. 29. Madison, WI: CSSA, pp. 1-13.

Unit-III
TEMPERATURE STRESSES

5

Heat Tolerance

5.1. INTRODUCTION

Temperature is the one of the most important environmental element influencing the plant growth and development limiting the productivity and adaptation of crops, especially when it coincides with critical ranges (Chen *et al.*, 1982; Paulsen, 1994). Nearly all commercial crops in the arid and semi-arid tropics often suffer from heat stress. Every plant has an optimal temperature for its normal growth, deviation from this may cause stress to plants. The adverse effects on plant of temperatures higher than the optimal considered as *heat stress*. Crops are exposed to periods of heat stress during their life-cycle. The optimum temperature for the most species ranges from 25 to 35 ^{0}C. Above this value, a decline in the photosynthesis rate is observed (Berry and Bjorkman, 1980; Pimentel, 1998). Even brief periods of heat-shock between temperature ranges of 45-50 ^{0}C induce marked change in plant growth processes. Under natural condition, a momentary water deficit at the daily warm hour is observed, and these can promote stomata closure. Consequently, the temperature of the leaves exposed directly to the sun can be equal or higher. This elevation of the leaf-temperature can result in the biochemical and biophysical disturbance in the mesophyl, which can be reversible or not (Berry and Bjorkman, 1980). High night temperatures during the growing season can have detrimental effects on reproductive development and yield of several crops (Hall, 1992).

5.2. INJURY MECHANISMS

High temperature would affect mainly in two different ways: (1) growth and development of plants, and (2) physiological process of plants. The nature and extent of heat injury may depend on the temperature, plant species and other meteorological parameters.

5.2.1. Effect of High Temperature on Growth and Development

High temperature is associated with death of cells, tissues, organs or whole plant. Heat stress is know to cause death of seedlings and flower abscission, pollen sterility and poor fruit set at reproductive stage and to cause yield reduction. Cell and tissue death usually occur in such organs that lack in transpirational cooling, and when their surface temperature ranges 48-50 ^{0}C. The extent of detrimental effects of heat depends on both the temperature and longibility of exposure at that temperature. Seedlings are more susceptible to heat stress.

Sensitivity to heat stress differs depending upon the stage of the plants. In general, seedlings are more prone to high temperature as they are contact with the soil surface. The temperature which kills 50% of the plants/seedlings could be heat killing temperature. When wheat seedlings are exposed to temperature in the range of 30-55 ^{0}C for 5-20 minutes, extent of growth, dry matter accumulation, chlorophyll synthesis, total RNA content and ^{32}P uptake reduce remarkably. Susceptibility to heat injury decreased toward maturity with the exception of a few species.

In many crops, pollination and fertilization are highly disturbed by heat during anthesis as pollen sterility caused by high temperatures (MacKill *et al.*, 1982). Different crop species and varieties respond differently to rise in temperature. Chowdhury and Wardlaw (1978) demonstrated how increase in day and night temperature reduces the duration of grain growth and thus the grin yield. Flowering stage, particularly anthesis in rice is the most sensitive stage to high temperature (IRRI, 1977). The maximum temperature, its duration and the temperature-rise pattern influence the incidence of unfertilized rice grain.

Damage during reproductive development occurs mainly during two distinct stages, early flowering and pod set stages (Warrag and Hall, 1984; Mutters *et al.*, 1989; Ahmed *et al.*, 1992; Ahamed and Hall, 1993). Flower initiation, floral bud development is suppressed by a combination of high night temperature and long photoperiods (Dew el-madina and Hall, 1986;

Patel and Hall, 1990). Hall and Kuo (1993) reported reduction in yield of cowpea under heat stress caused by damage to reproductive development. Genetic studies have demonstrated that heat tolerance during reproductive development is conferred by a set of major recessive genes, while heat tolerance during pod set also involved a major dominant gene. Two or more weeks of consecutive or interrupted high night temperature during the first 4 weeks after germination cause complete suppression of floral buds and prevents flowering (Ahmed and Hall, 1993). Many cowpea cultivars are susceptible to high temperatures during reproductive development (Patel and Hall, 1990) and can exhibit 13.5% reduction in first-flush grain yield per degree centigrade increase in daily minimum night temperature above 16.5 ^{0}C (Ismail and Hall, 1998). Heat-induce reduction in grain yield of cowpea under field conditions is mainly due to reduction in pod set and harvest index.

High night temperature during later stage of flower development can impair pod set by causing another indehiscence and low pollen viability (Warrag and Hall, 1984). Flowers are susceptible to heat for 7-9 days before anthesis, and the injury involves premature degeneration of the tapetum and lack of endothecial development (Ahmed *et al.*, 1992). Damage in pod is caused mainly by high night temperature between midnight and dawn (Mutters and Hall, 1992). Heat injury during flower development was associated with decreased proline accumulation in pollen and greater accumulation of proline in the anther wall (Mutters *et al.*, 1989). The most critical growth stage in green beans was found to be 2-3 days before anthesis (Dickson and Petzoldt, 1989).

High temperature in wheat from floral initiation onwards adversely affects floret development and the continued high temperature after ear emergence results in complete failure of grain production (Owen, 1971). When high temperature and low humidity prevail at the time of flowering, spikelets become sterile. In addition to yield, high temperature during the grain filling period also affects quality (Blumenthal *et al.*, 1991). They observed significant correlations of heat stress with protein content (positive) and with grain yields (negative). High temperature during endosperm cell division reduces grain yield of maize (Commuri and Jones, 2001) by reducing the endospermic cells.

5.2.2. Physiological Effects of High Temperature

High temperature affects plants' physiological process one way or other. Heat injury results in denaturizing of proteins, enzymes, metabolic imbalances, respiratory depletion of substrates, translocation and reduction in chloroplast photochemical activity. The cellular heat stress perturbations are to a large extent a result of membrane disturbance (Blum, 1988).

5.2.2.1. Respiration

Respiration is positively correlated with temperature. Total respiration increases with temperature. During growth and development, plant may avoid high temperature injury by mechanisms such as dramatically increased transpiration. Thus, high temperature may lead to utilization of more ATP.

5.2.2.2. Photosynthesis

Photosynthesis is activated after exposure to hardening temperature but inhibited at temperatures causing damages (Kurets and Popov, 1988). Photosynthetic processes are extremely sensitive to the temperature higher than the optimum range. Photosystem II (PSII), viz., photosynthesis of water and reduction of CO_2, is far more readily inactivated by heat than is photosystem I (PSI). The main effect of high temperature on photosynthesis results from alteration in thylakoid physical-chemical properties (Gilmore and Govindjee, 1999). The various enzymes located outside the thylakoid membrane can be destabilized by heat, and thereby, result in inhibition of photosynthesis. Besides inducing an increase in the lipid matrix fluidity (Raison *et al.*, 1982) with consequent formation of unilayer structure, high temperature provokes disturbances in the photosynthetic apparatus organization, and has been mentioned as follows:

1. Destruction of the oxygen evaluation complex;
2. Dissociation of the light harvesting complex of PS II accompanied by variation in energy distribution between PS II and PS I; and
3. Inactivation of PS II reaction centre (P680) that disturbs the grana stacking (Gounaris *et al.*, 1983; Enami *et al.*, 1994; Yamane *et al.*, 1998).

All these events result in photochemical and carboxylative efficiency losses, and serious metabolic restriction in the Calvin cycle, such as inactivation of Ribulose-5-bisphosphate carboxylase/oxygenase (Rubisco) and variations in the metabolic pool, especially ATP and NADPH availability (Pastenes and Horton, 1996a, b).

The relationship and genotypic variability among photosynthesis, thylakoid activity and productivity during high temperature stress at vegetative and reproductive stages were investigated by Al-Khatib and

Paulsen (1990) in wheat. They observed that high temperature increased mean photosynthetic rates by 32 and 11% in seedling and mature plants, respectively. Decreased photosynthetic rates and diminished productivity were significantly correlated at both growth stages. Relative grain yield were strongly influenced by decreased duration of photosynthetic activity.

In beans, photosynthesis shows a negative response to rising temperature (Masaya and White, 1991). According to Jones (1971), the optimum temperature for photosynthesis in bean leaf is 25 ^{0}C, showing decay with its continuous increase. This is particularly interesting because studies have shown that optimum temperature for bean plant is above 25 ^{0}C may be due to genetic breeding. A study presented by Pastenes and Horton (1996a, b) showed that potential photosynthetic rate of the *Barbucho* recombinant inbred line reaching maximum value at 35 ^{0}C.

5.2.2.3. Protein denaturation

High temperature induces conformational changes in proteins. Thus, it may affect the specific function of protein causing their denaturation, which may enhance their susceptibility to proteolytic enzymes.

5.2.2.4. Membrane composition and stability

Cell membrane of plants composts of protein and phospholipids. High temperature may affect membrane stability and function. In hot environment, the lipid becomes increasingly liquid. Heat stress can affect the structural integrity of protein in cytoplasms and cause membrane protein denaturation and aggregation (Levitt, 1969).

5.3. TOLERANCE MECHANISMS

The availability of plants to withstand high temperature and perform better than the others under the same environmental condition is termed as heat tolerance. Based on response of the plant to high temperature, the various mechanisms of heat tolerance/resistance may be classified as (1) heat avoidance, and (2) heat tolerance.

5.3.1. High Temperature Avoidance Mechanisms

High temperature avoidance is the ability of the plants to scatter the radiation energy causing cooling effect at stress temperature. The heat avoidance should be measured at a temperature just below the heat killing temperature and may be measured in the growth chambers as the ratio of the air temperature to the leaf temperature. The primary mechanism of dissipation of high temperature is transpiration. The other structural mechanisms of heat avoidance are pubescence, glaucousness, insulation by bark etc.

5.3.1.1 Transpiration

At high temperature rate of transpiration increases, causing cooling down the micro-environment. Vigorous growth seems essential to supply adequate water to the leaves so that these can maintain lose due to transpiration and good turgor at high temperature. Transpirational cooling is the link between dehydration and heat avoidance so that under conditions of high solar radiation and water stress the effect of the heat avoidance become inseparable from those of dehydration avoidance. Heat tolerance varieties of Chinese cabbage (*Brassica campestris* spp. *Pekinesis* Rupr.) had greater water uptake than heat sensitive ones at the on set of head formation (Kuo *et al.*, 1988).

5.3.1.2. Leaf Pubescence

Pubescence is a characteristic of leaves, a protective adaptation of some plant originating in regions with a long dry season. It is a covering a soft down or short hairs on leaves or soft stems. This texture is produced by microscopic hairs called trichomes that help plants to stand up to brutally dry or hot conditions. The trichome hairs help to shade the leaf surface. Each tiny hair casts a microscopic shadow to protect the leaf's outer layer from the direct sunlight. These hairs also reduce the moisture loss through stomata. The trichome hairs help to baffle the air at the leaf surface so it slows down. Even a minor slowing can have a significant effect on reducing moisture loss.

5.3.1.3. Reduction in the amount of foliage

The wheat variety Biancone dell'sElba possessing smaller leaf blades, which tend to reduce in size towards the apex of the plant, resembling rye, has a high resistant to hot spells of short duration (Azzi, 1950). Reduction in the amount of foliage with the microcellular structure of the leaf blades contributes to the resistance to drought and hot spells. In addition, reduction in foliage towards the apex decreases the area of the transpiration, thus preventing damage by heat, without the kernels losing plant its productivity.

5.3.1.5. *Adaptation to warm temperature*

Heat hardening of leaf tissue can occur within hours or even minutes when critical high-temperature thresholds are suppressed (Alexandrove, 1977). *Heat hardening* may be defined as an improved ability of a genotype to withstand a period of high temperature as a consequence of an earlier exposure to a high temperature for a given period of time. Exposure to sub-lethal high temperature can increase the thermotolerance of the plant (Howarth and Ougham, 1993; Dat *et al.*, 1998; Lopez-Delgado *et al.*, 1998). Potential mechanisms of response include synthesis of HSPs and isoprene and antioxidant production to protect the photosynthetic apparatus and cellular metabolisms. Buchner and Neuner (2001, 2003) reported that heat tolerance increased under warmer micro-site conditions and in warmer years, while long term artificial heating by +3 °C lead to a significant increase in heat tolerance by +0.6 °C. They also reported that higher the mean minimum leaf temperature at a particular growing site, the higher the mean heat tolerance in leaves. Buchner and Neuner (2001) also noted that under the warmer conditions of 1998 in Northern Hemisphere, all species investigated had higher heat hardening to overcome high temperatures. To this observation it has been postulated that any future concomitant increase in atmospheric CO_2 concentration would likely also enhance earth's plants' abilities to survive higher temperature (Taub *et al.*, 2000).

5.3.2. High Temperature Tolerant Mechanisms

High temperature tolerance generally associated with cellular and sub-cellular components. Bio-molecules are also involves in heat tolerance of plants. The following components and bio-molecules may be associated with heat tolerance.

5.3.2.1. *Molecular chaperones interacted to protect against heat*

Plants are known to synthesize, under heat shock, a large diversity of *heat shock proteins* (HSPs) that function as protectors on the biochemical level (Howarth and Ougham, 1993). HSPs are low molecular weight proteins produced during stress. These proteins have been classified into a number of families based on there molecular mass, and most have chaperon in function (Jaenicke and Creighton, 1993). All organism produce HSPs from all the major families (HSP90s, HSP70s, and small HSPs), but plants are unique in the number of different small HSPs that they produce (Jakob and Buchner, 1994).

Most studies investigating heat stress in plant have focused on HSPs (Schoffle *et al.*, 1997; Gurley, 2000). Evidences are accumulating to suggest an important role for heat shock proteins/molecular chaperones in stress resistance in plant (Gustavsson *et al.*, 1999; Wehmeyer and Vierling, 2000; Hardahl *et al.*, 2001). It may helps to prevent protein degradation, disassembly of aggregation.

Blumenthal *et al.*, (1990) reported that the exposure of wheat coleoptiles to a transient high temperature stress resulted in synthesis of a HSP. Appearance of these proteins was associated with a reduction in normal protein synthesis and was correlated with the acquisition of thermotolerance. Analysis of HSP synthesis in the green seedling leaves showed that tolerant wheat cultivars Mustang exhibited a greater degree of HSP synthesis (Krishnan *et al.*, 1987). They suggested that HSPs are intimately involved in the genetic control of thermal tolerance. Ristic *et al.* (1998) found that 45 kDa HSP(s) was associated with the drought and heat tolerance phenotype in F_2 plants of a cross between heat tolerance and heat susceptible cultivars of maize. They observed that synthesis of 45 kDa HSP(s) in F_2 plants increased ability to recover from soil drying and heat stresses. And they suggested that 45 kDa HSP(s) play a role in recovery from drought and heat stresses or their gene(s) are in close proximity to gene(s), which encode tolerance to drought and heat stresses.

The HSPs vary during different developmental stages and various tissues. Baszczynski (1988) reported that the various tissues- leaves, stems, flower buds, flowers, petals, hypocotyls and siliquas of three *Brassica* species showed characteristically unique pattern of polypeptide synthesis and the general pattern was maintained following heat shock. One HSP, ubiquitin, may help to degrade and remove heat-damaged proteins. Baszezynski *et al.* (1982) demonstrated the induction of HSP in an intact plant. They considered that this phenomenon may lead to breeding for tolerance to suboptimal environmental conditions. HSP70 is known to occur in glyoxysomes, infact the glyoxysomal protein is encoded by the same gene as the chloroplast form of the protein (Wimmer *et al.*, 1997). The HSPs that are localized to the cytosol appear to respond to specific developmental signals associated with the aquisition of desiccation tolerance that occur during seed development (Wehmeyer and Vierling, 2000). A small HSP (*HSP18.1*) has been shown to interact with HSP70 to reactivate heat denatured *luciferase* (Lee and Vierling, 2000).

The a-crystalline-related low-molecular-weight heat shock proteins (LMW HSPs) ranging in size from approximate 17-30 kDa and share a conserved C-terminal domain common to all eukaryotic LMW HSPs and to the a-crystalline proteins of the vertebrate eye lens. LMW HSPs act *in vitro* as molecular chaperones to bind partially denatured proteins, preventing irreversible protein inactivation and aggregation. It is known that LMW HSPs play an important role in protection of the organelles from hyperthermia damage. Chaperone activity of organelle LMW HSPs contributes to the development of thermotolerance. Mejority of LMW HSPs stabilizes protein structure and prevents damage and resultant turnover. LMW HSPs also influence different proteins or recognize various transitional states of partly denatured proteins. In maize mitochondria have a more abundant composition of chaperone than wheat and rye, and enhanced capability for preventing damage to the enzymes of the electron transport chain.

HSP101 is a molecular chaperone that is required for the development of thermotolerance in plants and other organisms. Hong and Vierling (2001) reported that *Arabidopsis thaliana* HSP101 is also regulated during seed development in the absence of stress, whereas, its activity is not essential in absence of stress.

5.3.2.2. *Amylopectin content*

It is suggested that the tolerant genotypes increased the portion of carbon preferably in the form of amylopectin, which is capable of holding more water. Pande *et al.* (1998) reported low amylase and high amylopectin contents compared with the starch from seeds of susceptible wheat genotypes.

5.3.2.3. *Membrane stability*

Heat tolerance species of plant tend to have a higher percentage saturated fatty acids in their membrane. Saturated fatty acids exhibit less increase in membrane fluidity associated with heat than unsaturated fatty acids. Cell membrane stability is an index in heat tolerance that bears considerable relationship with plant performance under stress environments. Cellular membrane-thermostability, measured as the conductivity of electrolytes leaking from leaf disks at high temperature, has been suggested as a screening technique for heat tolerance in plants (Sullivan, 1972). The relationship between the degree of injury and temperature at which that the injury is induced as a sigmoidal response. Genotypic differences in heat tolerance are associated with differences in the relative position of the response curve with respective to the treatment temperature. This technique could be used for selecting heat-tolerant genotypes in crop plants and in identifying forms with different degrees of adaptive resistance to extreme environmental conditions.

With a view of assessing the membrane thermostability of wheat cultivars under high temperature, Saadhalla *et al.* (1990a) exposed seedling and flag leaf materials from the same cultivars-grown to anthesis to various durations at 34 ^{0}C, with results exposed relative injury. Further, Saadhalla *et al.* (1990b) reported that relative injury varied significantly among the wheat genotypes with a range of 31-78%.

Based on membrane stability and photosystem function, Srinivasn *et al.* (1996) ranked from heat tolerant to sensitivity in the order: groundnut > soybean > pigeon pea > chick pea. They also found that membrane injury was negatively associated with specific leaf weight in groundnut and soybean, whereas electrolyte leakage and fluorescence ratio were negatively correlated in all the legumes under study.

5.3.2.4. Osmoregulators content

Osmoregulators, viz. polyamines have a protective role in plant stress responses (Evans and Malmberg, 1989; Flores, 1991). The diamine puterscine, the triamine spermidine, and tetramine spermine are ubiquitous in plant tissue and have been implecated in an overhelming array of plant growth and developmental process (Bangi, 1989). Proline and glycine-betaine protect several enzymes from heat inactivation *in vitro*.

Being cationic in nature, proline can associate with anionic components of membrane, such as phospholipids thereby stabilizing the bilayer surface and retarding membrane deterioration (Robert *et al.*, 1986; Barsa *et al.*, 1997). It also has radical scavenging property (Drohlert *et al.*, 1986). Protection of membrane from peroxidation by polyamines could involve both their ability to interact with phospholipids and their antioxidant activity (Kramer and Wang, 1989). It has also been suggested that under stress conditions, polyamines may partially replace calcium in maintaining integrity by binding to phospholipids components of the membrane (Naik and Srivastav, 1978).

Exogenous effects of polyamines (putrecine, spermidine, and spermine) on the recovery growth and membrane integrity of seedling tissues were studied by Barsa *et al.* (1997). They observed that application of polyamines either as a co-treatment (50 ^{0}C, 2h) during the heat shock period itself, enhance the recovery growth of the roots and hypocotyls, but specially the former with the other effectiveness being putrescine, spermidine, and spermine. Pertinently, the important role of polyamines in root formation and growth has also been demonstrated in the other studies (Chaterjee *et al.*, 1983; Palavan-Unsal, 1987). Heat stressed-induce polyamine accumulation has been reported

(Das *et al.*, 1987) and linked to their thermotolerance ability of resistance cell (Shevyakova *et al.*, 1994). Putrescine and spermine treatments also decrease leakage from embryonic axes, root and hypocotyls tissue sections (Basra *et al.*, 1997), and suggesting protection of membrane integrity. Further, they observed a remarkable decrease in lipid peroxidation of root tissue.

Pretreatment with polyethyle glycol (PEG) reduced damage by high temperature as measured by electrolyte leakage, dry weight and water content (Kocsy *et al.*, 2004). A great increase in the glutathione (GSH) and thioredoxin h (Trx h) levels induced by heat stress in maize, var. Pen.

5.3.2.5. Plant growth regulators

Various pre- and post-harvest chemicals treatments aimed at protecting leaf tissue integrity during and after supra-optimal heating have shown promise for increasing transplant success. Calcium ions (Sanders *et al.*, 1999; Knight, 2000), salicylic acid (Dat *et al.*, 1998), abscisic acid (Annamalai and Yanaghiara, 1999; Gong *et al.*, 1998a, b) and ethylene (Foyer *et al.*, 1997) are involved in several stress responses. Concentration of the some the growth regulators increase during high temperature, which contribute towards tolerant ability of plants. Plant hormone appears to induce chimeric genes in sunflower (Rojas *et al.*, 1999). Larkindal and Knight (2002) also reported the involvement of calcium, ABA, ethylene and salicylic acid in the protection against heat induced oxidative damage in *Arbidosis thaliana*. Role of few growth regulators have been briefed below.

Abscisic acid: The plant hormone, abscisic acid (ABA) could involve in some pathways resulting in survival of heat stress in plants. It induces thermotolerance in maize (Gong *et al.*, 1998a), a fact also observed in bromegrass (Robertson *et al.*, 1994). ABA has shown to induce a little amount of HSP70 at ambient temperature in plants (Wu *et al.*, 1994). The effect of heating at 38 °C of whole cucumber (*Cucumis sativa* L.) seedling or local heating of their shoots or roots on ABA content and heat tolerance of leaves and roots were investigated by Talanova *et al.* (2003) They observed that the initial period of the high temperature treatment of whole plant seedlings, the ABA concentration in the leaves and root increased considerably. Local heating of shoot or root resulted in an increase in the ABA concentration not only in the heated organ, but also in the unheated seedling parts. Kocsy *et al.* (2004) reported replacement with ABA reduce the heat induced electrolyte leakage in maize.

Humic acid: Zhang *et al.* (2003) investigated the influences of selected plant growth regulators on *sod* tolerant to stress during storage by examining the relationship between pre-harvest photochemical efficiency and transplant rooting of *Poa pratensis* L. Foliar application of extract at 0.5 kg/ha plus humic acid at 105 kg/ha, propiconazole at 0.44 kg/ha alone, or a combination of seaweed extract and humic acid with 0.22 kg/ha propiconazole, enhance the photochemical efficiency of pre-harvest *sod*. All plant growth regulators treatment combinations reduced visual injury. Their finding suggested that foliar application of seaweed extract, humic acid, propiconazole alone or in combination, may improve self life and transplant rooting of Kentucky bluegrass.

Ethylene: High temperature during flowering reduces yield of *Phaseolus vulgaris*, when ethylene production from plant tissue increases as a consequence of heat stress (Saulter *et al.*, 1990). The inheritance of leaf ethylene, evaluation rate of high temperature stressed (35/30 ^{0}C day/night) progenies from cross of *Phaseolus vulgare*.

Salicylic acid: Possibly, salicylic acid (SA) acts independently during the establishment of thermotolerance and particularly is effective when given as pretreatment before subjecting to heat shock. The thermotolerance ability linked with increased membrane thermostability of cells. Clarke *et al.* (2004) have tested thermotolerance and expression of pathogenesis-related and heat shock protein in *Arbidopsis thaliana* genotypes with modified SA signaling. They found evidence for SA-dependent signaling in thermotolerance of heat shock without prior heat acclimation in all the stages from seed to 3-week old plants. SA promoted thermotolerance during the heat shock itself and subsequent recovery. It reduced electrolyte leakage, indicating that it induced membrane thermoprotection.

SA enhanced the preharvest canopy photochemical efficiency (Dat *et al.*, 1998; Ervin *et al.*, 2005). SA increased canopy photochemical efficiency by 2% for *Poa prantensis* and 14% for *Festuca arundiacea*. SA reduced visible injury and enhanced post-harvest root strength. SA also increased transplant root strength by 26% for *P. pratensis* and 9% for *F. arundinacea*. Finally, Ervin *et al.* (2005) suggested that pre-harvest foliar application of SA might improve self-life and transplant successes of supra optimal heated cool season *sod*. He *et al.* (2005) studied the influence of SA on the production of active oxygen species (AOS) superoxide anion and hydrogen peroxide, and activities of antioxidant enzymes, superoxide dismutase (SOD) and catalase (CAT). SA (0.25 mmol) was effective in enhancing heat tolerance in Kentucky bluegrass. The SA application suppressed the increase of O_2^- generating rate and enhanced SOD activity. The SA application also decreased H_2O_2 level and increased CAT activity significantly. Their findings suggested that SA application enhanced heat tolerance in Kentucky bluegrass and could be involved in the scavenging of AOS by increasing SOD and CAT activities under heat stress. He *et al.* (1997) also suggested that high peroxidase activity may be beneficial with regard to heat tolerance.

Hydragen peroxide: Pretreatment with H_2O_2 (at a rate of 0.1 mmol/l) heat tolerance of maize seedlings greatly improved (Li and Gong, 2003). The activities of glutathione reductase, SOD, ascorbate peroxidase, and CAT increased remarkably. Their results indicated that after pretreatment with H_2O_2, the maize seedlings maintained higher antioxidant enzyme activities.

5.3.2.6. Thermo-stability of photosystem II

Thermo-stability of photosystem II enhances high temperature tolerance. Thus, effect of heat stress on photosynthesis and chloroplast damage can be assayed as variable chlorophyll fluorescence at 685 nm (Smillie and Gibbons, 1981). This heat is reasonably suitable for developing a mass screening test for heat tolerance. The measures of variable chlorophyll fluorescence are an indication of the capacity for electron flow through photosystem II. With the progressive heat-injury of leaves, their variable chlorophyll fluorescence decreased. Crop genotypes adapted to hot climate showed reduced sensitivity of photosystem II to heat stress. Different plant parts may show different tolerant to heat stress in terms of photosynthesis.

Moffatt *et al.* (1990a) suggested that chlorophyll fluorescence may be a useful tool for screening wheat of similar maturity for high temperature tolerance during the reproductive growth. Further, Moffat *et al.* (1990b) suggested that recurrent selection may be an appropriate method of accumulating genes that favour high temperature tolerance based on chlorophyll fluorescence measurements.

Costa *et al.* (2002) found that increased temperature promoted an increased in initial fluorescence at 45 °C, possibly associated with dissociation of the light harvesting complex from the reduction centre of PS II, but a decrease was observed at 48 °C in all cultivars. In some situations, initial fluorescence can be used as an indicator for invariable damage in PS II (Pastenes and Horton, 1996a), associated to light harvesting complex (LHC II) dissociation II (Braintais *et al.*, 1996; Yamane *et al.*, 1998) and blocking of the electron transference in the reductant side of PS II. According to the findings of Costa *et al.* (2002), the photosynthetic apparatus of Epace 10, a bean cultivar, widely distributed in warm and dry regions of Brazil showed differential tolerance to heat stress. The photosynthetic apparatus of Epace

10 makes use of mechanisms of tolerance to heat stress (McDonald and Paulsen, 1997), and this may be due to major capacity of *D1* protein regeneration of high capacity of the xanthophylls cycle and high capacity for energy dissipation via the protein gradient in thylakoids.

Temperature producing heat injury in leaves of 30 species were measured following heat induce changes in fluorescence of the lethal chlorophyll (Smillie and Nott, 1979). They observed that the chlorophyll fluorescence began to increase between 30-40 °C and reach a peak around 50 °C in an alpine plant *Platago glaciais*, at 51.2 °C in lettuce (*Lactuca sativa*) and at 57.6 °C in pawpaw (*Carica papaya*). Similarly, in creals, the peak of fluorescence reached at lower temperatures in the temperate ones (49.5 °C in barley and 48.9 °C in oat) than in the two from warm climate (55.2 °C in maize and 53.5 °C in sorghum). They have concluded that the mean temperature for the initial rise in fluorescence as well as for the peak fluorescence occurred at a low temperature in the alpine group of plants and at a high temperature in the tropical group of plants as compared with the temperate group.

5.3.2.7. Repairing of injured cells

The repair of plant cells exposed to heat stress after having returned to an optimal temperature has been reported (Bauer and Senger, 1979). But the degree of recovery depends upon the severity of the stress (Berry and Bjorkman, 1980).

5.4. GENETICS OF HEAT TOLERANCE

The development of high temperature tolerant genotype of crop plants, it is necessary to improve plant productivity under high temperature stress environments. The quantification high temperature-tolerance and the characterization of its genetic control are necessary for germplasm enhancement efforts. Findings of Xu *et al.* (1996) indicated that the chromosomes *1A, 2A, 2B, 2D, 3A, 3B, 3D, 5D* and *6B* were associated with heat tolerance in wheat cv. Hope. Heat tolerance appeared partially dominant in the cross of tomato Shelly *et al.* (1978). However, Hanna and Hernandez (1979) showed that in tomato heat tolerance is influenced mainly by additive gene effects. Additive gene action was more important than non-additive efforts for fruit set, flower drop and under-developed ovaries at high temperature (Hanna *et al.*, 1982; Beshir El-Ahmedi and Stevens, 1979).

Blumenthal *et al.* (1991) reported high significant correlation of heat stress with protein content (positive) and with grain yield (negative) in wheat.

Correlation analysis in bread wheat showed that the grain yield was positively correlated with tillers per plot and photosynthetic rate under heat stress (Munjal and Dhanda, 2004). Path analysis showed that tillers per plot had direct and positive effect on grain yield under normal conditions, while 1000-grain weight had the highest positive and direct effect on grain yield under high temperature.

Saulter *et al.* (1990) found that the ethylene evaluation rate is genetically controlled, with additive-dominance and epistatic effects present for ethylene evaluation rate. Ethylene production of bean (*Phaseolus vulgaris*) plant increases under heat environment. The findings of Porter *et al.* (1995) suggested that the enhancing the level of higher temperature tolerance in wheat germplasm are feasible utilizing existing levels of genetic variability and exploiting additive gene effects associated with high temperature tolerance. Near narrow sense of heritability estimates for seed yield per plant, number of pods per plant, and number of seed per pod were low to intermediate, suggesting the screening for tolerance in bean to higher temperature should be conducted using advance generation lines in replicated trails (Roman-Aviles and Beaver, 2003). Additive genetic correlations between seed yield per plant and number of pods per plant, number of seeds per pod, and 100-seed weight were positive and significant. Given the large additive genetic correlations between 100-seed weight and seed yield per plant and high narrow sense heritability for 100-seed weight, indirect selection for large seed size could have been used to select for heat tolerance.

The heat tolerance in cowpea was conferred by a recessive gene during floral bud development and by a dominant gene during pollen and anther development (Hall, 1990). Inheritance of tolerance to heat-induced low pod set is consistent with the effect of a single dominant nuclear gene (Marfo and Hall, 1992). Further, Hall (1993) found that inheritance of tolerance to heat induced floral bud expression is consistent with the effect of a single recessive nuclear gene.

Significant heterosis over better parent was observed for grain yield, heat injury, chlorophyll content and stomatal frequency in common wheat (Deshpande and Nayeem, 1999). The F_1 of cross Hindi 62 ´ Ajantha exhibited significant heterosis over mid- and better parent for heat injury and grain yield. Further, Nayeem and Veer (2000) identified that the crosses Hindi 62 ´ Ajantha, Hindi 62 ´ CC 464, and Kalyansona ´ NI 5439 were heat tolerant. Their study also reveled that the parents OC 464, NI 5439, Prabhani 51, PBN 1607-2 and MACS 2496 could be utilized in multiple cross programme and biparental mating for selection of high yielding progenies for heat tolerance.

Mean squares due to general combining ability (GCA) and specific combining ability (SCA) were significant, indicating the importance of both additive and non-additive type of gene action for the inheritance of seedling heat tolerance in *Pennisetum glaucum* L. (Singh *et al.*, 2003; Singh and Sharma, 2005). However, narrow sense of heritability was low, suggesting that simple mass selection would be not effective in improving the heat tolerance. Genetic analysis in cotton (*Gossypium hirsutum* L.) showed significant SCA effects, but GCA effect was non-significant (Azhar *et al.*, 2005). Significant effect due to SCA indicated that the importance of non-additive gene effects controlling heat tolerance.

5.5. SOURCES OF HEAT TOLERANCE

Breeding against heat injuries in crop plants, tolerant breeding materials, such as tolerant cultivars, inbreds, advanced linens, wild relatives are most essential. In this biotechnological era, the tolerant genes once identified can be incorporated from one species/genus to another, therefore heat tolerant genes available in other species/genus also important. Variability for heat tolerance has been reported among the varieties within the species (Table 5.1) as well as among the species within the genus. Tischler and Burson (1995) reported the Bahiagrass species, *Paspalum notatum* is greater heat tolerance than the other species.

5.6. SELECTION FOR HEAT TOLERANCE

Many characters have been reported in many literatures as contributing to heat tolerance and they can be classified as (1) easy to screen in breeding nurseries, (2) easy to screen but requiring special effort, and (3) unsuitable for mass screening due to increased laboratory and greenhouse requirements. The first two classes may be merged and may be categorized as direct selection and class-3 may be categorized as indirect selection for heat tolerance in crop plants.

5.6.1. Direct Selection

Heat stress is associated with soil and atmospheric (hot spell and wind) drought. Heat tolerance can be determined in respect of direct effect on yield and indirect measurement of high temperature tolerant criteria. The visual direct parameters for direct selection against heat injury are seminal root growth, coleoptile length, shoot/root ratio, first leaf area, internodes elongation etc. High night temperature during reproductive development can reduce yields. Visual assessment of early vigour is considered highly underused by breeders in hot environments.

Table 5.1. Heat tolerant genotypes of different crop species

Crop	Genotype	Reference
Rice *Oryza sativa*	Intermediate Jamaica Red	Roman-Aviles and Beaver, 2003
	BG33-2	Zhang *et al.*, 2004
Wheat *Triticum aestivum*	HD2501	Hanchinal *et al.*, 1994
	Hope	Xu *et al.*, 1996
	Erget	Stone and Nicolas, 1998
	Ventnor	Yang *et al.*, 2002
	WH730, WH781, C306, PBW373, AP1074 and Raj3765	Munjal *et al.*, 2004
Bean *Phaseolus vulgaris*	G122, G5273	Shonnard and Gepts, 1994
	GUNI 59	Udomprasert *et al.*, 1995
	Haibushi	Suzuki *et al.*, 2001
	Cornell 503	Rainey and Griffiths, 2005
Tomato *Lycopersicon esculentum*	Moneymaker, Red Cherry	Johijma *et al.*, 1995
	Sonali, Hotset, Kewalo, Saladette, NDTVR-60	Nainar *et al.*, 2004
Chinese cabbage *Brassica pekinensis*	Qngyan 1	Li *et al.*, 1999
Mustard/rape *Brassica* spp.		
Cotton (Upland)	FH-900, MNH-552, CRIS-19, Karishma	Hafeez-ur-Rahaman *et al.*, 2004
Turf grass *Poa pratensis*	Wabash	He *et al.*, 1997

In an experiment, Mann and Hettel (1994) reported significant positive phenotypic correlation of early vigour with grain yield, biomass, heads per m^2, days to anthesis, days to maturity and plant height in wheat. The relative changes in internodes elongation are also related to thermal tolerance, and it was suggested that this character may be used as a selection criterion for heat tolerance in potato (Nagarajan *et al.*, 1995). Screening whole plants for degree of flowering in hot environments has been effective in breeding for heat tolerance. Based on the morphological traits, varieties with shorter internodes, and thicker and erect leaves with short petioles are better survivor in extreme temperature. Thus, flower, fruit, seed formation and pollen fertility etc. may be used as selection criteria for heat tolerance (Table 5.2). Growth under heat environment is almost always used as selection criterion. Growth would reflect the integrated consequences of various cellular components involved in heat tolerance and ultimately it influences the yield.

Table 5.2. Selection criteria for high temperature tolerance in crop plants

Criteria Characteristics	Parameters for selection	Remarks	References
Germination	Percent germination under stress	It is performed when the crop faces high temperature during germination	Jensen and Williams, 1972
Growth and development	Longer hypocotyls	This was described as a rapid and useful technique for screening of heat tolerance in many crops	AVRDC, 1976; Nel, 1998
	Internodes elongation	Direct selection criterion for heat tolerance	Nagarajan *et al.*, 1995
	Early vigour of seedlings	Indirect selection, which may correlate with yield	-
	Yield	A significant positive correlation was recorded between heat tolerance and yield. And it was most commonly used selection criterion	Sulivan *et al.*, 1973; Ehdaie *et al.*, 1988; Shpiler and Blum, 1990
	Dry matter content	Recurrent selection for 6 cycle for high dry matter content significant increase survival rate, tuberization and yield under high temperature	Haynes, 1981; Hossain and Longan, 1983
Sensitivity of reproductive phase	Fruit setting	Displayed high level of heat tolerance based on fruit setting	Villareal *et al.*, 1978; Levy *et al.*, 1978; Weaver and Timm, 1989
	Pollen fertility	Pollen from heat tolerant genotypes expressed higher fertility than the heat sensitive cultivars	Rodriguez-Gray and Barrow, 1988; Suzuki *et al.*, 2001
Membrane stability	Solute leakage as measured by conductivity test	Membrane stability bears considerable relationship with plant performance under stress environments	Saadalla *et al.*, 1990a, b; Srinivasan *et al.*, 1996
Photosynthesis sensitivity	Chlorophyll-fluorescence at 685 nm	It is a potential method *in vitro* to screen heat-tolerant plants efficiently	Smillie and Nott 1979

Selection for heat tolerance can be carried out under natural high temperature environment (field), induced conducive environment required for high temperature selection. Sometimes, when natural field environment does not provide the suitable heat stress conditions, 'abnormal' field environments available at certain locations or during the off-season may be used.

5.6.1.1. Natural environment

Conventionally, natural environment, that is, field condition is suitable for the characters, which require a optimum temperature during a particular growth stage. Any deviation from critical temperature may cause heat injury to the susceptible plants. The natural field environment is the simplest, cheapest and suitable for screening of large population.

5.6.1.2. Artificially created environment

The programmed environment may be created in glasshouse or in growth chamber. The critical temperature range can be amended as per requirement at specific growth stage of the plant. A high relative humidity is maintained using humidifiers, misting or fogging facilities in the growth chamber or greenhouse. It is important not to allow any water stress during screening for heat tolerance. Created environments are very effective in selection for high temperature tolerance. It is quite costly and suitable only for small populations.

5.6.2. Indirect Selection

Selection of crop plants in high temperature environments is effective in breeding of heat tolerance, but suitable screening environments often are not available. In order to have a better understanding of plant responses to heat-shock, it is important to know the regulatory factors involve in imparting protection and growth recovery processes. A technique based on physiological characteristics is therefore required to screen the cultivars for heat-tolerance. Several indirect selection techniques were evaluated based on biochemical tests, viz. membrane thermostability, phospholipids content, chlorophyll fluorescence, osmoregulation etc. *In vitro* assay is also a selection criterion for high temperature tolerance. These criteria singly or in combination provide tolerance to plants under stress or recovery after stress.

5.6.2.1. Membrane thermostability

An indirect screening technique is evaluation of germplasm/breeding materials involving relative electrolyte leakage from leaf tissue sample. The major effect of high temperature stress is cellular membrane modification, causing what is termed as '*leaky membrane*'. This is expressed as an increased permeability for ions and electrolytes, which can be readily measured by outflow of electrolytes. Hence cellular membrane thermostability, measured as the conductivity of electrolyte leaking from leaf disks at high temperature, has been suggested as screening technique for heat tolerance in plants (Sullivan, 1972; Martineau *et al.*, 1979a, b). The cellular membrane thermostability assay indirectly measures integrity of cellular membranes through quantifying electrolyte leakage following heat treatment. This technique based on the observation that when leaf tissue is injured by exposure to high temperatures, cellular membrane permeability is increased and electrolytes diffuse out of the cell into the bathing solution. The amount of electrolytes leaking from injured cells can be evaluated by measuring the electrical conductivity of the solution. Several studies suggested the effectiveness of this technique in detecting genetic variability in heat tolerance in warm season crops (Saadalla *et al.*, 1990a, b; Marcuem, 1998). This method is simple, quicker and less expensive than the whole plant screening. Membrane thermostability may offer and advantage to growth at defined stress temperatures, and would be of considerable value when selection is carried under abnormal field or programmed environments. The technique could be used with early vegetative stage leaf tissue from plants grown in field nursery environments used by breeders for selecting plants and advancing generations.

Genotypic differences in electrolyte leakage in cowpea were obtained from plants grown in the growth chamber or field environment (Ismail and Hall, 1999). Genotypes with heat tolerance during flowering and pod set had less electrolyte leakage than either genotypes with heat susceptibility during flowering or pod set or genotypes having heat tolerance only during early flowering. Thus, they had suggested that the leaf electrolyte leakage, as a measure of leaf membrane thermostability, may provide an efficient indirect screening technique for reproductive stage heat tolerance genes that can be used with plants growth in a range of field nursery environments. Thiaw and Hall (2004) achieved enhanced grain production by selecting for reproductive-stage heat tolerance under hot field condition in cow pea. They found that slow leaf-electrolyte-leakage selection had high pod set, and high pod set selections had slow leaf-electrolyte-leakage, indicating that these traits are closely associated. Under very hot conditions, grain yield was positively correlated with pod production, pod set and harvest index. According to them, heat-resistant cowpea may be more efficiently bred by combining selection for slow leaf-electrolyte-leakage under heat stress.

Xu *et al.* (1997) observed positive correlation between membrane injury and grain weight in wheat, suggesting that membrane thermostability may be a better indicator of heat tolerance. Munjal *et al.* (2004) also reported variability in membrane stability under high temperature in wheat. The mean values for membrane thermostability indicated that WH730, had the minimum injury to the plasma membrane followed by WH781, C306, PBW373, AP1074 and Raj3765, while WH416, HD2329 had a high level of injury in their plasma membrane due to heat stress at seedling stage.

Method to determine cell membrane stability: Under heat stress the cell membrane may be damaged due to production of free radicals and this may be measured using the following protocol.

1. The plant must be exposed to heat stress for appropriate period before the test in order to allow for hardening (acclimation). The capacity for hardening is a major component of the capacity for tolerance. Hardening can be achieved in the natural field environment, if heat stress occurs, or in the greenhouse or a programmed chamber.

2. Leaf discs or pieces of leaf tissue cut with scissors or even whole small leaves are detached and placed in standard glass vials that can accommodate a conductivity electrode. The total area of leaf material per vial is about 15 to 25 cm^2. The sample is then washed for 2-3 times with de-ionized water. The water is drained off but samples remain wet so that they would not desiccate. In the case of screening, at least 10 vials (samples) are prepared for each genotype. In that case 5 pairs are taken from five different plants (replicates). For each pair, one vial is designated as treatment (T) and the other as control (C).

3. The treatment vials are subjected to the heat stress treatment *in vitro*. They are placed in racks and covered (not stopper) with '*Saran*' wrap so as to drying the samples. Racks are placed in thermostated water bath so that the leaf samples will be completely below the water surface level. Temperature is set to a predetermined stress (treatment) temperature and the samples remain in the bath for 1h. The control vials are placed in a rack, covered with *Saran* wrap and placed at room temperature (18-25 ^{0}C). The treatment temperature should be such that it will result in average population cell membrane stability values around 50%-60% for full separation of the accessions.

4. After treatment 20cc of deionized water is added to each vial making certain that all leaf materials are submerged. All vials are then placed for incubation at about 10 ^{0}C (typically, on the lowest refrigerator shelf) for 24h. After incubation samples are equilibrated for 1h to room temperature and the conductivity of the medium is measured by inserting a conductivity electrode into each vial.
5. All vials covered with *Saran* wrap are placed in an autoclave for 15 minutes to kill all tissues. Conductivity of all samples is measured after samples are equilibrated to room temperature. Calculation: where T1 and T2 are treatment conductivities before and after autoclaving and C1 and C2 are the respective control conductivities.

   CMS (cell membrane stability) % = [1-(T1/T2)]/[1-(C1/C2)] ´ 100%

   Injury = 100-CMS

5.6.2.2. Chlorophyll fluorescence

Chlorophyll fluorescence is another screening method, developed by Smillie and Gibbons (1981), which is an indication of the capacity for electron flow through photosystem II. Sensitivity of the photosynthetic processes and of chloroplast to heat stress can be assayed as variable chlorophyll fluorescence at 685 nm. With progressive heat injury in leaves their, variable chlorophyll fluorescence decreased. A positive correlation has been observed between electrolyte leakage and chlorophyll fluorescence in groundnut (Chauhan and Senboku, 1997) suggesting that chlorophyll fluorescence may provide an alternative means of screening for heat tolerance. Measurement of chlorophyll fluorescence *in vitro* provides a rapid method to monitor the onset of cellular heat injury, to rank plants in order of heat tolerance and to follow heat hardening. This method is suitable for mass screening of genotypes for high temperature tolerance.

In wheat and barley plants, identification of high temperature tolerance can be positively correlated with maximum initial fluorescence (Havaux *et al.*, 1988). However, Yamane *et al.* (1997) suggested that the inactivation of PS II reaction centre, caused by denaturation of chlorophyll-protein complexes in response to high temperature, correlates with reduction in maximum fluorescence values. Change in this fluorescence variable cause alteration in the PS II photochemical efficiency ratio, indicating a disturbance in photochemical activity of photosynthesis. The PS II photochemical efficiency ratio has been inferred as an indicator of environmental stresses, such as high temperature, water insufficiency, light excess and others, because it is easy and fast to measure (Maxwell and Johnson, 2000). Smillie and Nott (1979) also drew a relationship between heat resistance and temperature of the environment in which the plant evolved, illustrated the potential use of chlorophyll fluorescence method for *in vitro* measures of heat sensitivity in crop plants.

5.6.2.3. Phospholipid content

Under high temperature environment, appropriate amount of lipid increase, especially the phospholipids. Komarenko *et al.* (1979) reported that at the heading stage of wheat caused a greater increase in total lipids and phospholipids in tolerant varieties than susceptible; galactolipid content fell in susceptible and rose in tolerant varieties. The release of lipids from the membranes at lower temperatures in the less-adopted varieties is a result of their lesser membrane thermostability. Okanenko *et al.* (1991) observed marked increase in the content of sulpholipids in chloroplasts of resistant varieties of durum wheat and reduced in susceptible ones.

5.6.2.4. SDS-PAGE and Western Blot analysis

Sodium-dedosyl polyacrylamide gel electrophoresis (SDS-PAGE) may be used in screening of heat tolerant crop genotypes. Zee *et al.* (1978) reported SDS-PAGE of seed proteins of several Chinese cabbage cultivars resulted in clear, reproducible protein banding patterns. The heat-tolerant cultivars and heat-sensitive cultivars could be easily distinguished by the relative intensities of the two distinct bands in 75000 dalton and 85000 dalton molecular weight regions.

The relationship between heat tolerance and phosphoglucomutase (PGM) dominant loci was studied with tolerant and sensitive cultivars of Chinese cabbage (*Brassica pekinesis*). Cui *et al.* (1998) isolated a heat tolerance related PGM isoenzyme and purified from etiolated Chinese cabbage (cv. Baiyang) seedlings was estimated by SDS-PAGE to be of 29 kDa (approx.). Zheng *et al.* (1998) found correlation between the heat tolerance and the PGM-2 locus in Chinese cabbage. Further, they studied the correlation between heat tolerance and expression of PGM-2 using F_1, F_2, and F_3 populations from crosses of heat tolerant and sensitive cultivars; and finally they concluded that PGM-2 loci can be used as a genetic marker of heat tolerance in Chinese cabbage.

Another signature physiological response of all living organisms to heat stress is decreased synthesis of normal proteins accompanied by an accelerated increase heat-shock proteins (HSP) which has been elaborated in section 5.3.2.1.

HSPs are produced when plants are exposed to temperature 5-10 ^{0}F above their optimal growing conditions. One unique feature of heat tolerant plants is the abundance of HSP during heat stress. They can be measured by their molecular weight using *Western Blot Analysis*.

5.6.2.5. Pollen selection for heat tolerance

Since a gametophytic-sporophytic genetic overlap has been demonstrated in many plant species, pollen selection at high temperature might be an efficient method to achieve heat-tolerance. The male gametophytic selection (MGS) based on the assumption that the sensitive trait is controlled by genes expressed both in gametophytic as well as sporopytic phases (Sari-Gorla and Frova, 1997). A gametophytic-sporophytic genetic overlap of 60-80% has been demonstrated in many plant species (Ottaviano and Mulcahy, 1989). Because of haploidy and the very large size of the pollen populations, MGS is expected to be extremely efficient. Selection pressure can be applied to the pollen before pollination (Frova *et al.*, 1995) or during the pollen tube growth in the style and fertilization (Petolino *et al.*, 1990). Mean percentage pollen viability of bean (*Phaseolus vagaris*) lines most tolerant to heat was significantly greater than pollen viability of the heat sensitive lines (Suzuki *et al.*, 2001; Roman-Aviles and Veaver, 2003). Suzuki *et al.* (2001) suggested that heat tolerance at an early reproductive stage could be evaluated by analyzing pollen stain-ability using flowers developed under high temperature.

Contradictory result was obtained by Miao *et al.* (2005) cucumber. Their finding showed that cucumber yield under high temperature strongly depends on its vegetative heat tolerance. Pollen heat tolerance was not a credible screening index for cucumber heat-tolerant lines. They observed linear correlation between cucumber vegetative heat tolerance and leaf thickness, palisade tissue thickness, leaf temperature, stomatal conductance, transpiration rate and root dry weight of cucumber seedlings growing under high temperature.

5.7. PROTECTION FROM HEAT

The most important inputs of agriculture are water, fertilizer and agricultural chemicals. Water is abundant in nature, but quality irrigation water is very much limited. The ground water is being depleted day-by-day due to intensive and extensive crop production and industrial uses. Already many districts of southern states have been declared as black area, where no more drilling for deep/shallow tube well is permitted for agricultural purpose.

Now it has been felt to reduce the water use without affecting the crop production. This efficiency and reducing the consumptive use of water by crops plants. The consumptive use of water includes:

1. Evaporation from soil surface
2. Transpiration through plant parts
3. Water used for metabolic activities

The water used by plants for metabolic activities is only 1-2%. The major portion of water is lost due to transpiration followed by evaporation from soil. Evaporation from soil can be reduced dense cropping of covering the soil by crop canopy. Transpirational water may also be reduced by covering the stomates with antitranspirants inducing stomatal closure with chemical compounds. But, a certain level of transpiration is required to achieve potential yield of crop, but thereafter, the transpiration require to control (Arnon, 1975) for efficient use of water. Since stomates are common gateways of entry of CO_2 and exit of water vapour, it is likely that stomata-closing antitranspirants may affect photosynthesis adversely while reducing transpiration.

5.7.1. Types of Antitranspirants

Antitranspirants are chemical compounds, their role is to practice plants of hardening to stress, as a method of reducing the impact of drought created by many abiotic conditions, such as salinity, heat or high temperature, reduced precipitation, reduced soil moisture etc. it is gradually reduced by applying very dilute solutions of chemicals that close stomates or by emulsions of wax or latex to form a thin film covers over the stomates or by reflect back a portion of the incident radiation. Thus, based on mode of action, antitranspirants have been classified as:

1. Film forming antitranspirants
2. Reflecting type of antitranspirants and
3. Metabolic inhibitors which reduce the transpiration

Use of antitranspirants to reduce the transpiration in plants can play a useful role in this respect by penetrating the luxurious loss of water to atmosphere via stomata. However, outmost care has to be taken to use appropriate antitranspirants, since the processes of transpiration and photosynthesis involves the passage of water vapour and CO_2 via the stomata, both processes may be affected when these stomata were narrowed or curtailed by antitranspirants application. It also improves the water use efficiency under drought conditions, reduce leaf transpiration rate by 87-93% (Bora and

Mathur, 1998; Gaballah and Mandour, 1999). Moreover, this may reduce water lose through transpiration; consequently the amount of used water and increased yield of cotton plants by 63% (Makus, 1997).

5.7.1.1. Film Forming Antitanspirants

The efficiency of uptake of foliar-applied solutes is reportedly increased by the addition of emulsified oils to leaf sprays (Hull, 1970; Sole *et al.*, 19972). Oily layer often reduce the rate of evaporation for air-water interfaces and may thus prolong the period of relatively rapid solute absorption which is known to occur while the leaf surface remains moist (Witter & Bukovag, 1959).

Film forming types of anti-transpirants stop almost all transpiration. It is a mechanical barrier, which reduce the loss of water vapour through stomates. A film, which effectively retards transpiration, increases the turgidity of leaves and stomata guard cells, resulting in wider stomatal aperature (Davenport *et al.*, 1972a). A water proof film of plastic or wax emulsion exhibits a certain degree of certain permeability. It is more permeable to CO_2 and O_2 than to water vapour. A disadvantage of oil emulsions and of most other film forming materials is that they reduce the rates of photosynthesis and respiration of treated areas by increasing the leaf diffusion resistance to CO_2 and O_2 (Davenport *et al.*, 1972a; Parkinson, 1970). Increase in dry weight with little or no reduction in photosynthesis is possible with film forming antitranspirants under specific environmental conditions (Davenport *et al.*, 1972b; Gale and Hagan, 1966). The influence of film forming antitranspirants in increasing sorghum grain yield under field conditions was more pronounced with limited irrigation than with full irrigation (Fuchring, 1973). A film forming antitranspirant was found to be more effective in increasing the photosynthesis under adverse environmental conditions than under favourable conditions (Gale and Pojakoff-Mayer, 1967).

Many chemical compounds used as film forming type of antitranspirants, namely silicone compounds, octadecanol, Mobileaf, follicle, vapour guard, hexadecanol, paraffin wax, etc. have been used as film forming antitranspirants in crop plants.

Silicone compounds: Silicone oils (dimethylpolysiloxane) are chemically inert, non-toxic, clear, colourless liquid. Substitution of some of the methyl groups by phenol or other groups alters the properties of the oils.

Silicone oils have relatively high permeability to O_2 and CO_2 (Heinlrin and Haigh, 1971; Parkinson, 1970), are less inhibitory to gas exchanges, and several in vestigations into the use of silicone oils or antitranspirants have

cellular metabolism (Angus and Bielorai, 1965; Heinlrin and Haigh, 1971; Parkinson, 1970).

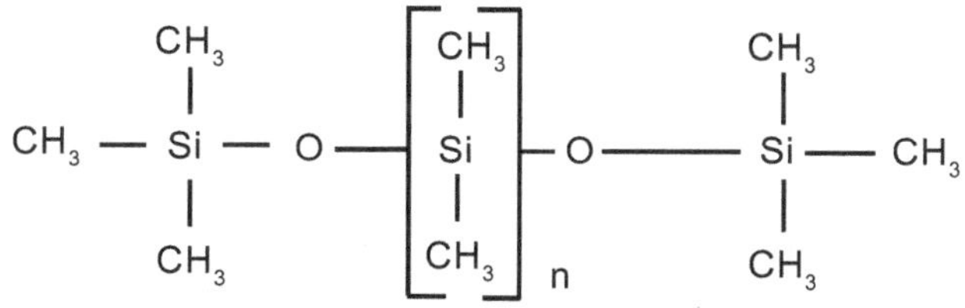

Fig. 6.1. Dimethylpolysiloxane, where 'n' determines the length of polymer chain and the viscosity

5.7.1.2. Reflectant Type of Antitranspirants

In reflecting type antitranspirants, the principle of light reflection is being used during hot and clear sky, which does not allow heating of the reflecting surface. The leaf surface is coated with chemical compounds for reflecting the light. The light reflection is multidirectional from the leaf surface within the crop canopy, subsequently increase photosynthesis (Gates *et al.*, 1965). There are many reflecting type of chemical compounds are being used in agriculture to reduce transpiration under water scarcity areas, such as kaolin, magnesium carbonate etc.

5.7.1.3. Kaolin (Aluminium silicate)

Kaolin is white reflecting type material used as antitranspirants in crop plants under drought condition. A reflectant coating of kaolinite was found to be decrease leaf temperature by reducing the energy input on leaves and to reduce the transpiration more than photosynthesis at high radiation levels in species which are light-saturated at low light intensities (Anoes and Bieloria, 1965). The kaoline remains effective for two weeks after application (Patil and De, 1976). Many investigators observed lower rate of transpiration on application of kaolin in crop plants (Moreshet *et al.*, 1977; Uppal, 1977). Saleh and El-Ashry (2006) reported increased yield by about 35.4% in orange var. Washington novel and 25.9 % in Succary by spraying of trees by 4% kaolin. Moreover, kaolin recorded the highest fruit weight, fruit set, number of fruits per tree and peel thickness.

5.7.1.4. Stomata-closing Antitranspirants

Application of certain chemical compounds on transpiring surfaces cause closing of stomates leading to reduction of water lose. These chemicals are known as stomata-closing antitranspirats (or metabolic inhibitors). Some of the chemicals of this category affect tugor of the guard cells by changing osmotic potentials (i.e. atrazine, diuron, hydroxylsulfonates, diaminozide, abscisic acid etc.) and other chemical compunds modify the permeability of the plasma membrane (e.g. phenyl mercuric acetate, alkenyl succinic acid etc.). It has been noted that while reducing transpiration, the stomata-closing antitranspiration, cause increase in leaf temperature from 1-4 °C in grass (Davenport, 1967), cotton (Simson, 1963) and Soybean (Poovaih and Wiefer, 1973).

Phenyl mercuric acetate (PMA): The mercury present in PMA combines with protein part and changes the permeability behaviour of guard cells plasma membrane, which results in efflux of water and thereby stomatal closure due to turgidity loss. Simson (1963) reported that PMA reduced transpiration more than photosynthesis in potted maize plants at both low and high moisture conditions. Transpiration was reported by 26% and 11% at high and low soil moisture, respectively, while photosynthesis was reduced by 3% at a low soil moisture regime. Patil and De (1976) reported increased water use efficiency due to reduction in transpiration by PMA. The use of PMA confined to forestry in the form of aerial spray and experimental purpose only. It can not be suggested for field application, particularly for agricultural crops, because of its mercury content.

5.7.2. Effect of Antitranspirants on Salt-Stress

Improving tolerance of crops grown under saline environment using antitranspirants has been an important practice to improve the yield in stress soils. Gaballah *et al.* (2007) reported that spraying plants with antitranspirants increased plant grown under saline condition. They observed highest value for growth characters for kaolin followed by $CaCO_3$. Many authors reported that a film which effectively retards transpiration increases the turgidity of the leaves and stomata guard cells, resulting in wider stomatal aperture. Davenport *et al.* (1972a) indicated that increase in growth characters of salt treated plants from antitranspirant treatments may be due to their effects on increasing plant eater potential at a time when the growth of particular plant part was more dependent on water status than on photosynthesis.

5.7.3. Spraying of Calcium (Ca^{2+})

Peanut (*Arachis hypogaea* L.) seedling were sprayed with CaCl solution (10 mmol/L) in successive days and then cultured at 40 ^{0}C in incubator for 12 hours (Zai *et al.*, 2001). They reported that Ca^{2+} effectively alleviated the disruption of chlorophyll, the decrease of soluble protein and the increase of membrane permeability in the heat stressed treatment. It also significantly reduced the O_2^- formation and lipid peroxidation induced by high temperature stress, protected and improves SOD and other antioxidant enzymes and helped keep the active oxygen metabolism of the seedlings at a relatively high level.

5.8. BIOTECHNOLOGY IN IMPROVEMENT FOR HEAT TOLERANCE

5.8.1. *In vitro* Screening

Tissue culture techniques for screening germplasm for heat tolerance have been used by many workers (Bochac *et al.*, 1988). They studied the tuberization of potato at high temperature in tissue culture system under a temperature gradient, which ranges from 14-34 ^{0}C, and suggested that the system was potentially useful for the study of potato tuberization under heat stress and *in vitro* screening of potato germplasm for heat tolerance. Rassadina (1988) screened heat-tolerant somaclones of potato. Tolerant genotypes were selected by introducing stress at various stages of callus growth. To identify heat tolerance ability of somaclones, Rassadina transferred the somaclones to a growth chamber. The somaclones were screened based on height, leaf size and number, the pigment system and rate of root development. *In vitro* response to the heat was investigated by Wang and Nguyen (1989) in wheat genotypes.

Regeneration and heat tolerance of lettuce cultivars Shanghai Butter, Summer and Guangdong Soft-tail from the cut stems were studied by Gao and Li (1994). They observed that the regenerated plants had higher activities of peroxidase, catalase and super oxide dismutase than seedlings, higher concentrations of IAA and iPA (isopentenyladenosine) in the root tips, lower leaf temperature, lower stomatal resistance and higher rate of transpiration during hot weather.

5.8.2. Marker Assisted Selection

Understanding of genetic basis of heat tolerance to high temperature is important for improving the productivity of crop plants in regions of heat stress. Molecular markers help in locating position of gene on chromosome, which can be use in improvement of the crop for desirable characters.

Ottaviano *et al.* (1991) identified six quantitative trait loci (QTL) in maize that accounted for a thermotolerance trait measured by cellular membrane stability. Recombinant inbred lines (RIL) of Chinese cabbage (*Brassica campestris* Subsp. *Pekinesis*) were analyzed using Isozyme, RAPD and AFLP techniques to find molecular markers that are liked to heat tolerance QTL (Zheng *et al.*, 2002, 2004). They located nine molecular markers closely linked with heat tolerance QTL including 5 AFLP markers, 3 RAPD markers, and 1 PGM isozyme marker. Total genetic contribution of these markers to heat tolerance was 46.7%. Five of the nine markers distributed one linkage group; the remaining 4 markers were located in separate groups. Thus the 9 heat tolerance linked markers distributed in 5 independent occasions in the genome of Chinese cabbage.

Cao *et al.* (2003) mapped QTLs for heat tolerance in rice (*Oryza sativa* L.) by using the constructed molecular linkage map and software QTL mapper 1.0. They detected six additive effect QTLs on chromosome 1, 3, 4, 8 and 11; and eight pairs of epistasis effect QTLs on chromosome 1, 2, 3, 4, 5, 7, 8 and 11 with variance explained being 2.27-8.13%. Total contribution of these markers to heat tolerance was 46.7%. Zhu *et al.* (2005) detected a total of three QTLs on chromosomes 1, 4 and 7, which confer heat tolerance during grain filling in rice. Of these, the QTL located in the C1100-R1783 region on chromosome 4 showed no QTL ´ environment interaction and epistatic effects, suggesting it could be stable express in different environments and genetic background, and would be valuable in rice breeding for heat tolerance.

Two markers, Xgwm11 and Xgwm293 were found to be linked to grain-filling duration in wheat (*Triticum aestivum*) by QTL analysis of F_2 population (Yang *et al.*, 2002) derived from the cross between cultivars- Ventnor (heat-tolerant) and Karl 92 (heat-susceptible) using simple sequence repeat pairs. Xgwm11-linked QTL had only additive gene action and contributed 11% to the total phenotype variation in grain-filling duration in F_2 population, whereas the Xgwm293-linked QTL had both additive and dominance action and contributed 12% to the total grain-filling duration. Their result demonstrated that heat tolerance of common wheat is controlled by multiple genes and suggested that marker-assisted selection with microsatellite primers might be useful for developing improved cultivars.

5.8.3. Transgenic Development

The successful development of transgenic has been presented in Table 5.3. The biochemical basis of heat tolerance was investigated by comparing the response of antisense and sense transgenic soybean plant containing L-Delta-1-pyrroline-5-carboxylate reductase gene (Ronde *et al.*, 2004) with non-transgenic wild-type plant. Upon stress, $NADP^+$ levels decreased in sense plants. Sense plants had the highest ability to metabolize proline after rewatering. They reported dissociation of oxygen-evolving complex (OEC) upon stress. In sense plants, which best resisted the stress, OEC dissociation was by passed by proline feeding electrons into PS II, maintaining an acceptable NADPH level, preventing further damage.

Table 5.3. Gene encoding for molecular chaperones and transgenic development

Gene	Gene action	Species	Phenotypic expression	Reference
hs	Heat shock transcription factor	*Arabidopsis thaliana*	Increased thermotolerance in transgenic plants	Lee *et al.*, 1995
Hsp70	Heat-inducible anti-sense HSP70	*Arabidopsis thaliana*	Increased thermotolerance in transgenic plants	Lee and Schoff, 1996
Hsp17.7	Heat shock protein	*Daccus carota* L.	Increased or decreased thermotolerance	Malik *et al.*, 1999
P5CR	Inducible heat shock promoter (IHSP)	*Glycin max*	Increased proline accumulation	De Ronde *et al.*, 2000
Hsp101	Heat shock protein	*Arabidopsis thaliana*	Decreased thermotolerance in Hsp101-deficient (hot1) mutant	Hong and Vierling, 2001
Hsp101	Heat shock protein	*Arabidopsis thaliana*	Manipulated thermotolerance in transgenic plants	Queitsch *et al.*, 2000
DcHSP17.7	Heat shock protein	*Solanum tubersum* L.	Improved cellular membrane stability and enhanced *in vitro* tuberization	Ahn and Zimmerman, 2006

Ahn and Zimmerman (2006) fused *DcHSP17.7* gene (a carrot heat shock protein gene encoding HSP17.1) to a 6XHistidine (His) tag to distinguish the engineered protein from endogenous potato proteins, and was introduced into the potato cultivar 'Desiree' under the control of the cauliflower mosaic virus (CaMV) 35S promoter. The integration was confirmed by Western Blot, which showed constitutive integration of *DcHSP17.7* in transgenic potato lines before heat stress. They observed improved cellular membrane stability at high temperature, compared with wild type and vector controlled plants. Transgenic potato lines also exhibited enhanced tuberization *in vitro*.

5.9. REFERENCES

Ahmed FE, Hall AE, DeMason DA. 1992. Heat injury during flower development of cowpea (*Vigna unguiculata*, Fabaceae). Am J Bot. 79: 784-791.

Ahmed FE, Hall AE. 1993. Heat injury during early floral bud development in cowpea. Crop Sci. 33: 764-767.

Ahn YJ, Zimmerman L. 2006. Introduction of carrot *HSP17.7* into potato (*Solanum tuberosum* L.) enhances cellular membrane stability and tuberization *in vitro*. Plant Cell Environ. 29: 95.

Alexandrove VY. 1977. Cells, Molecules and Temperature. Springer, Berlin, Germany.

Al-Khatib K, Paulsen GM. 1990. Photosynthesis and productivity during high temperature stress of wheat genotypes from major world regions. Crop Sci. 30: 1127-1132.

Annamalai P, Yanaghiara S. 1999. Identification and characterization of heat stress induced gene in cabbage encodes at kunitz type protease inhibitor. J Plant Physiol. 115: 226-233.

Anoes DE, Bielorai H. 1965. Transpiration reduction by surface films. Aust J Agri Res. 16: 107-113.

Arnon I. 1975. In Physiological Aspects of Dryland Farming. Gupta US (ed.), Oxford and IBM Publishing Co., New Delhi.

AVRDC (Taiwan). 1976. Annual Research Highlights, 1975. Shanhua, Taiwan, ROC. p. 23.

Azhar MT, Khan AA, Khan IA. 2005. Combining ability analysis of heat tolerance in *Gossypium hirsutum* L. Czech J Genet Plant Bred. 41(1): 23-28.

Azzi G. 1950. Biancone dell'Elba and its specific resistance to hot spells. Riv Ecol. 1: 231-237.

Bagni N. 1989. Polyamine and plant growth regulator and development. In: The physiology of polyamines, Bachrach U, Heimer YM (eds.), Vol II. CRC Press, Boca Raton. pp. 107-120.

Basra RK, Basra AS, Malik CP, Grover IS. 1997. Are polyamines involve in the heat-shock protection of mung bean seedlings? Bot Bull Acad Sin. 38: 165-169.

Baszczynski CL, Walden DB, Atkinson BG. 1982. Regulation of gene expression in corn (*Zea mays* L.) by heat shock. Can J Biochem. 60: 569-579.

Baszczynski CL. 1988. Gene expression in *Brassica* tissue and species following heat shock. Biochem Cell Biol. 66: 1303-1311.

Bauer H, Senger N. 1979. Photosynthesis of ivy leaves (*Hedera helix* L.) after heat stress. II. Activities of RuBP carboxylase, Hill reaction and chloroplast ultrastructure. Z Pflanzenphysiol. 95: 359-369.

Berry J, Bjorkman O. 1980. Photosynthetic response and adaptation to temperature in higher plants. Annu Rev Plant Physiol. 31: 491-543.

Beshir El-Ahmedi A, Stevens MA. 1979. Genetics of high temperature fruit set in the tomato. J Am Soc Hort Sci. 104: 691-696.

Blum A. 1988. Breeding crop varieties for stress environments. CRC Critical Review in Plant Science. 2(3): 199-238.

Blumenthal CS, Bekes F, Batey IL, Wrigley CW, Moss HJ, Mares DJ, Barlow EWR. 1991. Interpretation of grain quality results from wheat varieties trials with reference to high temperature stress. Aust J Agric Res. 42: 325-334.

Blumenthal CS, Bekes F, Wrigley CW, Barlow EWR. 1990. The acquisition and maintenance of thermo-tolerance in Australian wheat. Aust J Plant Physiol. 17: 37-47.

Bochac JR, Miller Jr JC, Borque JE. 1988. Tuberization response of potato to high temperature in tissue culture system. Am Potato J. 65: 47.

Bora KK, Mathur SR. 1998. Some plant growth regulator as antitranspirants in soybean. Ann Plant Physiol. 12: 175-177.

Briantais JM, Dacosta J, Goulas Y, Ducuet JM, Moya I. 1996. Heat stress in leaves an increase of the minimum levels of chlorophyll fluorescence, F_0: a time-resolved analysis. Photosynth Res. 48: 189-196.

Buchner O, Neuner G. 2001. Determination of heat tolerance: a new equipment for field measurements. J Appl Bot. 75: 130-137.

Buchner O, Neuner G. 2003. Plant heat tolerance: Adaptation to warmer temperature. Arctic, Antarctic, and Alpine Res. 35: 411-420.

Cao LY, Zhao JG, Zhang XD, Li DL, He LB, Chen SH. 2003. Mapping QTLs for heat tolerance and correlation between heat tolerance and photosynthetic rate in rice. Chinese J Rice Science. 17(3): 223-227.

Chaterjee S, Choudhari MM, Ghosh B. 1983. Changes in polyamine contents during root and nodule growth of *Phaseolus mungo*. Phytochemistry. 22: 1553-1556.

Chauhan YS, Senboku T. 1997. Evaluation of groundnut genotypes for heat tolerance. Ann Appl Biol. 131(3): 481-489.

Chen HH, Shen ZY, Li PH. 1982. Adaptability of crop plants to high temperature stress. Crop Sci. 22: 719-725.

Chowdhury S, Wardlaw IF. 1978. The effect of temperature on kernel development in cereals. Aust J Agric Res. 29: 205-223.

Clarke SM, Luis LAJ, Wood JE, Scott MS. 2004. Salicylic acid depended signaling promotes basal thermotolerance but is not essential for aquired thermotolerance in *Arabidopsis thaliana*. The Plant J. 38: 432.

Commuri PD, Jones RJ. 2001. High temperatures during endosperm cell division in maize: a genotypic comparison under *in vitro* and field conditions. Crop Sci. 41(4): 1122-1130.

Costa ES, Bressan-Smith R, Oliveira JG, Campostrini E, Pimentel C. 2002. Photochemical efficiency in bean plants (*Phaseolus vulgaris* L. and *Vigna unguiculata* L. Walp) during recovery from high temperature stress. Braz J Plant Physiol. 14(2): 105-110.

Cui HC, Li L, Zheng X, Samue SS. 1998. Isolation and purification of heat-tolerance related isozyme of phosphoglucomutase (PGM) from Chinese cabbage. Acta Agriculturae Boreali Sinica. 13(4): 86-92.

Das S, Basu R, Ghosh B. 1987. Heat stress- induced polyamine accumulation in cereal seedlings. Plant Physiol Biochem. 14: 108-116.

Dat JF, Foyer CH, Scott IM. 1998. Change in salicylic acid and antioxidants during induction of thermotolerance in mustard seedlings. Plant Physiol. 118: 1455-1461.

Davenport DC, Fisher MA, Hagan RM. 1972a. Some counteractive effects of antitranspirants. Plant Physiol. 49: 722-724.

Davenport DC, Martin PE, Hagan MR. 1972b. Antitranspirants for conservation of leaf water potential of transplanted citrus trees. Hort Sci. 7: 511-520.

Davenport DC. 1967. Effects of chemical antitranspirants on transpiration and growth of grass. J Exp Bot. 18: 332-347.

De Ronde JA, Spreeth MH, Cress WA. 2000. Effects of antisense D- 'pyrroline-5-carboxylate reductase transgenic soybean plants subjected to osmotic and drought stress. Plant Growth Regulators. 32: 13-26.

Deshpande DP, Nayeem KA. 1999. Heterosis for heat tolerance protein content, yield and yield components in bread wheat (*Triticum aestivu* L.). Indian J Genet Plant Bred. 59(1): 13-22.

Dew el-madina IM, Hall AE. 1986. Flowering of contrasting cowpea (*Vigna radiata* L. Walp) genotypes under different temperature and photoperiods. Field Crops Research. 14: 89-104.

Dickson MH, Petzoldt R. 1989. Heat tolerance and pot set in green beans. J Am Soc Hort Sci. 114: 833-836.

Drohlert G, Dumroff EB, Legge RL, Thompson JE. 1986. Radical scavenging properties of polyamines. Phytochemistry. 25: 367-371.

Ehdaie B, Waines JG, Hall AE. 1988. Differential responses of land race and improved spring wheat genotypes to stress environments. Crop Sci. 28: 838-842.

Enami I, Kitamura M, Tomo T, Isokawa Y, Ohta H, Katch S. 1994. Is the primary cause of thermal inactivation of oxygen evaluation in spinach PS II membranes release of the extrinsic 33 kDa protein or of Mn? Biochem Biophys Acta. 1186: 52-58.

Ervin EH, Zhang X, Schmidt RE. 2005. Exogenous salicylic acid enhances post transplant success of heated Kentucky bluegrass and Tall Fescue sod. Crop Sci. 45: 240-244.

Evan PT, Malmberg RL. 1989. Do polyamines have roles in plant development? Annu Rev Plant Physiol Plant Mol Biol. 40: 235-269.

Flores HE. 1991. Changes in polyamine metabolism in response to abiotic stress. In: Biochemistry and Physiology of Polyamines in Plants, Slocum RD, Flores HE (eds.). CRC Press, Boca Raton. pp. 214-228.

Foyer CH, Lopez-Delgado H, Dat JF, Scott IM. 1997. Hydrogen peroxide and glutathione associated mechanism of acclamatory stress tolerance and signaling. Physiol Plant. 100: 241-254.

Frova C, Portaluppi P, Villa M, Sari-Gorla M. 1995. Sporophytic and gametophytic components of thermotolerance affected by pollen selection. J Heredity. 86: 50-54.

Fuehring HD. 1973. Effects of antitranspirants on yield of grain sorghum under limited irrigation. Agron J. 65: 348-351.

Gaballah MS, Leila A, El-Zeiny HA, Khalil S. 2007. Estimating the performance of salt-stressed sesame plant treated with antitranspirants. J Appl Sci Res. 3(9): 811-817.

Gaballah MS, Mandour MS. 1999. Effect of irrigation interval and glycinebetaine application on growth, yield and seed quality of soybean. Egyptian J Appl Sci. 14: 142-154.

Gale J, Hagan RM. 1966. Plant antitranspirants. Annu Rev Plant Physiol. 17: 269-282.

Gale J, Poijakoff-Mayber A. 1967. Plastic film as antitranspirants. Science. 156: 650-652.

Gao LH, Li SJ. 1994. Physiological basis of heat tolerant in regenerated lettuce. J Nanjing Agril Univ. 17(2): 23-27.

Gates DM, Keegan HJ, Schelter JC, Weidner VR. 1965. In Annual Reviews of Plant Sciences. Malik CP (ed.), Kalyani Publishers, New Delhi.

Gilmore AM, Govindjee. 1999. How higher plants respond to excess light: Energy dissipation in Photosystem II. In: Concepts in Photobiology: Photosynthesis and Photomorphogenesis, Singhal GS, Renger G, Sopory SK, Irrgang K-D, Govindjee (eds.). Narosa Publ House, New Delhi, India. pp. 513-548.

Gong M, Li YJ, Chen SZ. 1998a. Abscisic acid induced thermotolerance in maize seedlings is mediated by Ca^{2+} and associated with antoxidant systems. J Plant Physiol. 153: 488-496.

Gong M, van der Luit A, Knight MR, Trewavas AJ. 1998b. Heat shock induced changes in intercellular Ca^{2+} level in tobacco seedlings in relation to thermotolerance. Plant Physiol. 116: 429-437.

Gounaris K, Brain APR, Quinn PJ, Williams WP. 1983. Structural and functional changes associated with heat-induced phase separation of no-bilayer lipids in chloroplast thylakoid membranes. FEBS Lett. 153: 47-52.

Gurley WB. 2000. HSP101: a key component for the acquisition of thermotolerance in plants. Plant Cell. 12: 457-459.

Gustavsson N, Harndahl U, Emanuelsson A, Roepstorff P, Sundby C. 1999. Methionine sulfoxidation of the chloroplast small heat shock protein and conformational changes in the oligomer. Protein Sci. 8(11): 2506-2512.

Hafeez-ur-Rahaman, Malik SA, Saleem M. 2004. Heat tolerance of upland cotton during the fruiting stage evaluated using cellular membrane thermostability. Field Crop Res. 85(2/3): 149-158.

Hall AE, Kuo CG. 1993. Physiology and breeding for heat tolerance in cowpea, and comparisons with other crops. *In*: Adaptation of food crops to temperature and water stress: Proc of an Intrnl Symp, Tawan, 13-18 August 1992. pp. 271-284.

Hall AE. 1990. Breeding for heat tolerance and approach based on whole-plant physiology. HortScience. 25: 17-19.

Hall AE. 1992. Breeding for heat tolerance. Plant Breed Rev. 10: 129-168.

Hall AE. 1993. Physiology and breeding for heat tolerance in cowpea and comparison with other crops. In: Adaptation of Food Crops to Temperature and Water Stress, George Kuo C (ed.). Asian Veg Res Dev Center, Proceeding of an International Symposium, Shanhua, Taiwan, 13-18 Aug, 1992. pp. 271-284.

Hanchinal RR, Tandon JP, Salimath PM, Hettel GP. 1994. Variation and adaptation of wheat varieties for heat tolerance in Peninsular India. In: Wheat in Heat-Stressed Environments: Irrigated, Dry Areas and Rice-Wheat Farmining System. Proc of the Intrl Conf, Saunders DA (ed.), heald at Wad Medani, Sudan, 1-4 February and Dinajpur, Bangladesh, 13-15 February 1993. pp. 175-183.

Hanna HY, Hernandez TP, Koonce KL. 1982. Combining ability for fruit set, flower drop, and under develop ovaries in some heat tolerance tomatoes. HortScience. 17: 760-761.

Hanna HY, Hernandez TP. 1979. Heat tolerance pistil size inheritance and heterosis in tomatoes. HortScience. 14: 458-459.

Harndahl U, Kokke BP, Gustavsson N, Linse S, Berggren K, Tjerneld F, Boelens WC, Sundby C. 2001. The chaperone-like of a small heat shock protein is lost after sulfoxidation of conserved methionine in a surface-exposed amphipathic alpha-helix. Biochem Biophys Acta. 1545(1-2): 227-237.

Havaux M, Ernez M, Lannoye R. 1988. Correlation between heat tolerance and drought tolerance in cereals demonstrated by rapid chlorophyll fluorescence. J Plant Physiol. 133: 555-560.

Haynes FL. 1981. Progress in adaptation and use in breeding of high land tropic diploid potatoes. Am Potato J. 58: 502.

He Y, Liu Y, Cao W, Huai M, Xu B, Huang B. 2005. Effect of salicylic acid on heat tolerance associated with antioxidant metabolism in Kentucky Bluegrass. Crop Sci. 988-995.

He YL, Shen J, Wang HL. 1997. Studies on heat tolerance mechanism of cool season turf grass. I. Changes of chlorophyll content and peroxidase activity of leaves of *Poa pratensis* L. under heat stress.

Heilrin JP, Haig WG. 1971. Transpiration control with silicon emulsion. Plant Physiol. 48: S47.

Hong SW, Vierling E. 2000. Mutants of *Arabidopsis thaliana* detective in the acquisition of tolerance to high temperature stress. Proc Acad Sci, USA. 97: 4392-4397.

Hong SW, Vierling E. 2001. Hsp101 is necessary for heat tolerance but dispensible for development and germination in the absence of stress. 27(1): 25-35.

Hossain M, Logan C. 1983. Recurrent selection for heat tolerance in diploid potatoes (*Solanum tubersum* subsp. *phureja* and *stenotonum*). Am Potato J. 60: 537-542.

Howarth CJ, Ougham HJ. 1993. Gene expression under temperature stress. New Phytol. 125: 1-26.

Hull HM. 1970. Leaf structure as related to absorption of pesticides and other compounds. Residue Rev. 31: 1-150.

IRRI. 1977. Climatic environment and its influence. High temperature. In: Annual Report 1976, Los Banos, Laguna, International Rice Research Institute, Philippines. pp. 264-266.

Ismail AM, Hall AE. 1998. Positive and potential negative effects of heat-tolerance genes in cowpea. Crop Sci. 38: 381-390.

Ismail AM, Hall AE. 1999. Reproduction stage heat tolerance, leaf membrane thermostability and plant morphology in cowpea. Crop Sci. 39: 1762-1768.

Jaenicke R, Creighton TE. 1993. Junior chaperones. Curr Biol. 3: 234-235.

Jakob U, Bucher J. 1994. Assiting spontaneity: the role of HSP90 and smHSPs as molecular chaperones. Trends Biochem Sci : 205-211.

Jensen SD, Williams NE. 1972. Selection for heat tolerance and drought resistance in maize. Agron Abstr, Madison, Am Soc Agron, USA. p. 12.

Johjima T, Fernandez-Munoz R, Cuartero J, Gomez-Guillamon ML. 1995. Inheritance of heat tolerance of fruit coloring in tomato. Acta Horticulturae. 412: 64-70.

Jones LH. 1971. Adaptive responses to tempreture in dwarf French beans *Phaseolus vulgaris* L. Ann Bot. 35: 581-596.

Knight H. 2000. Calcium signaling during abiotic stress in plants. Int Rev Cytol. 195: 269-324.

Kocsy G, Kobrehel K, Szalai G, Duviau MP, Buzas Z, Galiba G. 2004. Abiotic stress induced changes in glutathione and thioredoxin high levele in maize. Environl Exp Biol. 52(2): 101-112.

Komarenko NI, Protsenko DF, Okanenko AA. 1979. Effect of high temperatures on the lipid complex of winter wheat. Fiziol I Biokim Kultur Rastenii. 11: 29-34.

Kramer GF, Wang CY. 1989. Correlation of reduced chilling injury with increased spermine and spermidine leaves in zucchini squash. Physiol Plant. 76: 479-484.

Krishnan M, Nguyen HT, Burke JJ. 1987. Genetic diversity of heat shock protein synthesis and its relation to thermal tolerance in wheat leaves. In: Plant Gene System and their Biology, Key JL, McIntosh L (eds.). New York, USA, Alan R. Liss, Inc. pp. 109-120.

Kuo CG, Shen BJ, Chen HM, Chen HC, Opena RT. 1988. Association between heat tolerance, water consumption, and morphological characters in Chinese Cabbage. Euphytica. 39: 65-73.

Kurets VK, Popov EG. 1988. Evaluating the requirements of a genotype in respect of envirommental conditions. In: Diagnostika urtoichivosti rastenii stressovym vozdeistviyam, Leningrad, USSR. pp. 222-227.

Larkindale J, Knight MR. 2002. Protection against heat stress-induced oxidative damage in *Arabidopsis* involving calcium, abscisic acid, ethylene and salicylic acid. Plant Physiol. 128(2): 682-695.

Lee GJ, Vierling E. 2000. A small heat shock protein 70 systems to reactivate a heat denatured protein. Plant Physiol. 122(1): 189-198.

Lee JH, Hubel A, Schoff F. 1995. Depression of activities of genetically engineered heat shock factor causes constitutive synthesis of heat shock proteins and increased thermotolerance in transgenic *Arabidopsis*. Plant J. 8: 603-612.

Lee JH, Schoff F. 1996. An *Hsp70* antisense gene affects the expression of *HSP70/HSc70*, the regulation of *HSF*, and the acquisition of thermotolerance in *Arabidopsis thaliana*. Mol Gen Genet. 252: 11-19.

Levitt J. 1969. Growth and survival of plants at extreme of temperature- a unified concept. Soc of Exptl Biol Symp. 23: 395-448.

Levy D, Robinowitch HD, Kedar N. 1978. Morphological and physiological characters affecting flower drop and fruit set of tomatoes at high temperature. Euphytica. 27: 211-218.

Li JK, Luan ZS, Wang DS. 1999. Chinese cabbage F_1 hybrid 'Qungyan No. 1' for heat tolerance and extreme early maturity. Chinese Vegetables. No. 1: 35-37.

Li ZG, Gong M. 2003. Involvement of antioxidant system in the formation of H_2O_2-induced heat tolerance in maize seedlings. Plant Physiol Comm. 39(6): 575-579.

Lopez-Delgado H, Dalt JF, Foyer CH, Scot IM. 1998. Induction of thermotolerance in potato microplants by acetylsalicylic acid and H_2O_2. J Exp Bot. 49: 713-720.

MacKill DGJ, Coffman WR, Rutger JN. 1982. Pollen shading and combining ability for high temperature tolerance in rice. Crop Sc. 22: 730-733.

Makus DK. 1997. Effect of an antitranspirants on cotton grown under conventional tillage systems. Proc Beltwide Cooton Conf, New Orleans, LA, USA, Jan 6-10, 1997. pp. 642-644.

Malik MK, Solvin JP, Hwang CH, Zimmerman JL. 1999. Modified expression of a carrot small heat shock protein gene, *Hsp17.7*, results in increased or decreased thermotolerance. Plant J. 20: 89-99.

Mann CE, Hettel GP. 1994. Early vigor in wheat: a useful character for heat tolerance selection. In: Wheat in Heat-Stressed Environments: Irrigated, Dry Areas and Rice-Wheat Farmining System. Proc of the Intrl Conf, Saunders DA (ed.), heald at Wad Medani, Sudan, 1-4 February and Dinajpur, Bangladesh, 13-15 February 1993. pp. 247-250.

Marcuem KB. 1998. Cell membrane thermostability and whole-plant heat tolerance of Kentucky bluegrass. Crop Sci. 38: 1214-1218.

Marfo KO, Hall AE. 1992. Inheritance of heat tolerance during pod set in cowpea. Crop Sci. 32: 912-918.

Martineau JR, Specht JE, Williams JR, Sullivan CY. 1979a. Temperature tolerance in soybeans. I: Evaluation for technique for assessing cellular thermostability. Crop Sci. 19: 75-78.

Martineau JR, Williams JR, Specht JE. 1979b. Temperature tolerance in soybeans. II: Evaluation of segregating population for membrane thermostability. Crop Sci. 19: 79-81.

Masaya P, White JW. 1991. Adaptation to photoperiod and temperature. In: Common Beans- Research for Crop Improvement, Van Schoonhoven A, Voysest O (eds.). CIAT, Cali, Colombia. pp. 445-500.

Maxwell K, Johnson GN. 2000. Chlorophyll fluorescence- A practical guide. J Exp Bot. 51: 659-668.

McDonald GK, Paulsen GM. 1997. High temperature effect on photosynthesis and water relations of grain legumes. Plant Soil. 196: 47-58.

Miao MM, Zhang YH, Jiang YH, Xue LB. 2005. The relationship between cucumber heat tolerance and some characters of leaves, roots and pollens. J Yangzhou University Agril Lief sci edition. 26(2): 83-85.

Moffatt JM, Sears RG, Paulsen GM. 1990a. Wheat high temperature tolerance during reproductive growth. I. Evaluation by chlorophyll fluorescence. Crop Sci. 30: 881-885.

Moffatt JM, Sears RG, Paulsen GM. 1990a. Wheat high temperature tolerance during reproductive growth. II: Genetic analysis of chlorophyll fluorescence. Crop Sci. 30: 886-889.

Moreshet S, Stanhill G, Fuchi. 1977. In Annual Reviews of Plant Sciences, Malik CP (ed.), Kalyani Publishers, New Delhi.

Munjal R, Dhanda SS, Rana RK, Singh I. 2004. Membrane thermostability as an indicator of heat tolerance at seedling stage in bread wheat. Natl J Plant Impro. 6(2): 133-135.

Munjal R, Dhanda SS. 2004. Association of heat tolerance in bread wheat. National J Plant Improv. 6(1): 71-73.

Mutter RG, Hall AE, Patel PN. 1989. Photoperiod and light quality effects on cowpea floral development at high temperature. Crop Sci. 29: 1501-1505.

Mutters RG, Ferreira LGR, Hall AE. 1989. Proline content of the anthers and pollen of heat tolerant and heat-sensitive cowpea subjected to different temperatures. Crop Sci. 29: 1497-1500.

Mutters RG, Hall AE. 1992. Reproductive responses of cowpea to high temperature during different night periods. Crop Sci. 32: 202-206.

Nagarajan-Shantha, Minhas JS, Nagarajan S. 1995. Internode elongation: a potential screening technique for heat tolerance in potato. Potato Res. 38(2): 179-186.

Naik BI, Srivastava SK. 1978. Effect of polyamines on tissue permiability. Biochemistry. 17: 1885-1887.

Nainar P, Ranjangam J, Mohamed SEN, Thangarajan T. 2004. Studies on screening of tomato (*Lycopersicon esculentum* Mill) genotypes of heat tolerance. South Indian Hort. 52(1/6): 59-64.

Nayeem KA, Veer MV. 2000. Combining ability for heat tolerance traits in bread wheat (*Triticum aestivum* L. Em. Thell). Indian J Genet Plant Bred. 60(3): 287-295.

Nel AA. 1998. The response of germinating sunflower seed to heat tolerance induction. South African J Plant Soil. 15(2): 90.

Okanenko AA, Taran NU, Musienko NN. 1991. Content of sulpholipids in the chloroplast as a diagnostic index of heat and drought resistance in winter wheat. Sel'skokhozyaistvennaya Biologiya. 1: 132-138.

Ottaviano E, Mulcahy DL. 1989. Genetics of angiosperm pollen. Adv Genet. 26: 1-64.

Ottaviano E, Sari Giorla M, Pe E, Erova C. 1991. Molecular markers (RFLPs and HSPs) or the genetic dissection of thermotolerance in maize. Theor Appl Genet. 81: 713-719.

Owen PC. 1971. Responses of a semidwarf wheat to temperatures responsive representing a tropical dry season. II: Extreme temperatures. Exp Agri. 7: 43-47.

Palavan-Unsal N. 1987. Polyamine metabolism in the roots of *Phaseolus valgaris*-interaction of the inhibitors of polyamine biosynthesis with putrescine in growth and polyamine biosynthesis. Plant Cell Physiol. 28: 565-572.

Pande PC, Pathak PC, Sachdeva P, Ruwali KN, Sastry LVS. 1998. Amylose content in wheat grain: a trait for heat tolerance. Indian J Plant Physiol. 3(3): 247-248.

Parkinson KJ. 1970. The effect of silicone coating on leaves. J Exp Bot. 21: 566-579.

Pastenes C, Horton P. 1996a. Effect of high temperature on photosynthesis in beans. I: Oxygen evaluation and chlorophyll fluorescence. Plant Physiol. 112: 1245-1252.

Pastenes C, Horton P. 1996b. Effect on high temperature on photosynthesis in beans. II. CO_2 assimilation and metabolic contents. Plant Physiol. 112: 1253-1260.

Patel PN, Hall AE. 1990. Genotypic variation and classification of cowpea for reproductive responses to high temperatures under long photoperiods. Crop Sci. 30: 614-621.

Patil BB, De R. 1976. Influence of antitranspirants on rapeseeds (*Brassica campestris*) plants under water-stressed and nonstressed conditions. Plant Physiol. 57: 941-943.

Paulsen GM. 1994. High temperature response of crop plants. In: Physiology and Determination of Crop Yield, Boote KJ, Benett JM, Sinclair TR, Paulsen EM (eds). American Society of Agronomy, Madison. pp. 365-389.

Petolino JF, Cowen NM, Thompson A, Mitchell JC. 1990. Gamet selection for heat-stress tolerance in maize. J Plant Physiol. 136: 219-224.

Pimentel C. 1998. Metabolism de carbono na aericultura tropical. EDUR, Seropedica, RJ, Brazil.

Poovaiah BW, Wiefer HH. 1973. Influence of hydrogen fluoride fumigation on the water economy of soybean plants. Plant Physiol. 51: 396-399.

Porter DR, Nguyen HT, Burke JJ. 1995. Genetic control of acquired temperature tolerance in winter wheat. Euphytica. 83(2): 153-157.

Queitsch C, Hong SW, Vierling E, Lindquist S. 2000. Heat shock protein 101 plays a crucial role in thermotolerance in *Arabidopsis*. Plant Cell. 12: 479-492.

Rainey KM, Griffinths PD. 2005. Inheritance of heat tolerance during reproductive development in snap bean (*Phaseolu vulgaris* L.). J Am Soc Hortl Sci. 130(5); 700-706.

Raison JK, Roberts JKM, Berry JA. 1982. Correlation between thermal stability of chloroplast (thylakoid) membrane and the composition and fluididty of their polar lipids upon acclimation of the higher plant *Neuium oleander*, to grow temperature. Biochem Biophys Acta. 688: 218-228.

Rassadina GV. 1988. Developing methods of selecting heat-resistance somaclones of tomato. In: Biologiya Kul'tiviuemykh Kletok I biotekhnologiya 1, Butenko RG (ed.). Novosibirsk, USSR. p. 171.

Ristic Z, Yang GP, Martin B, Fullerton S. 1998. Evidence of association between specific heat shock protein(s) and the drought and heat tolerance phenotype in maize. J Plant Physiol. 153: 497-505.

Roberts DR, Dumbroff EB, Thompson JE. 1986. Exogenous polyamines alter membrane fluidity- a basis for potential misinterpretation of their physiological role. Planta. 167: 395-401.

Robertson AJ, Ishikawa M, Gusta LV, MacKenzie SL. 1994. Abscisic acid induced heat tolerance in *Bromus inermis* Lees. Cell suspension cultures. Plant Physiol. 105: 181-190.

Rodriguze-Gray B, Barrow JR. 1988. Pollen selection for heat tolerance for cotton. Crop Sci. 28: 857-859.

Rojas A, Almogurea C, Jordano J. 1999. Transcriptional activation of a heat shock gene promoter in sun flower embryos.: Synergism between ABI3 and heat shock factors. Plant J. 20: 601-610.

Roman-Aviles B, Beaver JS. 2003. Inheritance of heat tolerance in common bean of Andean origin. J Agri University Puerto Rico. 87(3/4): 113-121.

Ronde JA de, Cress WA, Kruger GHJ, Strasser RJ, Staden Jvan. 2004. Photosynthetic response of transgenic soybean plants, containing an *Arabidopsis* P5CR gene, during heat and drought stresses. J Plant Physiol. 161(11); 1211-1224.

Saadalla MM, Shanahan JF, Quick JS. 1990a. Heat tolerance in winter wheat. I: Hardening and genetic effects on membrane thermostability. Crop Sci. 30: 1243-1247.

Saadalla MM, Shanahan JF, Quick JS. 1990b. Heat tolerance in winter wheat. II: Membrane thermostability and field performance. Crop Sci. 30: 1248-1251.

Saleh MMS, El-Ashry SM. 2006. Effect of some antitranspirants on leaf mineral content, fruit set, yield and fruit quality of Washington navel and Succary orange trees. J Appl Sci Res. 2(8): 486-490.

Sanders D, Brownlee C, Harper JF. 1999. Communicating with calcium. Plant Cell. 22: 691-706.

Sari-Gorla M, Frova C. 1997. Pollen tube growth and pollen selection. In: Biotechnology and Crop Improvement, Sawgney and Shivanna (eds.). pp. 323-351.

Saulter KJ, Davis DW, Li PH, Wallerstein IS. 1990. Leaf ethylene evaluation level following high temperature stress in common bean. HortScience. 25: 1282-1284.

Schoffle F, Rossol I, Angermullar S. 1997. Regulation of transcription of heat shock genes in nuclei from soybean (*Glycine max*) seedlings. Plant Cell Environ. 10: 113-119.

Shelly RA, Greenleaf WH, Peterson CM. 1978. Comparative floral fertility in heat-tolerant and heat-sensitive tomatoes. J Am Soc Hort Sci. 103: 778-780.

Shevyakova NI, Roschupkin BV, Paramonova NV, Kuznetsov VV. 1994. Stress responses in *Nicotiana sylvestris* cells to salinity and high temperature. I: Accumulation of proline, polyamines, betaines and sugar. Russian J Plant Physiol. 41: 490-496.

Shonnard GC, Gepts P. 1994. Genetics of heat tolerance during reproductive development in common bean. Crop Sci. 34(5): 1168-1175.

Shpiler C, Blum A. 1990. Heat tolerance for yield and its components in different wheat cultivars. Euphytica. 51: 257-263.

Singh RV, Sharma TR, Khedar OP. 2003. Combining ability for seeling heat tolerance in pearl millet [*Pennisetum glaucum* (L.) R. Br.]. Indian J Genet Plant Bred. 63(4): 349.

Singh RV, Sharma TR. 2005. Combining ability for seedling heat tolerance in pear millet. Intrl Sorghum Newsl. 42: 70-72.

Simson D. 1963. Effect of oil moisture and phenylmurcuric acetate upon stomatal aperture, transpiration and photosynthesis. Plant Physiol. 38: 703-721.

Smillie RM, Gibbons GC. 1981. Heat tolerance and heat hardening in crop plants measured by cholorophyll fluorescence. Carlsberg Res Commu. 46: 395-403.

Smillie RM, Nott R. 1979. Heat injury in leaves of alpine, temperate and tropical plants. Australian J Plant Physiol. 6(1): 135-141.

Sole Z, Pixhas J, Loedenstein G. 1972. Evaluation of systemic fungicide and mineral oil adjuvants for the control of Mal Secco diease of lemon plants. Phytopathology. 62: 1007-1013.

Srinivasan A, Takeda H, Senboku T. 1996. Heat tolerance in food legumes as evaluated by cell membrane thermostability and chlorophyll fluorescence technique. Euphytica. 88: 35-45.

Stone PJ, Nicolas ME. 1998. The effect of duration of heat stress during grain filling on two wheat varieties differing in heat tolerance: grain growth and fractional protein accumulation. Aust J Plant Physiol. 25(1): 13-20.

Sullivan CY, Ross WM, Eastin JD, Clegg MD. 1973. In: Plant Physiology of Yield Mesurement of Sorghum in Relation to Genetic Improvement. University of Nebraska, USDA, ARS, North Central Region, The Rockefeller Foundation Annu Rep No. 7. pp. 43-57.

Sullivan CY. 1972. Mechanisms of heat and drought resistance in grain sorghum and methods of measurement. In: Sorghum in the Seventies, Rao NGP, House RL (eds.). New Delhi, India, Oxford & IBH Publishing Co. pp. 247-264.

Suzuki K, Tsukaguchi T, Takeda H, Egawa Y. 2001. Decrease in pollen stability of green bean at high temperatures and relationship to heat tolerance. J Am Soc Hortl Sci. 126(5): 571-574.

Talanova VV, Akinova TV, Titov AF. 2003. Effect of whole plant and local heating on the ABA content in cucumber seedling leaves and roots and on their heat tolerance. Russian J Plant Physiol. 50(1): 90-95.

Taub DR, Seeman JR, Coleman JS. 2000. Growth in elevated CO_2 protects photosynthesis against high temperature damage. Plant Cell Environ. 23: 649-656.

Thiaw S, Hall AE. 2004. Comparison of selection for either leaf-electrolyte-leakage or pod set in enhancing heat tolerance and grain yield of cowpea. Field Crop Res. 86(2/3): 239-253.

Tischler CR, Burson BL. 1995. Evaluating different bahaigrass cytotypes for heat tolerance and leaf epicuticular wax content. Euphytica. 84(3): 229-235.

Udomprasert N, Li PH, Davis DW, Markhart AH III. 1995. Root cytokinin level in relation to heat tolerance of *Phaseolus acutifolius* and *Phaseolus vulgaris*. Crop Sci. 35(2): 486-490.

Uppal HS. 1977. In Annual Reviews of Plant Sciences, Malik CP (ed.), Kalyani Publishers, New Delhi.

Villareal RL, Lai SH, Wong SH. 1978. Screening for heat tolerance in genus *Lycopersicon*. HortScience. 13: 479-481.

Wang WC, Nguyen HT. 1989. Thermal stress evaluation of suspension cell cultures in winter wheat. Plant Cell Rep. 8: 108-111.

Warrag MOA, Hall AE. 1984. Reproductive responses of cowpea (*Vigna unguiculata* L. Walp.) to heat stress II. Response to night air temperture. Field Crop Research. 8: 17-33.

Weaver ML, Timm H. 1989. Screening tomatoes for high temperature tolerance through pollen viability test. HortScience. 24: 493-495.

Wehmeyer N, Vierling E. 2000. The expression of small heat shock protein in seeds responds to discrete developmental signals and suggests a general protective role in desiccation tolerance. Plant Physiol. 122(4): 1099-1108.

Wimmer B, Lootspeich F, van der Klei I, Veenhuis M, Gietlc C. 1997. The glyoxysomal and plastide molecular chaperones (70 kDa heat shock protein) of watermelon cotyledones are encoded by a singal gene. Proc Natl Acad Sci USA. 94(25): 13624-13629.

Witter SH, Bukovag MJ. 1959. The uptake of nutrient through leaf surface. In: Handbuch der Pflsnzenemilhrung and Dungung, Linser H (ed.). Springer-Verlag, Berlin. Pp. 235-261.

Wu SH, Wong C, Chen J, Lin BC. 1994. Isolation of a cDNA encoding a 70 kDa heat shock cognate protein expressed in the vegetative tissue of *Arabidopsis*. Plant Mol Biol. 25: 577-583.

Xu RQ, Sun QX, Zhang SZ. 1996. Chromosomal location of genes for heat tolerance as measured by membrane thermostability of common wheat cv. Hope. Hereditas Beijing. 18(4): 1-3.

Xu RQ, Sun QX, Zhang SZ. 1997. Screening methods and indices of heat tolerance in spring wheat. Acta Agriculturae Boreali Sinica. 12(3): 22-29.

Yamane Y, Kashino Y, Koike H, Satoh K. 1997. Increase in the fluorescence F_0 level reversible inhibition of Photosystem II reaction center by high-temperature treatments in higher plants. Photosynth Res. 52: 57-64.

Yamane Y, Kashino Y, Koike H, Satoh K. 1998. Effects of high temperatures on the photosynthetic system in spinach: oxygen-evolving activities, fluorescence characteristics and the denaturation process. Photosynth Res. 57: 51-59.

Yang J, Sears RG, Gill BS, Paulsen GM. 2002. Quantitative and molecular characterization of heat tolerance of hexaploid wheat. Euphytica. 126(2): 275-282.

Zai XM, Wu GR, Lu CM, Gu GP, Rui HY. 2001. Effect of Ca^{2+} on heat tolerance and active oxygen metabolism of peanut seedlings. Chines J Oil Crop Sci. 23(1): 46-50.

Zee SY, Tso MYW, Tso WW. 1978. Electrophoretic analysis of seed proteins of heat-tolerant and heat-sensitive cultivars of Chinese cabbage. HortScience. 13: 547-548.

Zhang HY, Huang YJ, Wang DH, Qi YX, Zhong PA, Li GH, Liu K, Kuang HY. 2004. A study on development near isogenic lines of heat tolerance at grain filling stage in rice. Acta Horticulturae universitatis Jianxiensis. 26(6): 847-853.

Zhang X, Ervin EH, Schmidt RE. 2003. Plant growth regulators can enhance the recovery of Kentucky Bluegrass *sod* from heat injury. Crop Sci. 43: 952-956.

Zheng XY, Wang YJ, Song SH, Li L, Klocke E. 2004. Detecting molecular markers associated with heat tolerance of Chinese cabbage. Acta Hortculturae. 637: 317-323.

Zheng XY, Wang YJ, Song SH, Li L, Yu SC. 2002. Identification of heat tolerance linked molecular markers of Chinese cabbage (*Brassica campestris* L. ssp. *Pekinensis*). Agril Sci China. 1(7): 786-791.

Zheng XY, Wu GS, Wang YJ. 1998. Study on relationship between heat tolerance and phosphoglucomutase loci of Chinese cabbage. Acta Horticulturae Sinica. 25(3): 252-257.

Zhu CL, Xiao YH, Wang CM, Jiang L, Zhai HQ, Wan JM. 2005. Mapping QTLs for heat tolerance during grain filling in rice. Chinese J Rice Sci. 19(2): 117-121.

6

Cold Tolerance

6.1. INTRODUCTION

Low temperature is one of the major environmental factors that limit the plant growth. Many plants of tropical origin suffer cold damage when exposed to temperature below 20 ^{0}C (Graham and Patterson, 1982, Andrews, 1987). Cold temperature in crop plants are compounded by *cold snap*- a lower than usual drop in temperature that causes the crop to fail. Low temperature in growing season may reduce germination, retard vegetative growth by inducing metabolic imbalances and can delay or prevent productive development. Each plant species has an optimum range of temperature for its normal growth and development. It varies among the genotypes within a species, the specific temperature also depends on the growth stage and development of the particular genotype.

Low temperature cold injury can be categorized into two parts: (1) *Chiling injury*- the injury caused when temperatures remain above freezing point (>0 ^{0}C), and (2) *Freezing injury*- injury caused at temperatures below freezing point (<0 ^{0}C). Freezing injury may be of intracellular or extracellular. Intracellular freezing damages the protoplasmic structure and the ice crystals kill the cells once they grow large enough to be detected microscopically. In extracellular freezing, the protoplasm of the plant become dehydrated because a water vapour deficit is created as ice crystals are formed intercellular space.

Chilling temperature can damage the tissues of sensitive plants while freezing temperature will damage most tissues during active growth. Chilling sensitive plants are typically tropical and subtropical whereas, freezing sensitivity includes all plants. A chilling temperature is defined as a temperature low enough to cause plant tissue damage but not low enough to cause freezing of tissue water (Levitt, 1980). For most chilling-sensitive plants, chilling temperatures are between 10 and 0 ^{0}C (Lyons *et al*, 1979). Chilling injury is not translocatable. For example, when a cucumber plant was divided so that one shoot was chilled, which the other parts of the plant remained of warm temperatures, the chilling injury was restricted to that one shoot (Saltveit and Morris, 1990), which concludes that chilling injury is perceived locally, probably by each individual cell.

6.2. SYMPTOMS OF CHILLING INJURY

The visual symptoms of cold injury vary depending on the temperature, duration of exposure commodity, its stage of development and tissue, the time of day and other environmental conditions, such as light, wind, water and nutrients (Salveit and Morris, 1990). A brief chilling event will not produce obvious symptoms which the plant is at low temperature; the symptoms gradually appear afterwards, especially when the plant is return to optimal growth temperatures. In chilled bell pepper fruit most damage occurred after rewarming (Whitakar, 1995). Longer durations of chilling and lower temperatures cause the symptoms to appear more quickly in a time-dose dependant manner. The major visual symptoms of cold injuries are as follows:

1. Most plants stop growth at low temperatures, chilling sensitive plants suffers injury that persists after the plant returns to warm conditions. In many cases, plant development is slowed after chilling and the plants may remain visually stunted for prolonged period.

2. Severe chilling stress promotes cellular autolysis and senescence (Salveit and Moris, 1990). The degenerative or senescence process may be localized, causing surface lesions, such as spitting or sunken areas of the epidermis due to cell collapse. The loss of chlorophyll, apparent as leaf yellowish, may occur in the light as a consequence of photo-oxidation.

3. Chilling can detrimentally affect flower induction, pollen production and germination, and in some sensitive species it will cause sterility. Cold stress caused delayed heading and yield reduction due to spikelet sterility in rice (Andaya and Mackill, 2003a, b). In soybean, where the lowest temperature for fertilization and pod formation varies among

cultivars between 9 ^{0}C and 18 ^{0}C (Hume and Jackson, 1981), a difference that has been traced to pollen abnormalities at low temperature (Lawn and Hume, 1985). Even on cold night at 8 ^{0}C was sufficient to inhibit pod formation in field grown plants.

6.3. INJURY MECHANISM

Many crops that originated from tropical or subtropical regions are susceptible to injury when air temperature fall below critical threshold. The temperature pattern in the crop growing season determines the possible types of cold damage that may occur; it differs from region to region. The types of cold injury important in one country/region may not be important in another, suggesting the need of developing location specific varieties that will be resistant to the types of cold damage prevalent.

6.3.1. Membrane Damage

Membrane damage is considered as the primary effect of chilling injury. Under chilled environment, the normal lipid crystalline membrane changes to a solid gel state. This leads in changes in membrane organization, which reflects in the function of membrane bound enzymes and leakage of solutes.

Many forms of membrane damage can occur as a consequence of freezing-induced cellular dehydration including expansion-induced-lysis, lamellar-to-hexagonal-II phase transitions, and fracture jump lesions (Steponkus *et al.*, 1993). Freeze-induced production of reactive oxygen species also contributes to membrane damage (McKersie and Bowley, 1997). At sub-zero temperature, inter cellular ice can form adhesion with cell walls and membranes and cell rupture (Olien and Smith, 1977). Apart from this, the chemical potential of ice is less than that of liquid water at a given temperature, the formation of extra-cellular ice results in a drop in water potential outside the cell. Thus, there is movement of unfrozen water down the chemical potential gradient from inside to the intercellular space results in physical change to cell.

6.3.2. Prevention of Pollen Formation or Pollen Germination

Rice is originally a tropical plant, and it does not have an inbuilt resistance to cold. It is very sensitive to cold during flowering. The pollen doesn't set, the flowers are not fertilized under cold stress. A cell layer surrounding the pollen, called tapetum, is responsible for feeding the pollen with sugar. The tapetum is only active for 1-2 days, so if, a cold snap occurs at this time then there is no further chance for pollen growth in rice. Sugar can't easily move into the tapetum and pass through it to the pollen. If the pollen grains get no sugar they die and as rice is self fertilizing this means that no rice grains are formed. Instead, it has to be broken down then transported in bits to the pollen. Invertase is the catalyst that helps break down the sugar to transport it into the tapetum before it is transported to the pollen. Quantities of invertase are less in conventional rice when it is exposed to cold temperatures, but they remain at normal levels in a cold tolerant variety when it experiences cold. Cold snaps cause a reaction in the plant that prevents sugar getting to the pollen. Without sugar there is no starch build-up which provides energy for pollen germination. And without pollen, pollination can't occur, so no grain is produced.

Cold stress during the reproductive development of spring wheat cause grain set failure at high altitude, >1500m (Subedi *et al.*, 1998). Cool temperature (<10 ^{0}C) around heading prolonged the time to anthesis and led to greater degree of sterility in cold susceptible cultivars in wheat. The cooler it was between heading and anthesis, the longer was the period between these two phases and the more grains failed to set.

6.3.3. Abnormal Hormone Metabolism

Under stress, abnormal metabolism affects the normal metabolism of phytohormones. In general, the growth stimulators decreased and the inhibitors increase as an adaptation. Khusainova (1974) observed higher contents of growth inhibitors in the tillering nodes of cold-resistant winter wheat mutant of Kazakhastankay 126, and the growth stimulators disappeared during harvesting under natural conditions.

The induction in freezing tolerance can be inhibited by cycloximide (Chen *et al.*, 1983), indicating that protein synthesis is required. It also has been shown that the induction of freezing tolerance by ABA is associated with *de novo* synthesis of protein (Johnson-Flanagan and Singh, 1987; Lang *et al.*, 1989; Tseng and Li, 1990, 1991). Cycloheximide is a cytoplasmic protein synthesis inhibitor (Krizek *et al.*, 1985). Xin and Li (1992) reported that cycloheximide (0.5-10.0mm) in the ABA-containing medium inhibited to various degrees the synthesis of protein and development of chilling tolerance. The inhibition of protein synthesis may weaken the cell's viability and result in a decreased survival. The decrease in chilling tolerance of ABA-treated cells with the presence of cycloheximide is conceivably the result of a lack of *de novo* synthesis of protein (Xin and Li, 1992).

6.3.4. Imbibational Chilling Injury

As water is introduced to seed, it may become chilling sensitive (Herner, 1990). Imbibitional chilling injury occurs in sensitive seeds. If soil temperature is very low at planting, the initial imbibition of water disrupts membrane integrity, increase electrolyte leakage, and blocks germination. Imbibitional chilling injury may also occur to pollen of sensitive species; because the membrane in pollen may be studied *in situ* this system has been used as a model to study imbibitional damage (Crowe *et al.*, 1989). The lipid phase properties of membrane lipids in pollen are sensitive to both hydration and temperature. Normally, membrane lipids are in a fluid or liquid-crystalline phase, but at either low moisture or low temperature, they form the more rigid gel phase. If rehydration occurs when the membrane lipids are locked in the gel phase because of low temperature, their reorganization is faulty and they become leaky and dysfunctional. If rehydration occurs when membrane lipids are in the liquid crystalline phase, membrane reorganization is successful and sensitive to a subsequent low temperature treatment. Systematical it can be drawn as following (Crowe *et al.*, 1989):

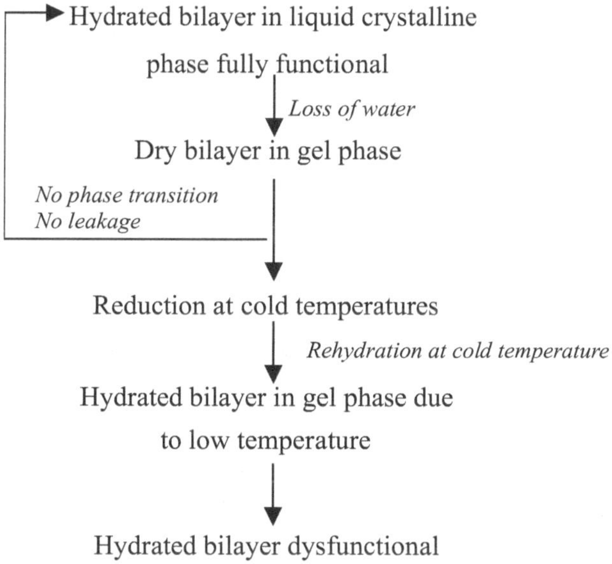

6.3.5. Reduction in Photosynthesis

Low temperature may lead to an accumulation of photosynthates in chloroplast by interfering with the export of assimilates. This may inhibit photosynthesis at temperatures higher than that at which photosynthesis would actually be inhibited. Chilling injury to photosynthesis is much more severe if exposure occurs coincident with moderate or high light intensities. The immediate consequences of exposure of leaves to low temperature are that the demand for chemical energy is automatically regulate water loss and CO_2 exchange is impeded. Stomatal aperture generally reduced at low temperatures, in part because water conductivity is decreased through the root plasma membrane (McWilliam *et al.*, 1982).

Chilling injury in night is not as severe as in the day. Although night temperature are important constraint in planting tropical legume crops, because photosynthesis is severely affected for several days following a single cool night (Bell, 1993). Chilling at night may disrupt electron transport, (Hallgren and Oquest, 1990), enzymes involved in CO_2 fixation (Sassenrath *et al.*, 1990), translocation of sugar from source to sinks leading to feedback inhibition of photosynthesis (Bagnall *et al.*, 1988) or altered water relations, due to slow stomatal responsiveness or reduced hydraulic conductivity of roots (McWilliam *et al.*, 1982).

At low temperatures, electron transport through photosystem II is inhibited. This photoinhibition may be explained as follows:

1. Low temperature promotes the excessive excitation of the photosystems by reducing the demand for chemical enzymes in process such as CO_2 fixation. Over excitation favours the transfer of excitation energy from light to oxygen causing photo-oxidative damage to the reaction *D1* protein.

2. The activity of enzymes that scavenge activated oxygen decreases at low temperature reducing these protective systems (Richter *et al.*, 1990), such as, catalase is photo-inactivated at low temperatures in chilling sensitive cucumber and maize (Feirabend *et al.*, 1992).

3. Low temperature prevents the repair of PS II. Turnover of the *D1* protein in the reaction centre is slow at low temperature (Gong and Nilson, 1989), therefore its assembly into new PS II complexes is blocked.

4. Low temperatures inhibit the formation of zeaxanthin (Bilger and Bjorkman, 1991). Zeaxanthin normally quenches excitation energy in the antenna of PS II and dissipates it as heat (Demmig-Adam, 1990). Zeaxanthin production from violaxanthin is induced under normal conditions by high light and low thylakoid lumen pH, but this process is blocked at low temperatures.

6.3.6. Oxidative Stress and Chilling Injury

Chilling injury is partly mediated by reactive oxygen species as agents causing the secondary injuries to membrane and photosynthesis. The development of chilling injury symptoms is frequently coincident with peroxidation of fatty acids (Parkinson *et al.*, 1989). Shewfelt and Erichson (1991) proposed that lipid peroxidation would alter the physical properties of membrane lipids, and thereby inhibit the function of membrane-bound proteins contributing to the development of visual symptoms of chilling injury.

Reactive oxygen species (ROS), such as superoxide, hydrogen peroxide and hydroxyl radicals increase under chilled environment (Omran, 1980; Hodgson and Raison, 1991; Prasad *et al.*, 1994a). These ROSs are degraded by catalase, and it is prevented in chilled-affected plants due to the induction of catalase inhibitor. ROSs, therefore, accumulates and acts as a source of free radical oxidants, which aggravate the chilling injury. Chilling tolerant plants are known to have more efficient antioxidant systems that chilling sensitive ones (Walker *et al.*, 1990; Dipierro and Leonardis, 1997).

6.3.7. Other Components of Cold Injury

Besides temperature, water availability, humidity, light intensity and time of day also influence the amount of chilling injury sustained. At low humidity, the effects of chilling are compounded by its effects on plant water relations because chilling reduces the ability of the roots to supply water (McWilliam *et al.*, 1982) and blocks stomata function (Guye and Wilson, 1987). High light intensity during day provides excess excitation energy to the photosystems promoting photoinhibition and photoxidation (Hallgren and Oquist, 1990). Diurnal fluctuations in metabolic pools, such as sucrose and glutathione, activation of enzymes by thiredoxin, or gene transcription and translation, may critical roles altering the cells sensitivity to low temperatures.

6.4. TOLERANCE MECHANISMS

Cold tolerance mechanism involves a number of biochemical and physical changes, which includes transient increase in abscisic acid levels, changes in membrane lipid composition, accumulation of compatible osmolytes, increase levels of antioxidants. Major metabolic changes in carbohydrates, have been documented during acquisition of cold tolerance. At plant or organ level, under low temperature there is reduction or cessation of growth, reduction in germination, poor seedling establishment, leaf discolourization may affect development of reproductive organs, delayed heading, incomplete excretion, prolonged flowering period due to irregular heading, degeneration of spike lets, irregular maturity, sterility, formation of abnormal grains and post harvest longibility, wilting, chlorosis, necrosis and ultimately death of the plant. Certain stages of life cycle of a palnt are sensitive to chilling than other (Sataka and Koike, 1983; Patterson *et al.*, 1987). Whether a particular temperature regima will cause cold damage depends on the variety and the growth stage of the rice plant (Nanda and Seshu, 1999).

Studying cold tolerance in the field is difficult. Field sites often exhibit either complete survival or complete winter kill. Because of this variability, laboratory procedures to measure freezing tolerance have been developed by a number of investigators. These include plant tissue, water content (Brule-Babel and Fowler, 1988), ion leakage from plant cells after a freezing stress (Teutonico *et al.*, 1993), and changes in luminescence (Brzestowicz and Barcikowska, 1987). Meristem growth after plants are subjected to freezing temperatures is also commonly estimates cold tolerance (Fowler *et al.*, 1981; Andrews and Morrison, 1992).

6.4.1. Biochemical and Physiological Basis of Tolerance

There are several biochemical and physiological processes and their end products influence the cold tolerance ability of plants; few of them briefed below.

6.4.1.1. Polyamines

There is a growing interest in the possible involvement of polyamines in the defence reaction of plants to various environmental stresses. It has been found that chilling tolerant plant increase endogenous polyamine levels in response to chilling to a much greater extent than chilling-sensitive ones (Guye *et al.*, 1986; Kramer and Wang, 1990; Lee, 1997). Polyamines such as spermidine (Spd) and spermine (Spm) occur ubiquitously in plants together with their diamine precursor putrescine (Smith, 1985). It is also synthesized by the decarboxylation of L-Arg by Arg decarboxylase (ADC), via agmative and N-carba-moglputrescine intermediates. Because of the polycationic nature at a physiological pH, polyamines can bind strongly to the negative charges in cellular components, such as nucleic acids, proteins and phospholipids. Interactions of polyamines with membrane phorpholipids may stabilize the membranes under condition of stress (Roberts *et al.*, 1986). *Spd* is involved in the chill-included increase of putrascine content in rice plants. It is known that abscisic acid and possibly ethylene are involved chilling tolerance of cucumber cultivar (Shen, 1999 a, b), Lee *et al.* (1995) have shown that the increase of abscisic acid content is responsible for the chilling tolerance of plants (Rikin and Richmond, 1976; Ciardi *et al.*, 1997; Morgan and Drew, 1997). Putrescine is primarily responsible for the chilling tolerance of bean (Guye *et al.*, 1986, Nadeau *et al.* 1987) and rice (Lee *et al.*, 1997).

6.4.1.2. Membrane-lipid unsaturation

The lipid composition of membrane is a key determinant for cold tolerance, and enzymes that modify membrane structure seem to be important for low temperature accumulation. Chilling tolerant varieties show a higher degree of membrane-lipid unsaturation than do susceptible varieties. The chilling sensitivity of higher plants is closely correlated with the degree of fatty acids unsaturation in the thylakoid membranes of these chloroplasts (Murata *et al.*, 1982). Thus membrane lipid fatty acid plays an important role in cold adaptation (Mastronicolis *et al.*, 2005). An increased production of highly unsaturated fatty acid, such as hexadecatrienoic acid at low temperature was observed in connection with cold acclimation in many plants (Graham and Patterson, 1982). A correlation between chilling sensitivity and degree of unsaturation of fatty acids in phosphotidylglycorol in plastid membranes has been reported in higher plants (Murata, 1983; Tasaka *et al.*, 1990).

The crucial role of polyunsaturated fatty acids in cold tolerance has been pointed out in studies with inhibitor of phosphatiglycerol synthesis (St. John *et al.*, 1979) and also with the *fod* mutants of *Arabidopsis thaliana*, which are effective in the desaturation of membrane lipids (Browse and Somerville, 1991). Cold tolerance was enhanced in cyanobacteria by an increase in amount of unsaturated fatty acids in the genetically engineered membrane lipids (Wada *et al.*, 1990). Makarenko *et al.*, (2003) also reported that cold tolerance was caused by a high content of unsaturated fatty acid residues in the total membrane lipid.

Qui *et al.* (1990) had noted that the seed oil content highly and significantly correlated with germination vigour. An association between cold resistance and the content of lipids in the embryo axis was found by Moreru and Syrku (1991). Albidum 114, a wheat variety differed from the other varieties in having a higher embryo to whole grain lipid weight and a higher lipid weight per embryo. Duhalde *et al.*, (1991) reported lower proportions of linolate (18:2) and higher of oleate (18:1) in all of the susceptible maize lines lipids than the cold-tolerant ones. The relative amounts of 18:2 and 18:1 in maize grain lipids are heritable traits, and may be important among the factors affecting the ability of maize to germinate at low temperatures. Novitskaya *et al.* (1990) reported that resistant species accumulate more linoleic acid than the susceptible ones. Reduced saturation of lipids in resistant plants forms the basis of adaptation to chilling temperatures. Li *et al.* (1989) observed that the high linolic acid: oleic acid ratio in dry seeds showed high cold tolerance. Plants with a high proportion of cis-unsaturated fatty acids are resistant to chilling and others with only a small proportion are not (Murata *et al.*, 1992).

Gomes *et al.*, (2000) had characterized *ALA1* (for aminophospholipid *ATPase1*), a novel P-type in *Arabidopsis* that belong to family *ALA1* to *ALA11*. The deduced amino acid sequence of *ALA1* is homologous with those of Yeast *DRS2* and Bovine ATPase II, both of which are putative aminophospholipid transtocases. *ALA1* complements the deficiency in phosphatidylserine internalization into intact cells that is exhibited by the *drs2* Yeast mutant, and expression of *ALA1* results in increased translocation of aminophospholipids in reconstituted yeast membrane vesicles. Gomes *et al.* (2000) also suggested that there is a link between regulation of transmembrane bilayer lipid arymmetry and the adaptation of plants to cold.

Novitskaya *et al.* (2000) studied the effect of a composition of polar lipids and their fatty acids into tomato (*Lycopersicon esculentum* Mitl., cv. *Sibirskie Skorospelye*) leaves. They demonstrated that chilling resulted in a decrease in the content of total polar lipids/mg of protein. The content of lipids in chloroplast membranes (monogalactosyl-diacylglycerols, digalactosydiacy-lglycerols, sulfoquinovosyldiacyl-glycerols, and phosphatidylglycerols) changed less substantially than then the content of phospholipids in other cell organelles and in the cytoplasm. Neutral lipids comprised only 1% of total lipids, and their content also decreased after chilling. The relative amount of unsaturated and saturated fatty acids in polar lipids was practically unchanged. These results suggest that the maintenance of a high level of chloroplast membrane lipids under low temperatures play an important role in the survival of cold tolerant important plants.

6.4.1.3. Phytohormones

There is a positive correlation between ABA accumulation and chilling resistance. ABA could be as a biochemical marker for cold tolerance in crop plants (Dorffling *et al.*, 1990). Application of exogenous ABA resulted in reduce electrolytes from left, retordation of leaf, decolouration, increase of soluble sugar content in rice seedling (Guo and Pan, 1984). Chen *et al.* (1983) also found that exogenous application of ABA increased freezing tolerance in *Solanum commersonii* plants. Some of the *COR* gene family- *COR78, COR47,*

COR15a, and *COR6-6* are highly expressed in response to the exogenous application of ABA (Hajela *et al.*, 1990; Kurkela and Franck, 1990; Gilmour *et al.*, 1992; Nordin *et al.*, 1993; Yamaguchi-Shinozaki and Shinozaki, 1993). The activation of those genes by exogenous ABA is consistent with observing an increase in freezing tolerance. ABA has been shown to reduce chilling injury in a number of other chilling sensitive plants (Eamus, 1987; Krizek *et al.*, 1985; Rikin *et al.*, 1983; Markhart, 1984; Duncan and Widholm, 1987). Pretreatment of seedling with ABA resulted in reduced chilling injury (Rikkin *et al.*, 1976). Similar result was observed by Rikin *et al.* (1979) in cotton. Galiba *et al*, (1983) studied the callus cultures of 3 wheat cultivars in frost resistance and also the *5A* and *5D* chromosome substitution times from the frost resistance variety Cheyenne into frost sensitive Chinese Spring. Following cold hardening, the increase in ABA level in the calli of two frost resistant cultivars was significantly higher than in those of the frost sensitive cultivars.

An abscisic acid (ABA) analogue under laboratory conditions is very effective in preventing leaf rolling and death of seedling during low temperature (Flores *et al.*, 1988a). The analogue is also effective in reducing transplanting shock. ABA analogue is effective in maintaining not only water balance, but also membrane stability (Flores *et al.*, 1991). Application of ABA analogue *LAB173711* decreased electrolyte leakage and minimized sterility in rice. The protective mechanism of *LAB173711* seems to involve the following processes:

1. Stomatal closure, which reduce water lose during chilling and drought stress, thus increasing leaf water potential and fresh weight and reducing retting (Flores and Dorffling, 1990).
2. Increases stability of membrane by a decrease in electrolyte leakage (Flores *et al.*, 1988b)
3. Prevention of chlorosis by an increase in chlorophyll fluorescence.
4. Stability in the root system
5. Decrease in spikelet sterility

ABA also improves water status by early closure stomata during chilling (Eamus, 1987; Rikin and Richmond, 1976). Because the water potential of a liquid medium changes little during culture was well as during chillig exposure, the complication of water stress during chilling is greatly in the maize suspension-cultured cell system (Xin and Li, 1992). Therefore, ABA induced chilling tolerance in maize suspension-cultured cell I likely to reflect a true tolerance to direct effect of chilling rather than amelioration of secondary stress during chilling exposure. It also facilitates uptake of exogenous chemicals into cells.

Singh and Jhonson-Elanagan (1989) reported that freezing tolerance was induced by ABA at 25 ^{0}C in a microspore derived from rape cultivar. *In vitro* labeling identified the appearance of a 20 KDa polypeptide concomitant with freezing tolerance induction. Poly(A) + mRNA isolated from ABA hardened cell programmed the *in vitro* translation of 20 KDa polypeptide but that from non-hardened cells did not.

Salicylic acid pretreatment provided protection of cell membrane of photosynthetic functions of cold-stressed banana seedlings, which enhances the cold tolerance of plants (Kang *et al.*, 2003).

6.4.1.4. Vernalization

Low temperature is one of the primary stress factors that limit growth and productivity of winter annual plants. The vernalization response and low temperature acclimation are among the most important for winter survival mechanisms. Laroche *et al.* (1992) estimated cell survival on excise leaves to determine freezing tolerance. Rapacz and Markowski (1999) found a significant correlation between vernalization requirement and both frost resistance and field survival when looking at older, high erucic acid cultivars of rape plants. Period of vernalization also plays an important role for reproductive growth phase. In cereals, most research findings indicated that although genotypes with longer vernalization requirements are generally more frost resistant, it was possible to find genotypes that require shorter vernalization period but exhibit high levels of cold tolerance (Brule-Babel and Fowler, 1988; Doll *et al.*, 1989). Fowler *et al.* (1996) reported that longer vernalization requirement delayed plant from entering the reproductive growth phase, a cold sensitive plant growth stage.

6.4.2. Morphological Bases of Tolerance

Chilling sensitivity is manifested by certain plants, whoes germination, growth, development of reproductive organs, and post harvest longevity are affected in the range of chilling temperature from 0 ^{0}C to about 15 ^{0}C (Wang, 1990).

6.4.2.1. Pollen fertility

Chilling temperatures at flowering cause floral abortion in most of the chick pea cultivars (Srinivasan *et. al.* 1999). Meiosis in anthers is the most suscepible to chilling. Distinct variation in flowering morphology, gamete development (viability and size of pollen and ovules) and function (pollen germination and tube growth, ovule viability and viability and fertilization). Srinivasna *et al.* (1999) found that the greater pod-setting ability of tolerant lines was associated with a higher pollen vigour (germination and tube growth) and ovule viability at low temperature in the sensitive lines. The number of ovules was not affected by cold stress in all cultivars/ lines but pollen size and viability were reducing in susceptible lines. The maturation of pollen is the most sensitive process in the entire life cycle of chilling-sensitive plants (Sataka and Koike, 1983; Patterson *et al.* 1987).

In cold tolerant varieties of rice, meiosis was disrupted at a lower temperature than in susceptible varieties. In case of sorghum, pollen sterility in chill-affected plant increased with a reduction in proline content of the pollen. Thus, high, proline content of pollen seems to confer on them chilling tolerance, proline may possibly be involved in chilling tolerance of other tissues as well.

Fertility in rice plants decreased with increasing length of cooling period and with decreasing temperature. Though, they were selected for tolerance at earlier growth stages. The number of pollen grains shade on the stigma and the number that germinated were reduced by cooling treatment. However, pollination with untreated pollen grains rested fertility in treatments of less than 12 days at more than 12 ^{0}C. This explains the fact that even when the rice plant grows well after adaptation to the environment, physiological process like pollination can be hampered (Satake and Koike, 1983).

6.4.2.2. Enhanced seed germination and seedling vigour

Genetic variation in chilling tolerance at germination is known in many crops, but nature of tolerance is little known. As a group dicots are less cold tolerant than the monocots. The C_4 crops which are essentially tropical are loss cold/frost-tolerant. To date, measurement of chilling tolerance, as germination under chilling stress has attained a high degree of sophistication, statistical, treatment, interpretation and use as a selection criterion.

In red kernelled rice varieties have adapted better than the white kernelled varieties, showing higher germinability at lower temperatures (Ichi and Tamai,

1988). In case of sorghum, cold tolerance refers to the ability of genotypes to germinate grow, and produce satisfactory grain yields under conditions of relatively cold (but above freezing) air and soil-temperature (Smith, 1985). Crobie *et al.* (1980) suggested percentage of emergence, emergence index (rate of emergence) and seedling dry weight as cold tolerant in maize.

Prostrate grain habit of legumes, has been found to be generally associated with cold tolerance. Lines showing cold tolerance usually take more time to emerge than those susceptible (Malhotra and Saxena, 1993).

6.4.2.3. Floral parts variation

Variations had been observed in fruit and seed setting in crop plants under chilling temperature, which is the result of floral structure and function. Tolerance in rice was associated with longer anthers and stigmas, while in brinjal it was related to longer styles, and flower position within a flower cluster.

6.4.2.4. Cell size

Direct significant positive correlation between size and chilling tolerance had been reported in crop plants. McMurphy and Rayburn (1992) noted that guard cell size of all cold-tolerant maize populations was longer than the cell size of the respective susceptible populations. Chloroplast number per guard cell was also higher in all the cold-tolerant populations than in their parental population. In controrary, the percentage of plant surviving winter correlated negatively with cell and vacuole size and positively correlated with cell-wall thickness. Findings of Baraskova *et al.* (1980) revealed that the hardy variety of wheat, albumin 114 has smaller cells and smaller growth rate than the other non-hardy varieties. Inanov (1980) also reported that cold hardy rye varieties differ in leaf epidermis characteristics, smaller cells with a thick cell wall, small vacuoles and smaller intercellular space.

6.4.3. Use of Chemicals to Induce Chilling Tolerance

The application of chemicals to plants prior to this exposure to chilling temperatures has had considerable success in enhancing chilling tolerance in laboratories (Wang, 1990). Vaccum-infiltration of calcium into avocado fruit before low temperature storage reduces their sensitivity to chilling injury (Chaplin and Scott, 1980). The application of the antioxidants ethoxyquin or sodium benzoate to cucumbers or sweet pepers reduced chilling injury (Wang and Baker, 1979). The exogenous application of ABA can substitute for low temperature to initiate acclimation in a number of species including cotton (Rikin *et al.*, 1979). The application chemicals in the triazole family including paclobutrazol, uniconizole and triadimefon protect plants from many stresses including chilling (Fletcher and Hofstra, 1985; Wang, 1985). The protective action of the triazoles may be related to the stimulation of ABA synthesis (Asare-Boamah and Fletcher, 1986), which has been suggested mode of action of another chemical, mefluidide, that also improves chilling tolerance (Tseng and Li, 1984).

6.5. GENETICS OF COLD TOLERANCE

Genetics of different characters that involve in chilling tolerance had been studied in major crops. Most of the cases showed that the additive gene effects were predominant, but dominance effects were also present. Significant effects were observed for germination at low temperature in all the crops that were studied. Electrolyte leakage after cold treatment and germination, seedling vigour, floret sterility and plant growth are all under polygenic control; only leaf discoulorization under chilling in rice and maize are governed by oligogenically.

Zeng and Deng (1982) reported that cold tolerance in rice is governed by single dominant gene. Genetic analysis of components of cold tolerance at seedling stage revealed non-additive gene action in expression of the character (Datta and Siddiq, 1983). Acharya and Sharma (1983) and Acharya (1987) observed additive and dominance effects; dominance predominant for all traits except panicle weight and spikelet fertility. However, Moon (1984) observed significant non-additive variance for cold induced sterility. Additive genetic variance was fairly consistent over generations; while dominance variance was less stable. A continuous uni-nodal distribution for the F_2 population suggested polygenic inheritance (Khan *et al.*, 1986). Shahi and Khush (1986) studied the inheritance in seven hybrids involving tolerant and non-tolerant parents. Cold tolerance measured by leaf yellowing was governed by a single dominant gene. A single dominant gene also controlled cold tolerance as measured by root generation ability.

Manishi *et al.* (1991) studied the inheritance of panicle exertion in rice under low temperature. They found that the well-exerted panicle character is dominant, whereas, partially exerted panicle character is recessive. Nagamine and Nakagahra (1989) have established an optimum method for evaluating chilling injury at the seedling stage and reported wide genetic variance and geographic distribution in the character among indigenous rice varieties. Genetic control of chilling injury was analysed at 5 ^{0}C for 4 days in a growth chamber (Nagamine and Nakaghara, 1991). They concluded that tolerance for chilling injury at the seedling stage is controlled by a single dominant gene. However, no linkage was found between chilling injury and low temperature chlorosis, indicating that the characters are controlled by different loci. Thus at least tow loci control tolerance for low-temperature stress during the seedling stage.

General combining ability (GCA) associated with the cold tolerant cultivars, cold tolerance was controlled by polygenic and both additive and dominance gene action were important with the major contribution being from dominance effects. But, Datta and Siddiq (1985) reported relatively high SCA effects in the F_1 and suggested a major role for non-additive gene action. With respect of yellowing, however, additive and additive × additive gene action appeared to predominant. Segregation analysis suggested that the cold tolerance of chilling is controlled by a single dominant gene (Nagamine, 1991; Xiong *et al.*, 1990 Mahishi *et al.*, 1991; Sthapit and Shrestha, 1991). However, Xu and Shen (1988) reported tow dominant duplicate gene for governing cold tolerance.

Cold tolerance is a complex quantitative character which is strongly influenced by environment. Dominant genes acted in the direction of lower frost resistance and recessive genes in the direction of higher levels of resistance. Three chromosomes viz. *4D*, *5D*, and *7A* of the cold resistant wheat variety Cappelle Desprez are involved and the action of three chromosomes is additive (Law and Jenkins, 1970). At least one dominant gene associated with cold hardiness differences between spring and winter wheat, while genes with mainly additives effects determined differences in cold hardiness among cultivars with the winter growth habit (Brule-Babel and Fowler (1988). Linin and Fowler (1993) also supported that the cold hardiness in wheat is controlled by both additive and dominant genes. The frost sensitive varieties had the largest number of dominant genes, while the frost resistance varieties had the highest number of recessive genes. At least 10 of the 21 chromosome pairs appear to be involved in frost resistance, with chromosomes 5A and 5D implicated most often and seed to carry major genes (Sutka, 1989).

Sutka (1981; 1983) found significant GCA and SCA in wheat, indicating additive and non-additive gene action in inheritance of frost resistance. The high GCA: SCA ratio revealed a predominance of additive genetic variance. The major portion of the genetic variability was associated with specific combining abining ability or specific heterosis effect due to dominance or additive × additive gene action (Parodi *et al.* 1983).

Cytoplasmic factors have been implicated in the control of low temperature tolerance (Cahalan and Law, 1979). However, Sutka (1981) have failed to show significant differences between reciprocals in F_1 hybrids. Khristolyubora *et al.* (1974) found that winter hardy forms have high adaptability connected with a rapid protective response of the cell mitochondria and endoplasmic reticulum. This adaptability is regulated by nuclear genes. *Triticum agropyron* Hybrids (THA) have different mitochondria in the coleoptile cells, if genome 'X' of *agropyron* is present, which TAH derives from male parent the related mitochondria become more resistant to low-temperature treatment.

Additive multiple gene factor genetic system affects cold tolerance in maize (Grogan, 1970). Epistatic gene effects contributed significantly to the variation observed for germination of maize at 72 ^{0}C in the laboratory and for emergence measured in field (McConnell and Gardner, 1979 a,b). Seedling vigour growth after emergence in field appeared to be conditioned predominantly by additive and dominance gene effects although some epistatic effects were detected. Seedling vigour or growth after emergence in the field is conditioned predominantly by additive & dominance gene effects (Ferance, 1980). Additive effects are more important in controlling low temperature response during germination, while non additive effects are important at the 3 leaf stage (Cabuba and Ochesann, 1978). The ratio of GCA and SCA indicated that both additive and nonadditive gene effects are involved in cold tolerance in maize (Chapman and Drolsom, 1983). The genetic nature of cold tolerance in maize is complex because of significant maternal and cytoplasmic effects associated with inheritance of germination and early growth (Grogan, 1970). Xie (1992) also stressed that the maternal effects has a greater influence on low temperature tolerance in hybrid than does the pollen parent.

6.6. VARIABILITY FOR COLD TOLERANCE

Difference in adaptation by some species and genotypes within the species to cold has resulted in genetic variability. Difference in variation for cold tolerance has also been noted between groups, dicots are less cold-tolerant than monocots, the C_4 plants which are essentially tropical, are less tolerant than C_3 plants. *Japonica* cultivars of rice are known to more tolerant than *indica* genotypes (Andaya and Mackill, 2003a). Sometimes cold tolerance ability may be introgressed from related wild species.

Stages of plant development also differ in their sensitivity to chilling. In vegetative stages, seedlings are generally more sensitive to chilling than mature plant. Variations to chilling tolerance also exist among plant parts. Roots, rhizomes and bulbs are more sensitive to chilling than vegetative shoot organs, because of the temperature moderating capacity of soil; these tissues in the field do not experience the same temperature as the shoots.

The most common method of reducing plants sensitivity is by the development and use of more chilling tolerant cultivars. Ample variability is present in well adopted breeding population and germplasm (Table 6.1); for example, maize (Meidema, 1982), soybean (Holmberg, 1973), tomato (Walker *et al.* 1990) and penut (Bell, 1993). Screening of germplasm collected from high altitudes and lower temperature margins of geographical distribution of the concerned crop may enable in isolation of cold tolerant genotypes. Variation may also be created by mutation and somaclonal breeding.

Table 6.1. Chilling tolerance crop genotypes

Crop	Genotype	Reference
Chickpea *Cicer arientium*	ICCV-88552, ICCV-88503, ICCV-88502	Srinivasn *et al.*,1999
	ILC-794, ILC-1071, ILC-1251, ILC-1256, ILC-1444, ILC-1455, ILC-1464, ILC-1875, ILC3465, ILC-3598, ILC-3746, ILC-3747, ILC-3791, ILC-3857, ILC-3861, FLIP-85-81C, FLIP-82-85C, 82-313C, -84-112C, -85-4C, -85-49C	Singh *et al.*, 1989
C. bijugum	ILWC-7-1, 7-2, 7-4-1, 7/S-1, 7/S-3, 7/S-4, 7/S-5, 7/S-11, 7/S-12, 7/S-13, 7/S-14, 7/S15, 7/S-17, 7/S-18, 8-3, 8-4, 8S-1-8 S3, 32-2, 34/S-1, 34/S-2,42/1,42/2	Singh *et al.*, 1990
C. pinnatifidum	ILWC-29/S-10	Singh *et al.*, 1990
C. echinospermum	ILWC-35/S-3	Singh *et al.*, 1990
Wheat *Triticum aestivum*	Albumin 114, Odesskara 51	Moreru and Syrku, 1991
	Albumin 24	Voinikov and Korytov, 1991
Rice *Oryza sativa*	Padi Sashal, Lambayaque 1, Mitak, Padi Labou Alumbis, Jumali, Pratao, Lengkwang, Silewah, Thagone, Dourodo Aguillia, G-21	IRRI, 1978
	KH 998, Calrose, KH1h351, Fujisaka 5	Coly, 1979

Contd...

Table 6.1. Contd...

	IET 6155, HPU 734, K 332, K-330, VL-191, VL-206, IR 781, IR 667, CR 126-42-1, CR 237-1, BG 34-8, K 28-1, K 46-15- B2-1, HK 32-162	Nanda and Mani, 1983
	Managala, Halubblu (S-317)	Manishi *et al.*, 1991
	K-39-96, Suweon 235	Kaw, 1991
	VL Dhan-163, Himalaya-741, Pant Dhan-6	Ram and Sigh, 1994
	RAU-1345-2, IR-56383-77-1-1-1-1, IR61608-28-70-2-2-1-2, Prabhat	Nilanjaya *et al.*, 2001
	Jinyou 402	Dai *et al.*, 2004
Maize *Zea mays*	MvTC290, MvSC429	Gupta and Kovacs, 1974
	BSSS2 (SCT), BSSS13 (SCT)	Mock and Bakri, 1976
	Gi3523	Stamp, 1979
	B73(VS × B14)-2-1, Mo17	Mock and McNeill, 1979
	C123, W117, B37, Mo17	Bocsi, 1988
	Pratap, WCYC3	Dhillon and Reddy, 1988
	CO255	Aidin *et al.*, 1991
	Labo78	Fascaroli, 1995
	SVP-PD7	Schans *et al.*, 1995
	CO255, CO304, CO308	Hodges *et al.*, 1997
	ZPBL1304	Ristic *et al.*, 1998
	B7cmsC, F564, Co158, W153R, Mo17	Paldi *et al.*, 1998
	Bozm855, PMS636, PoblacionD, Poblacion E, BOZM696	Bradolini *et al.*, 2000
Pea *Pisum sativum*	PI-102888, -125673, -163131, -167253, -173054, -173780, -180867, -180868, -184128, -189168, -210641, -220184, -251051, -269818	Auld *et al.*, 1983
	K-1053, -5284, -6140, -3198, Orlovski-29, Izum rud, Raman, Mutant-57, Dalibor	Balachkova *et al.*, 1986
P. arvense	PI-358620, -377693, -383730, -383731	Auld *et al.*, 1983
Pigeon pea *Cajana cajan*	NDA. 88-2, Bahar, IPC-4193, BR-111, NDA 91-1, NDA 91-2, MA-2	Singh *et al.*, 1997a

6.7. SCREENING AND SELECTION

The selection criteria for further developmental programme for chilling tolerance can be based on the parameters used by various workers and may be cited as germination, growth of radicle, hypocotyls or seminal roots, seedling growth rate and survival membrane stability, pollen variability, pollen fertility, lipid content/unsaturated fatty acids, protein/enzyme content, proline content, total soluble solids/carbohydrate (sugar) content, chlorophyll loss under chilling stress, photosynthesis seed/fruit set (Table 6.2).

Table 6.2. Selection for cold tolerance based on morpho-physiological characters

S No.	Selection criteria	Screening technique	Remarks
1.	Chlorophyll loss under stress	Leaf/seedling colour	Effect of light intensity
2.	Growth under stress	Dry weight, stem/leaf dry weight ratio	Commonly used to compare lines Unstable for segregating generation
3.	Germination	Field emergence Germination in incubators under stress	Extensively used
4.	Membrane stability	Leakage of solute is determined conductometrically	Quite usefull
5.	Photosynthesis	Variable chlorophyll florescence at 685 nm	Promising technique
6.	Seedling mortality	Seedling survival under stress	Simple
7.	Seed/fruit set	Fruit/seeds produced under stress	Simple
8.	Pollen fertility	Pollen sterility under stressed plant	Indirect selection

6.7.1. Germination and Seedling Vigour under Chilling Environment

Normal rate of germination in some crops is affected by chilling temperature, and variation in germination among the varieties of crop had also been reported. In such case, germination in the field/controlled environment is used as the indices of selection for chilling tolerance; which is usually quit effective. But chilling tolerance at germination may not reflect tolerance ability of the plant in the subsequent developmental stages.

Rice seeds germinated at 10 ^{0}C, showed significant differences in the rate of germination (Sreedharan and Rao, 1983). Li *et al.* (1989) emphasized relative germination rate determined between percentage germination at low and optimum temperature as a selection criterion for rice. Regression analysis of 4 cycles of recurrent selection indicated that germination of maize seeds under cold laboratory conditions improved in pioneer cold tolerant and Iowo Stiff Stalk (McConnell, 1979, McConnell and Gardner, 1979 b). Genotypes those are tolerant during germination not always necessary to continue its tolerant ability through the different stages of growth. Selvakumar *et al.* (1987) reported that rice varieties tolerant during germination, but none were tolerant at all stages vegetative growth and flowering. Cold tolerance at different stages inherited independently (Dickson and Petzoltt, 1987).

Scott and Jones (1982) screened tomato varieties for low temperature germination. Based on germination in cold environments, selections were made in parental and F_2 families derived from a diallel cross of 2 cold-tolerances and 4 intolerant tomatoes inbreed lines (Vos *et al.*, 1982). Average response to selection was significant among the families. Li *et al.* (1990) selected S78 TE as cold tolerant based on germination in laboratory at 10 ^{0}C and it was noted that the seedling vigour after chilling increased substantially.

Nilanjaya *et al.* (2001) screened 21 rice varieties based on germination percentage, seedling vigour and cold tolerance score. Based on the mean value of rice cultivars of the mentioned parameters, RAU-1345-2, IR-56383-77-1-1-1-1, IR-61608-28-70-2-2-1-2 and Prabhat showed optimum tolerance to cold with least damage to the seedling.

6.7.2. Growth and Development under Chilling Stress

Gyue *et al.* (1987) assessed the chill sensitivity of three *Phaseolus* species by measuring pigment loss, leaf diffusion resistant, relative growth rate recovery, change in leaf water and severity of leaf necrosis on return to the warmth. Growth of excised roots may also be used for screening cold tolerant varieties (Burbanova, 1980). Xu and Chen (1989) found it more suitable to select rice varieties at seedling stage.

The differences in vigour, growth and development have been observed among the rice varieties under cold environment. The optimum time to make observations for screening tolerant genotype would be seedling, tillering, flowering and mature stages. Standard Evaluation System for Rice as outlined by International Rice Research Institute (IRRI, 1988) can be used for assessment of rice genotypes for cold tolerance at seedling and adult active growth stages (Table 6.3).

Table 6.3. Scoring of cold tolerance at seedling and adult plant stages in rice (IRRI, 1988)

Seedling stage		Adult plant stage	
Scale	**Symptoms**	**Scale**	**Symptoms**
1	Seedling dark green	1	Plant show a natural colour, rate of growth and flowering
3	Seedling light green	3	Plant slightly stunted and growth slightly stunted and growth slightly delayed
5	Seedling yellow	5	Plant moderately stunted leaves pinkish and development delayed
7	Seedling brown	7	Plant severely stunted, leaves yellow, development delayed and panicle poorly exerted
9	Seedling dead	9	Plant severely stunted with leaves brown, development much delayed and, panicles not exerted

6.7.3. Pollen Fertility

The potential of pollen selection as part of the breeding efforts to increase chilling tolerance investigated in many crops. Male gametogenesis in flowering plant, from archesporal cells to mature pollen, involves a series of complicated events (Goldberg *et al.*, 1993). The differentiation and development of the male gametophyte of angiosperms depend on the expression of the haploid genome after meiosis. Mature pollen grains contain mRNA whos protein products appear to function during the late maturation stages of pollen tube growth and germination (Mascarenhas, 1975). Meiotic stages of PMCs is extremely sensitive to chilling; this is the stage at which chilling must be applied if pollen fertility is to be used as a selection criterion.

Pollen grains from maize lines have higher percentage germination and a better capacity to develop normal pollen tubes at 12 ^{0}C. Lyakh and Soroka (1990) have suggested this as a parameter of assessing cold tolerance in maize. Cold storage of pollen for a week 1-4 ^{0}C and then applied on female flowers is effective for selection of cold tolerant plants (Barnabas and Kovacs, 1988).

The seed set after pollination with cold treated pollen showed a higher germination frequency at low temperature than seed set after pollination with untreated pollen (Kovacs, 1989)

Haploid selection for traits related to pollen cold tolerance in tomato was performed by Domiguez *et al.* (2004) in segregating population derived from a *Lycopersicon esculentum, L. pennelli* hybrid. Back cross, populations were obtained by combining normal and low temperature treatments on two stages of pollen development: pollen formation, and germination and pollen tube growth. F_1 hybrids were cultivated under low and normal temperatures and pollen was used to pollinate *L. esculatum* plants at low and normal temperature. The four BC, populations obtained were tested for the quality and quantity of pollen produce at low temperatures. The population obtained by cold treatment at both stages had significantly improved pollen germination ability at low temperatures. The two other cold selected BC, populations showed no differences compared with the unselected BC population. A second cycle of pollen selection, corresponding to BC_2, was applied in order to test its persistence in the subsequent generations and the possibility to further improvement of the character. This second cycle showed no improvement, although some plants retained the high pollen germination ability at low temperatures that was observed in the first cycle. Hence gametophytic selection of some characters related with tomato pollen performance may be feasible, at least for the first cycle of selection.

Clarke *et al.* (2004) improved chilling tolerance in crossbreds of chickpeas compared with the parental genotypes, the progeny derived without pollen selection. They reported that the pollen selection resulted in lower threshold temperature for podding, which leads to pod setting 2-4 weeks earlier.

6.7.4. Membrane Stability

Membrane stability is directly correlated with the solute leakage from tissues, such as, leaf discs and it is measured conductometrically. Patterson *et al.* (1976) used leaf electrolyte leakage as a method of distinguishing chilling tolerance (low leakage rate) and chilling sensitive (high leakage rate) *Passiflora* genotypes. Leaves from plants of *Passiflora* spp., which had been grown less than 25 ^{0}C day/18 ^{0}C nights was chilled at 0 ^{0}C using conductivity. Potassium was the principal cation lost from the tissue and its leakage was proportional to that of total electrolyte during the time of leakage. Tolerance to chilling in the species and hybrids examined showed a gradual increase in the series *P. deulis* forma *flavicarpa* $<$ *P. malformis* $<$ *P. cincinnata* Masters $<$ *P. edulis* $<$ *P. caerulea*.

Several workers have determined electrical conductivity of plant sap and have found a direct correlation between frost resistance and electrical conductivity of wheat varieties (Melniskii and Bidyukova, 1980); Ding, 1984; Brzostowicz *et al.* 1985)

6.7.5. Acclimation

To increase the crops consistency in the cold effected environments, cultivars with improved winter hardiness and management recommendations that utilize all of genotypes winter hardiness need to be developed. Better understanding of the variability in cold tolerance over the winter season need to be determined to help screening for cold tolerance in crop plants.

Winter plants have evolved survival mechanisms that are temperature related to cope with low temperature stress. Cold acclimatization of plant is a higly active process resulting from the expression of a number of physiological and metabolic adaptations to low temperature. Many plants increase in freezing tolerance upon exposure to low nonfreezing temperatures a phenomenon known as cold acclimation. Therefore, chilling injury can be alleviated if the plants are exposed first to a low non-chilling temperature. In chilling sensitive tomato acclimation is associated with accumulation of ABA (Doie and Campbell, 1981) and exogenous application of ABA can substitute for low temperature to initiate the acclimation response (Rikin *et al.* 1979). Cold acclimation includes the expression of certain cold induced genes that function to stabilized membranes against freezing induced injury. Plants acclimate to survive metabolic lesions because of ice formation, as well as to survive the dehydration effects of frost (Kacperska, 1984).

Rife and Zeineli (2003) achieved cold tolerance by 3 days of acclimation in spring cultivars and between 6 and 9 days in winter cultivars. All tested lines were able to recover cold tolerance to the same level as before the deacclimation treatment after exposure to 5 ^{0}C conditions for 7 days. Cold tolerance did not tend to decrease with prolonged acclimation durations. Subjecting chilling sensitive genotypes to various moderate stresses can cause them to acquire resistance to chilling injury. Such a chilling causes an accumulation of H_2O_2 in both the coleoptile-leaf and the mesocotyl but not roots of dark grown maize seedling (Anderson *et al.*, 1995). Fewer symptoms were observed in those seedlings given cold pre-treatment (Prasad *et al.*, 1994 a, b). Elevated levels of glutathione in these acclimated seedlings may have contributed to an enhanced ability to scavange H_2O_2 (Anderson *et al.*, 1995). The increased activity of the enzymes of glutathione synthesis at low temperatures contributed to increased glutathione levels during acclimation

(Burnner *et al.*, 1995). Other research in maize has also implicated glutathione in protection against chilling damage (Kecsy *et al.*, 1996). Molecular evidences also suggested that there may be a relationship between chilling, free radical production and acclimation, because the promoter of cytosolic Cu/Zn -SOD was induced in tobacco after chilling (Herouart *et al.*, 1994).

6.7.6. Chlorophyll Fluorescence

Chlorophyll fluorescence have been applied in breeding programmes (Walker *et al.*, 1990), in crop breeding, which is the most common and successful approach has been to growth the populations in a marginal environment and select based on yield.

6.8. HYBRIDIZATION

Hybridization is an excellent option for creating variability for improvement of crops and subsequently selection in the segregating generations. Yang *et al.* (2003) developed an early maturing, good quality and cold tolerant rice cultivar, 'Inweolbyeo'. This genotype was derived from a cross between Fukei 127 × Unbongbyeo. The milled rice kernel is non-glutinous and clear with excellent taste. Milled rice yield was ~5.33t/ha.

6.9. LIMITATION IN CONVENTIONAL BREEDING

1. Low temperature tolerance is usually found in distant wild relatives of commercial crops. When these are crossed to combine tolerance with other agronomic characteristics, there is a lot of unwanted characters associate with cold tolerance, that cause a loss of commercial quality.
2. Inheritance studies indicated that the variability for chilling tolerance is a quantitative polygenic trait with additive effects among the genes.
3. The identification of plants with chilling tolerance in segregating population is technically difficult.

6.10. BIOTECHNOLOGY IN COLD TOLERANCE

6.10.1. *In Vitro* Screening

The potential of tissue culture for effective genetic modification has been stressed in number of research papers. One approach to such genetic modification involves the isolation of stable variant cell lines from established cell cultures either directly or following mutagen treatment; regeneration of plants from such times, and study of the phenotypic expressions of the genetic variants. The isolation of variant cell lines requires that the cells be subjected to appropriate selection pressures. Cold tolerant cell lines can be selected relatively easily and tolerant plants can be regenerated from them (Gerasimova, 1985). Templeton-Somers and Sharp (1980) have described a cell culture technique for improving cold tolerance of carrot. Low temperature tolerant plants have been detected and isolated from irradiated cuttings/ pedicels and from *in vitro* cell suspension cultures at low temperature of *Chrysanthemum* (Jung-Heiliger and Horn, 1980; Broertjes *et al.*, 1983; Preil *et al.*, 1983; Broertjes and Lock, 1985; Hu *et al.*, 1989).

Cold tolerant lines have been produced from rice anther culture (Zapata *et. al.*, 1986). Bouharmont *et al.* (1991) cultured mature embryos from a large number of rice varieties. Plant regenerated from the callus obtained together with their progeny, showed somaclonal attempted by long-term culture of callus at sublethal temperature (11-13 ^{0}C). Some plants that were regenerated from selection and tested under hydrophonic conditions showed improved tolerance.

6.10.2. Marker Assisted Selection (MAS)

The quantitative trait loci (QTLs) for cold tolerance at bud-burst period (CTBT) were identified in rice (Qiao *et al.*, 2004). A molecular linkage map of 97 simple sequence repeats (SSR) markers were constructed using interval mapping. They identified three QTLs (qCTBP2, qCTBP4 and qCTBP7) on chromosome 2, 4 and 7, which were associated with CTBT, were detected on RM6-RM240, RM273-RM303 and RM214-RM11, repeatedly.

Oh *et al.* (2004) identified that chromosomal region responsible for cold tolerance in rice. Cold tolerance was measured as days to heading, culm length, spikelet fertility, leaf discolourization, and panicle exertion. They identified 14 QTLs for 7 characters, the number of traits ranged from 1-3. Suh *et al.* (2004) used single simple repeat linkage map and a total of 22 significant QTLs were identified. Dai *et al.* (2004) also constructed linkage map of rice to identify QTLs controlling cold tolerance at reproductive stage using DNA markers. Zhang *et al.* (2004) mapped three putative QTLs on chromosome 1, 3 and 9 for seedling height, four QTLs on chromosome 2, 8 and 11 for chlorophyll content, and four QTLs on chromosome 5, 9 and 12 for chilling injury of rice seedling under low temperature. They also found loci RM160 on chromosome 9 affected seedling growth, chlorosis, withering and dying. Andaya and Mackill (2003 a,b) identified a major QTL on chromosome 12, designated as qCTS12a, that was closely associated cold induce necrosis and wilting tolerance, and accounted for 41% of the phenotypic variation.

Lee *et al.* (2004) identified 19 QTLs associated with cold tolerance using RILs (Recombinant Inbred Lines) derived from a cross between Hwayeongbyeo (*japonica*) and *Oryza rufipogon*. *O. rufipogon*-derived allele contributed a desirable agronomic effect, and fevourable wild alleles were detected for panicle exertion, days to heading and culm length associated with the cold water treatment. They have concluded that *Oryza rufipogon* is useful as a source of valuable alleles for rice improvement. Zhang *et al.* (2003) found that one QTL each in chromosome 1, 3, 4, 5, 6, 8, 9 and 12 and two QTLs in chromosome 11, which govern seedling height at low temperature. They also found seven QTLs located in chromosome 1, 2, 5, 6 and 8 govern temperature chlorosis resistant and five QTLs in chromosome 3, 4, 7, 8 and 11 govern resistance to chilling injury. Andaya and Mackill (2003a, b) measured cold tolerance as the degree of spikelet sterility of treated plants and mapped QTLs on chromosomes 1, 2, 3, 5, 6, 7, 9 and 12, which confer cold tolerance at booting stage.

Han *et al.* (2005a) evaluated seedling cold tolerance with constant cold and identified three QTLs on chromosome 1, 5 and 9 using a F_2 and F_3 population derived from a cross between Milyang 23 (indica) and Jileng (japonica) rice. Using root characters for as indices for cold tolerant at seedling stage, Han *et al.* (2005b) found another 17 QTLs on chromosome 1, 2, 6, 11 and 12 with the same population and cold treatment. Zhan *et al.*, (2005) detected 11 QTLs on chromosomes 1, 2, 3, 5, 6, 7, 8, 9, 12 and five QTLs on chromosomes 3, 4, 7, 8, 11 using chlorophyll content and chilling injury as indicators for cold tolerance at seedling stage, respectively. Lou *et al.* (2007) identified five main effective QTLs with LOD > 4.0 on chromosome 1, 2, 8 using a composit interval mapping approach. The accumulated contribution of the five QTLs was 62.28% and a major QTL (LOD = 15.09) was identified on chromosme 2 flanked by RM561 and RM341, which explained 27.42% of the total phenotypic variation.

Planting maize earlier than the recommended guidelines, would give great contribution to ecological and sustainable agriculture, where maize is a major crop. In order to early planting, maize varieties with good cold tolerance and strong earlier vigour are required, both the characters with quantitative genetic background and are not easily introduced in modern maize varieties. therfore, MAS can improve the efficiency of breeding activities. Engelen *et al.* (2003) identified nine QTLs for cold tolerance and six QTLs for early vigour.

6.10.3. Cold Regulated Genes

In response of low temperature, plants induce relevant gene expression to increase cold tolerance, which is called *cold acclimatization*. The nature of cold-inducible genes and their role in cold tolerance have been established. A large number of these genes encode protein with known enzyme activities that potentially contribute to cold tolerance. Many cold regulated genes (COR) were isolated and characterized in *Arabidopsis thaliana* and other cold tolerant plant species. Studies on regulation of COR in *A. thaliana* have regulated in the discovery of a family of transcriptional activators, of which, CBF1, a member of the gene family, controls expression of a battery of COR in *A. thaliana* and other cold tolerant plants. The molecular mechanism of cold acclamitization has been studied, suggesting that the CBF transcription activators are critical in the signal transduction of cold acclimatization (Zhong *et al.*, 2006). Few cold tolerant genes have been detailed bellow:

6.10.3.1. Osmotin

Osmotin is stress responsive cation protein with a pI^2 value of > 8.2 and it accounts for 10-12% of total cellular protein (Singh *et al.*, 1997b). It was reported to be involved in multiple responses against abiotic stresses such as excess salt, drought etc. Osmotin produces a 27 kD pathogenesis related protein. Furthermore osmotin was found to interact with a number of other proteins to activate signal transduction pathway leading to defense gene expression through changing the balance of death cell and survival proteins in the plant system. Osmotin synthesis is regulated by abscisic acid, but its accumulation is affected by cells, and by other stresses. The synthesis of protein could induce synthesis and accumulation of certain solutes or could be involved in metabolic or structural changes.

6.10.3.2. COR15a

It is a cold induce gene, having a role in freezing tolerance. *COR15a* enconds a 15 kDa polypeptide and targated to the chloroplast. Inside the organelle, *COR15a* convert into a 9.4 kDa polypeptide and know as *COR15am*. Expression of *COR15a* in transgenic *Arabidopsis* plants increased freezing tolerance (Artus *et al.*, 1996). They revealed that it involves in stabilization of membrane.

6.10.3.3. CAP (cold acclamation protein) gene

These genes are induced during cold acclamation. *CAP85* (Guy *et al.*, 1992; Neven *et al.*, 1993) and *CAP160* also induced in response of dehydration and ABA, and they encode hydrophilic polypeptides. Kaye *et al* (1998) developed transgenic tobacco plants that constitutively expressed the spinach *CAP85* and *CA160* protein. They found that the both proteins have a detectable effect on freezing tolerance.

6.10.4. Transgenic Development

Cold temperature is major limiting geographical locations suitable for growing crop and horticultural plants and periodically account for significant losses in plant productivity. Classical plant breeding has limited success in imparting cold hardiness to crop plants. Biotechnology with its powerful tolls however may provide answer through the isolation of cold fighting genes and thus may help in the development of crop plants that can withstand frigid temperature.

Several comprehensive genomic studies based on DNA micro array performed in the field of plant cold acclimation, identified large number of cold responsive genes (Seki *et al.*, 2001; Fowler and Thomashow, 2002; Kreps *et al.* 2002; Seki *et al.*, 2002), such as *COR*, lipid transfer protein and β-amylase (Table 6.4). A large number of these genes encode proteins with known enzyme activities, that potentioally contribute to freezing tolerance. Studies revealed that the cold acclimation is controlled by many genes and that cell membranes are particularly vulnerable to cold damage. Lee and Lee (2003) also identified many cold inducible alcohol dehydrogenase, b-amylase, and many novel genes.

Many of the known cold regulated genes were under control of a primary master regulator, CBF/DREB1. Dehydration response element (DRE) plays an important role in the response to low temperature (Fig. 6.1). The transcription factor *DREB1A* specifically interacts with DRE and induces the expression of stress tolerance genes in plant.

	Signal perception
Transduction	Temperature change
	↓
Transcription	ABA independent
	↓
	DREB1/CBF genes
	↓
Trans-elements	CBF/DREB1
	↓
	rd29A promotor
cis elements	DRE/CRT

Fig. 6.1: Stepwise gene expression in molecular responses to low temperature (Datta, 2002)

Table 6.4. Cold tolerance transgenic plants

Genes	Gene action	Species	Phenotypic Expression	Reference
Sod	Cu/Zn-SOD	Tobacco	Retained 90% photosynthesis under chilling and heat stress	Gupta *et al.*, 1993
Sod	Mn-SOD	Alfalfa	Increased tolerance to freezing stress	McKersie *et. Al.*, 1993
Nt107	Glutathion Stransferase	Tobacco	Sustained growth under cold and salinity stress.	Roxas *et al.*, 1997
Wx	Controls amylase synthesis	Rice	Increased amylose content at low temperature	Hirano and Sano, 1998
CBF1	Transcription factor	*Arabidopsis*	Increased cold tolerance	Jaglo-Otlosen *et al.*, 1998
Cod A	Choline oxised (Glycine betaine synthesis)	*Arabidopsis*	Seedling tolerant of salinity stress and increased germination under cold.	Hayashi *et al.*, 1997 Alia *et al.*, 1998
COR15a	Cold induced gene	*Arabidopsis*	Increased freezing tolerance	Steponkus *et al.*, 1998
DREB	Transcription factor	*Arabidopsis*	Increased tolerance to cold, drought and salinity	Kasuga *et al.*, 1999
AB13	Transcription factor	*Arabidopsis*	Increased freezing tolerance	Tamminen *et al.*, 2001
Gs2	Chloroplastic glutamine synthetase	Rice	Increased salinity resistance and chilling tolerance	Hoshida *et al.*, 2000
OsCDPK7	Transcription factor	Rice	Increased tolerance to cold, salinity and drought	Saijo *et al.*, 2000
SCOF1	Transcription factor	*Arabidopsis*	Increased winter tolerance	Kim *et al.*, 2001
Sod	Cu, Mn, Fe, Zn-SOD	Alfalfa, rye grass	Increased winter hardiness	McKersie, 2001
Osmotin	induce praline accumulation	Tomato	More tolerant to cold	Sarad *et al.*, 2004

An alteration of chilling sensitivity has been reported in transgenic tobacco, *Nicotiana tabacum* (Murata *et al.* 1992) and *Arabidopsis thaliana* (Wolter *et al.* 1992) into which the glycerol-3-P acyltransferase gene was introduced. However, the increase in amount of polyunsaturated fatty acids was slight in these transgenic plants. Kodama *et al.* (1994) reported transgenic tobacco plants in which the level of trienoic fatty acids is considerably increased by introduction of the *fad7* desaturase gene. They observed that the low-temperature induced chlorosis was observed that the low temperature induced chlorosis, reduced in transformed with the *fad7* gene. These results indicate that increased levels of trienoic fatty acid in genetically engineered plants enhance cold tolerance.

Transgenic plants expressing the *COR45a* gene, in *Arabidopsis* showed reduced damage to freezing temperatures when compared to those from the control plants (Personal communication of Thomashow MI, Michigan 48824). The *COR15a* gene enhanced the freezing tolerance of chloroplasts in engineered plants by almost 2 °C, which was nearly one third of the increase seen due to cold acclimation. While this might not appear as large increase, an improvement of freeze tolerance by 2 °C could potentially benefit certain crop plants (Artus *et al.*, 1996).

A transgene encoding a Cu/Zn- SOD from pea was introduced into *Nicotiana tabaccum* (Sen Gupta *et al.*, 1993). When the transgenic plants were subjected to chilling temperatures (3° C) under moderate or high irradiance, they retained higher rates of photosynthesis compared to the nontransgenic plants. Presumably the transgenic plants have higher SOD activity and lower levels of superoxide than the control plants leading to reduced oxidative damage.

Sarad *et al.* (2004) developed transgenic tomato with *osmotin* gene. Their preliminary tests revealed that the transgenic plants are more tolerant of cold than wild types.

Over expression of *DREB1A* in transgenic *Arabidopsis* plants activated the expression of many of these stress tolerance genes and resulted in tolerance to freezing (Liu *et al.* 1998; Kasuga *et al.*, 1999; Seki *et al.* 2001; Urao *et al.* 1998, 2000). Liu *et al.* (1998) and Jaglo-Ottosen *et al.* (1998) activated gene expression by overexpressing a homolog of CBF1, designated *DREB1A*. Their result indicated the expression of the *CRT/DRE* regulation also increase drought tolerance. Liu *et al.* (1998) observed that the overexpression of *DREB1A* in transgenic *Arabidosis* resulted in dwarf phenotype. Fan *et al.* (2002) cloned an antifreeze gene from carrot and successfully transferred it into tobacco.

To elucidate the contribution of dehydrins (*DHNs*) to freezing stress tolerance, Puhakainen *et al.* (2004) developed transgenic *Arabidopsis* plants over-expressing multiple *DHN* genes. The transgenic plants exhibited lower LT50 values and improved survival when express to freezing stress compared to control plants. Thus they concluded that dehydrines contribute to freezing stress tolerance in plants and this could be partially due to dehydrines protective effect on membrane.

Parvanova *et al.* (2004) transferred tobacco to accumulate different compatible solutes (proline, fructans or glycine-betaine) in order to improve its tolerance to low temperature. The elevated parameters of transgenic plants successfully survived freezing stress. Mukhoopadhyay *et al.* (2004) developed transgenic with *OSISAP1* gene conferred tolerance to cold, dehydration and salt stress. The gene *OSISAP1* was isolated from rice encoding a zinc-finger protein induced under abiotic stresses. Hur *et al.* (2004) also developed transgenic rice with *OsP5CS2* gene. This gene encodes for a protein that is highly homologous to δ-1-proline-5-caboxylate synthetase (*P5CS*), a proline biosynthesis enzyme. Their result indicated that the *OsP5CS2* gene is necessary for plant tolerance to salt and cold tolerance.

Potato plants (*Solanum tuberosum* cv. Desiree) transferred with east invertase gene acquired a higher tolerance to cold temperature as compared to the control plants, apparently due to the changes in sugar ratio produced by the foreign invertase. The expression of *Osmyb4* in *Arabidopsis thaliana* plants showed a significance increase cold and freezing tolerance, measured as membrane or photosystem-II stability and as a whole plant tolerance (Vannini *et al.*, 2004). They demonstrated by transient expression of *Myb4* transactivater the *PAL2, ScD9, SAD* and *COR15a* cold inducible promoters.

Li *et al.* (2003) suggested that *Hsf* (heat shock factor) gene may play a pivotal role in heat-shock-induced chilling tolerance, and constitutive expression of the transcriopton regulated gene in chilling sensitive crops may be useful in improving tolerance against chilling stress. They transferred *Arabidopsis thaliana Hsf1b* (*AtHsf1a*) gene into tomato. The transgenic tomato plants harbouring this gene showed increased chilling tolerance.

6.11. REFERENCES

Acharya S, Sharma KD. 1983. Genetics of cold tolerance at rice reproductive stage. Int Rice Res Newsl. 8(2): 10-11.

Acharya S. 1987. Genetic parameters and their implication in breeding cold tolerant varieties of rice (*Oryza sativa* L.). Crop Improvement. 14: 100-103.

Aidin VL, Migus WN, Hamilton RI, 1991. Use of inbred seedling cold tolerance to predict hybrid cold tolerance in maize (*Zea mays* L.). Canadian J Plant Sci. 71(3): 663-667.

Alia HH, Chen THH, Murata N. 1998. Transformation with a gene for choline oxidase enhances the cold tolerance of *Arabidopsis* during germination and early growth. Plant Cell Environ. 21: 232-239.

Andaya VC, Mackill DJ. 2003a. Mapping of QTLs associated with cold tolerance during the vegetative stage in rice. J Expt Bot. 54(392): 2579-2585.

Andaya VC, Mackill DJ. 2003b. QTLs conferring cold tolerance at the booting stage of rice using recombinant inbred lines from japonica × indica cross. Theor Appl Genet. 106(6): 1084-1090.

Anderson MD, Prasad TK, Stewart CR. 1995. Changes in isozyme profiles of catalase peroxidase, and glutathione reductase during acclimation to chilling in mesocotyls of maize seedlings. Plant Physoils. 109: 1247-1257.

Andrews CJ, Morrison MJ. 1992. Freezing and ice tolerance for winter *Brassica*. Agron J. 84: 960-962.

Andrews CJ. 1987. Low temperature stress in field and forage crop production an-overview. Canadian J Plant Sci. 67: 1121-1133

Artus NN *et al.* 1996. Constitutive expression the cold-regulated. *Arabidopsis thaliana* COR15a gene effects both chloroplast freezing tolerance. Proc Natl Acad Sci. 93: 13404-13409.

Asare-Boamah NK, Fletcher RA. 1986. Protection of bean seedlings against heat and chilling injury by triadimefon. Physoil Plant. 67: 353-358.

Auld DI, Adams KJ, Swensen JB, Murray GA. 1983. Screening peas for winter-hardiness in peas. Crop Sci. 23: 763-766.

Bagnall DJ, King RW, Farquhar GD. 1988. Temperature dependent feed back inhibition photosynthesis in peanut. Planta. 175: 348-354.

Balachkova NE, Lakhanov AP, Zaitsev VN. 1986. Resistance of pea and French bean breeding material to unfavourable temperature (in Russian). Nauchno-Tekni-Cheskii-Byulleten Vsesoyusnogo. Nauchno-Inssledvatel Skogo Instituta Zernobovykhi Krupanykh Kul tur 1986. No. 35: 66-71.

Baraskova EA, Gradchaninova DD, Gudkova GN, Smirnova VS. 1980. Morphological anatomical analysis of winter wheats differing in frost resistance during over wintering and after chilling. Fiziologiya I Biokkiya Kulturnykh rastenii. 12: 115-119.

Barnabas B, Kovacs G. 1988. Perspective of pollen and male gamete selection in cereals. In: Plant Sperm Cells as Tools for Biotechnology, Wilms HJ, Keijzer CJ (eds.). Waganingen, Netherlands, Pudoc. pp. 137-147.

Bell MJM. 1993. Low night temperature response in peanut (*Arachis hypogaea* L.) Ph. D. Thesis, University of Guelph.

Bilger W, Bjorkman O. 1991. Temperature dependence of violaxanthin de-epoxidation and non-photochemical fluorescence quenching in intact leaves of *Gossypium hirsutum* L. and *Malva parvifkia* L. Planta. 184: 226-234.

Bocsi J. 1988. *In vitro* study of cold tolerance in maize (*Zea mays* L.) Novenytermeles. 37(5): 421-429.

Bouharmont J, Dekeyser A, Van Sint Jan V, Dogbe YS. 1991. Application of somaclonal variation and *in vitro* selection to rice improvement. In: Rice Genetics II. Proc Second Int Rice Genetics Symp, 14-18 May, 1990. IRRI, Manila, Philippines. pp. 271-277.

Brandolini A, Landi P, Monfredini G, Tano F. 2000. Variation among Andean races of maize for cold tolerance during heterophic and early atotrophic growth. Euphytica. 111(1): 33-41.

Broertjes C, Koene P, Pronk T. 1983. Radiation induced low temperature tolerant cultivars of *Chrysanthemum morifolium* Ramat. Euphytica. 32(1): 97-101.

Broertjes C, Lock CAM. 1985. Radiation induced low temperature tolerant solid mutants of *Chrysanthemum morifolium* Ramat. Euphytica. 34: 97-103.

Browse J, Somerville C. 1991. Glycerolipid synthesis: biochemistry and regulation. Annu Rev Plant Physiol Plant Mol Biol. 42: 467-506.

Brule-Babel AL, Fowler DB. 1988. Genetic control of cold hardiness and vernalization requirement in winter wheat. Crop Sci. 28: 879-884.

Bunner M, Socsy G, Ruegsegger A, Schmutz D, Brunold C. 1995. Effects of chilling on assimilatory sulfate reduction and glutathione synthesis in maize. J Plant Physoil. 146: 7743-747.

Brzostowicz A, Barcikowska B. 1987. Possibility of frost resistance testing of *Brassica napus* with the help of delayed luminescence intensity. Cruciferae Newsl. 12: 27.

Brzostowicz A, Prokowski Z, Grabikowski E. 1985. Use of the delayed luminescence test for evaluation of changes in frost resistance of winter wheat. Acta Agrobotanica. 38: 5-10.

Burbanova RS. 1980. The growth response as an index of plant resistance to low temperature. In: Fiziologiya utoichivostt rastenii k nizkim temperatum i zamonzkam. 12: 125-130.

Cabuba I, Ochesann C. 1978. Additive and nonadditive genetic action in the inheritance of low temperature response in maize. Probleme de Genetica Toeretica Si Aplicata. 10: 175-194.

Cahalan C, Law CN. 1979. The genetic control fo cold resistance and vernalization requirement in wheat. Heredity. 42: 125-132.

Chaplin GR, Scott KJ. 1980. Association of calcium in chilling injury susceptibility of stored avacadoes. Hort Sci. 15: 541-515.

Chapman MA, Drolsom PN. 1983. Combining ability of cold tolerance traits of corn. In: Agron Abstr Am Soc Agron. p. 59.

Chen HH, Li PH, Brenner ML. 1983. Involvement of abscisic acid in potato cold acclamation. Plant Physiol. 71: 362-365.

Ciardi JA, Deikman J, Orzolek MD. 1997. Increased ethylene synthesis enhances chilling tolerance in tomato. Physoil Plant. 101: 333-340

Clarke HJ, Khan TN, Siddique KHM. 2004. Pollen selection for chilling tolerance at hybridization lead to improve chickpea cultivation. Euphytica. 139(1): 65-74.

Coly A. 1979. Varietal screening for cold tolerance in the Sahelian Zone I. Some promising rice varieties selected for dry cold season. Int Rice Commu Newsl. 28(2): 29-31.

Crobie TM, Mock JJ, Smith OS. 1980. Comparison of grains predicted by several selection methods for cold tolerance traits of two maize populations. Crop Sci. 20: 649-655.

Crowe JH, Hoekstra FA, Crowe LM. 1989. Membrane phase transitions are responsible for imbitional damage in dry pollen. Proc Natl Acad Sci, USA. 86: 520-523.

Dai YC, Deng XL, Jiang XC, Chen LB. 2004. Identification and exploitation of chilling-resistant physiology in different rice seedlings. J Natural Sci Hunan Univ. 27(3): 86-89.

Datta D, Siddiq EA. 1983. Genetic analysis of cold tolerance at seedling phase in rice. Indian J Genet. 43: 345-349.

Datta D, Siddiq EA. 1985. Genetic analysis of cold tolerance at seedling phase in rice. Indian J Genet Plant Breed. 43: 345-349.

Datta SK. 2002. Recent development in transgenic for aboitic stress tolerance in rice. JIRCAS Working Report (2002). pp. 43-53.

Demmig-Adams B. 1990. Carotinoids and photoprotection in plants: a role for the xanthophyll, zeaxanthin. Biochem Biophys Acta. 1020: 1-24.

Dhillon BS, Reddy AN. 1988. Response to selection for cold tolerance in maize. Indian J Genet. 48: 129-133.

Dickson MH, Petzoldtt R. 1987. Inheritance of low temperature in beans at several growth stages. Hort Science. 22: 481-483.

Ding ZR. 1984. Study on the use of electrical resistance measurement in the assessment of cold resistance in winter wheat cultivars. Zhiwn Shenglixue Jongum, Plant Physiology Communication. No. 1: 26-28.

Dipierro S, Leonardis SD. 1997. The ascorbate system and lipid peroxidation in stored potato (*Solanum tubersum* L.) tubers. J Exp Bot. 48: 779-783.

Doie J, Campbell WF. 1981. Response of tomato plants to stress full temperatures: Increase in abscisic acid concentrations. Plant Physoil. 67: 26-29.

Doll H, Haahr V, Sogaard B. 1989. Relationship between vernalization requirement and winter hardiness in doubled haploid of barley. Euphytica. 42: 209-213.

Dominguez E, Cuartero J, Fernandez-Munoz R. 2004. Breeding tomato for pollen tolerance to low temperatures by gametophytic selection. Euphytica. 142(3): 253-263.

Dorffling K, Schulenburg S, Lesselich G, Dorffling H. 1990. Abscisic acid and proline levels in cold hardened winter wheat levels in relation to variety specific differences in freezing resistance. J Agron Crop Sci. 165: 230-239.

Duhalde MC, Aveldano MI, Careeler M. 1991. Major fatty acid and cold tolerance during germination in *Zea mays*. Maydica. 36: 251-256.

Duncan DT, Widholm JM. 1987. Proline accumulation and its implication in cold tolerance of generable maize callus. Plant Physoil. 83: 703-708.

Eamus D. 1987. Stomatal behavior and leaf water potential of chilled and water stressed *Solanum melongena* as influenced by growth history. Plant Cell Environ. 10: 649-654.

Engelen S, Reheul D, Gheysen G. 2003. Molecular markers for cold tolerance and early vigour in maize (*Zea myse* L.). In: 17th Forum for Applied Biotechnology and Genetics, Belgium, 18-19 Sep 2003. pp. 367-374.

Fan Y, Liu B, Wang H, Wan S, Wang J. 2002. Plant Cell Reports. 21: 296-301.

Fascaroli E. 1995. Pollen cold tolerance of maize genotypes with different ability to germinate at low temperature. Maydica. 40: 229-232.

Feierabend J, Schaan C, Hertwig B. 1992. Photoinactivation of catalase occurs under both high-and low-temperature stress conditions and accompanies photoinhibition of photosystem II. Plant Physiol. 100: 1554-1561.

Feranec P. 1980. Investigation of the inheritance of cold hardiness in maize. Vedecke Prac Vyskumsecho ustarv Kururice No. 9: 61-70.

Fletcher RA, Hofstra G. 1985. Triadimefon a plant multi-protectant. Plant Cell Physoil. 26: 775-780.

Flores A, Dorffling K, Vergra S. 1988a. Influence of new terpenoid analogue of abscisic acid on chilling resistance of rice seedlings. Int Rice Res Newsl. 13(5): 29.

Flores A, Dorffling K. 1990. A comparative study of the effect of abscisic acid analogues on plant physiological processes. Plant Growth Regulation. 9: 133-139.

Flores A, Grau A, Laurich E, Dorffling K. 1988b. Effect of new terpenoid analogue of abscisic acid on chilling and freezing resistance. J Plant Physoil. 132: 362-369.

Flores AA, Groffling K, Vergara BS. 1991. Increasing cold and drought tolerance in rice with a new phytohormone analogue. In: Rice Genetics II, Proc Second Intr Rice Genetics Symp, 14-18 May, 1990. IRRI, Manila, Philippines. pp. 134-136.

Fowler DB, Gusta LV, Tyler NJ. 1981. Selection for winter hardiness in wheat III. Screening method. Crop Sci. 21: 896-901.

Fowler DB, Limin AE, Wang S-Y, Ward RW. 1996. Relationship between low-temperature tolerance and vernalization response in wheat rye. Can J Plant Sci. 76: 37-42.

Fowler S, Thomashow ME. 2002. *Arabidopsis* transcriptome profiling indicates that multiple regulatory pathways are activated during cold acclimation in addition to the CBF cold response pathway. Plant Cell. 14: 1675-1690.

Galiba G, Ruberosa R, Kocsy G, Sutka J. 1983. Involvement of chrosome 5A and 5D in cold induced abscisic acid accumulation in and frost tolerance of wheat calli. Plant Breeding. 110: 237-242.

Gerasimova SI. 1985. Use of tissue culture methods in breeding potato for cold resistances. Referativrujii Zhurnale. 10: 65.257.

Gilmour SJ, Artus NN, Thomashow MF. 1992. cDNA sequence analysis and expression of two cold-regulated genes of *Arabidopsis thaliana*. Plant Mol Biol. 18: 13-21.

Goldberg RB, Beals TP, Sanders PM. 1993. Anther development : basic principles and practical applications. Plant Cell. 5: 1217-1229.

Gomes E, Jakobsen MK, Azelsen KB, Geisler M, Palmgren MG. 2000. Chilling tolerance in *Arabidopsis* involves ALA1, a membrane of a new family of putative aminophospholipid trenslocases. Plant Cell. 12: 2441-2454.

Gong H, Nilson S. 1989. Effect of temperature photoinhibition of photosynthesis, recovery and turnover of the 32 kD chloroplast protein in *Lemma gibba*. J Plant Physiol. 135: 9-14.

Graham D, Patterson BD. 1982. Response of plant to low, non-freezing temperatures: Proteins, metabolism and acclimation. Annu Rev Plant Physiol. 33: 347-372

Grogan CO. 1970. Genetic variability in maize (*Zea mays* L.) for germination and seedling growth at low temperatures. Annu Corn and Sorghum Res Cong Proc. 25: 90-98.

Guo Q, Pan CC. 1984. Effect of ABA on the resistance of rice seedling to chilling injury. Acta Phytophysiologiica Sinica. 10: 275-303.

Gupta AS, Heinen JL, Holaday AS, Burke JJ, Allen RD. 1993. Increased resistance to oxidative stress in treangenic plant that over express chloroplastic Cu/Zn superoxide dismutase. Proc Natl Acad Sci, USA. 90: 1629-1633.

Gupta D, Kovacs I. 1974. Cold wave tolerance of maize seedlings. Zeitschrift fuer Acker and Pflanzenbau. 140(4): 306-311.

Guy C, Haskell D, Neven L, Klein P, Smelser C. 1992. Hydration-state-responsive proteins link cold and drought stress in spinach. Planta. 188: 265-270.

Guye MG, Vigh L, Wilson JM. 1986. Polyamine titre in rotation to chilling sensitivity in *Phaseolus* sp. J Exp Bot. 37: 1036-1043.

Guye MG, Vigh L, Wilson JM. 1987. Recovery after chilling: An assessment of chilling tolerance in *Phaseolus* spp. J Exp Bot. 38: 691-701.

Guye MG, Wilson JM. 1987. The effect of chilling and chill-hardening temperatures on stomatal behavour in chilling sensitive species and cultures. Plant Physoil Biochem. 25: 717-721.

Hajela RK, Horvath DP, Gilmour SJ, Thomashow MF. 1990. Molecular cloning and expression of *cor* (cold-regulated) genes in *Arabidopsis thaliana*. Plant Physiol. 93: 1246-1252.

Hallgren JE, Oquest G. 1990. Adaptation to low temperatures. In: Stress Responses in Plants: Adaptation and Acclimation Mechanisms, Alscher RG, Cumming JR (eds.). Wiley-Liss Inc. New York. pp. 265-293.

Han LZ, Qiao YL, Cao GL, *et al.* 2005b. QTL analysis on cold tolerance during early growth period in rice (in Chinese with English abstract). Chinese J Rice Sci. 19(2): 122-126.

Han LZ, Zhang SY, Qiao YL, *et al.* 2005a. QTL analysis of root traits at the seedling stage in rice under cold irrigation (in Chinese with English abstract). Acta Agronom Sin. 31(11): 14151421.

Hayashi HA, Mustardy L, Deshnium P, Ida M, Murata N. 1997. Transformation of *Arabidopsis thaliana* with the *Cod A* gene for choline oxidase accumulation of glycine betaine and enhanced tolerance to salt and cold stress. Plant J. 12: 133-142.

Herner RC. 1990. The effects of chilling temperatures during seed germination and early seedling growth. In: Stress Responses in Plants: Adaptation and Acclimation Mechanisms, Alsher RG, Cumming JR (eds.). Wiley-Liss Inc. New York. pp. 265-293.

Herouart D, Montagen MV, Inze D, Van Montagu M. 1994. Developmental and environmental regulation of the *Nicotiana plumbaginifolia* cystosolic Cu/Zn superoxide dismutase promoter in transgenic tobacco. Plant Physoil. 104: 873-880.

Hirano Hy, Sano Y. 1998. Enhancement of *Wx* gene expression and the accumulation of amylase in response to low temperatures during seed development in rice. Plant Cell Physiol. 39(8): 807-812.

Hodges DM, Andres CJ, Johnson DA, Hamilton RI. 1997. Sensitivity of maize hybrids to chilling and their combining abilities at two developmental stages. Crop Sci. 37: 850-856.

Hodgson RA, Raison JK. 1991. Superoxide production by thylakoid during chilling and its implication in the susceptibility of plants to chilling induced photo inhibition. Planta. 183: 222-228.

Holmberg SA. 1973. Soybeans for cool temperate climates. Agrihortique Genetica. 31: 1-20.

Hoshida H, Tanaka Y, Hibino T, Hayashi Y, Tanaka A, Takabe T. 2000. Enhanced tolerance to salt stress in transgenic rice that overexpress chloroplast glutamine synthetase. Plant Mol Biol. 43: 103-111.

Hu Y, Yin DC, Yu QC. 1989. Selection and breeding of a new *Chrysanthemum mutant* cv. Hanrong No.1 Jiangsu. Agric Sci. 6: 31-32.

Hume DJ, Jackson AKH. 1981. Pod formation in soybeans at low temperature. Crop Sci. 21: 933-937.

Hur JH, Jung KH, Lee CH, An GH. 2004. Stress-inducible *OsP5CS2* gene is essential for salt and cold tolerance in rice. Plant Sci. 167(3): 417-426.

Ichi M, Tamai K. 1988. Low temperature adaptability in red kernelled rice. I. Preliminary test of heterosis in F_1 intervarietal hybrids of rice. Taiqan Agric Res. 18(3): 1-17.

IRRI. 1978. Program Report for 1978. International Rice Research Institute, Los Banos, Philippines.

IRRI. 1988. Standard Evaluation System for Rice. International Rice Research Institute, Manila, Philippines. pp. 54.

Ivanov MI. 1980. Winter hardiness of short strawed forms of winter rye in: Selekts Semenovod i Sortor aggrotekhn Zern Kutur i Mnogolet trav Set-Zap Nechernozen. Zony. Leningrad, USST. pp. 11-14.

Jaglo-Ottosen KR, Gilmour SJ, Zarka DG, Schabenberger O, Thomashow MF. 1998. *Arabidopsis* CBF1 overexpression induces COR genes and enhances freezing tolerance. Science. 280: 104-106.

Johnson-Flanagan AM, Singh J. 1987. Alteration of gene expression during the induction of freezing tolerance in *Brassica napus* suspension culture. Plant Physoil. 85: 699-705.

Juag-Heiliger H, Horn W. 1980. Variation nach mutgener Behandlung von Stecklingen und in intro-kulturen bei chrysanthemum. Z pflanzenzuchtg. 85: 185-199.

Kacperska A. 1984. Mechanisms for cold acclimation in winter rape plants. In: Proc Intr Rapseed Congr, 6th GCIRC, Paris. pp. 78-82.

Kang GZ, Ou ZY, Wang ZX, Sun GC. 2003. Salicylic acid alleviated the damage caused by low temperature to cell membrane and some photosynthetic functions of banana seedlings. Acta Horticultura Sinica. 30(2): 141-146.

Kasuga M, Liu Q, Miura S, Yamaguchi-Shinozaki K, Shinozaki K. 1999. Improving plant drought, salt and freezing tolerance by gene transfer of a single stress inducible transcription factor. Nature Biotechnol. 17: 287-291.

Kaw RN. 1991. Phenotypic stability for reproductive stage cold tolerance in rice. Indian J Genet. 51: 66-71.

Kaye C, Neven L, Hofig A, Li QB, Haskell D, Guy C. 1998. Characterization of a gene for spinach CAP160 and expression of two spinach cold acclimation proteins in tobacco. Plant Physiol. 116: 1367-1377.

Kecsy G, Burnner M, Ruegsegger A, Stamp P, Brunold C. 1996. Glutathione synthesis in maize genotypes with different sensitivities to chilling. Planta. 198: 365-370.

Khan DR, Mackill DJ, Vergara BS. 1986. Selection for tolerance to low temperature-induced spikelet sterility at anthesis in rice. Crop Sci. 26: 694-698.

Khristolyubora NB, Khvostova VA, Safonova VT, Usova TK. 1974. Studies of the functional activity of mitochondria and cell ultrastructure of the coleoptile in wheat, *Agropyron* and their hybrids (incomplete amphidiploids) during decreasing temperature. Thor Appl Genet. 44: 255-261.

Khusainova GK. 1974. Content of growth substances in wheat cv. Kazakhstanskaya 126 and its mutant. In: Vaprosy Povysheniya Produktivnosti Zernovykh Kulter, Irkutks, USSR. pp. 146-149.

Kim JC, Lee SH, Cheng YH, Yoo CM, Lee SI, Chun HJ, Yun DJ, Hong JC, Lee SY, Lim CO, Cho MJ. 2001. A novel cold-inducible zinc finger protein from soybean. SCOF-1, enhances cold tolerance in transgenic plants. Plant J. 25: 247-259.

Kodama H, Hamada T, Horguchi G, Nishimura M, Iba K. 1994. Genetic enhancement of cold tolerance by expression of a gene for chloroplast w-3 fatty acid desat urease in transgenic tobacco. Plant Physoil. 105: 601-605.

Kovacs G. 1989. Selection for hardiness in pollen population of maize. Novenytermeles. 38: 9-13.

Kramer GE, Wang CV. 1990. Effects of chilling and temperature preconditioning on the activity of polyamine biosynthetic enzymes in gucchinic. J Plant Physiol. 136: 115-119.

Kreps JA, Wu Y, Chang HS, Zhu T, Wang X, Harper JF. 2002. Transcription changes for *Arabidopsis* in response to salt, osmotic and cold stress. Plant Physoil. 130: 2129-2141.

Krizek DT, Semeniuk P, Moline HE, Mirechi RM, Abbott JA. 1985. Chilling injury in coleus as influenced by photosynthetically active radiation, temperature and abscisic acid pretreatment I. Morphological and physiological response. Plant Cell Environ. 8: 135-142.

Kurkela S, Franck M. 1990. Cloning and characterization of cold-and ABA-induced *Arabidopsis* gene. Plant Mol Biol. 15: 137-144.

Lang V, Heino P, Palva ET. 1989. Low temperature acclimation and treatment with exogenous abscisic acid induce common polypeptides in *Arabidopsis thaliana* (L.) Heynh. Their Appl Genet. 77: 729-734.

Laroche A, Geng XM, Singh J. 1992. Difference of freezing tolerance and vernalization responses in cruciferae exposed to low temperature. Plant Cell Environ. 15: 439-445.

Law CN, Jenkins G. 1970. A genetic study of cold resistance in wheat. Genet Res. 15: 187-208.

Lawn RJ, Hume DJ. 1985. Response of tropical and temperate soybean genotypes to temperature during early reproductive growth. Crop Sci. 25: 137-142.

Lee JY, Lee D-H. 2003. Use of serial analysis of gene expression technology to reveal changes in gene expression in *Arabidopsis* pollen undergoing cold stress. Plant Physoil. 132: 517-529.

Lee SJ, Oh CS, Choi YH, Suh JP, Ju HG, Yoon DB, Ahn SN. 2004. Mapping quantitative trait loci controlling cold tolerance in an *Oryza sativa* × *O. rufipogon* BC1F7 population. Korean J Breed. 36(1): 1-8.

Lee TM, Lur HS, Che C. 1995. Abscisic acid and putrocine accumulation in chilling tolerant rice cultivars. Crop Sci 35: 502-508

Lee TM, Lur HS, Che C. 1997. Role of abscisic acid in chilling tolerance of rice (*Oryza sativa* L.) seedlings. II. Modulation of free polyamine levels. Plant Sci. 126: 1-10

Lee, TM. 1997. Polyamine regulation of growth and chilling tolerance of rice (*Oryza sativa* L.) roots cultured *in vitro*. Plant Sci. 122: 111-117

Levitt J. 1980. Responses of plants to environmental stress, edition 2, Vol 1. Academic Press, New York. pp. 23-64.

Li HY, Chang CS, Lu LS, Liu CA, Chan MT, Charng YY. 2003. Overexpression of *Arabidopsis thaliana* heat shock factor gene (AtHsf1b) enhance chilling tolerance in transgenic tomato. Botanical Bulletin Academia Sinica. 44(2): 129-140.

Li Y, Chang RZ, Zheo ZT. 1989. Response of soybean to low temperature during germination. Soybean Genetics Newsl. 16: 33-37.

Li YJ, Chang RZ, Zhao YT, Sun JY. 1990. Application studies in the field on screening soybeans. Soybean Genetics Newsl. 17: 32-42.

Linin AE, Fowler DB. 1993. Inheritance of cold hardiness in *Tricum aestivum*´ synthetic hexaploid wheat crosses. Plant Breeding. 110: 103-108.

Liu Q, Kasuga M, Sakuma Y, Abe H, Miura S, Yamaguchi-Shinozaki K, Shinozaki K. 1998. Two transcription factor, DREB1 and DREB2, with EREBP/AP2 DNA binding domin sepatate two cellular signal transduction pathways in drought and low temperature responsive gene expression, respectively in *Arabidopsis* expression. Palnt Cell. 10: 1391-1406.

Lou Q, Chen L, Sun Z, Xing Y, Li J, Xu X, Mei H, Luo L. 2007. A major QTL associated with cold tolerance at seedling stage in rice (*Oryza sativa* L.). Euphytica. 158: 87-94.

Lyakh VA, Soroka AI. 1990. Cold resistance of male gametophytes in different genotypes of maize. Izvestiyta Akademii Naiik Moldavkoi SSR. Biologicheskie i Khimicheskie No. 2: 27-32.

Lyons JM, Raison JK, Steponkus PL. 1979. The plant membrane in response to low temperature : an overview. In: Low Temperature Stress in Crop Plants, Lyons JM, Graham D, Raison JK (eds.). Academic Press, New York. pp. 1-24.

Makarenko SP, Konstantinov YM, Khotimchenko SV, Konenkia TA, Arziev ASH. 2003. Fatty acid composition of mitochondrial membrane lipids in cultivated (*Zea mays*) and wild (*Elymus sibiricus*) grass. Russian J Plant Physiol. 50(4): 487-491.

Malhotra RS, Saxena MC. 1993. Screening for cold and heat tolerance in cool season food legumes. In: Breeding for Stress Tolerance in Cool-season Food Legumes, Singh KB, Saxena MC (eds.). Jhon Wiley, UK. pp: 227-244.

Manishi DM, Mahadevappa M, Gopala Reddy P. 1991. Inheritance of panicle exertion in *Oryza sativa* under low temperature. In: Rice Genetics II, Proc of the Scond Intr Rice Genet Symp, 14-18 May, 1990, IRRI, Manila, Philippines. pp. 175-177.

Markhart AH III. 1984. Amelioration of chilling induced water stress by abscisic acid induced changes in root hydraulic conductance. Plant Physoil. 74: 81-83.

Mascarenhas JP. 1975. The biochemistry of angiosperm pollen development. Bot Rev. 41: 259-314.

Mastronicolis SK, Arvantis N, Karaliota A, Litos C, Stavroulakis G, Moustaka H, Tsakirakis A, Heropoulos G. 2005. Cold dependence of fatty acid of different lipid structure of *Listeria monocytogenes*. Food Microbiol. 22(2/3): 213-219.

McConnell RL, Gardner CO. 1979a. Inheritance of several cold tolerance traits in corn. Crop Sci. 19: 847-852.

McConnell RL. 1979. Inheritance and improvement of cold tolerance traits in maize (Astr.). Diss Abstr Int B. 39: 1074B.

McConnell RL., Gardner CO. 1979b. Selections for cold germination in two corn populations. Crop Sci. 19: 765-768.

McKersie BD, Bowley SR. 1997. Active oxygen and freezing tolerance in transgenic plants. In: Plant Cold Hardiness, Molecular Biology, Biochemistry and Physiology, Li PH, Chen THH (eds.) New York, Plenum. pp. 203-214.

McKersie BD, Chen Y, Debens M, Bowley SR, Bowler C, Inze D, D'Halluin K, Botterman J. 1993. Superoxide dismutase enhances tolerance of freezing stress in transgenic alfalfa (*Medicago sativa* L.). Plant Physiol. 103: 1155-1163.

McKersie BD. 2001. Stress tolerance of transgenic plants over expressing superoxide dismutase. J Expt Bot. 52: 8.

McMurphy LM, Rayburn AL. 1992. Chromosomal and cell size analysis of cold tolerance maize. Theo Appl Genet. 84: 798-802.

McWilliam JR, Kramer PJ, Musser RL. 1982. Temperature induced water stress in chilling sensitive plants. Australian J Plant Physoil. 9: 343-352.

Melniskii VN, Bidyukova GF. 1980. Determining the relative frost resistance of cold hardened seedlings of winter wheat by registering electrical resistance. Fiziologiya i Bioghimiya Kulturnykh rastenii. 27: 1110-1114.

Miedema P. 1982. The effects of low temperature on *Zea mays*. Adv Agron. 35: 93-128.

Mock JJ, Bakri AA. 1976. Recurrent selection cold tolerance in maize. Crop Sci. 16(2): 230-233.

Mock JJ, McNeill MJ. 1979. Cold tolerance in maize inbred lines adapted to various latitudes in North America. Crop Sci. 19(2): 239-242.

Moon HP. 1984. Inheritance of low temperature induced-sterility and its relationship to agronomic characters in rice (*Oryza sativa*). Diss Abstr Int B. 44: 2065B.

Moreru KV, Syrku VL. 1991. Lipids in the grain and embryos of winter wheat varieties differing in frost resistance. Iz vestiya Akademii Nauk SSR moldava BiologichesKie I Khimicheskie Nauki No. 1: 70-71.

Morgan PW, Drew MC. 1997. Ethylene and plant response to stress. Physoil Plant. 100: 620-630.

Mukhoopadhyay A, Vij S, Taygi AK. 2004. Overexpression of a zinc-finger protein gene from rice confers tolerance to cold, and salt stress in transgenic tobacco. Proc Nat Acad Sci, USA. 101(16): 6309-6314.

Murata N, Ishizaki-Nishizawa O, Higashi S, Hayashi H, Tasaka Y, Nishda I. 1992. Geneticallly engineered alteration in the chilling sensitivity of plants. Nature. 356: 710-713.

Murata N. 1983. Molecular species composition of phosphatidylglycerols from chilling sensitive and chilling resistant plants. Plant Cell Physoil. 24: 81-86.

Murata S, Sato N, Takahashi N, Hamazaki Y. 1982. Plant Cell Physoil. 23: 1071-1079.

Nadeau P, Delaney S, Chouinard L. 1987. Effects of cold hardening on the regulation of polyamine levels in Wheat (*Triticum aestivum* L.) and alfalfa (*Medicago sativa* L.). Plant Physiol. 84: 73-77.

Nagamine T, Nakagahra M. 1989. Genetic variations and genetic control of chilling injury in young seedlings of worldwide rice vrieties. In: Proc 6th Intr Cong SABRAO, August 21-25. 1989, Iyama S, Takeda G (eds.), Organizing Committee of the 6th International Congress of SABRAO, Tokyo.

Nagamine T, Nakagahra M. 1991. Genetic control of chilling injury in rice seedlings detected by low temperature treatment. in: Rice Genetics II, Proc Second Rice Genetics Cymp, 14-18 May, 1990, Intr Rice Res Inst, Manila, Philippines. pp. 737-739.

Nagamine T. 1991. Generic control of tolerance to chilling injury at seedling stage in rice *Oryza sativa* L. Japan J Breed. 41: 35-40.

Nanda JS, Mani SC. 1983. Problems and prospects of breeding cold tolerance varieties of rice (*Oryza sativa* L.) in India. In: Rice in West Bengal, Mukherji DK (ed.). Directorate of Agriculture, Govt. West Bengal. pp. 55-72.

Nanda JS, Seshu DV. 1999. Breeding strategy for cold-tolerance rice. In: Report of Rice Cold Tolerance Workshop. International Rice Research Institute, Manila, Philippines. pp. 91-99

Neven LG, Haskell DW, Hofig A, Li QB, Guy CL. 1993. Characterization of a spinach gene responsive to low temperature and water stress. Plant Mol Biol. 21: 291-305.

Nilanjaya Kumar M, Choudhary VK, Singh JRP. 2001. Screening of Boro rice genotypes for cold tolerance. J Appl Biol. 11(1-2): 98-100.

Nordin K, Vahala T, Palva ET. 1993. Dofferential expression of two related, low-temperature-induced gene in *Arabidopsis thaliana* (L.) Heynh. Plant Mol Biol. 21: 641-653.

Novitskaya GV, Salnikova EB, Suvorova TA. 1990. Change in the level of unsaturated fatty acids in lipids of winter and spring wheat plants during the hardening process. Giziologiya I Biokhimiya Kulturrykh Rastenii. 22: 257-263.

Novitskaya GV, Suvorava TA, Trunova TI. 2000. Lipid composition of tomato leaves as related to plant cold tolerance. Russian J Plant Physoil. 47(6) : 728-733.

Oh CS, Choi YH, Lee SJ, Yoon DB, Moon HP, Ahn SN. 2004. Mapping of quantitative traits for cold tolerance in weedy rice. Breed Sci. 54(4): 373-380.

Olien CR, Smith MN. 1977. Ice adhesions in relation to freezing stress. Plant Physiol. 60: 499-503.

Orman RG. 1980. Peroxidase levels and the activities of catalase, peroxidase and indolacetic acid oxidase during and after chilling cumcumber seedling. Plant Physiol. 65: 407-408.

Paldi E, Szalai G, Janda T, Marton LC. 1998. Effect of low temperature on the synthesis of N-containing compounds in inbred maize lines with varying degrees of cold tolerance. Novenytermeles. 47(5): 483-490.

Parkinson KL, Marangoni A, Jackman R, Yadav R, Stanley D. 1989. Chilling injury. A review of possible mechanisms. J Food Biochem. 13: 127-153.

Parodi PC, Nyquist WE, Patterson FL, Hodges HF. 1983. Traditional combining ability and Gardner Eberhart analysis of a diallel for cold resistance in winter wheat. Crop Sci. 23: 314- 318.

Parvanova D, Povova A, Zaharivena I, Lambrev P, Kostantinova T, Taneva S, Atnassov A, Goltsev V, Djilianov D. 2004. Low temperature tolerance of tobacco plants transferred to accumulate proline, fructans, or glycine betaine. Variable chlophyll fluorescence evidence. Phytosynthetica. 42(2): 179-185.

Patterson BD, Mutata T, Graham D. 1976. Electrolyte leakage induced by chilling *Pasciflora* species tolerant to different climates. Australian J Plant Physiol. 3: 435-442.

Patterson BD, Mutton L, Paull RE, Nguyen VQ. 1987. Tomato pollen development: stage sensitive to chilling and a natural environment for the selection of resistant genotypes. Plant Cell Environ. 10: 363-368

Prasad TK, Anderson MD, Martin BA, Steward CR. 1994a. Evidence for chilling induced oxidative stress maize seedling and a regulatory role for hydrogen peroxide. Plant Cell. 6: 65-75.

Prasad TK, Anderson MD, Stewart CR. 1994b. Acclimation, hydrogen peroxide and abscisic acid protect mitochondria against irreversible chilling injury in maize seedlings. Plant Physoil. 105: 619-627.

Preil W, Engelhardt M, Walter F. 1983. Breeding for low temperature tolerant poinsettia (*Euphorbia pulcherrima*) and *Chrysanthemum* by means of mutant induction in *in vitro* cultures. Acta Horticulture. 131: 345-351.

Puhakainen T, Hess MW, Makela P, Svensson J, Heino P, Palva ET. 2004. Overexpression of multi-dehydrin gene enhances tolerance to freezing stress in *Arabidopsis*. Plant Mol Biol. 54(5): 743-753.

Qiao YL, Han LZ, An YP, Zhang YY, Cao GL, Koh HJ. 2004. Molecular mapping of QTLs for cold tolerance at the budburst period in rice. Agril Sci China. 3(11): 801-806.

Qui N, Liu ZT, Han YZ, Guo T, Nara. 1990. Fat and protein content of soybean seeds as related to their cold hardiness during germination. Oil Crops of China, No. 3: 39-41.

Ram HH, Singh HG. 1994. Rice. In: Crop Breeding and Genetics, Kalyani Publisher, New Delhi. pp. 58-103.

Rapacz M, Markowski. 1999. Winter hardiness, frost resistance and vernalization requirement of European winter oilseed rape (*Brassica napus oleifera*) cultivars within the last 20 years. J Agric Crop Sci. 183: 243-253.

Richter M, Rühle M, Wild A. 1990. Studies on the mechanism of photosystem II photoinhibition. II. The involvement of toxic oxygen species. Photosyn Res. 20: 237-243.

Rife CL, Zeinali H. 2003. Cold tolerance in oilseed rape over varying acclimation durations. Crop Breeding Genet Cytol. 43: 96-100.

Rikin A, Atsmon D, Gitler C. 1979. Chilling injury in cotton (*Gossypium hirsutum* L.) : prevention by abscisic acid. Plant Cell Physoil. 20: 1537-1546.

Rikin A, Atsmon D, Gitler C. 1983. Quantitation of chill-induced release of a tubulin-like factor and its prevention by abscisic acid *Gossypium hirsutum* L. Plant Physoil. 71: 747-748.

Rikin A, Blumenfeld A, Richmond AE. 1976. Chilling resistance as affected by stressing environments and abscisic acid. Bot Gaz. 137: 307-312.

Rikin A, Richmond AE. 1976. Amilioration of chilling injuries in cucumber seedlings by abscisic acid. Physiol Plant. 38: 95-97.

Ristic Z, Yang GP, Sterzinger A, Zang L, Yang GP, Zang LQ. 1998. Higher chilling tolerance in maize is not always related to ability for greater and faster abscisic acid accumulation. J Plant Physiol. 153(2): 154-162.

Roberts DR, Dumdroff EB, Thompson JE. 1986. Exogenous polyamines alter membrane fluidity in bean leaves: a basis for potential misinterpretation of their true physiological role. Planta. 167:395-401.

Roxas VP, Smith RK Jr, Allen ER, Allen RD. 1997. Overexpression of glutathione-S-transferase /glurathione peroxidase enhances the growth of transgenic tobacco seedlings during stress. Nature Biotechnol. 15: 988-991.

Saijo Y, Hata S, Kuozuka J, Shimamoto K, Izui K. 2000. Over-expression of a single Ca^{+} dependent protein kinase confer both cold and salt/drought tolerance on rice plants. Plant J. 23: 319-327.

Saltveit ME Jr, Morris LL. 1990. Overview of chilling injury of horticultural crops. In: Chilling Injury of Horticultural Crops, Wang CY (ed.). CRC Press, Boca Raton. pp. 3-15.

Sarad N, Rathore M, Singh NK, Kumar N. 2004. Genetically engineered tomatoes: New vista for sustainable agriculture in high altitude regions. In: New Directions of Diverse Planet: Proceedings of 4th International Crop Science Congress, Brisbane, Australia, 26 Sep-1 Oct, 2004.

Sassenrath GE, Ort GE, Portis AR. 1990. Impaired reductive activation of stomatal biophosphatases in tomato leaves following low temperature exposure at high light. Arch Biochem Biophy. 282: 302-308.

Satake T, Koike S. 1983. Sterility caused by cooling treatment at the flowering stage in rice plants. I. The stage and organ susceptible to cooling temperatures. Jap J Crop Sci. 52: 207-214.

Schans DA, VanDer, Van Dijk DA, Dolstra O. 1995. Influence of plant spacing sowing time and cold tolerance on nitrogen efficiency of maize during the juvenile stage. Verslag Proefstation Voor de Akkerbouw en de Groenteteelt in de Vollegrond. 1991: 113.

Scott SJ, Jones RA. 1982. Low temperature seed germination of *Lycopesircon* species evaluated by survival analysis (abstr.) Hort Science. 17 (3, II): 477.

Seki M, Marusaka M, Abe H, Kasuga M, Yamaguchi-Shinizaki K, Carninei P, Hayashzaki Y, Shinozaki K. 2001. Monitoring the expression pattern of 1,300 *Arabidopsis* gene under drought and cold stresses using a full-length cDNA microarry. Plant J. 31: 279-292.

Seki M, Marusaka M, Ishida J, Nanjo T, Fujita M, Dona Y, Kamiya A, Nakajima M, Enju A, Sakuraj T *et. al.* 2002. Monitoring the expression profiles of 7,000 *Arabidopsis* genes under drought cold and high salinity stresses using full length cDNA microarray. Plant Cell. 13: 61-72.

Selvakumar KS, Sundarapandian G, Vaaithialingam R. 1987. Physiological indices for cold tolerance in rice. Madras Agric J. 74: 34-41.

Sen Gupta, A, Heimen JL, Holoday AS, Burke JJ, Allen RD. 1993. Increased resistance to oxidative stress in transgenic plants that overexpress chloroplastic Cu/Zn superoxide dismutase. Proc Natl Acad Sci, USA. 90:1629-1633.

Shahi BB, Khush GS. 1986. Genetic analysis of cold tolerance in rice. In: Rice Genetics. Intr Rice Res Inst, Manila, Philippines. pp. 429-435.

Shen W, Nada K, Tachibana S. 1999a. Oxygen radical generation in chilled leaves of cucumber (*Cucumis sativus* L.) cultivars with different tolerance to chilling temperature. J Japan Soc Hortic Sci. 68: 780-787

Shen W, Nada K, Tachibana S. 1999b. Effect of chilling treatment on enzymic and non-enzymic antioxidant activities in leaves of chilling tolerant and chilling sensitive cucumber (*Cucumis sativus* L.) cultivars. J Japan Soc Hortic Sci. 68: 967-973.

Shewfelt RL, Erickson ME. 1991. Role of lipid peroxidation in the mechanism of membrane associated disorders in relationship to antioxidant content. J Plant Physoil. 133: 56-61.

Singh BB, Singh DP, Singh NP. 1997a. Genetic variability in pigeon pea germplasm for cold tolerance. Indian J Genet. 57: 425-430.

Singh J, Jhonson-Elanagan AM. 1989. Alterations of gene expression during the induction of freezing tolerance in a *Brassica napus* cell suspension culture. In: Environmental Stress in Plants Biochemical and Physiological Mechanisms, Cherry JH (eds.). Berlin, Springer-verlag. pp. 291-301.

Singh KB, Malhotra RS, Saxena MC. 1989. Chickpea evaluation for cold tolerance under field conditions. Crop Sci. 29: 282-285.

Singh KB, Malhotra RS, Saxena MC. 1990. Source of tolerance to cold in *Cicer* sp. Crop Sci. 30: 1136-1138.

Singh NK, Bracker CA, Hasegawa PM, Handa AK, Buckel S, Hermodson MA, Pfankoch E, Regnier FE, Bressan RA. 1997b. Characterization of *Osmotin*. Plant Physiol. 85: 529-536.

Smith TA. 1985. Polyamines. Annu Rev Plant Physoil. 36: 117-143.

Sreedharan PN, Rao GJN. 1983. A new screening technique for isolation of donors for cold tolerance in rice. Oryza. 20: 156-158.

Srinivasan A, Saxena NP, Johanson C. 1999. Cold tolerance during early reproductive growth of chickpea (*Cicer arientinum* L.) : generic variation in gamete development and function. Field Crops Res. 60: 209-222

St. Johan JB, Christiansen MN, Ashworth EN, Gentner WA. 1979. Effect of BASE 13-338, a substituted puridazinone, on linolenic acid levels and winter hardiness of cereals. Crop Sci. 19: 65- 69.

Stamp P. 1979. Pigment content and activities of photosynthetic enzymes in the leaves of young maize in relation to the temperature at grain ripening. Zeitschrift fuer Acker und Pflanzenbau. 148(3): 230-238.

Steponkus PL, Uemura M, Joseph RA, Gitmour S, Thomashow ME. 1998. Mode of action of the COR15a gene on the freezing tolerance of *Arabidopsis thaliana*. Proc Natl Acad Sci, USA. 95: 14570-14575.

Steponkus PL, Uemura M, Webb MS. 1993. A contrast of the cryostability of the plasma membrane of winter rye and spring oat- two species that widely differ in their freezing tolerance and plasma membrane lipid composition. In : Advances in Low Temperature Biology, Steponkus PL (ed). London, JAI press. 2: 211-312.

Sthapit BR, Shrestha KP. 1991. Breeding for cold tolerance at reproductive phase in the high hills of Nepal. Intr Ric Res News;. 16(5): 14.

Subedi KD, Floyd CN, Budhathoki CB. 1998. Cold temperature induced sterility in spring wheat (*Triticum aestivum* L.) at high altitudes in Nepal: Variation among cultivars in response to sowing date. Field Crop Res. 55: 141-151.

Suh JP, Cho YC, Jeon EG, Choi IS, Choi YS, Baek MK, Yea JD, Ahn SN, Kim YG, Hwang HG. 2004. SSR analysis of heterosis for cold tolerance in rice. Korean J Breed. 36(3): 157-164.

Sutka J. 1981. Genetic studies of frost resistance in wheat. Theor Appl Genet. 59: 145-152.

Sutka J. 1983. Genetic analysis of frost resistance of hexaploid wheat in the F_2 generation of a 6 × 6 diallel cross. Narenytermeles. 32: 385-391.

Sutka J. 1989. Genetic control of frost resistance in wheat. Sveriges Utsadesforenings Tidskrift. 99: 135-142.

Tamminen I, Makela P, Heino P, Palva ET. 2001. Ectopic expression of AB13 gene enhances freezing tolerance in response to abscisic acid and low temperature in *Arabidopsis thaliana*. Plant J. 25: 1-8.

Tasaka Y, Nishida I, Higashi S, Beppu T, Muratu N. 1990. Fatty acid composition of phosphatidglycerols in relation to chilling sensitivity of woody plants. Plant Cell Physiol 31: 545-550.

Templeton-Somers K, Sharp WR. 1980. Selection of cold tolerance cell lines of carrot. Plant Physiol. 65 (6, Suppl.): 38.

Teutonico RA, Palta JP, Osborn TC. 1993. *In vitro* freezing tolerance in relation to winter survival rape seed cultivars. Crop Sci. 33: 103-107.

Tseng MJ, Li PH. 1984. Mefluidide protection of severely chilled crop plants. Plant Physoil. 75: 249-250.

Tseng MJ, Li PH. 1990. Alteration of gene expression in potato (*Solanum commersonii)* during cold acclimation. Physoil Plant. 78: 535-547.

Tseng MJ, Li PH. 1991. Changes in protein synthesis and translatable messenger RNA populations with ABA-induced cold hardiness in potato. Physoil Plant. 81: 349-358.

Urao T, Yakubov B, Yamagnchi-Shinozaki K, Shinozaki K. 1998. Stress responsive expression of gene for two componenets response. Regulator like protein in *Arabidopsis thaliana.* FEBS Lett. 427: 175-178.

Urao T, Yamaguchi-Shinozaki K, Shinozaki K. 2000. Two component systems in plant signal transduction. Trends Plant Sci Rev. 5(2): 67-74.

Vannini C, Locatelli F, Bracale M, Magnani E, Marsoni M, Osnato M, Mattana M, Baldeni E, Coraggio I. 2004. Overesxpression of the rice *Osnmyb4* gene increase chilling and freezing tolerance of *Arabidopsis thaliana* plants. Plant J. 37(1): 115-127.

Voinikov VK, Korytov. 1991. Low temperature induction of cold shock protein synthesis in seedlings of maize lines and hybrids. Tsitologiya i Genetika. 25(6) : 64-69.

Vos DA de, Hill Jr. RR, Hepler RW. 1982. Response to selection for low temperature sprouting ability in tomato populations. Crop Science. 22: 876-879.

Wada H, Gombos Z, Murata N. 1990. Enhancement of chilling tolerance of a cyanobacterium by genetic manipulation of fatty acid desaturation. Nature. 347: 200-203.

Walker MA, Smith DM, Pauls KP, McKersie BD. 1990. Development of a chlorophyll fluorescence screening test to evaluate chilling tolerance in tomato. HortSci. 334-349.

Wang CY, Baker JE. 1979. Effects of two free radical scavengers and intermittent warning on chilling injury and polar lipid composition of cucumber and sweet pepper fruits. Plant Cell Physoil. 20: 243-251.

Wang CY. 1985. Modification of chilling susceptibility in seedlings of cucumber and zucchini squash by the bioregulator pactobutrazol. Sci Hort. 26: 293-298.

Wang CY. 1990. Alleviation of chilling injury of horticultural crops. In: Chilling Injury of Horticultural Crops, Wang CY (ed.). CRC Press, Boca Raton. pp. 281-301.

Whitaker BD. 1995. Lipid changes in mature green bell pepper fruit during chilling at 2 degrees C and after transfer to 20 degrees C subsequent to chilling. Physoil Plant. 93: 683-688.

Wolter FP, Schmidt R, Heinz E. 1992. Chilling sensitivity of *Arabidopsis thaliana* with genetically engineered membrane lipids. EMBO J. 11: 4685-4692.

Xie ZY. 1992. Low temperature germination ability in inbred lines and hybrids of maize. Crop Genetics Resources. No. 3: 27-28.

Xin Z, Li PH. 1992. Abscisic acid induced chilling tolerance in maize suspension-cultured cells. Plant Physoil. 99: 707-711.

Xiong ZM, Min SK, Wang GL, Chen SH, Cao LY. 1990. Genetic analysis of cold tolerance at the seedling stage of early rice (*Oryza sativa* L. sub sp. *indica*). Chinese J Rice Sci. 4(2): 75-78.

Xu YB, Chen ZT. 1989. Genetic study on cold tolerance in crosses between *indica* and *japonica* rice at seedling stage. Scientia Agricultura Sinica. 22(5): 14-18.

Xu YB, Shen ZT. 1988. Inheritance of seedling stage cold tolerance in 6 *indica* and *japonica* crosses. Intr Rice Res Newsl. 13(4): 15-16.

Yamaguchi-Shinozaki K, Shinozaki K. 1993. Characterization of expression of a desiccation-responsive *rd29* gene of *Arabidopsis thaliana* and analysis its promoter in transgenic plants. Mol Gen Genet. 236: 331-340.

Yang BK, Noh TH, Park HK, Ko JK, Seo JH, Kim TS, Kim KY, Sim IS, Lee DJ, Cho KJ, Shin HT, Cho SY, Lee SY, Kim CH. 2003. A new early-maturing good quality and cold tolerant rice cultivar 'Inweolbyeo'. Korean J Breed. 35(4): 249-250.

Zapata EJ, Torrizo LB, Aldemita RR, Noverno AU, Mazaredo AM, Visperas RV, Moon HP, Heu MH. 1986. Rice, anthers culture: a tool for production of cold tolerant lines. In: Rice Genetics. Proc Int Rice Genet Symp, 27-31 May, 1985. Manila Philippines, IRRI. pp. 77-78.

Zeng DH, Deng YC. 1982. Cold tolerance of rice at different seedling stages. J Agic Res, China. 31: 255-264.

Zhan QC, Zhu KY, Chen ZW *et al*. 2005. Studies on the QTL for cold tolerance related characters of rice seedlingby molecular markers (in Chinese with English abstract). Hybrid Rice. 20(1): 50-55.

Zhang QC, Koji S, Akira K, Zhan QC. 2003. Molecular mapping for seedling cold tolerance QTLs in rice. Hunan Agril Sci Technol Newsl. 4(1): 8-15.

Zhang QC, Zhu KY, Chen ZW, Zeng SZ. 2004. Mapping of QTLs controlling seedling cold tolerance in rice using F_2 population. J Hunan Agril Univ. 30(4): 303-306.

Zhong KY, Ye MS, Hu XW, Guo JC. 2006. Role of transcription factors CBF in plant cold tolerance. Hereditas Beijing. 28(2): 249-254.

Unit-IV
OXIDATIVE STRESS

7

Oxidative Stress Tolerance

7.1. INTRODUCTION

Abiotic stress is a major limiting factor in crop productivity. Several stresses, such as high temperature, high light intensity, osmotic stress, heavy metals, and several herbicides and *toxin* lead to over production of reactive oxygen species (ROS), causing extensive cellular damage and inhibition of photosynthesis (Allen, 1997). ROS is produced in both unstressed and stressed cells. Plants have well developed mechanisms against ROS, involving both limiting the formation of ROS as well as instituting its removal. Under unstressed conditions, the formation and removal of O_2 are balance. Introduction of molecular oxygen (O_2) into atmosphere by O_2^- evolving photosynthetic organisms, ROS has been the unwelcome companion of anaerobic metabolism. In contrast to O_2, these partially reduced or activated derivatives of oxygen (O_2^1, O_2^-, H_2O_2, and HO) are highly reactive and toxic and can lead to the oxidative destruction in tissues during stress.

The toxic superoxide radical has a half-life of less than one second. The steady state level of ROS in different cellular compartment is determined by interplay between multiple ROS-producing pathways and ROS-scavenging mechanisms. There are many potential sources aerobic metabolism, such as photosynthesis and respiration. Some other new sources of ROS are NADPH oxidases, amine oxidase, and cell wall bound peroxidase.

Under normal growth conditions, the production of ROS in cell is estimated at a constant rate of 240 mMs^{-1} O_2^- and a steady state level of 0.5 mM H_2O_2. However, stresses that disrupt the cellular homeostasis of cells result in the enhanced production of ROS (upto 720 mMs^{-1} O_2^- and steady state level of 5-15 mM H_2O_2). This includes drought, desiccation, salt, chilling, heat shock, heavy metals, UV-radiation, air pollutants such as ozone and SO_2, mechanical stresses, nutrient deficiency, pathogen attack, and highlight intensity. It is also thought that during stress ROS are actively produced by cells and acts as signals for the induction of defense pathways. Thus, ROS may be viewed as cellular by-products of stress metabolism, as well as secondary messengers involved in the stress response signal transduction pathway.

7.2. INJURY MACHANISMS

7.2.1. Origins of ROS in Plants

The circumstance in which cellular redox homeostasis is disrupted can lead to oxidative stress or the generation of ROS (Asada, 1994). Production of ROS during environmental stress is one of the main causes for decrease in productivity, injury and ultimately death of plant accompany these stress in plants. ROSs are produced in both unstressed and stressed cells and in various locations (Halliwell and Gutteridge, 1989). These ROSs are generated endogenously during certain developmental transitions, such as seed maturation, photosynthesis and respiratory metabolism. An initial oxyradical product, the superoxide radical (O_2^-), upon further reaction within the cell, can form more ROS such as hydroxyle radicals and singlet oxygen.

Cell organelles, such as peroxisome and chloroplast, produce ROSs at comparatively high-rate. In chloroplast, changes in light intensity and temperature or limitation in the substrates of the photosynthesis occur frequently, resulting in increased production of ROS. ROSs are produced at high levels in peroxisomes. Hydrogen peroxide is produced peroxisomal respiratory pathway by flavin oxide. Fatty acid oxidation and glycolate oxide action are another sources of hydrogen peroxide production in peroxisome. Developmental transitions, such as seed maturation, in which peroxisome play and important role, also involve oxidative stress (Leprince *et al.*, 1990; Walter, 1998).

7.2.2. Injury by ROS in Plants

Highly energetic reactions of photosynthesis and an abundant supply of oxygen make the chloroplasts a rich source of ROS in plants. The potential sources of ROS have been presented in Table 7.1. High light intensity can lead to reduction of PSI so that CO_2 fixation cannot keep pace and $NADP^+$ pools are reduced. Under these conditions, O_2 can complete for electrons from PSI, leading to the generation of ROS. When CO_2 fixation is limited by environmental conditions such as cold temperatures or low CO_2 variability (closes stomata) excess PSI reduction and increased ROI production can occur even at moderate light intensities. Efficient removal of ROIs from chloroplast is critical, since H_2O_2 concentrations as low as 10mM can inhibit photosynthesis (Kaiser, 1979). When too much of oxygen radicals accumulate, peroxidation of fatty acid occur, and this can alter the functions of the cell membrane proteins.

Table 7.1. The potential source of ROS production in plants

Source of ROS	Production site (organ/organelle)	Primary ROS
Photosynthetic electron transport and PSI and II	Chloroplast	O_2^1
Respiration electron transport	Mitochondria	O_2^1
Glycolate oxidase	Peroxisome	H_2O_2
Excited chlorophyll	Chlorophyll	O_2^1
NADPH oxidase	Plasma membrane	O_2^1
Fatty acid β-oxidation	Peroxisome	H_2O_2
Oxalate oxidase	Apoplast	H_2O_2
Xanthine oxidase	Peroxisome	O_2^1
Peroxidases, Mn^{2+}, and NADH	Cell wall	H_2O_2, O_2^1
Amine oxidase	Apoplast	H_2O_2

PS: Photo system

Molecular O_2 (dioxygen) at its ground state upshots and sometimes raised to lethal reactive states containing free radicals and derivatives. Utilization of O_2 proceeds readily via a complete stepwise, four-electron reduction to water during which partially reduced reactive intermediates are generated. The reactive species of reduced oxygen include the superoxide radicals (O_2^-), hydrogen peroxide (H_2O_2), and hydroxyl radical (OH). If the production rate of superoxide radical under abiotic stress exceeds SOD (superoxide dismutase) activity, then oxidative damage results (Casano *et al.*, 1997).

$$3O_2 + 2H_2O \rightarrow O_2^{\cdot -} + 2H_2O^{\cdot} + H_2O_2$$

Activated oxygen species are extremely reactive and cytotoxic to all organisms. A common consequence of most abiotic stressed is that they result, at some stage of stress exposure, in a serial production of ROS (Polle and Rennenberg, 1993). The success reduction of molecular oxygen to H_2O yields the intermediates O_2^-, HO., and H_2O_2, which are potentially toxic, because they are relatively reactive oxygen compared with O_2^- ROS may lead to interspecific oxidation of proteins and membrane lipids or may cause DNA injury. Hagar *et al.* (1996) also suggested that ROS play an important role in endonuclease activation and DNA damage. As a consequence, tissues injury by oxidative stress generally contains increased concentration of carbohydrated proteins and malondialdehyde and show an increased production of ethylene (Dean *et al.*, 1993; Ames *et al.*, 1993). ROS can react with unsaturated fatty acid to cause peroxidaton of essential membrane lipids in the plasmalemma or intercellular organelles. Peroxidation damage of the plasmolemma leads to leakage of cellular contents, rapid desiccation, and cell death. Intercellular membrane damage can affect respiratory activity in mitochondria, cause pigment breakdown, and cause loss of carbon fixing ability in chloroplasts.

Superoxide is a charged molecule and cannot cross biological membranes. Subsequently compartmentation of defense mechanisms is, therefore, crucial for efficient removal of superoxide anions at their site of generation through out the cells. Hydrogen peroxide, on other hand, which is formed as a result of SOD action, is capable of diffusing across membrane and thought to fulfill a signaling function in defense response (Mullineaux *et al.*, 2000).

Several Kalvin-cycle enzymes within chloroplasts are extremely sensitive to H_2O_2 (the product of superoxide dismutation) directly inhibit CO_2 fixation (Kaiser, 1979). H_2O_2 was found to be active with mixed function oxidase in marking several types of enzymes for proteolytic degradation. The hydroxyl radical indiscriminately and rapidly attacks virtually all macromolecules leading to serious damage in cellular components, DNA lesions, and mutations, and often leading to irreparable metabolic disfunctions and cell death. Thus, O_2, although essential for the existence and survival of aerobic life, presents leaving organisms with a variety of physiological challenges collectively known as *"Oxidative stress"*.

In plants, the superoxide radicals and singlet oxygen are commonly produced in illuminated chloroplasts by occasional transfer of an electron from an excited chlorophyll molecule or PSI component under conditions of high NADPH/NADP ratios to molecular O_2^- carotenoids, which are essential components of thylakoid membranes, can effectively quench the excited triplet state of chlorophyll and/or single oxygen. However, various environmental perturbations (e.g. intense light, drought, temperature stress, herbicides, etc.) can cause excess reactive O_2 species, overwhelming the system and necessitating additional defense. For example, stomatal closure resulting from drought conditions limits CO_2 availability for photosynthetic carbon assimilation. Under such conditions and in intense sunlight, excess superoxide production in the chloroplast can result in photoinhibition damage. In addition to normal metabolic activity, reactive O_2 species can result from cellular exposure to various environmental stimuli. Such factors as UV light and other from of radiations, herbicides (e.g. Paraquat, diquat), pathogens, certain injuries, *hypoxia*, ozone, temperature fluctuations, and various other stresses are known to induce free radical formation in most aerobic organisms. Lipid peroxidation occurs mainly in membranes, where the content of unsaturated fatty acid is relatively high. Peroxidation of membrane lipids arising out of the oxidative damage intact cells results in decrease fluidity, inactivation of membrane bound enzymes and receptors and changes in non-specific ion permeability (Bast, 1993).

7.3. TOLERANCE MECHANISMS

All aerobic organisms must posses the means to protect themselves from the toxic effects of reduced oxygen species. While oxygen is vital for all aerobic life forms, its reduced derivatives or activated oxygen species such as superoxide, hydrogen peroxide, hydroxyl radicals, singlet oxygen are associated with a number of physiological disorders in plants causing lethal reactions (Asada and Takahashi, 1987; Monk *et al.*, 1987). Although activated oxygen is produced as a by-product of normal cell metabolism, its levels are enhanced by exposure to chemical and environmental stress. Increased production of activated oxygen species was found to be associated with the development of injury symptoms resulting from diverse environmental stress, including chilling (Wise and Naylor, 1987), drought (Burke *et al.*, 1985; Price and Hendry, 1991), desiccation (Senaratna *et al.*, 1985; Leprince *et al.*, 1990), flooding (Van Toai and Bolles, 1991), freezing (Kendall and McKersie, 1989; McKersie *et al.*, 1998), ozone (Decleire *et al.*, 1984), calcium deficiency (Monk and Davies, 1989), hypoxia (Monk *et al.*, 1987), pathogens (Bowler *et al.*, 1992).

ROS are toxic, but also participate in key signaling events, plant cell require different mechanisms to regulate their intercellular ROS concentrations by scavenging of ROS. One of the most crucial functions of the plant cells is their ability to fluctuation to their environment. Plant cells responds defensively to oxidative stress by removing the ROS and maintaining antioxidant defense compounds at level of that reflect ambient environmental conditions (Scandalios, 1997). The mechanisms that act to adjust antioxidant levels to afford protection include changes in antioxidant gene expression (Cushman and Bohrnet, 2000). The problem incurred by aerobic organism is the need to effectively eliminate toxic O_2 intermediate species generated during normal metabolic activity or as a consequence of various exogenous environmental insults. Through selective pressure and evaluation, numerous defense mechanisms, both enzymatic and non-enzymatic, were emerged to protect cells against oxidative injury. Antioxidant system in plant chloroplasts include enzymes such as SOD, peroxidases, catalases, GP (glutathione peroxidase) and APX (ascorbate peroxidase), which are capable of removing, neutralizing, or scavenging free radicals and oxyintermediates (Asada, 1999). Food derived antioxidants, such as α-tocopherol and β-carotene are considered to play a significant role in quenching ROS (Halliwell and Gutteridge, 1990; Shahidi *et al.*, 1992). Without these defenses, plants could not efficiently convert solar energy to chemical energy. The balance between SOD and APX (and/or CAT) actively in cells is considered to be crucial for determining the steady-state levels of O_2^- and H_2O_2. Antioxidants, such as ascorbic acid and glutathione, found at very high concentrations in chloroplast and other cellular components, are also important for defense of plants against oxidative stresses. Over-expression of ROS-scavenging enzymes increases tolerance of plants to abiotic stresses. ROS production can also be decreased in cells by the chloroplast and mitochondria by a group of enzymes called alternative oxidase.

The maintenance of ascorbates in its reduced form is achieved by monodehydroascorbate radical reductase (MDAR) and NAD(P)H or feredoxin as reductant or by the operation of the ascorbate-glutathione pathway (Foyer and Halliwell, 1976; Borraccino *et al.*, 1986; Miyake and Asada, 1994). In later pathway, the reduction of dehydroascorbate is coupled to the oxidation of glutathione, which in turn, is reduced by glutathione reductase by oxidation of NAD(P)H (Foyer and Halliwell, 1976).

Antioxidant defense includes ascorbate peroxidase and GSH reductase, which scavenges H_2O_2 in chloroplasts and mitochondria, respectively; the catalase and peroxidase that remove H_2O_2 very efficiently; and SODs that scavenge the superoxide anions. Superoxide radicals are produced by the reduction of molecular oxygen at photo system I. This $.O_2^-$ is rapidly dismuted to H_2O_2 by SOD that is associated with the thylakoid. The superoxide and H_2O_2 that diffuses away from the membrane-associated enzymes can be scavenged in the storma. The CATs and SODs are the most efficient antioxidant enzymes. Their combined action converts the potentially dangerous superoxide radicals (O_2^-) and hydrogen peroxide H_2O_2 to water (H_2O) and molecular oxygen (O_2), thus causing cellular damage. These hydroxyl radicals can cause damage to all classes of biologically important macromolecules, especially nucleic acids. It can also modify protein so as to make them more susceptible to proteolytic attack (Davis, 1987). The combined action of SOD and CAT abate the formation of most toxic and highly reactive oxidant, the hydroxyl radical (OH), which can react indiscriminately with all macromolecules.

Antioxidant enzymes are critical components in preventing oxidative stress in plant. The activity of one or more of these enzymes is generally increased in plants exposed to stressful conditions, and this elevated activity correlates with increased stress tolerance. Pre-treatment of plants under one form of stress can increase tolerance to different stresses. The plant lines selected for resistance to ROS-inducing herbicides such a methyl viologen (MV) was found to inflate levels of one or more of these enzymes and exhibit cross-tolerance (Gressel and Galun, 1994). The above observations indicate that exposure to environmental stresses can stimulate plants to enhance their ROS scavenging systems and this enhancement can be apparently provide generalized stress protection.

The mechanisms to detoxify oxygen radicals are varied, and the complex interactions among the antioxidants in different subcellular compartments, cells and tissues are now being elucidated. A number of enzymes are involved in the selective detoxification of activated oxygen species. One enzyme, SOD (EC 1.15.1.1), catalyzes the dismutation of super oxide to hydrogen peroxide and molecular oxygen (Bowler *et al.*, 1992). Asada (1994) proposed reactive oxygen intermediate scavenging pathway of plant chloroplasts. Superoxide radicals are dismuted by SOD associated with PSI and the resulting H_2O_2 is scavenged by the thylakoid-bound APX(tAPX). Reactive oxygen species the escape destruction at the thylakoid are scavenged by stomatal SOD and Stromal APX(sAPX).

$$2H + 2O_2^- \ (\textit{in presence of SOD}) \rightarrow H_2O_2 + O_2^-$$
$$2\,H_2O_2 \ \ (\textit{in presence of APX}) \rightarrow 2\,H_2O$$

Over years, SOD enzyme was variably referred as erythrocuprein, indophenol oxidase, and tetrozolium oxidase. The catalytic function of the enzyme was discovered by McCord and Fridovich (1969) distributed among O_2^- consuming organisms, aerotolerant anaerobes, and some obligate anaerobes (Fridovich, 1986). The hydrogen peroxide is removed by a number of different peroxidase enzymes such as those in the ascorbate-GSH cycle present in the chloroplast (Salin, 1991).

7.4. ANTIOXIDANTS

The ROSs generate due to the biotic and abiotic stresses are need to be removed or deactivate to maintain the normal metabolic processes in plants. Here, antioxidants play major role. Antioxidant system in plant chloroplasts include enzymes such as SOD and APX and other enzymes, which are capable of removing, neutralizing, or scavenging free radicals and oxyintermediates. Without these defenses, plants could not efficiently convert solar energy to chemical energy.

7.4.1. Superoxide Dismutase (SOD) Isozymes

Within a cell, the SODs constitute the first line of the defense against ROS. It is found in all subcellular locations, like mitochondria, chloroplast microsomes, glyoxysomes, peroxysomes, apoplast and the cytosol. SODs are metaloproteins found in various compartments of plant cells and contain Cu and Zn, Fe or Mn cofactors. These metaloenzymes are classified into three distinct types according to their metal cofactor: Cu/Zn (Cu/Zn-SOD), Mn (Mn-SOD) and Fe (Fe-SOD). Plant SODs are products of nuclear genes, and cDNAs that encode chloroplastic Cu/Zn-SODs include amino-terminal chloroplast transit peptide sequences (Scioli and Zilinskas, 1988). A full-length Fe-SOD cDNA from soybean also includes an apparent chloroplast-targeting domain. All SODs irrespective of source are multimeric metallic proteins that are very efficient at scavenging the superoxide radical. These metal enzymes catalyze the disproportion of superoxide (O_2^-) radicals, and they play an important role in protecting cells against toxic effect of superoxide radicals produced in different cellular loci (Fridovich, 1986; Halliwell and Guteridge, 1989). The Cu/Zn-SODs, as well as most prokaryotic Mn-SODs and Fe-SODs are dimeric, whereas, the Mn-SODs from mitochondria and certain thermophilic bacteria are tetrameric.

The Cu/Zn-SODs is found almost exclusively in eukaryotes in the cytosol and/or chloroplast. The Mn-SOD group is widely distributed in prokaryotes and in the mitochondrial matrix of eukaryotes. Fe-SOD was found in prokaryotes and recently within the chloroplast of some plants families (Bannister *et al.*, 1987; Bowler *et al.*, 1992). The presence of different types of SOD was demonstrated in peroxisomes from several plant species (Dell Rio *et al.*, 1983; Sandalio *et al.*, 1987; Sandalio and Dell Rio, 1987, 1988; Lopez-Huertas *et al.*, 1994; Bueno *et al.*, 1995). There is abundant information on SODs from cytosolic, mitochondrial and chloroplastic origin (Steinman, 1982; Parker *et al.*, 1984; Bannister *et al.*, 1987, 1991; Halliwell and Guterridge, 1989;

Asada, 1992; Bueno and dell Rio, 1992). The evolutionary reason for the separation of SODs with different metaloproteins is probably related to the different availability of soluble transition metal compounds in the biosphere in relation to the O_2 content of atmosphere in different geological eras (Bannister *et al.*, 1991).

The SODs with different metal-cofactors have been briefed individually with their functions and achievement in the field of transgenic development.

7.4.1.1. Iron superoxide dismutase (Fe-SOD)

It is probably the most ancient SOD group. Fe-SOD is found both in prokaryotes and eukaryotes. Fe-SOD is inactivated by H_2O_2 and is resistant to KCN inhibition. It is located in the chloroplast (Salin, 1988; Kliebenstein *et al.*, 1998). It has been suggested that Fe was probably first metal used as metal-cofactor at the active site of the first SOD because of an abundance of iron in soluble Fe (II) form (Bannister *et al.*, 1991).

Fe-SOD is absent in animal, which suggested that Fe-SOD gene originated in the plastid and moved to nuclear genome during evaluation. This theory has been supported by Bowler *et al.* (1994). They observed the existence of several conserved regions that are present in plant and cyanobacterial Fe-SOD sequences, but absent in non-photosynthetic bacteria.

Two distinct groups of Fe-SOD are found in plants. One group is a homodimer formed from two identical 20 kDa sub-units of proteins, with 1-2 g atom of iron in the active centre (Yost and Fridovich, 1973; Puget and Michelson, 1974; Baldensperger, 1978; Kanematsu and Asada, 1978; Salin and Bridges, 1980).

7.4.1.2. Manganese superoxide dismutase (Mn-SOD)

Increased levels of O_2 in the environment, the mineral compounds are being oxidised, which decrease the amount of available Fe (II) causing a shift to the use of a more available metal Mn (III). Mn-SODs are found in mitochondria (del Rio *et al.*, 1983; 1998; Sandalio and del Rio, 1987; del Rio *et al.*, 1992; Zhu and Sandalio, 1993; Zhu *et al.*, 1999) and peroxisomes (Jackson *et al.*, 1978; Foster and Edwards, 1980; Reddy and Venkaiah, 1982; Sandalio and del Rio, 1987; Droillard and Paulin, 1990; Bowler *et al.*, 1994) of the cells. Mn-SODs carry only one metal atom per subunit. These enzymes can not function without the Mn atom present at the active site. Catalysis by Mn-SODs is through the attraction of negatively charged O_2^- molecules to a site formed positively charged amino acid present at the active site of the enzyme. The metal present at the active site then denotes an electron directly to the O_2^-, reducing O_2^- molecule, which in turn forms H_2O_2 by reacting with a proton (Asada, 1994; Bowler *et al.*, 1994). Plant Mn-SODs have approximately 65% sequences similarity to one another, and these enzymes also have high similarities to bacterial Mn-SODs (Bowler *et al.*, 1994).

In green algae and cyanobacteria, Mn-SOD is found in the thylakoid membrane (Kanematsu and Asada, 1979; Okada *et al.*, 1979). Although thylakoid bound form of Mn-SOD was reported in spinach plants (Hayakawa *et al.*, 1985).

Mn-SOD was reported to respond positively to salt stress (Hernandez *et al.*, 1993, 1995; Gomez *et al.*, 1999), manganese toxicity (Gonzalez *et al.*, 1998), chilling stress (Lee *et al.*, 1999; Lee and Lee, 2000) and drought (Wu *et al.*, 1999).

7.4.1.3. Copper-Zinc superoxide dismutase (Cu/Zn-SOD)

When the environment is completely replenished with oxygen, Fe (II) is almost completely unavailable in environment and insoluble Cu (I) is converted into Cu (II). At this stage, Cu (II) began to use as the metal cofactor at the active site of SODs. It was proposed that Cu/Zn-SOD in apoplast function in lignification and that in the nucleus it protects the cells against fatal mutation caused by O_2^- molecules (Ogawa *et al.*, 1996, 1997). Cu/Zn-SODs are found throughout the plant cells.

7.4.1.4. Cambialistic superoxide dismutase (Fe/Mn-SOD)

Cambialistic SOD (Fe/Mn-SOD) is an enzymatically active protein that can accommodate either metal ligand (Fe/Mn). Some bacterial species use a common SOD for both Fe and Mn and utilize either of these two metals as the active metal cofactor, according to the availability of the metal. The similar electrical properties of Fe and Mn allow the enzyme to function with either metal without a conformation change in its protein structure.

7.4.2. Other Antioxidants

Ascorbic acid, glutathione, and α-tocopherol have each been shown to act as antioxidants in the detoxification of ROS (Table 7.2). These compounds have central and interrelated roles, acting both non-enzymatically and as substrates in enzyme catalyzed detoxification reaction. Metallic cycles located within the aqueous phase of the peroxisome, chloroplast, cytosol, and the mitochondrion successively oxidize and re-reduce glutathione and ascorbate, using NADPH as the ultimate electron donor. Ascorbate reduces glutathione (GSH), ascorbate peroxidase (APX), glutathione reductase (GR), and monodehydroascorbate reductase (MDHAR) are involved in several contexts in antioxidants regeneration throughout the plant cell.

Table 7.2. Major ROS scavenging enzymes in plant

Enzyme/metabolite	Localization	Primary ROS
Superoxide dismutase (SOD)	Chl. Cyt, Mit, Per, Apo	O_2^1
Ascobic peroxidase (APX)	Chl. Cyt, Mit, Per, Apo	H_2O_2
Catalase (CAT)	Per	H_2O_2
Glutathione peroxidase (GPX)	Cyt	H_2O_2, ROOH
Peroxidases	CW, Cyt, Vac	H_2O_2
Thioredoxin peroxidase	Chl, Cyt, Mit, Per, Apo	H_2O_2, O_2^1
Ascorbic acid	Chl, Cyt, Mit, Per, Apo	H_2O_2, O_2^1
Glutathione	Chl, Cyt, Mit, Per, Apo	H_2O_2
α-Tocopherol	Membrane	ROOH, O_2^1
Caretenoids	Chl	O_2^1
Alternative oxidase (AOX)	Chl, Mit	O_2^1

Chl: Chloroplast **Mit:** Mitochondria **Per:** Peroxisome. **Apo:** Apoplast
O_2^1: Singlet oxygen **Vac:** Vacuol **Cyt:** Cytosol

7.4.2.1. Ascorbate peroxidase (APX)

APXs have high substrate specity ascorbate and are the primary H_2O_2 scavenging enzyme in the chloroplast and cytosol of plant cells (Asada, 1992). A membrane associated APX has been described in the peroxisome, and also in the chloroplast (Mullen and Trelease, 2000). Peroxisomes are subcellular respiratory organelles that contain catalase and H_2O_2 producing flavin oxidizes as basic enzymatic constituents, these organelles posses an essentially oxidative type of metabolism. Peroxisomes carry out essential function in almost all eukaryot cells. Cytosolic and chloroplastic forms of APX have been identified in plant cells. These enzymes differ in substrate specificity, pH optimum, and sensitivity to ascorbate depletion (Asada, 1992). Distinct thylakoid-bound and stomatal forms of APX also apparently exist in chloroplast (Miyake and Asada, 1992; Miyake *et al.*, 1993).

7.4.2.2. Glutathione reductase (GR)

Glutathione reductase has been reported in chloplast, mitochondria, peroxisome and cytosolic cellular components of *Pisum sativum* (Edwards *et al.*, 1990; Creissen *et al.*, 1991). Glutathione acts as redox sensor of environmental cues and forms part of the multi-regulatory circuitry coordinating defense expression. Glutathione has been reported to regulate rates of cell division (Sanchez-Fernandez *et al.*, 1997) and the induction of antioxidant defense (Herourt *et al.*, 1993). GR has been suggested as an intermediary in a redox sensing signaling pathway in plant involving the ROS-mediated oxidation of membrane lipids to oxylipins as the initial step (Ball *et al.*, 2001).

Plants that expressed high levels of bacterial GR showed no decrease in ozone activity but were some what more tolerant of methyl viologen and were able to maintain the reduction state of their ascorbate pools more efficiently than control plants (Aono *et al.*, 1991; Foyer *et al.*, 1991).

Higher GR activities have been reported in transgenic tobacco plants than the control. These plants were reported to have increased tolerance to methyl viologen and sulfur oxide, but not to ozone (Aono *et al.*, 1993). Transgenic tobacco plants that overexpress pea GR have also been developed (Creissen *et al.*, 1994).

Seed develop trough a series of stages that may be tracked at morphological, physiological and molecular levels. During the final stage of embryogenesis, seed must acquire desiccation tolerance to survive. As both fruit and seed dry, metabolism decline rapidly as water is lost. As respiratory activity slows during desiccation, electrons leak and react with oxygen to generate reduced oxygen species, so antioxidants are necessary to protect mitochondria from damage (Vertucci and Farrant, 1995). Free radical scavengers may provide additional protection during desiccation, because the development of desiccation tolerance, which involves a period of water stress, coincides with an increase in free radical scavengers in seeds (Vertucci and Farrant, 1995; Haslekas *et al.*, 1998). Water stress involves the production of ROS and their containment by antioxidants (Bohnert and Jenssen, 1996; Bohnert and Shevelava, 1998). Thus, availability of antioxidants such as GSH may be essential for seed maturation. Decrease in GR activity also associated with aging of seed (De Vos *et al.*, 1994; Bailly *et al.*, 1996).

7.4.2.3. Methionine

Methionine residues act as antioxidants protein reservoir (Hoshi and Heinemann, 2001). Surface exposed methionine residues are available for oxidation by molecules such as H_2O_2, thus effectively lower the degree of the damage caused by free radicals. In case of glutathione synthatase from *E. coli*, it was found that 8 of the 16 methionine residues present in the protein could be oxidized with little effect in overall enzyme activity (Levine *et al.*, 1999). They are re-reduced through the action of protein methionine sufoxide reductase, using thioredoxine as a source of reductant (Brot and Weissbach, 2000). The oxidation of surface methionine residues, which is mediated by the hydroxyl radical, ozone, and poeroxynitrile as well as H_2O_2, has been proposed to function in an antioxidant capacity in animal and yeast system (Moskovitz *et al.*, 1997; Preville *et al.*, 1999).

7.5. ROLE OF 'ROS' IN PLANT DEFENSE MECHANISM

ROSs were mainly known for their dangerous molecules, whose levels need to be kept as low as possible. Now this opinion is changing gradually. It has been realized that ROSs play important roles in defense system against pathogens, and it is termed as *'oxidative burst'* (Alvarez and Lamb, 1997; Doke, 1997; Bolwell *et al.*, 2002), make certain developmental stages, such as tracheary element formation, lignification and other cross-linking process in the cell in the cell wall- 'programmed cell death' (Jacobson, 1996; Teichmann, 2001; Fath *et al.*, 2002) and act as intermediate signaling molecules to regulate the expression of genes (May *et al.*, 1998; Karpinski *et al.*, 1999; Neill *et al.*, 2002; Vranova *et al.*, 2002).

Against pathogen attack, ROSs are produced by plant cells via the enhanced enzymatic activity of plasma membrane-bound NADPH oxidases, cell wall-bound peroxidase and amine oxidases in the apoplast. Hydrogen peroxide produced during these responses is thought to diffuse into cells through aquaporins and together with salicylic acid and nitric oxide activate many of the plant defenses, including the induction of programmed cell death. The activity of APX and CAT is suppressed during this response by salicylic acid and nitric oxide, the expression of APX is post-transcriptionally supressed, and the expression of the CAT is down regulated at the steady-state mRNA level. Thus, the plant simultaneously produces more H_2O_2 and the activation of programmed cell death.

Hydrogen peroxide is a substrate for peroxide-dependent reaction in biosynthesis of lignin. In spinach leaves, up to 40% of the cytosolic isoforms of

Cu/Zn-SOD associated with the lignifying cells in the leaf and hypocotyls tissues (Ogawa *et al.*, 1995). Further more they have demonstrated a preliminary evidence for an association of sites of generation of superoxide and of lignifying tissues in spinach hypocotyls (Ogawa *et al.*, 1996, 1997). Thus, Cu/Zn-SOD is essential to lignification since it supplies H_2O_2 and protects the peroxidase from inactivation by superoxide.

ROSs also play a role in intercellular sensing, activating antioxidant resistance mechanisms (Karpinski *et al.*, 1997; May *et al.*, 1998). A number of redox sensitive transcription factors have been identified in animals, bacterial, and plant cells (Pastori and Foyer, 2001).

7.6. TRANSGENIC DEVELOPMENT

Molecular analyses to be carried out to asses the presence of intruded transgenes in the regenerated plantlets and its stable integration in recipient genome. The regenerated plantlets are analyzed at molecular levels using PCR and Southern blot techniques other techniques, which have been briefly described here.

7.6.1. Screening Techniques

7.6.1.1. SOD Activity and Oxidative Stress Tolerance

SOD activity can be measured following standard protocols described by many researchers (Arisi *et al.*, 1998; McKersie *et al.*, 2000)

1. To measure SOD activity, collect leaf samples from greenhouse-grown plants at the mid-vegetative stage of development (Kalu and Fick, 1983).
2. Whole leaves are grounded with liquid nitrogen in a mortar, and add 0.4 mL of homogenizing buffer, 50 mM KH_2PO_4, pH 7.8, to extract the soluble proteins.
3. SOD activity is determined using the *in situ* staining technique (Beauchamp and Fridovich, 1971). The protein samples are separated by native PAGE on a separating gel of 13% (w/v) polyacrylamide in a tank buffer containing 25 mM Tris (pH 8.3).
4. The gels are stained for 30 min in the dark using a 1:1 mixture of: (a) 0.06 mM riboflavin and 0.651% (w/v) TEMED, and (b) 2.5 mM nitroblue tetrazolium (NBT), both in 50 mM phosphate buffer at pH 7.8, and then developed for 20 min under moderate light conditions.

5. The gels may be digitally photographed and measure the intensity of each band. The area of individual SOD isozymes can be expressed relative to a standard of *Escherichia coli* Fe-SOD (Sigma Chemical, St. Louis) to calculate each enzyme's activity. One unit of enzyme activity is the amount that will inhibit the rate of reduction of cytochrome *c* by 50% in a coupled system with xanthine and xanthine oxidase at pH 7.8 at 25 ^{0}C in a 3 mL reaction volume (McCord and Fridovich, 1969).
6. The protein content can be determined (Bradford, 1976), and SOD activity for each individual isozyme is expressed as units per milligram of protein.

7.6.1.2. Methyl viologen (MV) test

The herbicide paraquat produces its cytotoxic effects via a free radical mechanism. Although it is a banned herbicide generally used as an indicator in O_2 toxicity studies in many organisms. In maize, the increase observed in total SOD activity which was correlated with increased levels of specific isozymes. Light dependant MV-mediated oxidative damage to membranes and *Chl* (measured by the accumulation of pheophytin were found to be significantly decreased in plants that were expressed the chloroplastic Mn-SOD gene compared to untransformed control plans). Level of these transgenic tobacco plants also had fewer lesions than control plants when fumigated under conditions that simulate natural diurnal ozone fluctuations in industrialized regions. Significant increases in resistance to the herbicide acifluorofen and to freezing were found to co-segregate with the expression of this transgene in alfalfa, *Medicago sativa* (McKersie *et al.*, 1998), and cotton (*Gossipum hirsutum*). Preliminary analysis has indicated that expression of the chloroplastic Mn-SOD gene may provide increased chilling tolerance.

Comparing the MV-sensitivity of Cu/Zn-SOD over expressing plants with that of transgenic tobacco plants of the same variety (cv. Xanthi) that express a chloroplast Mn-SOD gene construct, it can be concluded that Cu/Zn-SOD-expressing plants showed similar levels of damage that were significantly lower than control plants. However, at a higher MV concentration (2.4 mM), protection in the Mn-SOD plants was maintained while protection in Cu/Zn-SOD plants was not significantly different from that in control plants.

A general method have been described for screening against methyl viologen test. At tillering stage, one tiller from each established plant was placed in test tubes containing 10 ml of 75 and 100 mM paraquat or methyl viologen (Roy, 2002). Following the protocol described by (Perl *et al.*, 1993), after 16 h incubation with MV, the solution is to be discarded and the test tubes are washed thoroughly and filled with tap water. Each experiment should be repeated three times. A control set (non-transgenic plants) was also maintained with each replication for comparative assessment (Fig. 7.1).

Fig. 7.1. Methyl viologen test by dipping roots in 75 mM methyl viologen to confirm oxidative stress tolerance of putative transformants. **Plant 1:** control plant grown in water, **Plant 2:** MV tolerant transformant- ATP-14; **Plant 3 and 4** (Designation: ATP1 and ATP 2) turned to be susceptible to MV as displayed by substantial leaf rolling. *(For colour version of this figure see page 535)*

The oxidative stress tolerance of the plants can also be assessed using the method described by Bowler *et al.* (1991):

1. Leaf slices, 0.5 cm across the largest width of the leaflet, are to be incubated in aqueous solutions of methyl viologen [Paraquat, Sigma Chemical; (1, 1 dimethyl 4,4 bipyridinium chloride, $C_{12}H_{14}N_2Cl_2$)], with concentrations ranging from 0 to 16 μM, overnight in the dark at 21 ^{0}C.
2. Then the samples are exposed to light for 2 h (200 μmol m^2 s^1 photosynthetically active radiation) at 26 ^{0}C, and were then allowed to develop in the dark for 20 h at 30 ^{0}C. The methyl viologen-dependent oxygen radical damage is estimated by chlorophyll fluorescence determination of photochemical yield, F_v/F_m (Genty *et al.*, 1989), using a portable chlorophyll fluorometer.
3. SOD and ascorbate peroxidase (APX) activity are determined on the same extracts from alfalfa leaves treated with methyl viologen.
4. To quantify APX activity, the extract (25 μL) and 150 μL of 0.03% (w/v) H_2O_2 are added to 2 ml of assay buffer, containing 0.5 mM ascorbate

and 0.1 mM EDTA in phosphate buffer at pH 7.0 (Nakano and Asada, 1981).

5. APX is determined by measuring the rate of oxidation of ascorbate at 290 nm using an absorbance coefficient of 2.8 mM^1 cm^1. One unit of APX is defined as 1 μmol of ascorbate oxidized per minute at pH 7.0 and 25 °C and is expressed per milligram of protein.

7.6.1.3. PCR analysis

Application of target DNA sequence using specific primer(s) may detect the transgenic plants. Forward primer 1 [5′ TGC GGA GCG GCG ATA CCG TA 3′] and reverse primer 2 [5′ GAG GCT ATT CGG CTA TGC TG 3′] were used by Roy (2002) for amplification of *npt*II using PCR. Individual PCR reaction was brought to a final volume of 25 μl consisting 1.5 units of *Taq* DNA polymerase, 10 mM Tris-HCl (pH 9.0 at RT), 50 mM KCl, 105 nM $MgCl_2$, 200 μM of each dNTP, stabilizer including BSA, 100 ng template DNA, 2μl (25 pm) of each forward and reverse primers. DNA amplification was performed in a Peltier Thermal Cycler (MJ Research, PTC-200). After initial heat denaturation of the DNA at 94 °C for 1 min, thermal cycling was performed for 30 times following the temperature regimes 95 °C for 1 min, 40 °C for 2 min, followed by 72 °C for 3 min. The final extension step at 72 °C for 5 min was followed by cooling at 8 °C to complete the PCR. PCR amplification of genomic DNA showed the occurrence of expected band size of 0.75 kb size in all selected putative transgenic plants of Annada selected on kanamycin-supplemented medium. However, this band is found to be absent in non-transgenic control plantlets (Fig. 7.2).

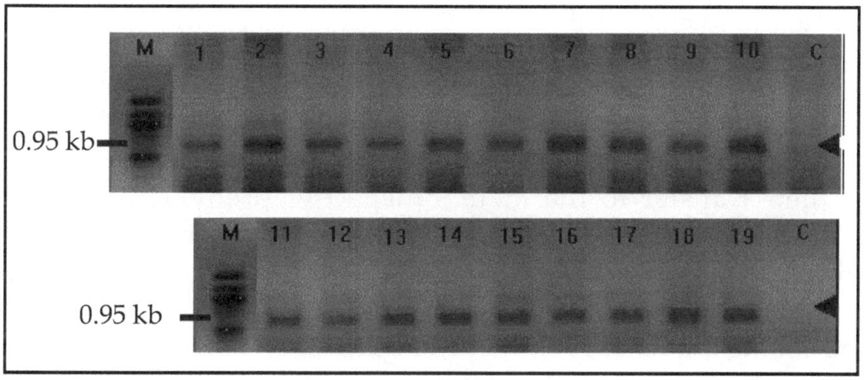

Fig. 7.2. PCR amplification of genomic DNA isolated from transformed rice plants a) Amplicon profile involving Annada transformed with Fe-SOD. Lanes M: *Hind*III digested λDNA, Lanes 1-19: putative transgenic plants (ATP 1-19), Lanes C: control plant.

PCR amplification confirmed the integration of *npt*II gene in all the plantlets. Since both *npt*II and Fe-SOD in plasmid pEXSOD10, *npt*II was ligated within the T-DNA border in *cis* orientation, presence of *npt*II is generally expected to indicate the presence of Fe-SOD. These plantlets derived from Annada and Pusa Basmati through *Agrobacterium*-mediated genetic transformation were assessed for further confirmation through Southern blot analysis.

7.6.1.4. Southern hybridization

PCR positive and kanamycin resistant nature of the plantlets derived from transformed calli tentatively advocates their transgenic character. To confirm the transgenic status, genomic DNA from putative plantlets is to be assayed by Southern blot to detect the presence of gene of interest. Isolated genomic DNA is generally digested with one or more RE and the resulting fragments are fractionated according to their size by electrophoresis in agarose gel. Subsequently, the DNA is denatured *in situ* and transferred from the gel to a solid support (generally NYLON membrane). The relative positions of the DNA fragments in the gel are preserved during their transfer to the membrane. The DNA attached to the membrane is generally hybridized to radiolabelled DNA of the sequence of insert (probe) to locate the positions of bands complementary to the probe.

Roy (2002) confirmed the presence of *FeSOD* gene in putative transgenic plants using Southern blot. The genomic DNA from transgenic plants of Annada was digested with *Bam*HI and *Eco*RI to release 954 bp fragment containing *FeSOD* gene from integrated T-DNA. The digested DNA fragment were subjected to electrophoresis on 0.8% (w/v) agarose at 50V for 3 h. Vacuum blotting was performed with the aid of vacuum blotting system, (Hyderabad, India), to transfer the DNA to nylon membrane (Hybond- N^+). The relative positions of the DNA fragments in the gel were preserved during their transfer to the membrane. After prehybridization, which reduces non-specific hybridization with the probe, the membrane was hybridized with desired radiolabelled nucleic acid probes to locate the positions of bands complementary to the probe. The probes were prepared by using coding sequence of the gene of interest. The membrane was washed to remove unbound and weekly bound probes followed by autoradiography on X-ray plates. The X-ray plates were developed following Sambrook *et al.* (1989).

PCR positive and kanamycin resistant nature of the plantlets derived from transformed calli tentatively advocates their transgenic character. To confirm the transgenic status, genomic DNA from putative plantlets was assayed by Southern blot to detect the presence of *Fe-SOD* gene (Roy, 2002). Southern blot analysis revealed the presence of *Fe-SOD* gene in the transformants. Out of 19 putative transgenic plant of Annada, three were found to display positive results (Fig. 7.3). They were designated as ATP 5, ATP 14 and ATP 20. Southern hybridization profile was found to be uniform among the transformants and each had a single band of 950 bp, which was lighter than the molecular size (954 bp) of *Fe-SOD* housed in pEXSOD10. Single copy insertion of gene of interest is desirable to avoid interaction among copies of the same or with others. This results in unexpected expression pattern of foreign genes (Jain and Jain, 2000). Rearrangement of bases in the transgene sequence before integration and deletion of the part of the T-DNA was occasionally observed as in case of *Agrobacterium*-mediated transformation in dicotyledons (Komari, 1990) and monocotyledons (Hiei *et al.*, 1994). Remaining 16 PCR positive putative transformants did not show integration of the transgene into the plantlets.

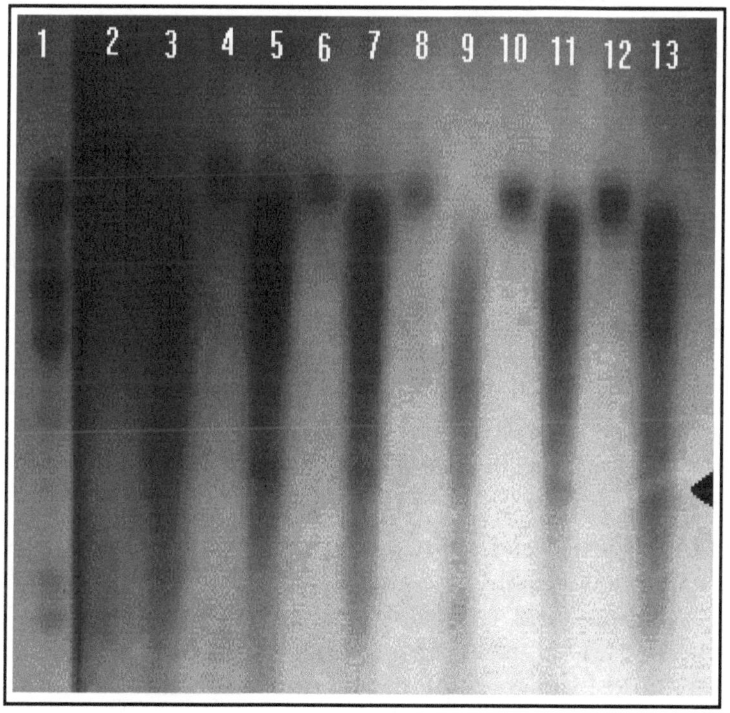

Fig. 7.3. Southern blot analysis of rice transformants a) DNA from putative transgenic plants of Annada probed with Fe-SOD gene; Lane 1: *Hind*III digested λDNA; Lanes 2, 4, 6, 8, 10 and 12: uncut genomic DNA obtained from putative transgenic plants (Designation: ATP3, ATP5, ATP14, ATP15, ATP19, and ATP20), Lane 3, 5, 7, 9, 11 & 13: genomic DNA of transgenic plants (Designation: ATP3, ATP5, ATP14, ATP15, ATP19, and ATP20) digested with *Hind*III and *Eco*RV

Fig. 7.4. Regeneration and maintenance of transgenic plants under *in vitro* and *in vivo* containments.

a) Plants turned to be albino on selection medium fortified with 100 mg $L^{-1}$of kanamaycin and 50 mg L^{-1} cefotaxime,

b) Fertile transgenic plant of Annada on cement pots

In essence, rice plantlets carrying Fe-SOD gene in Annada was produced using *Agrobacterium*-mediated transformation system. This result also confirmed successful *Agrobacterium*-mediated genetic transformation in rice and ample potential of this technique in genetic engineering for crop its improvement. Percent regeneration (%) of transformants as obtained in this study is found to be corroborative with the findings of Lin *et al.* (1995), Datta *et al.,* (1999, 2000).

7.6.2. Achievements in Transgenic Development

Protection of sensitive metabolic reaction through maintaining the structures of protein complexes or membrane by an increased capacity for hydroxyl radical scavenging may be an important strategy for engineering tolerance to oxidative stress.

Unlike most other organisms, plant posses multiple enzymatic forms (isozymes) of SOD. The existence of SOD isozymes in plants and genetic basis was first demonstrated in maize, and first plant *sod* gene was cloned from maize. Subsequently several SOD cDNAs were cloned from plants (Bowler *et al.,* 1992; Perl *et al.,* 1993) and transgenic plants were produced that exhibit enhanced SOD activity (Table 7.3). In some cases, these transgenic plants displayed enhanced tolerance of oxidative stress (Bowler *et al.*, 1991; Perl *et al.*, 1993; Sen Gupta *et al.*, 1993), whereas in other cases, they did not (Tepperman and Dumsmuir, 1990; Pitcher *et al.*, 1991).

Table 7.3. List of transgenic research for oxidative stress in plants

Gene	Gene action	Species	Phenotypic expression	References
sod	Cu/Zn-SOD	Tobacco, Tomato	No protection seen against superoxide toxicity	Tepperman and Dusmunir, 1990
sod	Mn-SOD	Tobacco	Reduced cellular damage under oxidative stress	Bolwer *et al.*, 1991
sod	Cu/Zn-SOD	Tobacco	Retained 90% photosynthesis under chilling and heat stresses	Gupta *et al.*, 1993
sod	Mn-SOD	Alfalfa	Increased tolerance to freezing stress	McKersie *et al.*, 1993
sod	Fe-SOD	Tobacco	Protected plants from ozone damage	Van Camp *et al.*, 1996
sod	Mn-SOD	Alfalfa	Increased tolerance to water deficit	McKersie *et al.*, 1996
Nt107	Glutathione s-transferase	Tobacco	Sustained growth under cold and salinity stresses	Roxas *et al.*, 1997
sod	Mn-SOD	Alfalfa	Increased winter survival	McKersie *et al.*, 1999
sod	Mn-SOD	Tobacco	Increased tolerance to Mn-deficiency	Yu *et al.*, 1999
Apx3	Ascorbic peroxidase	Tobacco	Increased protection against oxidative stress	Wang *et al.*, 1999
MsFer	Feratin (Iron storage)	Tobacco	Increased tolerance to oxidative damage caused by excess iron	Deak *etal.*, 1999
GST/GPX	Glutathione s-transferase with glutathione peroxidase	Tobacco	Increased abiotic stresses tolerance	Roxas *et al.*, 2000
par	Glutathione s-transferase	*Arabidopsis*	Protects aginst Al-toxicity and oxidative stress	Ezaki *et al.*, 2000
sod	Cu, Mn, Fe, Zn-SOD	Alfalfa, rye grass	Increased winter hardiness	McKersie, 2001
sod	Fe-SOD	Rice	Increased tolerance to methyl viologen	Roy, 2002

Cu/Zn-SOD gene was successfully introduced in many crop plants to manipulate the level of Cu/Zn-SODs. Transgenic potato (*Solanum tuberosum*) plants expressing a tomato Cu/Zn-SOD were found to be more resistant to paraquat (Perl *et al.*, 1993). On the other hand, tobacco plants harbouring a petunia (*Petunia hybrida*) chloroplastic Cu/Zn-SOD cDNA showed a 30-50 fold increase in choloroplastic Cu/Zn-SOD activity, but plant did not displayed enhanced tolerance to oxidative stress imposed by the herbicide paraquat (Tepperman and Dunsmuir, 1990). Tomato (*Lycopersicon esculentum*) transformed with the same construct did not exhibit increased tolerance to photo-inhibition at the low temperature (4 ^{0}C) and high light (Tepperman and Dunsmuir, 1990). SenGupta *et al.* (1993) demonstrated increased resistance to photo-inhibition in tobacco plants that expressed a cDNA for pea (*Pisum sativum*) Cu/Zn-SOD. Transgenic tomato plants showed photosynthetic rate approximately 20% higher than non-transformed tobacco plants when subjected to chilling temperature (3 ^{0}C) under moderate light intensity. These differences among the experiments on transgenic SOD planted are possibly due to difference in the level of expression. Over expression was found to be detrimental. Expression of pea (*Pisum sativum*) chloroplstic Cu/Zn-SOD in tobacco (cv. Xanthi) resulted in a 3 folds increase in total SOD activity, which led to a significant increase in resistance to light mediated MV-induced membrane damage. This protective effect was found only over a relatively narrow range of MV concentrations. It could activate Cu/Zn-SOD which was deactivated by H_2O_2 at higher MV concentrations. This observation could partially explain the discrepancies between these results and those of Tepperman and Dunsmuir (1990).

The presence of SOD was also demonstrated in peroxisomes from watermelon (*Citrulles vulgaris* Schrad.) cotyledons, but in this case two isozymes were detected, a Cu/Zn and Mn-containing SOD (Sandalio and del Rio, 1987). The peroxisomal Cu/Zn-SOD represented about 18% of the total SOD isozyme in these organelles (Sandalio and del Rio, 1987). The study of the intraorganellar distribution of SOD in plant peroxisomes showed that the Cu/Zn-SOD was present in the matrix of these organelles. The generation of O_2^- radicals, the substrate of SOD, in matrix and membranes of plant peroxisomes was reported and the generating systems of superoxids were partially characterized.

Chloroplasts in wheat (*Triticum aestivum* L.) was found to reduce oxygen to superoxide under water deficiency due to impaired electron transport, which promoted McKersie *et al.*, (1998) to hypothesize that overexpression of SOD might have improved tolerance of water deficit. Poplar trees exposed to relatively low levels of SO_2 proved to be more resistant to subsequent damage on exposure to high levels of SO_2; the increased resistance was correlated with increased SOD activity.

Bowler *et al.*, (1991) developed transgenic tobacco plants (cv. SRI) that expressed a chimeric gene derived from a Mn-SOD cDNA from *Nicotiana plumbaginifolia* in which the native mitochondria transit peptide was replaced with a chloroplastic transit peptide from the *Arabidopsis* RuBP-Case gene. Expression of this transgene was controlled by CaMV 35S promoter. Plants that carried this novel chloroplastic Mn-SOD gene expressed a unique Mn-SOD isoform that was correctly targeted to the chloroplasts. The Mn-SOD activity in these plants was estimated to be between 1.5 and 2-folds higher than in untransformed plants. Transgenic tobacco plants containing oxidative stress- related genes showed elevated levels of glutathione reductase, superoxide dismutase, and ascorbate peroxidase, resulting in enhance drought tolerance (Van Rensburg and Kruger, 1994). However, the results are often not correlated with the expectations as has been shown by Tepperman and Dunsmuir (1990) in transgenic tobacco and tomato with the Cu/Zn- SOD gene, which failed to show any protection against superoxide toxicity. Overexpression of Fe-SOD in transgenic tobacco plants did not show any tolerance for the salt stress, though expression elevated levels of oxidative stress related oxidative enzymes (Van Camp *et al.*, 1996). However, Fe-SOD in transgenic maize (Van Breusegem *et al.*, 1999) and in rice (Roy, 2002) showed enhanced tolerance in presence of metyl viologen. Another report in transgenic tobacco containing the *Nt107* gene encoding glutathione-d-transferase showed better seedling growth under cold and salt stress (Roxas *et al.*, 1997). A Mn-SOD cDNA from *Nicotiana plumbaginifolia* was introduced into alfalfa, the primary transformants and their F_1 transgenic progeny showed increased survival and vigour after exposure to sub-lethal freezing stress (McKersie *et al.*, 1998).

Over expression of SOD in chloroplasts provides increased protection from oxidative stress. Elevated activity of $H_2O_2^-$, resistance Mn-SOD in the chloroplast stroma apparently provides superior protection from oxidative stress caused by chemical exposure (MV, aciflurofen, ozone), but this is enzyme was found to be much less effective than Cu/Zn-SOD in protection from photo oxidation under photo-inhibitory conditions. These differences showed that elevated SOD activity in the chloroplast stroma might not alone be sufficient in imparting tolerance. Rather, it appears that the type of SOD that is increased is also a critical factor. Since the Cu/Zn-SOD is associated with the surface of the thylakoid membrane in close association with PSI, it likely that the different protective activities provided by Cu/Zn-SOD and Mn-SOD in chloroplasts are due to differences in the biochemical characteristics and sub-organellar localization of these enzymes.

Transgenic tobacco plants that overexpressed cytosol APX have been developed by Pitcher *et al.* (1994). Another gene construct that encodes chimeric chloroplastic APX isoforms, which consists of a 5′chloroplast transit peptide sequence fused to cytosolic APX cDNA have been developed and successfully expressed in tobacco plants. Increased tolerance to methyl viologen damage has been reported in tobacco plants that overexpress cytosolic APX, but not those that express a chloroplast targeted isoform (Pitcher *et al.* 1994). SenGupta *et al.* (1993) have reported a 3-fold increase in APX activity and mRNA in transgenic tobacco plants that overexpress chloroplastic Cu/Zn-SOD. Induction of APX activity to similar levels could also be achieved by treating untransformed tobacco leaf discs with H_2O_2 (Allen *et al.*, 1994).

7.7. REFERENCES

Allen RD, SenGupta A, Webb RP, Holaday AS. 1994. Protection of plants from oxidative stress using SOD transgenics: Interactions with endogenous enzymes. In: Frontiers of Reactive Oxygen Species in Biology and Medicine, Asada K, Yoshikawa T (eds.). Excerpta Medica, Amsterdam. pp. 321-322.

Allen RD. 1997. Use of transgenic plants to study antioxidant defenses. Free Radic Biol Med. 23: 473-479.

Alvarez ME, Lamb C. 1997. Oxidative burst-mediated defense response in plant disease resistance. In: Oxidative Stress and Molecular Biology of Antioxidant Defense, Scandalios JG (ed.). Cold Spring Harbor Laboratory Press. pp. 815-839.

Ames B, Shigenaga M, Hagen T. 1993. Oxidants, antioxidants, and the generative diseases of aging. Proc Natl Acad Sci USA. 90: 7915-7922.

Aono M, Kubo A, Saji H, Natori T, Tanaka K, Kondo N. 1991. Resistance to active oxygen toxicity of transgenic *Nicotiana tabaccum* that express the gene for glutathione reductase from *Escherichia coli*. Plant Physiol. 32: 691-697.

Aono M, Kubo A, Saji S, Tanaka K, Kondo N. 1993. Enhanched tolerance to photo-oxidative stress in transgenic *Nicotiana tabaccum* with high chloroplastic glutathione reductase activity. Plant Cell Physiol. 34: 129-135.

Arisi ACM, Cornic G, Jouanin L, Foyer CH. 1998. Overexpression of iron superoxide dismutase in transformed poplar modifies the regulation of photosynthesis at low CO_2 partial pressures or following exposure to the pro-oxidant herbicide methyl viologen. Plant Physiol. 117(2): 565–574.

Asada K, Takahashi M. 1987. Production and scavenging of active oxygen in photosynthesis. In: Photoinhibition. Elsevier Science Publishers, Kyle DJ, Osmond CB, Arentzen CJ (eds.). Amsterdam. pp. 227-287.

Asada K. 1992. Ascorbate peroxidase- a hydrogen peroxide scavenging enzyme in plant. Physiol Plant. 85: 235-241.

Asada K. 1994. Production and action of active oxygen species in photosynthetic tissue. In: Causes of photooxidative stress and amelioration of defense systems in plants. Boca Raton, Foyer CH, mullineaux PM (eds.). CRC Press. pp. 77-104.

Asada K. 1999. The water-water cycle in chloroplast: Scavenging of active oxygen and dissipation of excess photons. Annu Rev Plant Physiol Biochem. 36: 689-694.

Bailly C, Benamaur A, Corbineau F, Come D. 1996. Changes in malondialdehyde content and in superoxide dismutase, catalase and glutathione reductase activities in sunflower seeds as related to deterioration during accelerated aging. Physiol Plant. 97: 104-110.

Baldensperger JB. 1978. An iron containing superoxide dismutase from the chemo lithopheric *Thiobacillus denitrifican* Rt strain. Archives Microbiol. 119: 237-444.

Ball L, Richard O, Behtold U, Penkett C, Reynolds H, Kular B, Creissen G, Karpinski S, Schuch W, Mullineaux P. 2001. Changes in global gene expression in response to excess excitation energy in *Arabidopsis thaliana*. J Exp Bot. 52 (Special issue).

Bannister JV, Bannister WH, Rotilio G. 1987. Aspect of the structure, function and applications of super oxide dismutase. CRC Crit Rev Biochem. 22: 111-180.

Bannister WH, Bannister JV, Barra D, Bond J, Bossa F. 1991. Evolutionary aspects of superoxide dismutase: The copper-zinc enzyme. Free Redic Commun. 12-13(Pt1): 349-361.

Bast A. 1993. In: Diet and Free Radicals, Halliwell B, Aruoma OI (eds.). Ellis Harwood, London. pp. 95.

Beauchamp C, Fridovich I. 1971. Superoxide dismutase: improved assays and an assay applicable to acrylamide gels. Anal Biochem 44: 276-287.

Bohnert HJ, Jensen RG. 1996. Strategies for engineering water-stress tolerance in plants. Trends Biotech. 14: 89-97.

Bohnert HJ, Shevelava E. 1998. Plant stress adaptations-making metabolism move. Curr Pin Plant Biol. 3: 267-274.

Bolwell GP, Bindschedler LV, Bolee KA, Butt VS, Devis DR, Gardner SL, Gerrish C, Minibayeva F. 2002. The apoplastic oxidative burst in response to biotic stress in plants: a three-component system. J Exp Bot. 53: 1367-1376.

Borraccino G, Dipierro S, Arrigoni O. 1986. Purification and properties of ascorbate free radical reductase from potato tubers. Planta. 167: 521-526.

Bowler C, Slooten L, Vandenbranden S, de Rycke R, Botterman J, Sybesma C, van Montagu M, Inzé D. 1991. Manganese superoxide dismutase can reduce cellular damage mediated by oxygen radicals in transgenic plants. EMBO J. 10: 1723-1732.

Bowler C, Van Camp W, Van Montogu M, Inze D. 1994. Superoxide dismutase in plants. Critica Rev Plant Sci. 13: 199-218.

Bowler C, Van Montage M, Inze D. 1992. Superoxide dismutage and tress tolerance. Annu Rev Plant Physiol Plant Mol. Biol. 43:83-116.

Bradford MM. 1976. A rapid and sensitive method for the quantitation of microgram quantities of protein utilizing the principle of protein-dye binding. Anal Biochem 72: 248-254.

Brot N, Weissbach H. 2000. Peptide methionine sulfoxide reductase: Biochemistry and physiological role. Biopolymers. 55(4): 288-296.

Bueno P, del Rio LA. 1992. Purification and properties of glyozysomal cuprozine superoxide dismutage from watermelon cotyledons (*Citrullus vulgaris* Schrad.). Plant Physiol. 98:331-336.

Bueno P, Varela J, Gimenez-Gallego G, del Rio LA. 1995. Peroxisomal Copper, Zinc Superoxide Dismutage: Characterization of Isoenzyme from watermelon cotyledons. Plant Physiol. 108: 1151-1160.

Burke JJ, Gamble PE, Hatfield JL, Quisenberry JE. 1985. Plant morphological and biochemical response to field water deficit. Response of regulation reductase activity and paraquat sensitivity. Plant Physisol. 79:415-419.

Casano L, Gomez L, Lascano C, Trippi V. 1997. Induction and degradation of Cu/Zn-SOD by active oxygen species in wheat chloroplast exposed to photo oxidative stress. Plant Cell Biol. 38: 433-440.

Creissen G, Edwards EA, Enard C, Wellburn A, Mullineaux P. 1991. Molecular characterization of glutathione reductase cDNAs from pea (*Pisum sativum* L.). Plant J. 2: 129-131.

Creissen GP, Edwards A, Mullineaux PM. 1994. Glutathione reductase as ascorbate peroxidase. In: Causes of Photooxidative Stress and Amelioration of Defense System in Plants, Foyer CH, Mullineaux PM (eds.). CRC Press, Boca Raton, FL. pp. 343-364.

Cushman JC, Bohrnet HJ. 2000. Genomic approaches to plant stress tolerance. Curr Opin Plant Biol. 3: 117-124.

Datta K, Kouklikova-Nicola Z, Baisakh N, Oliva N, Datta SK. 2000. Agrobacterium-mediated engineering for sheath blight resistance of indica rice cultivar from different ecosystem. Theor Appl Genet. 100: 832-839.

Datta K, Velazhahan R, Oliva N. 1999. Over expression of cloned rice thaumatin like protein (PR-5) in transgenic rice plant enhance environmental friendly resistance to *Rhizoctonia solani* causing sheath blight disease. Theor Appl Genet. 98: 1138-1145.

Davis KJA. 1987. Protein damage and degradation by oxygen radicals. J Bio Chem. 262: 9895-9901.

De Vos CHR, Kraak HL, Bino RL. 1994. Aging of tomato seeds involves glutathione oxidation. Physiol Plant. 92: 131-139.

Deak M, Horvath GV, Davletova S, Torok K, Sass L, Vass I, Barna B, Kiraly Z, Dutis D. 1999. Plants ectopically expressing the iron-binding protein ferritin are tolerant to oxidative damage and pathogens. Nature Biotechnol. 17: 192-196.

Dean RT, Gieseg S, Davis M. 1993. Reactive species and their accumulation on radical-damaged proteins. Trends Biotechnol Sci. 18: 427-441.

Decleire M, De Cat W, De Temmerman L, Baeten H. 1984. Change of peroxidase, catalase and superoxide dismutase activities in ozone-fumigated spinach leaves. J Plant Physiol. 116: 147-152.

Del Rio LA, Lyon DS, Ohal I, Glick B, Salin ML. 1983. Immunocytochemical evidence for a peroxisomal localization of manganese superoxide dismutase in leaf protoplasts from a higher plant. Planta. 158: 216-224.

Del Rio LA, Pastori GM, Palma JM, Sandalio LM, Corpas FJ, Jimenez A, Lopez-Huertas E, Hernandez AJ. 1998. The activated oxygen role of peroxysomes in senescence. Plant Physiol. 116: 1195-2000.

Del Rio LA, Sandalio LM, Palma JM, Bueno P, Corpas FJ. 1992. Metabolism of oxygen radicals in peroxisome and cellular implications. Free Radicals in Biology and Medicine. 13: 557-580.

Doke N. 1997. The oxidative burst: role in signal transduction and plant stress. In: Oxidative Stress and Molecular Biology of Antioxidant Defense, Scandalios JG (ed.). Cold Spring Harbor, Cold Spring Harbor Laboratory Press. pp. 785-813.

Droillard MJ, Paulin A. 1990. Isozymes of superoxide dismutase in mitochondria and peroxisomes isolated from petals of carnation (*Dianthus caryphyllus*) during senescence. Plant Physiol. 94: 1187-1192.

Edwards A, Rawsthorne S, Mullineaux PM. 1990. Subcellular distribution of multiple isoforms of glutathione reductase in leaves of pea (*Pysum sativum* L.). Planta. 180: 278-284.

Ezaki B, Gardner RC, Ezaki Y, Matsumoto H. 2000. Expression of aluminium-induced gene in transgenic *Arabidopsis* plants can ameliorate aluminium stress and/or oxidative stress. Plant Physiol. 122: 657-665.

Fath A, Bethke P, Belligni V, Jones R. 2002. Active oxygen and cell death in cereal aleuron cells. J Exp Bot. 53: 1273-1282.

Foster JG, Edwards GE. 1980. Localization of superoxide dismutase in leaves from C_3 and C_4 plants. Plant Cell Physiol. 21: 895-906.

Foyer CH, Halliwell B. 1976. The presence of glutathione and glutathione reductase in chloroplasts: a proposed role in ascorbic acid metabolism. Planta. 133: 21-25.

Foyer CH, Lalandais M, Galap C, Kunert KJ. 1991. Effects of elevated cytosolic glutathione reductase activity on the cellular glutathione pool and photosynthesis in leaves under normal and stress conditions. Plant Physiol. 97: 863-872.

Fridovich I. 1986. Superoxide dismutases. In: Advances in Enzymology and Related Areas of Molecular Biology, Merister A, (ed.). John Wiley & Sons, New York. pp. 61-97.

Genty B, Briantais JM, Baker NR. 1989. The relationship between quantum yield of photosynthetic electron transport and quenching of chlorophyll fluorescence. Biochem Biophys Acta. 990: 87-92.

Gomez JM, Hernandez JA, Jimenez A, dell Rio LA, Sevilla F. 1999. Differential response of antioxidative enzymes of chloroplasts and mitochondria to long-term NaCl stress of pea plants. Free Radical Res. 31: S11-18.

Gonzalez A, Steffen KL, Lynch JP. 1998. Light and excess manganese. Implications for oxidative stress in common bean. Plant Physiol. 118: 493-504.

Gressel J, Galun E. 1994. Causes of photooxidative stess and amelioration of defence systems in plants. In: Genetic Controls of Photooxidant Tolerance, Foyer CH, Mullineaux PM (eds). CRC Press, Boca Raton, F.L. pp. 237-273.

Gupta AS, Heinen JL, Holada AS, Burke JJ, Allen RD. 1993. Increased resistant to oxidative stress in transgenic plant that overexpressed chloroplastic Cu/Zn-superoxide dismutase. Proc Natl Acad Sci USA. 90: 1329-1633.

Hagar H, Ueda N, Shah SV. 1996. Oxygen metabolites in DNA damage and cell death in chemical hypoxic injury to LLC-PK-1 cells. Am J Physiol. 271: 209-215.

Halliwell B, Gutteridge JMC. 1989. Free Radicals in Biology and Medicine, Ed 2. Clarendon Press. Oxford, U. K.

Halliwell B, Gutteridge JMC. 1990. The antioxidant of human extracellular fluids. Arch Biochem Biophys. 280:1.

Haslekas C, Stacy R, Nygaard V, Culianez-Macia F, Aalen R. 1998. The expression of a periredoxin antioxidant gene *AtPer1*. In: *Arabidopsis thaliana* is seed specific and related to dormancy. Plant Mol Biol. 36: 833-845.

Hayakawa T, Kanemastu S, Asada K. 1985. Purification and characterization of thylakoid-bound Mn-superoxide dismutase in spinach chloroplast. Plant. 166: 111-116.

Hernandez JA, Corpas FJ, Gomez M, del Rio LA, Sevilla F. 1993. Salt-induced oxidative stress mediated by activated oxygen species in pea leaf mitochondria. Physiologia Plantarum. 89: 103-110.

Hernandez JA, Olmos E, Corpas FJ, Sevilla F, del Rio LA. 1995. Salt-induced oxidative stress in chloroplast of pea plants. Plant Sci. 105: 151-167.

Herourt D, Van Montagu M, Inze D. 1993. Redox-activated expression of the cytosolic copper/zinc superoxide dismutase gene in *Niccotiana*. Proc Acad Sci, USA. 90: 3108-3112.

Hiei Y, Ohta S, Komari T, Kumashiro T. 1994. Efficient transformation of rice (*Oryza sativa* L.) mediated by *Agrobacterium* and sequence analysis of the boundaries of the T-DNA. Plant J. 6(2): 271-282.

Hoshi T, Heinemann S. 2001. Regulation of cell function by methionine oxidation and reduction. J Physiol. 531: 1-11.

Jackson C, Dench J, Moore AL, Halliwell B, Foyer CH, Hall DO. 1978. Subcellular localization and identification of superoxide dismutase in the leaves of higher plants. European J Biochem. 91: 339-344.

Jacobson MD. 1996. Reactive oxygen species and programmed cell death. Trends Biotechnol Sci. 21: 83-86.

Jain R, Jain S. 2000. Transgenic strategies for genetic improvement of Basmati rice. Indian J Exptl Biol. 38: 6-17.

Kaiser WM. 1979. Reversible inhibition of the Calvin cycle and activation of the oxidative pentose phosphate cycle in isolated chloroplast by hydrogen peroxide. Planta. 145: 377-382.

Kalu BA, Fick GW. 1983. Morphological stage of development as a predictor of alfalfa herbage quality. Crop Sci. 23: 1167-1172.

Kanematsu S, Asada K. 1978. Superoxide dismutase from an anaerobic photosynthetic bacterium, *Chromatium visosum*. Archives Biochem Biophysics. 185: 473-482.

Kanematsu S, Asada K. 1979. Ferric and manganese superoxide dismutase in *Euglena gracilis*. Archives of Biochem and Biophysics. 1995: 535-545.

Karpinski S, Escobar C, Karpinska B, Creissen G, Mullineaux PM. 1997. Photosynthetic electron transport regulates the expression of the cytosolic ascorbate peroxidase genes in *Arabidopsis* excess light stress. Plant Cell. 9: 327-640.

Karpinski S, Reynolds H, Karpinska B, Wingsle G, Cressen G, Mullineaux P. 1999. Systemic signaling and acclamation in response to excess excitation energy in *Arabidopsis*. Science. 284: 654-657.

Kendall EJ, McKersie BD. 1989. Free radical and freezing injury to cell membranes of winter wheat. Physiol Plant. 76: 86-94.

Kliebentein DJ, Monde R, Last RL. 1998. Superoxide dismutase in *Arabidopsis*: An electric enzyme family with disparate regulation and protein localization. Plant Physiol. 118: 637-650.

Komari T. 1990. Genetic characterization of a double- flowered tobacco plant obtained in transformation experiment. Theor Appl Genet. 80: 167-171.

Lee H, Xiong L, Ishitani M, Stevenson B, Zhu JK. 1999. Cold-regulated gene expression and freezing tolerance in an *Arabidopsis thaliana* mutant. Plant J. 17: 301-308.

Lee YB, Lee SM. 2000. Effect of *S*-adenosylmethionine on hepatic injury from sequential cold and warm ischemia. Archives of Pharmacological Research. 23: 250-258.

Leprince O, Thorpe PC, Deltour R, Atherton NM, Hendry GAF. 1990. The role of free radicals and radical processing system in loss of desiccation tolerance in germinating maize. New Physiol. 116: 573-580.

Levine RL, Bertt BS, Moskovitz J, Mosoni L, Stadtman ER. 1999. Methionine residues may protect proteins from critical oxidative damage. Mech Ageing Dev. 107(3): 323-332.

Lin W, Anuratha CS, Datta K, Potrykus I, Muthukrishnan S, Datta SK. 1995. Genetic engineering of rice for resistance to sheath blight. Bio/Technology. 13: 686-691.

Lopez-Huertas E, Sandalio LM, del Rio LA. 1994. Priliminary study of the superoxide generating system of membranes from plant peroxisomes. In: Frontiers of Reactive Oxygen Species in Biology and Medicine, Asada K, Yshikawa T (eds.). Elsevier, Amsterdam. pp. 37-38.

May M, Vernoux T, Leaver C, van Montagu M, Inze D. 1998. Glutathione homeostasis in plants: Implication for environmental sensing and plant development. J Exp Bot. 49: 649-667.

McKersie BD, Bowley SR, Harjanto SR, Leprince O. 1996. Water deficit tolerance and field performance of transgenic alfalfa overexpressing superoxide dismutase. Plant Physiol. 111: 1177-1181.

McKersie BD, Bowley SR, Jones KS. 1999. Winter survival of transgenic alfalfa overexpressing superoxide dismutase. Plant Physiol. 119: 839-848.

McKersie BD, Chen Y, Debens M, Bowley SR, Bowler C, Inze D, D'Halluin K, Botterman J. 1993. Superoxide dismutase enhance tolerance of freezing stress in transgenic alfalfa (*Medicago sativa* L.). Plant Physiol. 103: 1155-1163.

McKersie BD, Murnaghan J, Jones KS, Bowley SR. 2000. Iron-superoxide dismutase expression in transgenic alfalfa increase winter survival without a detectable increase in photosynthetic oxidative stress tolerance. Plant Physiol. 122: 1427-1438.

McKersie BD, Senaratna T, Walker MA, Kendall EJ, Hetherington PR. 1998. Deterioration membrane during aging in plants: evidence for free radical mediation. In: Senesence and Aging in Plants, Nooden L, Leopold AC (eds.). Academic Press, New York. pp. 441-463.

McKersie BD. 2001. Stress tolerance of transgenic plant overexpressing superoxide dismutase (Abstr.). J Exp Bot. 52 (supplement): 8.

Miyaka C, Asada K. 1992. Thylakoid bound ascorbate peroxidase in spinach chloroplasts and photoreduction of its primary oxidation product, monodehydroascorbate radicals in thylakoids. Plant Cell Physiol. 33: 541-553.

Miyake C, Asada K. 1994. Ferridoxin-dependent photoreduction of the monodehydro-ascorbate radical in spinach thylakoids. Plant Cell Physiol. 35: 539-249.

Miyake C, Cao WH, Asada K. 1993. Purification and molecular properties of the thylakoid-bound ascorbate peroxidase in spinach chloroplasts. Plant Cell Physiol. 34: 881-889.

Monk LS, Davies HV. 1989. Antioxidant status of the potato tuber and Ca^{2+} deficiency as a physiological stress. Physiol Plant. 75: 411-416.

Monk LS, Fagerstedt KV, Crawford RMM. 1987. Superoxide dismutase as an anaerobic polypeptide- a key factor in recovery from oxygen deprivation in Iris Pseudocous? Plant Physiol. 85: 1016-1020.

Moskovitz J, Berlett BS, Poston JM, Stadtman ER. 1997. The yeast peptide-methionine sulfoxide reductase functions as an antioxidant *in vitro*. Proc Natl Acad Sci USA. 94(18): 9585-9589.

Mullen RT, Trelease RN. 2000. The sorting signals for peroxisomal membrane-bound ascorbate peroxidase are within its C-terminal tail. J Biol Chem. 275(21): 16337-16344.

Mullineaux P, Ball L, Escobar C, Karpinska B, Creissen G, Karpinski S. 2000. "Are diverse signaling pathways integrated in the regulation of A*rabiodopsis* andtioxidant defense gene expression in response to excess excitation energy?" Philos: Trans R Soc Lond B Biol Sci. 355(1402): 1531-1540.

Nakano Y, Asada K. 1981. Hydrogen peroxide is scavenged by ascorbate specific peroxidase in spinach chloroplasts. Plant Cell Physiol. 22: 867-880.

Neill SJ, Desikan R, Clarke A, Hurst RD, Hancock JT. 2002. Hydrogen peroxide and nitrate oxide as signaling molecules in plants. J Exp Bot. 53: 1237-1247.

Ogawa K, Kanematsu S, Asada K. 1995. Spinach chloroplastic and cytosolic CuZn-SODs are localized at the site of superoxide generation. In: Photosynthesis from Light to Biosphere, Vol. IV, Mathis P (ed.). Kluwer Acad Publishes, Dordrecht. pp. 339-342.

Ogawa K, Kanematsu S, Asada K. 1996. Intra- and extra-cellular localization of "cytosolic" CuZn-superoxide dismutase in spinach leaf and hypocotyl. Plant Cell Physiol. 37: 790-799.

Ogawa K, Kanematsu S, Asada K. 1997. Generation of superoxide anion and localization of CuZn-superoxide dismutase in the vascular tissue of spinach hypocotyl: Their association with lignification. Plant Cell Physiol. 38(10): 1118-1126.

Okada S, Kanematsu S, Asada K. 1979. Intercellular distribution of manganese and ferric superoxide dismutase in blue-green algae. FEBS Letters. 103: 106-110.

Parker MW, Schnina ME, Bossa F, Bannister JV. 1984. Chemical aspects of the structure, function and evaluation of superoxide dismutase. Inorg Chim Acta. 9d1: 307-317.

Pastori GM, Foyer CH. 2001. Indentifying oxidative responsive genes by transposom tagging. J Exp Bot. 52 (Special Issue).

Perl A, Perl-Treves R, Gatili G, Aviv D, Shalgi E, Malkin S, Galun E. 1993. Enhanced oxidative stress defense in transgenic potato expressing tomato Cu, Zn superoxide dismutase. Theor Appl Genet. 85: 568-576.

Pitcher LH, Brennan E, Hurley A, Dunsmuir P, Tepperman JM, Zilinskas BA. 1991. Over production of *Petunia* chloroplastic coper zinc superoxide dismutase does not confer ozone tolerance in transgenic tobacco. Plant Physiol. 97: 452-455.

Pitcher LH, Repetti P, Zilinskas BA. 1994. Over-production of ascorbate peroxidase protects transgenic tobacco plants against oxidative stress. Plant Physiol. 105: S-169.

Polle A, Rennenberg H. 1993. Significance of antioxidants in plant adaptation to environmental stress. In: Plant Adaptation to Environmental Stress, Mansfield T, Fowden L, Stoddard F (eds.). Chapman & Hall, London. pp. 263-273.

Preville X, Salvemini F, Giraud S, Chaufour S, Paul C, Stepien G, Ursini MV, Arrigo AP. 1999. Mammalian small stress protein preotects against oxidative stress through their ability to increase glucose-6-phosphate dehydrogrnase activity and by maintaining optimum cellular detoxifying machenary. Exp Cell Res. 247(1): 61-78.

Price AH, Hendry GAF. 1991. Iron-catalyzed oxygen radical formation and its possible contribution to drought damage in nine native grasses and three cereals. Plant Cell Environ. 14: 477-484.

Puget K, Michelson AM. 1974. Iron containing superoxide dismutase from luminous bacteria. Biochem. 56: 1255-1267.

Reddy CD, Venkaiah B. 1982. Studies on isozymes of superoxide dismutase from mung bean (*Vigna radiata*) seedlings. J Plant Physiol. 116: 279-284.

Roxas VP, Lodhi SA, Garrett DK, Mahan JR, Allen RD. 2000. Stress tolerance in transgenic tobacco seedlings that overexpress glutathione-d-transferase/glutathione peroxidase. Plant Cell Physiol. 41: 1229-1234.

Roxas VP, Smith RK Jr, Allen ER, Allen RD. 1997. Overexpression of glutathione-d-transferase/glutathione peroxidase enhance the growth of transgenic tobacco seedlings during stress. Nature Biotechnol. 15: 988-991.

Roy B. 2002. Genetic Studies on Seed Components and Standardization of Genetic Transformation in Rice. Ph. D. Thesis, Bidhan Chandra Krishi Viswavidyalaya, Mohanpur, Nadia, West Bengal, India. pp. 246-259.

Salin ML, Bridges SM. 1980. Localization of superoxide sdismutase in chloroplast from *Brassica campestris*. Zeitschrift fur Pflanzenphysiologie. 99: 37-47.

Salin ML. 1988. Toxic oxygen species and protective systems of the chloroplast. Physiologia Plantarum. 72: 381-689.

Salin ML. 1991. Chloroplast and mitochondrial mechanism for protection against oxygen toxicity. Free Radical Res Commun. 12: 851-858.

Sambrook J, Fritsch EF, Maniatis T. 1989. Molecular cloning: In: A Laboratory Manual, 2nd Edn. Cold Spring Harbor Laboratory Press.

Sanchez-Fernandez R, Friker M, Corben LB, White NS, Sheard N, Leaver CJ, Van Montagu M. 1997. Cell proliferation and hair tip growth in the *Arabidopsis* root are under *mechanistically* different from of redox control. Proc Natl Acad Sci, USA. 94: 2745-2750.

Sandalio LM, del Rio LA. 1987. Localization of superoxide dismutase in glyoxysomes from *Citrullus vulgaris*. Functional implications in cellular metabolism. J Plant Physiol. 127: 395-409.

Sandalio LM, del Rio LA. 1988. Intraorganellar distribution of superoxide dismutase in plant peroxisomes (glyoxysomes and leaf peroxesomes). Plant Physiol. 88: 1215-1218.

Sandalio LM., Palma JM, del Rio LA. 1987. Localization of manganese peroxidase dismutase in peroxisome isolated from *Pisum sativum* L. Plant Sci. 51:1-8.

Scandalios JG. 1997. Molecular genetics of superoxide dismutase in plants. In: Oxidative Stress and the Molecular Biology of Antioxidative Defense, Scandalios JG (ed.). Plainview, Cold Spring Harbor. pp. 527-568.

Scioli JR, Zilinskas BA. 1988. Cloning and characterization of a cDNA encoding the chloroplastic copper/zinc-superoxide dismutase form pea. Proc Natl Acad Sci, USA. 85: 7661-7665.

Senaratna T, McKersie BD, Stinson RH. 1985. Simulation of dehydration injury to membranes from soybean axes by free dadicals. Plant Physiol. 77: 472-474.

SenGupta A, Webb RP, Holaday AS, Allenn RD. 1993. Overexpression of superoxide dismutase protects plant from oxidative stress. Plant Physiol. 103: 1067-1073.

Shahidi F, Janitra PK, Wanasmdara PD. 1992. Phenolic antioxidant. Crit Rev Food Sci Nutr. 32: 67-

Stieinnman HM. 1982. Superoxide dismutase protein chemistry and structure-function relationship. In: Superoxide dismutage, Oberley LW (ed.)., Vol. 1. CRC Press, Boca Raton, FL. pp. 11-68.

Teichmann T. 2001. The biology of wood formation: Scientific challenges and biotechnological perspective. Recent Res Dev Plant Physiol. 2: 269-284.

Tepperman JM, Desmuir P. 1990. Transformed plants with elevated levels of chloroplasts SOD are not more resistant to superoxide toxicity. Plant Mol Biol. 14: 501-511.

Van Breusegem F, Slooten L, Stassart JM, Moens T, Botterman J, Van Montagu M, Inze D. 1999. Overproduction of *Arabidopsis thaliana* Fe-SOD confers oxidative stress tolerance to transgenic maize. Plant Cell Physiol. 40: 515-523.

Van Camp W, Capiau K, Van Montagu M, Inze D, Slooten L. 1996. Enhancement of oxidative stress tolerance in transgenic tobacco plants overproducing Fe-superoxide dismutase in chloroplast. Plant Physiol. 112: 1703-1714.

Van Rensburg L, Kruger GHJ. 1994. Applicability of ascorbic acid and (or) proline accumulation as selection criteria for drought tolerance in *Nicotiana tabaccum*. Can J Bot. 72: 1535-1540.

Van Toai T, Boller CS. 1991. Postanoxia injury in soybean (Glycine max) seedling. Plant Physiol. 97: 588-592.

Vertucci CW, Farrant JM. 1995. Acquisition and loss of desiccation tolerance. In: Seed Development and Germination, Kigal J, Galili G (eds.). New York, Marcel Dekker. pp. 237-271.

Vranova E, Inze D, Van Breusengen F. 2002. Signal transduction during oxidative stress. J Exp Bot. 53: 1227-1236.

Walters C. 1998. Understanding the mechanisms and kinetics of seed aging. Seed Sci Res. 8: 223-244.

Wang J, Zhang H, Allen RD. 1999. Overexpression of *Arabidopsis* peroxisomal ascorbate peroxidase gene in tobacco increases protection against oxidative stress. Plant Cell Physiol. 40: 725-732.

Wise RR, Naylor AW. 1987. Chilling- enhanced photo-oxidation. Evidence for role of singlet oxygen and superoxide in the breakdown of pigments and endogenous antioxidants. Plant Physiol. 83: 278-282.

Wu G, Wilen RW, Robertson AJ, Gusta LV. 1999. Isolation of chromosomal localization, and differential expression of mitochondrial manganese superoxide dismutase and chloroplast Cu/Zn superoxide dismutase genes in wheat. Plant Physiol. 120: 513-520.

Yost Jr RJ, Fridovich I. 1973. An iron-containing superoxide dismutase from *Escherichia coli*. J Biological Chem. 248: 4905-4908.

Yu Q, Osborne LD, Renzel Z. 1999. Increased tolerance to Mn deficiency in transgenic tobacco over-producing superoxide dismutase. Ann Bot. 84: 543-547.

Zhu D, Scandalios JG. 1993. Maize mitochondrial manganese superoxide dismutases are encoded by a differentially expressed multigene family. Pro Natl Acad Sci, USA. 90: 9310-9314.

Zhu YL, Philon-Smith EAH, Tarun AS, Weber SU, Jounanin L, Terry N. 1999. Cadmium tolerance and accumulation in Indian mustard is enhanced by overexpressing gamma-glutamylcyseine synthatase. Plant Physiol. 121: 1169-1178.

Unit-V
METAL TOXICITY TOLERANCE

Introduction

Heavy metals are important environmental pollutants and their toxicity is a problem of increasing significance for ecological, evolution, nutritional and environmental reasons. Metals present naturally in soils, which may be beneficial or harmful to plants. Usually, excess amount of metals adversely affect plant growth and development. Heavy metals are more harmful to plants. Heavy metals are defined as metals with a density higher than 5 g/cm^2. Fifty three of 90 naturally occurring elements are heavy metals (Weast, 1984), but not all of them are of biological importance. Based on their suitability under physiological conditions, 17 heavy metals may be available for living cells and of importance for organisms and ecosystems (Weast, 1984). Heavy toxic metals include Fe, Al, Cd, Cu, Co, Hg, Mn, Mo, Ni, Pb, and Zn. Nutrient toxicity or deficiency affects the soil quality. Among these metals, Fe, Mo, and Mn are important as micronutrients. Zn, Ni, Cu, V, Co, W, and Cr are toxic elements with high or low importance as trace elements. As, Hg, Ag, Sb, Cd, Pb, and U are have no known functions as nutrients and seen to be more or less toxic to plants and microorganisms (Breckle, 1991; Nies, 1999).

Some regions naturally contain high levels of heavy metals. Anthropogenic releases of heavy metals into the environment continuously increase soil contamination. In most terrestrial ecosystems, there are two main sources of heavy metals: the underlying parent materials and the atmosphere.

The concentrations of heavy metals in soil depend on the weathering of the bedrock and on atmospheric imputes of metals. Natural resources are volcanoes and continental dusts. Anthropogenic activities like mining, combustion of fossil fuels, metal-working industries, phosphate fertilizers, etc., lead to the emission of heavy metals and accumulation of these compounds in ecosystems (Lantzy and MacKersie, 1979; Galloway *et al.*, 1982; Angelones and Bini, 1992). In the soil, mobile and immobilized fractions have to be distinguished since heavy metals bind to the inorganic and organic soil compounds and to the humus. The solubility and mobility of the metals is affected by adsorption, and complexion processes, which in turn are dependent on the soil type. Nutrient toxicity or deficiency affects the soil quality. N, P, K and Mg are major elements normally the subject of routine fertilization, but many secondary elements may be limiting (Vose, 1984), and this is due to deficiency or toxicity.

Soil characteristics associated with metal toxicity have been presented in Table U.V-1. Heavy metal ions play essential role in many physiological processes. In trace amounts, several of these ions are required for metabolism, growth and development. However problem arises when cells are confronted with an excess of these vital ions or with non-nutritional ions that lead to cellular damage (Avery, 2001; Schutzendubel and Polle, 2002; Gaetke and Chow, 2003; Polle and Schutzendubel, 2003). Heavy metal toxicity comprises inactivation of biomolecules by either blocking essential functional groups or displacement of essential metal ions (Goyer, 1997). It has been shown to affect chlorophyll content and biosynthesis (Somashakaraiah *et al.*, 1992), germination and seedling growth (Hsu and Chang, 1992), and membrane damage (Hendry *et al.*, 1992).

Table U.V-1. Soil characteristics associated with common major metal toxicity effects on crop plants

Metal	Soil characteristics
Aluminium	Toxicity occurs on acid soils ($pH < 5.0$), such as ultisols, oxisols, acid sulfate soils and flooded soils with $pH < 4$ before Fe-toxicity symptoms appear.
Iron	Fe is toxic on acid ferralsols and acrisols and application of urban or industrial sewage with high Fe content.
Cadmium	The effect of Cd salt on the growth of seedlings was weaker in loamy soil and stronger in sandy soil
Lead	The absorption of Pb in soil increases with increasing pH between 3.0 and 8.5

Environmental pollution by metal has increased due to industrialization and anthropogenic activities (Foy *et al.*, 1978). Toxic ions are often added indiscriminately as pesticides, sewage sludge, factory disposal and mine wastes. In addition to toxic, it also causes an ion imbalance in the soil. Toxic metals are also unknowingly added to the soil by human beings, such as, lead from leaded gasoline, industrial smoke, etc. Removal or fixation of the toxic ions into forms unavailable to plants is often not feasible. Such problem soil may be brought under crop cultivation through development and cultivation of tolerant varieties, which would be less expensive, eco-compatible and more permanent than more expensive transient soil reclamation by physicochemical means.

Accumulation of toxic ions varies among the plant organs (Cole *et al.*, 1968). Generally, roots and leaves take up the most toxic ions, and stems and inflorescences the least. Accumulation also depends on plant species and metal concentrations. The chemical forms of heavy metals in soil solution is dependant of the metal concerned in plant in the presence of excess amount of heavy metals may be due to a range of interactions at the cellular level.

Interactions among the elements present on the root surface and within the plants affect their uptake and accumulation in crop plants (Nan *et al.*, 2002). It had been reported that addition of Cd decrease the Zn concentration in corn (Root *et al.*, 1975) and barley (Wu and Zhang, 2002). Cheng *et al.*, (2006) had analyzed correlations among the concentrations of five toxic heavy metals and Fe and Zn in rice grain (Table U.V-2). They found a significant positive correlation between Cd and As, Cr and Ni, As and Pb or Zn, and Fe and Zn. On the other hand, there was a significant negative association between Zn and Ni.

Table U.V-2. Correlations among concentrations of toxic heavy metal and two nutrients in rice grains (Cheng *et al.*, 2006)

Metals	Cd	Cr	As	Ni	Pb	Fe	Zn
Cd	1.0000						
Cr	0.1665	1.0000					
As	0.4162**	-0.0823	1.0000				
Ni	0.0320	0.2008*	0.0089	1.0000			
Pb	0.1613	0.0672	0.1967*	0.0384	1.0000		
Fe	0.1761	0.1460	0.0512	-0.0477	0.0869	1.0000	
Zn	0.1655	0.0234	0.2763**	-0.2241	0.0154	0.5202**	1.0000

REFERENCES

Angelone M, Bini C. 1992. Trace elements concentrations in soils and plants of Western Europe. In: Biogeochemistry of Trace Metals, Adriano DC. Boca Raton, FL: Lewis Publishers. pp. 19-60.

Avery SV. 2001. Metal toxicity in yeast and the role of oxidation stress. Adv Appl Microbiol. 49: 111-142.

Breckle CW. 1991. Growth under heavy metals. In: Plant Roots: The Hidden Half, Waisel Y, Eshel A, Kafkafi U (eds.). New York, Marcel Dekker. pp. 351-373.

Cheng W, Zhang G, Yao H, Wu W, Xu M. 2006. Genotypic and environmental variation in cadmium, chromium, arsenic, nikle, and lead concentrations in rice grains. J Zhejiang University Science. 7(7): 565-571.

Cole MM, Provan DMJ, Tooms JS. 1968. Geobotany, biochemistry and geochemistry in mineral exploration in the Bulman-Waimura Springs area, Northern territory, Australia. Trans Instn Min Metal. 77: 81-104.

Foy CD, Chaney RL, White MC. 1978. The physiology of metal toxicity in plant. Annu Rev Plant Physiol. 29: 511-566.

Gaetke LM, Chow CK. 2003. Copper toxicity, oxidative stress, and antioxidant nutrients. Toxicology. 189: 147-163.

Galloway JN, Thornton JD, Norton SA, Volcho HL, McLean RA. 1982. Trace metals in atmospheric deposition: A review and assesment. Atmospheric Envi. 16: 1677.

Goyer RA. 1997. Toxic and essential metal interactions. Annu Rev Nutr. 17: 37-50.

Hendry GAF, Baker AJM, Ewart CF. 1992. Cadmium tolerance and toxicity, oxygen radical process and molecular damage in Cd tolerant and Cd sensitive clones of *Hocus lanatus* L. Acta Bot Nether. 40: 271-281.

Hsu FH, Chang HC. 1992. Inhibitory effect of the heavy metals on seed germination and seedling growth of *Miscanthus* species. Bot Bull Acad. 33: 335-342.

Lanstzy RJ, MacKersie FT. 1979. Atmospheric trace metals: Global cycles and assessment of man's impact. Geochica et Cosmochica Acta. 43: 511.

Nan ZR, Zhao CY, Li JJ, Chen FH, Sun W. 2002. Relation between soil properties and selected heavy metals concentrations in spring wheat (*Triticum aestivum* L.) grown in contaminated soils. Water, Air and Soil Pollut. 133(1/4): 205-213.

Nies DH. 1999. Microbial heavy-metal resistance. Appl Microbiol Biotechnol. 51: 730-750.

Polle A, Schutzendubel A. 2003. Heavy metal signaling in plants: Linking cellular and organismic responses. In: Plant Responses to Abiotic Stress, Vol. 4, Hirt H, Shinozaki K (eds). Springer-Verlag, Berlin. pp. 187-215.

Root RA, Miller RJ, Koeppe DE. 1975. Uptake of cadmium its toxicity, and effect on the iron ratio in hydroponocally grown corn. J Environ Qual. 4: 473-476.

Schutzendubel A, Polle A. 2002. Plant response to abiotic stress: Heavy metal-induced oxidative stress and protection by mycorrhization. J Exp Bot. 53: 1351-1365.

Somashakaraiah BV, Padmaja K, Prasad ARK. 1992. Phytotoxicity of cadmium ions on germinating seedlings of mung bean (*Phaselus mungo*): Involvement of lipid peroxidase chloropyhyll degradation. Physiol Plant. 85: 85-89.

Vose PB. 1984. Effects of genetic factors on nutritional requirement of plants. In: Crop Breeding and Contemporary Basis, Vose PB, Blixt SG (eds). Pergamon Press, Great Britain. pp. 67-114.

Weast RC. 1984. CRC Handbook of Chemistry and Physics, 64th edn., Boca Raton, CRC Press.

Wu FB, Zhang GP. 2002. Genotypic differences in effect of Cd on growth and mineral concentration in barley seedling. Bull Environ Contamination Toxicol. 69(2): 219-227.

8

Aluminium Toxicity Tolerance

8.1. INTRODUCTION

Aluminium (Al) is not considered as essential nutrient, but at low concentrations it may increase plant growth or induce other desirable effects (Bollard and Butler, 1966; Foy, 1988; Huang *et al.*, 1992a, b; Roy and Mandal, 2005). Al affects about 40-70% of the world arable land, which has the potential for food crops biomass production (Haug and Caldwell, 1985). Acid soils constitute about 3.95 billion hectares globally. About 38% of those are present in Tropical Asia (Datta, 2002). It is believed to cover more than 800 million hectares of forest and tropical Savana ecosystems of tropical America (Jaffe and Rojar, 1994; Prakash 2000). Nearly 75% of Amazon Basin contains acid and infertile soil classified as oxisols and utisols. About 383 million hectares (79%) of the Amazon area, suffers from Al toxicity. Acid soil occurs extensively in the some of the Mediterranean countries and Southern Australia. Low laying acid sulfate of Bay Islands in India chronically suffer from Al-toxicity (Chowdhury *et al.*, 2001).

Al present in all soils, but its toxicity impairs productivity in soils having low pH (< 5.0). Free Al-ions are solublized at low pH. An acid soil is commonly formed of acidic sediments from rocks or evolved from previous soils or supply of basic ions to the soils. The main process of their formation is the reduction of clay content in the upper horizon. Acidity increase the availability of

Fe, Mn and Al and these elements are often toxic to plants. However, other factors, such as Mn toxicity and deficiency of P, Ca and Mg also interplay in manifestation of Al toxicity. Toxic concentration of Al is greatly available in acid soils (Gregorio *et al.*, 2002).

8.2. Al-TOXICITY SYMPTOMS IN PLANT

1. Poor or stunted growth of the affected plant.
2. Orange yellow to white interveinal chlorosis and mottling on leaves followed by leaf tips death and margin scorch. Rice suffering from Al-toxicity shows interveinal white to yellow discolorization of the tips of older leaves, which may later turn necrotic.
3. Excess Al induces nutrient deficiency symptoms. It cause iron deficiency *symptoms* in rice (*Oryza sativa* L.), sorghum and wheat (Clark *et al.*, 1981; Foy and Fleming, 1982; Furlani and Clark, 1981). In some cases Al-toxicity in plants appears as Ca deficiency resulting curling or rolling of leaves and collapse of growing points or petioles.
4. Al was found to have deleterious effect on leaf area, increase chlorophylase activity followed by a decline in pigment content and Hill activity. Such effects on physiological and biochemical changes resulted in decrease in seed yield of mung (*Vigna radiata*) bean (Neogy *et al.*, 2001).
5. Stunted and deformed roots in susceptible cultivars. Al-toxicity is clearly expressed in retardation of root growth and inhibition of plant growth. It is not observed before root symptom development. Root growth inhibition was detected 2-4 days after the initiation of seed germination (Bennet *et al.*, 1991). It also induces abnormalities in root system either by reduction or inhibition of growth of the main axis or by inhibition of numerous lateral roots, the growth of which is subsequently inhibited. Root growth at toxic levels of soluble Al do not develop so fine and branch in a pattern as they do in the absence of Al. Roots become stubbier and thicker and have a lower specific surface.
6. Lateral roots system become thickened and turn brown (Kinraide, 1988; Roy *et al.*, 1988). The root system as a whole is coralloid in appearance with many stubby lateral roots but lacks fine branching (Foy *et al.*, 1978).
7. Callose formation is one of the major symptoms of Al-toxicity (Nichol *et al.*, 1993).

8.3. SITE OF ALUMINIUM ACCUMULATION IN CROP PLANTS

The primary site of Al accumulation and toxicity target in plants was reported to be the root meristems, which consist of actively dividing cells (Rincon and Gonzalez, 1992; Ryan *et al.*, 1993). Al accumulated in large amount in the younger cells in the roots of pea (Wagatsuma *et al.*, 1995) and cell destruction was extensive in the region with high Al concentration. At cellular level, Al is mainly localized in the cell wall, although some are also found in the nucleus. Al absorbed by roots of wheat penetrated the boundary between root apexes and accumulated in the nuclei and cytoplasm of cells adjacent to this zone (Henning, 1975). It is evident that mass signal spread throughout the intercellular area and it is not intensive in the cell wall as evident through secondary ion mass spectrometry involving Al (Lazof *et al.*, 1996). The specific site of Al-toxicity is the plasma membrane of younger and outer cells in roots and that Al tolerance depends largely on the integrity of the plasma membrane (Wagatsuma *et al.*, 1995). An increase in the rigidity of the action net work (Grabski and Schindler, 1995) and the inhibition of phospholipase C (Jones and Kochian, 1995) were proposed as possible intercellular Al-target sites that might be involved in root growth.

8.4. INJURY MECHANISMS

A wide range of results such as less uptake of some cations (Renegel, 1990; Robinson and Renegel, 1991), reduced root growth, nutrient deficiency symptoms (Foy, 1984), photosynthesis by lowering chlorophyll content (Roy *et al.*, 1988), blocking Mg^{2+} transport (MacDiarmind and Gardner, 1996), inhibiting Hill reaction and photophosphorrelation of chloroplast (Hao and Lfuo, 1989), and low yield (Delhaize and Ryan, 1995) responses were amply recorded in different plant species/genotypes under Al stressed environments (Slootmaker, 1974; Lafever *et al.*, 1977). Some of the injury mechanisms have been elaborated in this section.

8.4.1. Reduced Root Growth

The principal effect of Al-toxicity in plants is severe restriction on root growth because Al directly inhibits cell division (Gunse *et al.*, 2003) in the root apical meristem and cell elongation. Inhibition of root elongation is generally observed because Al^{3+} is a toxic ionic species with high binding ability in cellular components of roots and usually shows marginal translocation to the upper parts of the plants. Al also acts directly on the cell, particularly in inducing lesion on the cell membrane possibly by interacting with membrane associated proteins (MacDiarmind and Gardner 1996) and on the whole plants by impairing intake of essential ions and water. Cell membrane damage may be due to the effect of Al on cell wall, as well as the toxic effects of Al on the plasma membrane of younger and outer cells in root or on the root symplasm. Wagatsuma *et al.* (1987) reported that the cells of the epidermis and outer cortex of maize in the root portion about 1 cm from the root-tip were damaged, and the walls of those cells were abnormal and partially detached in barley. Al also decreases the extensibility of cell wall by cross-linking proteins in the middle lamella (Rorison, 1958). Bennet *et al.* (1985) reported that an anisotropic growth response of cortical cells with 20 hours root exposure to Al were associated with collapse of the conducting tissues of the style and disintegration of the outer cell of the roots. Al affects plasma-membrane function and decreases the influx of Ca^{2+} and Mg^{2+}.

'H' ion increases permeability of cell membranes and allows cell constituents to leak out. At higher temperatures 'H' ions are much more damaging to the tissues. Al increases viscosity of protoplasm in plant root cells and decreases overall permeability to salts and water. This is due to cross-linking between adjacent protein molecules (Clarkson and Sanderson, 1969). Al also accumulates in the root cell nuclei (Morimura *et al.*, 1978). Microscopic observation of the meristematic region shows that the number of cell in metaphase decrease sharply after one-hour of Al treatment (10^{-3} M $AlCl_3$) to onion roots and become undetectable after 10 hours of treatment, whereas the control tissues maintain a level of 50-60 metaphase cells/1000 cells. Frye and Anderson (1978) also observed that in wheat variety Eagle (Al-sensitive) inhibition of elongation and DNA synthesis increased with the duration of exposure to Al. However, the elongation of DNA synthesis in the wheat var. Atlas 66 (Al-tolerant) were unaffected by Al treatment.

Positive correlation existed between the extent of growth inhibition and the amount of Al accumulation in the cells (Yamamoto *et al.*, 1996). Since, the cells in the stationary phase did not take up Al, this appears that the accumulation of Al is a prerequisite for the manifestation of toxicity (Yamamoto *et al.*, 1994). The location of Al in tobacco cells was estimated by staining with haematoxylin, which indicated the presence of Al over the entire cell surface and in the nucleus (Yamamoto *et al.*, 1996). Al was either observed in the entire cell or found to be tightly bounded to the plasma membrane or some minor parts of the cell wall (Yamamoto *et al.*, 1994).

Whole root response to Al toxicity showed discernible effect within the first few hours of uptake of Ca^{2+} (Lindberg, 1990; Huang *et al.*, 1995), K^+ (Nichol *et al.*, 1991), and NH_4 (Durieux *et al.*, 1993; Nichol *et al*, 1993).

However, comparison of direct effects on Al-induced nutrient uptake among the developmental regions of a root is scanty. In response to 5 mM $^{45}Ca^{2+}$ to the root of the wheat, subsequent to absorption in 0-5 mm root tip was decreased by 75%, but the figure was only 20% in case of mature root region, 5 cm behind the tip (Huang *et al.*, 1993). An immediate inhibition of K^+ uptake in Al-sensitive wheat was demonstrated in a zone 2 cm behind the root apex (Miyasaka and Kochian, 1989). Uptake of $^{15}NO^{3}$- was found inhibited by 2 h exposure to Al in four regions of soybean roots, as well as in the whole roots (Lazof *et al.*, 1994c).

8.4.2. Effect of Al-toxicity on Nutrient Uptake

Al-toxicity interfere uptake and transport of essential nutrients and water supply to plants (Fleming *et al.*, 1974; Rufty *et al.*, 1995). Long-term exposure of plants to Al also inhibits shoot growth by inducing nutrient (Mg, Ca, P and Fe) deficiencies, drought stress and phytohormones imbalance. Phosphorous transport between root and shoot diminished with increased Al concentration in roots. Poor plant growth with Al-toxicity was a result of P starvation (Ligon and Fierre, 1932). Foy *et al.* (1969) reported decreased Ca uptake in Soybean (*Glycine max*). The Ca influx in the root apex was strongly inhibited by Al in barley (Nichol and Oliveira, 1995). Rufty *et al.* (1995) showed that NO_3^- uptake by soybean decreased in high Al concentrations. Al-toxicity inhibits shoot growth by reducing transport of Ca from roots to shoots (Strid, 1996).

8.4.3. Al-toxicity Affects Pollen Germination

Many workers reported reduced pollen germination and pollen tube growth under Al-toxic environment. Zhang *et al.* (1999) observed that pollen germination was inhibited by micromolar concentrations of Al^{3+}.

8.4.4. Chromosomal Abnormality and Reduced DNA Synthesis

Al-toxicity cause chromosomal abnormalities. Mohanty *et al.* (2004) reported chromosomal abnormalities in rice (*Oryza sativa* L.) var. Lalat, which includes chromosome stickiness, laggards, sticky bridge, occurance of micronuclei, and binucleate and multinucleat cells as Al treatment. The mitotic and meiotic indexes were significantly low. They observed significant variation in interphase nuclear volume between treated and non-treated plants. The occurrence of different types of chromosomal aberrations, reduction in amount of nuclear DNA, and persistence of phytotoxic effects at the post-treatment stage suggest carcinogenic effects of Al on rice plants. The presence of Al in acidic soils might thus be extremely hazardous and might cause permanent cytotoxic disorder in rice plants. Al also affects the movement of chromosome carried out by spindle.

Al-toxicity reduces DNA replication by increasing the rigidity of the double helix. Minocha *et al.* (1992) reported that Al treatment resulted in a severe inhibition of DNA synthesis within 16-24 h. Matsumoto *et al.* (1977) suggested that binding of Al tDNA was a potential cause of inhibition of cell division. Bennet *et al.* (1985) reported that nuclear changes were obtained with a low level of Al due to chromatin condensation of the nucleus and an increase in size and frequency of vacuoles in the nucleus.

8.4.5. Effects of Al-toxicity on Cell and Cellular Organelles

Plant cells respond to a certain stress factor in different ways depending on their developmental stage and type of tissue. Structural damage may be severe of even lethal in individual cells within a tissue that exhibits moderate or no effects of stress. Two different structural changes had been observed by Ciamporova and Baluska (2000) within the root epidermis just behind the root cap of *Zea mays*. Cells with dark and shrunken cytoplasm occurred next to swollen cells with preserved cellular compartments. Within the root cortex, individual cells of file had severely damaged cytoplasm, in contrast to almost undisturbed cytoplasm of adjacent cells.

Disruption of golgi apparatus had been reported in the peripheral cap cells of *Zea mays* (Bennet *et al.*, 1987). Al-toxicity also caused redistribution of amyloplasts to proximal halves of central cap cells as well as alterations in the linear arrangement. Al-toxicity cause extensive vacuoleation, cell distortion, decreased viability and increased cytoplasmic density of many rhizodermal cells of the root tip were found probably due to excessive Al accumulation in root cap (Budikova and Uzik, 1999).

8.4.6. Effects of Al-toxicity on Metabolisms

Al can bind either proteins or lipids, depending upon pH and other conditions. Al decreased lipid fluidity in membranes of *Termoplasma acidophilium* (Vierstra and Haug 1978). Al-toxicity closely related to nitrogen metabolism (Foy and Fleming, 1982). Al-stress increases peroxidase activity, which reduce the plant growth (Cakmak and Horst, 1991). Severi (1997 a,b) reported that the presence of Al had the tendency to decrease the multiplication rate of *Lemna minor* L. with significantly increasing guaiacol peroxidase activity.

8.4.7. Al-toxicity Injury Depends on Cell Growth Stages

The sensitivity of tobacco (*Nicotiana tabccum* L.) cells to Al was largely depended upon the phase of growth (Lazof *et al.*, 1994b; Yamamoto *et al.*, 1994; Ono *et al.*, 1995). Cell at the logarithmic phase they showed resistance. After ~10 h of exposure, cells started accumulating Al in the logarithmic phase.

8.4.8. Other Physiological Effects of Al-toxicity Injury

Al interacts strongly with proteins, polynucleotides, lipids and glycosides (MacDonald *et al.*, 1987; Akeson *et al.*, 1989). Al generally enters in the symplasm of the root cells very quickly within 1-2 hours of inhibition of exposure (Yamamoto *et al.*, 1996) and Al photo-toxicity might be the result of a symplastic Al interaction (Lazof *et al.*, 1994a &b; Vitorello and Haung, 1996). Although concentration of free Al^{3+} was reported to be decreased to less than 10^{-10} M at pH 7.0 due to formation of in soluble $Al(OH)_3$, such low concentrations are potentially phytotoxic because of the strong affinity of Al for O_2 donor compounds such as inorganic phosphate, ATP, RNA, DNA, protein, carboxulic acids and phospholipids (Martin, 1988). Plucinska and Zeigler (1995) indicated that the longer exposure to Al ions led to a drastic decrease in total adenylate and ATP pool-levels with a corresponding rise in ADP and AMP content and great depression both in ATP/ADP and AEC (adenylate energy charge) and inhibition of metabolic activity (Plucinska and Zeigler, 1996). Woolhouse (1969) reported that Al inhibited the activity of ATPase in plants. He also suggested that structural changes in these enzymes might be responsible differential Al tolerance of the ecotypes. Prolonged Al-stress induced an enhancement of lipid peroxidation (Severi, 1997 b) and caused formation of high toxic oxygen free radicals (Cakmak and Horst, 1991). Therefore, internal detoxification mechanisms are required for building tolerance to high Al in Al-accumulating plants.

8.5. TOLERANCE MECHANISMS

Considerable research has been done in under standing physiological (Kochian, 1995; Pellet *et al.*, 1995; Ryan *et al.*, 1995a &b), molecular (Snowden and Gardner, 1993; Snowden *et al.*, 1995; Ezaki *et al.*, 1996) and genetic (Aniol,

1995) mechanisms governing Al-toxicity tolerance. Tolerance is manifested at cellular level too (Taylor, 1995). An element present in excess can interfere with metabolism through competition for uptake, inactivation of the structure of water. Many of the elements involve in modification of membrane structure and function. Several potential mechanisms of tolerance for uptake of Al in the symplasm were also advocated (Taylor, 1991; Kochian, 1995). These include chelation of the vacuoles, involvement of Al binding proteins, elevation of Al-tolerant enzyme activities etc.

8.5.1. Plant Exudation

It was proposed that organic acid exudation acts in maintaining low Al^{3+} in the rhizosphere, thereby to Al exclusion from the root cells (Kochian, 1995). Al stress in barley reduces the concentrations of citric acid, succinic acid and total organic acid in the roots (Foy *et al.*, 1987). It was proposed that organic acid exudation acts in maintaining low Al^{3+} in the rhizosphere, thereby leading to Al exclusion from the root cells (Kochian, 1995).

Exudation of malate was suggested as an important mechanism governing Al tolerance in wheat (Delhaize *et al.*, 1993; Delhaize and Ryan, 1995). Malate exudation was correlated with Al tolerance in the whole root (Basu *et al.*, 1994; Huang *et al.*, 1996) and in mature regions even through exudation per unit length was much greater for 4 mm root tips. Malate may remove Al from either the apoplast or symplast, possibly accounting for Al efflux (Rincon and Gonzalez, 1992).

In some plants, Al concentration in the shoots are not consistently different from those of Al sensitive plants, but in the root the Al concentrations are lower in certain cultivars of wheat, barley, soybean (Foy, 1974a, b) and pea (Klimashevskii *et al.*, 1976). Here, Al-tolerant apparently involves an exclusion mechanism. Pintro *et al.* (1995a, b) also reported existence of Al exclusion mechanism, which confer tolerance ability to maize plants.

8.5.2. Cell-wall Composition Modulates Al-Toxicity

There is close positive correlation between Al contents and relative callose induction in the maize cells (Schmohl *et al.*, 2001). Investigating the spatial Al-sensitivity of root apices of *Zea mays* and *Vicia faba* and combining the data for both species they found close positive relationships between pectin and Al contents, pectin contents and relative callose induction. Thus, binding of Al to cell-wall pectin-matrix represents an important step in the expression of Al-toxicity.

8.5.3. Biochemical Mechanisms of Al-tolerance

Proposed biochemical mechanisms governing Al-toxicity tolerance, such as the effects on calmodulin (Siegel and Haug, 1983) or on phospholipase (Haug *et al.*, 1994) might be altering the metabolism in mature cells as well as at the tip. On the other hand, the proposed mechanisms for biochemical mediated Al phytotoxicity assumes that Al must enter cells and should remain somewhat active within the cells (Lazof *et al.*, 1996).

Under Al-stress, a calcium-dependent multifunctional regulatory protein, calmodulin is synthesized (Haug, 1984). Binding Al to calmodulin is supposed to induce functional modification of this protein and cause changes of calmodulin-assciated activation of target enzymes (Haug and Caldwell, 1985). There are two kinds of NAD-kinase present in pea root apex: a calmodulin-dependent and calmodulin independent forms. The high calmodulin concentration was detected in the root cap and the cell division area, therefore the effects of Al ions on this enzyme and calmodulin might be connected with Al tolerance mechanism. Aniol (1991) showed a shift from calmodulin-dependent NAD-kinase activity to calmodulin-independent activity of this enzyme observed under Al stress in Al-tolerant wheat genotypes.

Resistance for Al-toxicity in calcifuges results from a binding or chelating mechanism, which also has an affinity for Fe, since many of the calcifuges were found to be highly susceptible to Fe chlorosis when grown in absence of Al. Tea plants contain appreciable amount of organic acids and poly-phenols, which detoxify Al by chelating and account for the Al tolerance of the species. Al stress in barley reduces the concentrations of citric acid, succinic acid and total organic acid in the roots (Foy *et al.*, 1987). The ability of certain plant species to tolerate Al in their tissues might also result because of the presence of Al tolerance enzymes.

In sensitive plants, reactive oxygen species formed due to Al-toxicity. Darko *et al.* (2004) reported formation of superoxide dismutase, catalase, ascorbate peroxidase and glutathione-S-transferase, which played important role in the detoxification of reactive oxygen species in Al-tolerant plants, since they were found to have higher activity than in the Al-sensitive plants.

8.5.4. Uptake and Accumulation of Al

Al-toxicity tolerance is directly associated with Al accumulation in plant tops, such plants have high internal tolerance to Al particularly pine trees, tea and mangroves (Foy, 1978). The majority of the plats growing in acid environments do not accumulate large concentration of Al in their foliage and majority of the Al they take up remains inside or in the roots. Tolerance in these species must result from a mechanism that restricts the entry of Al into cell metabolism within the root. Under *in vitro* culture condition, quantum of Al present in selected and non-selected calli and plants substantiated the concept of different mechanisms being involved in governing Al resistance at both cellular and *in planta* levels. Organic acid in plants may also be responsible for maintaining high foliar concentration of Al.

Most plant species accumulate Al ~ 0.2 mg/g dry weight. They may also contain Al more than 10 times of this level without displaying any Al-injury. Tea plants are found to be typically Al accumulators. The Al content in their old leaves could reach as high as 30 mg/g dry weight, although young leaves contained only about 0.6 mg/g dry weight (Matsumoto and Hirasawa, 1976). Some trees in the tropical rain forests, such as *Richeria grandish*, were reported to accumulate Al more than 1 mg/g dry weight in their leaves (Cuenca *et al.*, 1990). The top of some Al-tolerant plants contain lower concentrations of Al than do those of Al-sensitive plants. The tea plants appear to protect itself against Al-injury by trapping most of the excess Al in older leaves. These may contain 2000 ppm Al or more, whereas young leaves may contain only 100 ppm and buds as little as 50 ppm. Hydrangea (*Hydrangea macrophylla*) is also a well known accumulator and relationship between Al and blue colourization of its sepals was thoroughly investigated (Takeda *et al.*, 1985a, b; Ma *et al.*, 1997). Total accumulation of intercellular Al in tobacco root tip was as high as 210 nmol/g fresh weight after 4 h (Lazof *et al.*, 1996).

Baggie (2002) reported that Al uptake was higher in tolerant genotypes of rice [WAC102531(2) and E425] than in susceptible genotypes (IDSA6, SMMG88-9 and Seniwa), this suggested that the mechanism of tolerance in WAC102531(2) and E425 may be related to their ability to tolerate high Al in the tissue, presumably by Al detoxification. Conversely, for the Al tolerant genotype Ngbongo nyenye, the lowest shoot Al concentration may suggest tolerance mechanism of Al exclusion in the root apoplasm.

8.5.5. Nutrients Associated with Al-tolerance

Plants that are able to maintain sizable concentrations of Al within their tissues with concurrent maintenance of adequate P levels must have endowed with a mechanism whereby, Al is prevented from precipitating with P at physiological pH. The plant chelate Al by organic acids are generally prevented from interfering P metabolism. Since, deficiency is often associated with Al-toxicity on acid soils, they ranked inbred with respect to Al-tolerance using P deficiency symptoms and P content of tops as their criteria. Al-tolerant cultivars of upland rice have higher concentration of both P and Ca and lower levels of Al in their shoots than do susceptible varieties (Howeler and Cadavid, 1976).

8.5.4.1. Silicon

Silicon (Si) concentrations are generally found to be much higher in monocot than dicot plants. Conversely, high mineral cation-anion uptake ratio is being observed in dicots. When large amount of anionic Si participate in cation-anion balance to add the excess of anion uptake, equivalent of hydroxyl ion (OH-) should be expelled from roots, which can increase rhizosphere pH and decrease the uptake of Al. Si-induced amelioration of Al-toxicity was investigated by many workers (Kidd *et al.*, 2001; Gu and Li 2002).

Si causes Al to accumulate in root and facilitated P transfer to shoots (Huang and Shen, 2003). Si alleviated Al toxicity to barley (*Hordium vulgare*) by forming soluble less toxic hydroxyl-aluminium silicate ion in the soil and solution, thus reducing the concentration of toxic Al by redistributing Al and P in shoots and roots. Morikawa and Saigusa (2002) suggested that the soil amelioration of Al toxicity by the waste porous hydrated calcium silicate may be due to the increase in soil pH rather than to the increase of Si concentration in the soil solution. The supply of various forms of Si effectively alleviates Al stress leading to normal root growth (Gu and Li, 2002), Si change the forms in the solution or medium. With Si addition, the ratio of exchangeable Al decreased from 78 to 48%, while exchangeable Si increased from 0 to 20%. Al in root generally accumulates in apoplastic tissues, especially in cell wall. They reported that, the addition of low-molecular Si increased total Al in root and Al content in apoplast, but decreased Al in cell walls. The addition of high-molecular Si significantly decreased total Al in roots and Al contents in apoplasts, symplast and cell walls. Al and Si triggered release of catechol flavonoid-type phenolics: catechin, and quercetin. Kidd *et al.* (2001) reported that Al-tolerant cultivar of maize, Sikuani, Al-exposed plants treated with Si exuded up to 15 times more phenolics than those plants not pretreated with Si. Thus, flavonoid-type phenolics appeared to play a role in the mechanism of Si-inducedc amelioration of Al-toxicity.

8.5.4.2. Calcium and magnesium

Calcium (Ca) and magnesium (Mg) are quite affective in protecting wheat root from Al-toxicity (Ali, 1973). Increasing the concentration of either Ca or

Mg in an Al-toxic solution greatly reduce the toxic effect of Al on root growth. Some varieties of wheat and barley are tolerant of high Al while others are tolerant of high Mn, but wheat variety April Red, known to be tolerant of lime-deficient conditions (Ca-efficient) is highly resistant to both elements.

8.5.6. Root-induced pH Change

Some Al-tolerant plants raise the pH of the rhizosphere, thus reduces the solubility and toxicity of Al. Al-tolerant genotypes of wheat, rice, barley, peas and maize follow this mechanism to avoid Al-toxicity in soil (Clark and Brown, 1974; Mugwira and Patel, 1977). This includes 'H' ion release resulting from excess cation over anion absorption, release and hydrolysis of CO_2, release of 'H' ions from carboxyle group of polygalacturonic acid residues of pectic acid and excretion of protons from microorganisms associated with roots.

8.5.7. Soil Microorganisms

Al toxicity limits the growth and nutrient acquisition of sensitive tree species in regions receiving acidic deposition. Symbiosis between tree roots and mycorrhizal fungi may offset the negative impacts of Al in the root zone. The availability of heavy metals to plants and thus, their toxicity depends on complex rhizospheric reaction involving not only exchange processes between soils and plant but also microbial activities. In this respect, *Rhizobium* and mychorrizal fungi appear to play a central modulating role. Scholl *et al.* (2005) observed better growth of *Pinus sylvestris* seedlings in soil colonized with ectomycorrhiza fungi. Ectomycorrhiza fungi improved the uptake of immobile nutrients such as P through a better soil exploration by the external mycelium or by detoxification of Al by organic anions excreted by the fungi. Patterns of labile Al in solution, nutrients, and Al accumulation in tissue suggest that arbuscular mycorrhizal fungal ecotypes may alter the form or compartmentation of Al within the rhizosphere and plant, thus protecting seedlings from the effects of exposure to Al in the soil solution. Non-mychorrizal plants appeared to be the result of the disruption of P translocation to leaves and Ca, Mg, P, Cu and Zn uptake in roots.

Considerable variations in acid tolerance of four strains of *Rhizobium* have been reported (Freire, 1976). The strain *R. phaseoli* 148-S, which survived well in unlimed soil, formed many nodules on plant grown in both limed and unlimed soils. Thorton and Daevy (1983) evaluated the effect of soil acidity on nodulation N_2-fixation, and dry matter production of *Trifolium subterranean* inoculation with 33 strains of *Rhizobium trifoli*. It exhibited a large and varied tolerance of soil acidity in symbiotic association with the host. Rice (1982) found variation in low pH tolerant ability of different strains of *R. meliloti* on the basis of dry matter production of Ladak. New and Kennedy (1989) have also reported variation in Al-toxicity tolerance of different strains of *Azospirillum* spp.

Banana is a cash crop and food crop in tropical regions and it is sensitive to Al-toxicity. Rufytikiri *et al.* (2000) reported a significant positive effect of arbuscular mycorrhzal fungi Glomus on plant growth with Al treatment. The mycorrhizal plants had increased shoot dry weight, uptake of water and of most nutrients, and Ca, Mg, and P content, particularly in roots; decrease in Al content in root and shoot; and delay in the appearance of Al-induced symptoms.

8.6. VARIATION IN Al-TOXICITY TOLERANCE ACROSS CROP PLANTS

The traditional method of liming is economically viable and sustainable only when lime is available locally. Further, amelioration of acid soils is impractical (sub-soil acidity) and temporary; breeding acid tolerant variety is the only tangible success. Selection and breeding of cultivars appears more promising than liming. Thus, identification of the adaptable crop genotypes for Al-stressed soils is the durable solution. It is a fact that plant species or cultivars within the same species greatly differ in their tolerance pattern to Al-stress (Khatiwada *et al.*, 1996). Genetic variability among species (Reid, 1970; Konazak *et al.*, 1976; Foy *et al.*, 1988) in relation to Al tolerance has been noted in different crop species (Table 8.1).

The traditional tall cultivars of rice have higher Al tolerance and require lower liming rate than the new semi-dwarf cultivars, which require high rate of liming (Howeler and Cadavid, 1976). Extensive field and laboratory experiments indicated that rice genotypes show sizable variability for Al-toxicity tolerance (Konzak *et al.*, 1976; Fageria and Zimmerman, 1979; Fargeria and Barbosa-Fillo, 1983; Chowdhury *et al.*, 2001). Several improved rice cultivars have been developed at IRRI that can survive partially under these adverse soil conditions (IRRI, 1990).

Despite the problems encountered in adapting culture media for *in vitro* selection for Al-toxicity resistance genotypes or somaclones (Conner and Meredith, 1985a; Wersuhn *et al.*, 1994; Taylor, 1995), Al-resistant cells were obtained in several species, such as, alfalfa (Coulombe and Scoyoe, 1990; Parrot and Button, 1990; Arihara *et al.*, 1991), carrot (Arihara *et al.*, 1991), soybean (Campbell *et al.*, 1987), tomato (Meredith, 1987a, b), sorghum (Conner and

Meredith, 1985a), tobacco (Petolino and Collins, 1983), *Nicotiana plumbaginifolia* (Conner and Meredith, 1985b, c) and rice (Van Sint Jan *et al.*, 1997; Chowdhury *et al.*, 1998; Roy and Mandal, 2005).

Al-tolerant among the tritical followed the order: Canada and Mexico > California > Albama > Oregon, while in the Bladen soil Albana triticals had higher Al-toxicity tolerance ability than Canadian triticales. Brazilian wheat had higher Al-toxicity tolerance than the USA wheat. Wheat varieties from Indiana and the Plains State were more sensitive to Al than those developed in Eastern States.

Table 8.1. Donor germplasm of different crops harbouring Al-toxicity tolerance ability

Crop	Genotypes	References
Rice *Oryza sativa*	BW196, Bhura Rata, Co37, Basmati 370	Sivaguru and Paliwal, 1993
	IR64, VN86G	Tang *et al.*, 1993a
	Azucena, IRAT104, Morobekeran, SiyamKuning, Gudabang Putih, Siyam, Lemo, Khao Daeng, Siyamhalus, Bjm-12, Ketan, Seribu, Gantang, Bayer Raden, Rati, Padi Kanji	Khatiwada *et al.*, 1996
	Atals 66, Shironsanjyaku	Andrade *et al.*, 1996
	FM404	Eckert *et al.*, 1996
	Guapore	Ferreira *et al.*, 1997
	IAC-227, CMH80A.747	Camargo, 1998
	Guarani, Guapore, IAC 25	Ferreira *et al.*, 1999
	IR43, Agulh Arroz, Vermelho, IAC3, IRAT109, Dinorado	Dobermann and Fairhurst, 2000
	WAC10253(2), E425, Ngbongo nyenye	Baggie, 2002
	Haskaima, Gara, Karjat23, Ajan146, Mohishadhan, Mattu, Panvel, Sendursail, Karjat18, Indryni, Dhurdras, Seetabhog	Mandal *et al.*, 2004
	Asominori, IR24	Xue *et al.*, 2005
Weat *Triticum* spp.	Fulz, Trubull, Fulcaster, Thorne, Seneca, Lucas, Butter, Fulton, Atlas66, Blue By, Pennoll	Lafever *et al.*, 1977; Cocker *et al.*, 1998
	Waalt	Wheeler, 1994
	BH1146, AIC60	Camargo *et al.*, 1995

Contd...

Table 8.1.Contd...

	Kadett	Strid, 1996
	BH1146, Embrapa 15, Embrapa 24, Embrapa 49, IAC5-Maringa, IAC 18-Xavantes, IAC27-Pantaneiro, IAPR46, IAPAR53, RS8-Westphalen, Trigo BR15, Tigo BR20, Trigo BR25, Trigo BR35, Trigo BR41-Ofaie	Sousa and Sousa, 1998
	BH1146	Camargo, 1999
Maize	B37, Virginia17	Lutz *et al.*, 1971
Zea mays	M162W, I137N, HL-2, TX24, SR52	Farina *et al.*, 1982
	Trinidad Group 1 and 2< PDMR Comp1	Jariel *et al.*, 1990
	C525-M	Pintro *et al.*, 1995a, b
	Sikuani	Kidd *et al.*, 2001
Sorghum *Sorghum bicolor*	PaB8A, B37, Oh43, W117, Mo17, Pa54, PaW703, A635, A554, C103	Clark, 1973
Barley *Hordium vulgare*	Aura	Ramaskeviciene *et al.*, 2000
Soybean *Glycine max*	No. 62, Lachang, Guangzhou Dali	Wan *et al.*, 2001
Groundnut *Arachis hypogea*	TAG-24, TCGP-5, TCGP-6, TCGP-7, TCGP-10, JL-220, TCGP–34, ICGS-44, ICGS-76, TPT-1	Prathap *et al.*, 2002

8.7. GENETICS OF Al-TOXICITY TOLERANCE

Al-tolerance is genetically determined (Reid, 1971), thus selection possible for better Al-toxicity tolerance in crop plants. Al-toxicity tolerance appeared to be determined by one or more major-dominant genes (Lafever, 1981; Wheeler *et al.*, 1992). Al-stress tolerance in rice is controlled by a complex multi-genic system (Campbell *et al.*, 1994; Sledge *et al.*, 2002). This is in contrast to that of mineral deficiency resistance, which is generally monogenically controlled. Tolerance to Al-toxicity in wheat is controlled by two genes pairs; each gene pair affecting the same characters, with complete dominance of both gene pairs, but their recessive homozygote is epistasis to effects of the other gene (Aniol, 1991). The tolerance of Al of disomic substitution lines having the chromosomes of the *D* genome of *Triticum aestivum* L. cv. Chinese Spring individually substitutated for their homologues in *T. turgidum* L. cv. Langdon was investigated by hematoxylin method (Luo and Dvorak, 1996). Studies of Al-toxicity tolerance in ditelosomic lines of Chinese Spring wheat cultivar revealed that genes controlling this character are located on the short arm of chromosome 5A and the long arm of chromosomes *2D* and *4D*. Further, it had been concluded that Al-toxicity tolerance genes were located on wheat chromosome *5A* (long arm), *6A* (short arm), *2D* (long arm), *4D* (long arm), *4B* (long arm), *5D* and *7D*.

Al tolerance in certain barley population is controlled by one major, dominant gene (Reid, 1971). In diploid *Hordium valgare* Al-toxicity tolerance is controlled by a single dominant gene located on chromosom-4 (Stolen and Anderson, 1978). In maize (*Zea mays*), additive gene action is important for Al-toxicity tolerance (Rhue and Grogan, 1977; Rhue *et al.*, 1978). Tolerance ability is controlled by single gene with multiple alleles; tolerance seems to be the pleotropic effects of heat resistant gene *te2*. Ferreira *et al.* (1999) reported in rice that inheritance was oligogenic with tolerance determined by dominant alleles. Rice cultivars, Guarani, Giapore and IAC25 had more dominant alleles and CNA5600, CNA5615 and IAC899 more recessive alleles. Whereas, the findings of Bunta (2000), demonstrated the complexicity of genetic mechanism involved in wheat tolerance of Al-toxicity. Their results suggested the presence of additive, dominant and epistatic gene actions, with positive and negative effects, for Al-tolerance. The heritability was high, but, the genotypes exhibited different hereditary capacity. Czembor and Wiewiora (2001) found that single gene pair controls tolerance to aluminium toxicity in wheat cultivars-Carazinho, Ruivo Tardio, Trigo des Acores and Barbela PI 185700. Again, Czembor (2003) reported a single gene pair controls tolerance o f Al-toxicity in wheat cultivars- Quality and Dong 120. It represents the source of tolerance as that found in the spring wheat BH-1146.

In sorghum, Al-toxicity tolerance was shown to be polygenic. The ratio of GCA and SCA showed that additive gene effects were more important than non-additive gene effects (Borgonovi *et al.*, 1985). Thus, GCA was more important than SCA.

8.8. SCREENING FOR Al-TOXICITY TOLERANCE

Screening for Al-toxicity tolerance can be performed in field condition having Al-toxic soil, creating toxicity in pots or tanks, hydroponics or sand culture. For initial screening or screening small populations in hydroponics, sand culture, pots, or tanks are suitable, whereas, Al-toxic soil in the field condition facilitates evaluation of large populations.

8.8.1. Field Evaluation

The tolerant mechanism of the toxic ions appears to keep them away from the active sites of metabolism by chelating the cell wall. When the plants were grown in soils containing high concentrations of heavy metals, several metabolic activities could be adversely affected. However, plants can be selected that can grow on normally toxic concentrations of metal ions. Bradshaw (1974) studied on natural vegetation which revealed that genes for metal tolerance are highly frequent and randomly distributed. Thus, selection for tolerance in contaminated site is very efficient (Gartside and McNeilly, 1974). Large population can be screened under field condition with relatively low cost. But, multiple metal toxicities may exhibits in the same field; for example, acid soils exhibit both Al and Mn toxicities. Therefore, selection for tolerance for Al-toxicity alone is not possible in the field condition.

The efficiency of natural selection for metal toxicity tolerance in metal toxic soil is related to the inherent genetic variation in those populations and the selection pressure of the toxic soils (Gartside and McNeillty, 1974; Wild and Bradshaw, 1977). Evaluation criteria for metal toxicity tolerance under natural toxic soil are yield and general performance rather than tolerance to metal toxicity.

8.8.2. Screening in Tanks or Pots

Tanks or pots are filled with problem soil. They are usually maintained under controlled environment, and they are allowed a considerable control on the soil medium. Foy (1996) screened barley (*Hordium valgare* L.) cultivars for Al-toxicity tolerance by growing them for 25 days in greenhouse in pots containing acid soil and Al-toxic Tatum subsoil. Seven days after transplanting the tanks are irrigated with 30 ppm of Al in the form of $Al_2SO_4.18H_2O$. The plants are irrigated with Al-toxic water for 83 days in case of rice (Costa *et al.*, 1997; Van Sint Jan *et al.*, 1997; Roy and Mandal, 2005). Tolerant plants are selected based on the appearance of symptoms on leaves. Screening in tanks or pots is not suitable to study root characteristics as well as for large scale evaluation.

8.8.3. Hydroponics and Sand Culture

Small pots or test tubes may be filled with pre-washed and air dried sands. Tubes filled with sands impregnated with nutrient solution. To impose stress to the plants selective nutrient solution containing 240 mg L^{-1} $MgSO_4.7H_2O$, 228.6 mg L^{-1} NH_4NO_3, 41.02 mg L^{-1}, Ca $(NO_3)_2$,.$4H_2O$, 27.8 mg L^{-1} $FeSO_4.7H_2O$, 16.09 mg L^{-1} KCl, 6.16 mg L^{-1} $NaH_2PO_4.7H_2O$ (Van Sint Jan *et al.*, 1997) and micronutrients (Yoshida *et al.*, 1976) may be applied. The pH of the solution is maintained below 5, preferably 3.85.

For screening of rice, pre-germinated seeds of 5-days old are grown for 28 days in test tubes filled with sands (Baggie *et al.*, 2002; Roy and Mandal, 2005). After 28 days, sand is removed from each test tube without disturbing the spread and orientation of roots to study the root characteristics (Fig. 8.1a).

Fig. 8.1. Screening for Al-toxicity tolerance. **a)** Sand culture in test tube, **b)** hydroponics culture on nutrient solution. *(For colour version of this figure see page 537)*

Nutrient culture can also be used as hydroponics. This system is suitable for seedling studies. To establish the stability of the Al-toxicity tolerance character, pre-germinated seeds (of 5-10 days old) are transferred on the selective nutrient solution for 80 days (Costa *et al.*, 1997; Van Sint Jan *et al.*, 1997). The selective nutrients are renewed weekly interval and maintained the solution pH. Depending upon the leaf necrosis, the plants are selected (Costa *et al.*, 1997). The resistant plants remain entirely green or exhibit very limited necrotic zones, whereas sensitive plants show total necrosis of their leaves. Sivaguru and Paliwal (1993) tested tolerant ability of 22 rice cultivars in nutrient solution and isolated six tolerant cultivars, which showed significant changes in their expression in the presence of Al compared to control on the basis of root tolerance index, shoot tolerance index and relative growth reduction in root and shoot.

Spehar (1994) selected Al-tolerant soybean genotypes in hydroponics experiments. Ma *et al.* (1997) conducted rapid hydroponics screening for Al tolerance in 600 barley lines and isolated 90 intermediate-tolerant genotypes. Krizek *et al.* (1997) also screened *Coleus blumei* in nutrient solution containing excess Al. Ramaskeviciene *et al.* (2000) had screened barley (*Hordium valgare* L.) in liquid culture based on relative root length. Among the cultivars they had tested, Aura was more tolerant to Al-toxicity.

Sand culture system is cheaper and less complicated than hydroponics systems. Hydroponics requires adequate aeration and greater expertise. Both the systems are based on same principles, and used properly designed nutrient solutions to generate the specific mineral toxicity stress.

8.8.4. Selection Criteria

There are several selection criteria for Al-toxicity tolerance in crop plants, and they have been described bellow.

8.8.4.1. Hematoxylin test

This test is suitable in such cases where Al-toxicity tolerance is based on Al-exclusion from roots. The staining solution consists of 2 g hematoxylin and 0.2 g of $NaIO_3$ dissolved in one litter of distilled water.

8.8.4.2. Relative root length (RRL)

Root length is often used as selection a criterion for Al-toxicity tolerance. RRL has been used to asses Al-tolerance of sorghum (Furlani, 1981; Ohki, 1987), wheat (Kerridge *et al.*, 1971), soybean (Hanson and Kamprath, 1979), rice (Stolen and Anderson, 1978; Zhang *et al.*, 2007), and legumes (Baligar *et al.*, 1988; Blamey *et al.*, 1990; Mackay *et al.*, 1991). Nutrient solution screening method for screening rice seedling was developed by Howeler and Cadavid (1976). Wheat seedlings were screened with solution containing aluminium-sulfate ranging from 5-100 ppm (Prestes *et al.*, 1975). The RRL of seedling grown at 30 ppm Al was used as an index of Al tolerance. RRL and root length also used at International Rice Research Institute, Philippines (IRRI, 1977) for screening rice varieties (Table 8.2).

Table 8.2. Root length and relative root length of some rice varieties at 0 and 30 ppm Al in culture solution (adopted from IRRI, 1977)

Variety	Root length (cm)		RRL
	30 ppm Al	0 ppm Al	
E425	9.1	19.5	0.47
OS6	7.8	16.6	0.47
Dark Mali	3.0	10.9	0.28
HR334	2.7	9.3	0.29
CICA4	2.4	13.0	0.18
IR22	2.0	11.9	0.17

Root growth study of rice (Roy and Mandal, 2005), var. IR72 somaclones at R_3 generation showed normal root growth and distribution in response (Fig. 8.2) to Al-toxicity at seedling stage in Al-toxicity tolerant lines (Table 8.3). Notably, root length of tolerant somaclones remains almost unaltered across stress gradients (Roy and Mandal, 2005). Toxic effects of Al on root growth in respect of length, spread and orientation in the sensitive cultivars were well documented (Doberman and Fairhurst, 2000).

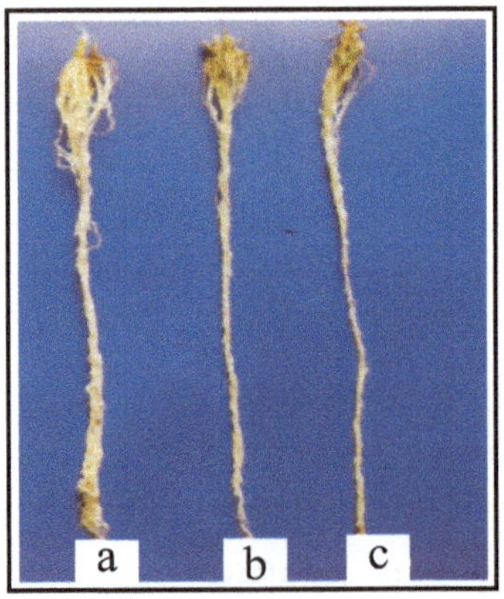

Fig. 8.2. Relative root length of Al-toxicity tolerant somaclone of IR72 under Al-toxic environment: a) Root of control plant, b) and c) Roots of under 30 ppm of Al $SO_4.7H_2O$

Xue *et al.* (2004) has coined the term the RRL as Relative Root Elongation (RRE), and they used it as an indicator for Al-toxicity tolerance. The RRE of each line was calculated as follow:

RRE = (RE1/RE2) × 100

Where,

RE1: Root elongation (average of replications) wih Al solution

RE2: Root elongation without Al solution

As Al-tolerance is a heritable character, and a long root regrowth after Al-stress is indicative of apparent Al-tolerance while a short or no regrowth is indicative of Al-intolerance, artificial selection of seedlings with longer regrown root length would be expected to produce plant with improved Al-tolerance. However, the effectiveness of the selective breeding depends on a range of factors, among which are the target character, gene effects involved, estimates of heritability as affected by the interaction between genotype and environment and the selection methods.

8.8.4.3. Root deformation and discolourization

Al-sensitive plants grown in stress media produced deformed and discolourized roots. Toxic concentration of Al generally found in acid soils (pH < 5.0), inhibit root growth (Lu-Ning and Hou-Tian, 1989), restricting water and nutrient uptake and leading to poor growth. This may be used as selection criterion for evaluation in hydroponics.

8.8.4.4. Leaf necrosis

Al-toxicity cause leaf necrosis and it is more reliable criterion for evaluation Al-toxicity tolerance (Costa *et al.*, 1997; Van Sint Jan *et al.*, 1997). Depending upon the extent of leaf necrosis, the plants may be classified into three categories:

1. Resistant plants entirely green or exhibiting very limited necrotic zones
2. Moderately tolerant plant shows necrosis on about half of their leaves
3. Sensitive plants totally necrosed

8.8.4.5. Dry matter production

Al mainly retards root growth, disturbed plant metabolism by decreasing water absorption and mineral nutrition. It ultimately reduces shoot growth too (Table 8.3). Therefore, shoot dry matter production is often used as a selection criterion in nutrient solution or sand culture (Campbell *et al.*, 1989; Roy and Mandal, 2005). Foy (1996) reported relative shoot dry weights averaged 28.6% for tolerants and 14.1% for sensitive cultivars.

Table 8.3. Root inhibition in IR72 somaclone for Al-sensitivity under sand culture (Roy and Mandal, 2005)

Genotype	Al-stress (AlSO4.18H2O ppm)	Root length (cm)	Root fresh weight (g)	Root dry weight (g)
IR72	Control	21.50 a*	0.29 a	0.08 a
	30	14.20 b	0.08 d	0.02 c
	60	12.60 c	0.05 e	0.02 c
Tolerant somaclone	Control	20.95 a	0.22 bc	0.06 ab
	30	20.50 a	0.24 b	0.05 b
	60	20.70 a	0.20 c	0.06 ab

*Means having same letter in the column are not significantly different by DMRT at P= 0.05

8.9. BREEDING FOR Al-TOXICITY TOLERANCE

Once the genetic control mechanism(s) of metal toxicity tolerance identified, it may be possible to combine metal tolerance with other desirable characters to produce plants that are better adapted to high metal-toxic soils. Breeding for acid-tolerant and/or Al-tolerant cultivars appears to be the most economically viable solution to the acid soils, because genetic selection for acid soil tolerance offers an avenue for increasing plant production and reducing production cost (Samac and Tesfaye, 2003). Moreover, genetic improvement of Al tolerance in crop is the only alternative to soil limiting, a traditional but short term and expensive agricultural cure to raise soil pH.

8.9.1. Mutation Breeding

Different mutagens may be used to induce mutation for rapid creation of variability for Al-tolerance in crop plants. This technique may also be used to correct the existing better performing cultivars for abiotic stresses. Tulmann *et al.* (1995) used gamma rays (25 and 35 kR) to create variation in wheat cv. IAC-24. They isolated a mutant line which was tolerant to Al-toxicity. Nawrot *et al.* (2001) created mutants in barley using N-methyl-N-nitrosourea (MNH) and sodium azide and developed few Al-tolerant mutants. These selected mutants also could be used in further breeding programmes to improve barley genotypes.

8.9.2. Recurrent Mass Selection

Diallel cross was made to study the inheritance of Al-tolerance ability in rice (Zhang *et al.*, 2007). Their finding suggested that the complex genetic nature and expression of Al tolerance in the diallel crossing system. A-tolerant parent GAAT'S', was the most promising parent, conferring the heist GCA effects for regrowth root length, root length and total root length. The existence of significant GCA variance in regrowth root length may also suggest the feasibility of improving Al tolerance through enhanced root regrowth using phenotypic recurrent mass selection to pyramid desirable Al-tolerant gene(s).

8.10. CROP MANAGEMENT FOR Al-TOXICITY

Crops can be grown in the Al-toxic soils following some cultural, varietal and fertilizer managements.

8.10.1. Cultural Practices

Al-toxicity is relatively rare especially in irrigated rice systems. It is more common in upland acid soil rice environments and can be a major source of yield loss. There are several measures to prevent loss due to Al-toxicity. Early ploughing is recommended right after the recession of floods at the end of the rainy season. Acid sulfate soils should have a shallow drainage system. Delay planting (to mobilize Al) provides crops with sufficient water to maintain reduced soil conditions and prevent the top soil from drying out.

8.10.2. Varietal Management

Al-tolerant cultivars, which accumulate less Al in their foliage and take up and efficiently use Ca and P in the presence of Al are recommended than going for soil reclamation. Tolerant varieties of different crops have been listed in Table 8.1.

8.10.3. Fertilizer Management

8.10.3.1. Liming the acid soils

Al-toxicity is an important growth limiting factor for plants in many acid soils. The problem is accounted by heavy application of acid-forming nitrogenous fertilizers. On acid upland soils with Al-toxicity, play special attention to Mg fertilizer. Grower should use domestic lime, if possible. It is necessary to apply phosphorus and potassium fertilizers. Acidifying nitrogenous sources should not be used. Al-toxicity decreases when sufficient Mg is supplied. Liming with $CaCO_3$ may not be sufficient, whereas, the application of dolomite instead of $CaCO_3$ not only raises the pH, but also supplies Mg. It also decreases the toxic Al ions in the soil and an increase in the amount of exchangeable Ca and available P (Badora and Filipek, 1998). Soil amended with wallboard-quality gypsum ($CaSO_4.2H_2O$) and $CaSO_3.5H_2O$ increased soil pH and exchangeable Al (Wendell and Ritchey, 1996). Sudangrass (*Sorghum bicolor*) seedlings grown in the amended soil showed higher root vigour.

Kieserite and Langbeinite can part of an integrated management strategy on acid upland soils to reduce Al-toxicity, but are less cost efficient than finely ground dolomite. Small amounts of Kieserite and Langbeinite (50 kg/ha) may have an effect similar to that of liming with more than 1000 kg $CaCO_3$. There are various options for treating Al-toxicity (Table 8.4).

Table 8.4. Materials for treating Al-toxicity soils (Dobermann and Fairhurst, 2000)

Name	Formula	Content	Remarks
Lime	$CaCO_3$	40% Ca	
Dolomite	$MgCO_3 + CaCO_3$	13% Mg, 21% Ca	Slow-acting, content of Ca and Mg varies
Gypsum	$CaSO_4.2H_2O$	23% Ca, 18% S	Slightly soluble, slow acting
Kieserite	$MgSO_4.7H_2O$	23% S, 16% Mg	Quick acting
Langbeinnite	$K_2SO_4.MgSO_4$	18% K, 11 Mg, 22% S	Quick acting
Partly acidulated rock phosphate	$Ca_2(PO_4)_2$	10-11% P	>1/3 water soluble
Rock phosphate finely powdered	$Ca_2(PO_4)_2$	11-17 % P, 33-36% Ca	Very slow acting (25-30% P2O5)
Single super phosphate	$Ca(H_2PO_4)_2.H_2O + CaSO_4.2H_2O$	12% S, 7-9% P, 13-20% Ca	Soluble, quick acting

8.10.3.2. Boron (B) application

Root growth inhibition is an early symptom of A-toxicity and B deficiency. Thus, application of B in the soil may promote root penetration in Al-stressed soils. Lenoble *et al.* (1996a, b) suggested that B concentrations may need to be increased under acidic soils by incorporating B in the soil. Application of B protects against Al-inhibition of root growth in *Medicago sativa* cv. Hy-Phy (Lenoble *et al.*, 1996b), and *Cucurbita pepo* cv. Sunbar (Lenoble *et al.*, 1996a).

8.10.3.3. Fluoride application

Crop production in acid soils may also be improved by small fluoride application in the form of fluoroapatite (Tang *et al.*, 1993b). They observed reduction of Al-toxicity in rice by addition of fluoride anions in the nutrient solutions, which resulted increase in available phosphate and pH. Fluoride in nutrient solution with high Al^{3+} concentration (20 ppm) induced an increase nutrient absorption and dry matter production in rice. High-F concentration (15 ppm) did not reduce root growth and development.

8.10.3.4. Phosphate application

Application of rock phosphate in acid soils increased root length of wheat seedlings (Hu *et al.*, 1995). They observed increased soil pH and Ca content, and decreased Al content in the amended soils.

8.10.3.5. Organic manuring

Green manuring acid soils may reduce the effect of Al-toxicity. Hairiah *et al.* (1996) identified eight species, *viz. Gmelina arborea, Melastoma* sp., *Syzygium aqueum, Peroema canescens, Calliandra calothysus, Peltophorum dasyrachis, Leucaena leucocephala,* and *Flemingia congesta* can be used as green biomass in acid soils to improve the soil status. Soil application of farm yard manure (FYM) and other organic composts detoxify Al-toxicity in acidic soils (Badora and Filipek, 1998; Castillo *et al.*, 1999). Combined application of lime and organic matter increased the amount of root, resulting in increasing dry weight of maize (Vannavanich and Vannavanich, 2003). Chen *et al.* (2002) also found that combined application of pig manure and lime increased soil pH, decreased the content of exchangeable Al and increased chlorophyll content and photosynthetic rate and dry weight of plant top.

8.11. BIOTECHNOLOGY

In spite of the explosive development, biotechnology in the last 20 years, its use for obtaining new abiotic stress tolerant varieties has been very limited.

The basic of scientific information of the molecular biology of the crops and animals is certainly the greatest constraint to the use of biotechnology for stress tolerance. Abiotic stress tolerance is the result of many genes expressed simultaneously or in a certain order. Often this expression concerns plant architecture and complex physiological responses control by group of genes. This limitation has not prevented some innovative biotechnology approaches to develop stress tolerance. Plant tissue culture, molecular markers and genetic engineering and transformation technologies facilitated in development of abiotic stress tolerant lines and varieties.

8.11.1. Plant Tissue Culture

Induction of calli or callus culture in growth medium with excessive Al, followed by plant regeneration was reported to induce metal toxicity tolerance (Van Sint Jan, 1997; Chowdhury *et al.*, 1998; Roy and Mandal, 2005), which might be heritable. Successful inheritance of tolerance was observed in tissue cultured derived plants for Al-tolerance (Coulombe and van Scoyoe, 1990; Van Sint Jan, 1997, Roy and Mandal, 2005). Plant cell cultures are known to undergo genetic instability. Although, the mechanism subtending such instability in many cases were partly elucidated in instability in many cases were partly elucidated in exploitation of somaclonal variation (Sibi, 1990) and callus selection for prospective production of lines tolerant to environmental stresses (Sun *et al.*, 1991, Chowdhury *et al.*, 2001).

Somaclonal variations induced *in vitro* were used to enhance tolerance towards aluminium (Al) toxicity in rice (Roy and Mandal, 2005; Van Sint Jan *et al.*, 1997), wheat (Dornelles *et al.*, 1997), Tomato (Taghian and Enany, 1996). Roy and Mandal (2005) exploited somaclonal variability and generated toxicity tolerant rice plantlets by applying selection pressure both at callus and plantlet regeneration levels (Fig. 8.3). Tolerant plants were obtained through *in vitro* screening of embryogenic calli derived from mature seeds on medium supplemented with toxic levels of Al in the form of $Al_2(SO_3)_4.18H_2O$. Seed germination (%), callus induction, plantlet regeneration (%) and callus health declined with increased levels of Al-stress (Table 8.5). Callus health deteriorated substantially at higher concentrations of Al and displayed partial or total necrosis. Plantlet regeneration varied greatly among the varieties as well as among the treatments. IR72 showed maximum plantlet regeneration among all. Higher concentration (60 ppm) of Al was found to be more toxic, which greatly reduced the plantlet regeneration from callus. The R_0 plantlets were grown in glasshouse condition. Based on appearance of bronzing symptoms, the tolerant R_1 plants were selected. The putative tolerant somaclones derived

R_1 and R_2 lines were evaluated in the fiberglass tank filled with Al-toxic soil. R_3 population was field evaluated in open range. A few lines derived from IR72 showed high yield and good plant type. Their progenies showed normal root growth under stressed environment in sand culture at R_3 generation. The studies prospects that *in vitro* screening would be an appropriate supplementary to conventional breeding in evolving Al-tolerant lines alike other abiotic stress tolerances as well as in creation of *de novo* synthesized Al-tolerance character in rice.

Fig. 8.3. *In vitro* screening at cellular level for increased Al-tolerance in rice (Roy, 2002): **(a)** Putative Al-tolerant calli proliferating on MS fortified with 1 mg L^{-1} 2,4-D (initially passed for 28 days on Al-stressed callus induction medium); **(b)** Calli redifferentiating into plantlets on RM consisting of MS with 1 mg L^{-1} kinetin, 1 mg L^{-1} BAP and 0.5 mg L^{-1} NAA; **(c)** Plantlets developed from Al-tolerant embryogenic calli on RM; **(d)** Full grown Al-toxicity tolerant plants on cement pot; **(e)** Selected IR72 somaclones at R_1 progeny grown under 30 ppm Al-stressed condition in fiberglass tank; **(f)** R_3 generation of selected IR72 somaclone in Experimental Farm. *(For colour version of this figure see page 538)*

Table 8.5. *In vitro* screening of mature seed derived calli for increased Al-toxicity tolerance in rice (Roy and Mandal, 2005)

Genotype	Al-stress $AlSO_4.18H_2O$ (ppm)	Seed germination (%)	Callus induction (%)	Callus health*		Plantlet regeneration	
				After 1st selection	After 2nd selection	Regeneration (%)	No. of plants developed
IR72	Control	92.00 abc#	81.01 abc	2	2	77.50 a	43
	30	90.25 abc	67.86 e	3	4	31.81 c	15
	60	92.25 abc	57.72 f	3	4	3.45 d	7
Taichung Sen Yu	Control	88.00 bc	75.64 cde	2	2	62.79 b	38
	30	89.75 abc	72.98 cde	3	4	4.87 d	6
	60	76.50 de	77.45 bcd	4	4	0.00 d	-
Annada	Control	98.00 a	85.00 ab	1	2	75.00 a	47
	30	95.25 ab	66.40 ef	3	4	5.77 d	11
	60	91.75 abc	68.12 de	3	4	0.00 d	-
S_1P_1 681032	Control	83.00 cd	86.74 a	2	2	71.69 a	33
	30	65.75 f	79.08 abc	3	4	7.14 d	12
	60	67.75 ef	77.86 bcd	4	4	2.17 d	6

*1-Excellent; 2-Good; 3-Medium; 4-Weak; 5&6-Poor

#Means having same letter in the column are not significantly different by DMRT at P= 0.058.11.2. F_1 anther culture

Somaclonal variation-induced multiple mutations were observed by Moon *et al.* (1997) in a progeny of the S1587 plant regenerated from type I calli of the aluminium-tolerant inbred maize line Cat100-6. The tolerance ability of Cat100-6 was due to Al exclusion from the cells. Their findings also indicated that the tolerance in Cat100-6 was controlled by a single nuclear, semi-dominant *Alm1* gene.

It is necessary to mention that *in vitro* screening for Al-toxicity tolerant variants at cellular level is obviously a difficult tusk, which requires standardization and adoption of efficient techniques/technologies to exclude artifacts (Wersuhn *et al.*, 1994). Regeneration of Al-toxicity tolerant rice plants was amply reported by many workers (Bouharmont *et al.*, 1991; Van Sint Jan, 1997; Chowdhury *et al.*, 1998; Roy and Mandal, 2005). *In vitro* screening for Al-toxicity tolerance in important field crops like rice, and possibly could become complementary to genotype improvement by classical methods. Field evaluation for acid soil is essentially required over years across locations for final recommendation. On the other hand, progeny testing and back crossing would be require establishing new resistant lines and restoration of all essential parental characters in an improved variety.

Double haploids increased the selection efficiency. Thus, it would be easier to develop quickly superior variety involving F_1 hybrid. The anther culture strategy also enables greater use of minor genes for resistance. The cumulative effect of which may provide more desirable resistance than provided by the major genes (Toenniessen and Khush, 1991). Rice lines have been developed through anther culture for saline soil. Thus, F_1 anther culture may be utilized in development of metal-toxicity tolerant in crop planrts.

8.11.2. Isozyme Analysis

Isozyme analysis in rice have progressed several lines of research involving gene mapping, gene regulation, developmental genetics and evaluation (Endo and Marishima, 1983). Chowdhury *et al.* (2001) demonstrated isozyme band-shift and activity changes in the tolerant plants recovered through *in vitro* screening in response to varying levels of Al-toxicity. Many bands showed declined activity in response to Al-stress, whereas, activity increased in many others. Disappearance of bands is indicative of their high susceptibility towards Al-toxicity. Appearance of new bands and disappearance of existing bands in stress condition possibly due to activation or inactivation of diverse domains in the genome seems to be involved in governing Al-toxicity tolerance. They also reported that some bands remained unaltered in the stressed environment. This might be due to their stability under Al-toxic condition and/or genesis from conservative domains of the genome through stable mRNA synthesis. It is assumed that several potential mechanisms for Al-toxicity tolerance involve Al-tolerant enzymes and evaluated enzyme activities (Taylor, 1991; Kochian, 1995). Chowdhury *et al.* (2001) also reported that dehydrogenase and esterase exhibited maximum stress responsive alteration, impaired pentose phosphate pathway and enhanced lipid peroxidation under Al-stress condition. It also highlights that Al-toxicity led to a wide variation in the isozyme profiles as evident in other abiotic stresses. The differential accumulation of isozymes with respect to varied stress intensities, thus, advocates their suitability as biochemical markers.

8.11.3. Marker Assisted Selection

Marker assisted selection (MAS) offers precision and efficient indirect selection of complex characters. Identification and integration of stress tolerant quantitative loci (QTL) into high yielding varieties would increase crop production in toxic soils. MAS can assist gene pyramiding to achieve multiple tolerances also. Molecular marker linked to Al-toxicity was mapped recently (Wu *et al.*, 2000). The use of MAS techniques in QTL analysis opened new opportunities to work with the tolerance for Al-toxicity. Refined mapping using near isogenic lines for the development of polymerase chain reaction (PCR)-based MAS are now possible and being perused. Rapid and reliable techniques were developed at International Rice Research Institute, Philippines to detect tolerance Al-toxicity. F_8 recombinant inbred line of IR1552/Azucena population was developed to map the genes for Al-toxicity tolerance using relative root growth (RRL) as a parameter.

The putative QTLs of rice for Al-toxicity tolerance were detected on chromosome 1, 5, 9 and 11 (Wu *et al.*, 2000; Nguyen *et al.*, 2002; Ma *et al.*, 2002; Nguyen *et al.*, 2003). A mapping population developed from the cross Kinmaze ´ DV85 was used by Xue *et al.* (2004, 2005) for mapping of QTL for Al-tolerance in rice. Five QTLs controlling Al-toxicity tolerance also have been detected by Xue *et al.* (2004) on chromosome 1, 5, 8, 9 and 11, respectively (Table 8.6). Individual QTL accounted for 8.64-18.60% of the phenotypic variance in the recombinant inbred lines (RILs) population. Direction of additive gene effect coincided with that predicted by phenotypes of the parents at 3 loci, *qRRE5*, *qRRE8*, and *qRRE11*. At these 3 loci, the DV85 alleles increased the tolerant to Al-toxicity, and at *qRRE1* and *qRRE9*, the Kinmaze alleles increased the tolerance to Al-toxicity. Some of best target QTLs identified in rice have been listed in Table 8.7.

Table 8.6. Putative QTLs detechted for Al-toxicity tolerance in recombinant inbred lines from the cross of Kinmaze and DV85 (Xue *et al.*, 2004)

Locus	Chromosome	Marker interval	LOD	R^2 (%)	Additive gene effect
qRRE-1	1	X317-X252	2.7	18.6	-0.10
qRRE-5	5	C246-X297	2.1	12.7	0.08
qRRE-8	8	C166-C1121	3.1	18.2	0.10
qRRE-9	9	R1751-C1121	2.3	8.6	-0.07
qRRE-11	11	C1172-G1465	3.2	16.8	0.09

LOD: Logodds, which is likelihood score for the presence of the QTL; **R^2:** which is the proportion (or percentage) of the phenotypic variation explained (PVE) by the QTL

Table 8.7. Target QTLs for marker-assisted backcrossing of Al-toxicity in rice

Trait	Chr	Marker	LOD	R^2	Population*	Reference
RRL	3	CDO1395-RG391	8.4	24.9	IR64/**O. refipogon**	Nguyen *et al.*, 2003
RRL	7	RZ629-RG650	5.4	22.5	IR64/**O. refipogon**	Nguyen *et al.*, 2003
RRL	1	CDO345	8.1	24.1	**CT9993**/IR62266	Nguyen *et al.*, 2002
RRL	8	C1121	8.2	28.7	**CT9993**/IR62266	Nguyen *et al.*, 2002
RRL	12	RG9	6.8	20.0	IR1552/**Azucena**	Wu *et al.*, 2000

Chr: Chromosome; **LOD:** Logodds, which is likelihood score for the presence of the QTL; **R^2:** which is the proportion (or percentage) of the phenotypic variation explained (PVE) by the QTL; **RRL:** Relative root length; *: Parent in bold is considered source of the desirable allele.

A major QTL accounting for 85% of the phenotypic variation for Al-toxicity tolerance in wheat was mapped on chromosome *4D* (Ried and Aderson, 1996). Using Chinese Spring disomic substitution lines, Luo and Dvorak (1996) also mapped a dominant Al-toxicity tolerance gene on chromosome *4D*. Brondani and Paiva (1996) reported association of RFLP markers with Al-tolerant and sensitivity and suggested to use this technique in marker-assisted selection in plant breeding for Al-tolerance.

8.11.4. Transgenic Development

Biotechnological applications, especially transgenic plants, probably hold the most promise in augmenting agricultural production in the first decades of the next millennium. However, the application of these technologies to the agriculture of tropical regions where the largest areas of low productivity are located, and where they are most needed, remains a major challenge. In this paper, some of the important issues that need to be considered to ensure that plant biotechnology is effectively transferred to the developing world are discussed. The subject may be considered to develop aluminium toxicity-tolerant crop plants.

At molecular level, identification of tolerant genes to Al-toxicity remains scanty. However, development of transgenic tobacco plants, tolerant to Al-toxicity was reported by Pan *et al.* (1994). Mexican scientist also have developed transgenic rice conferring tolerance to high concentration of Al (Conway and Toenniessen, 1999), which could have immense prospects in genetic management of Al-toxicity. Some of other works on Al-tolerant transgenic development have been given in Table 8.8.

Table 8.8. Transgenic plants with gene conferring Al- and Fe-toxicity tolerance

Gene	Gene action	Species	Phenotypic expression	Reference
MsFer	Ferritin (Iron storage)	*Nicotiana tubaccum*	Increased tolerance of oxidative stress caused by excess Fe	Deak *et al.*, 1999
parβ	Glutathione S-transferase	*Oryza sativa*	Tolerant to higher concentration of Al-toxicity	Conway and Toenniessen, 1999
parβ	Glutathione S-transferase	*Arabidopsis thaliana*	Protects against Al-toxicity and oxidative stress	Ezaki *et al.*, 2000

8.12. REFERENCES

Akeson MA, Munns DW, Baru RG. 1989. Absorption of Al^{3+} to phosphatidylcholine vesicle. Biochem Biophys Acta. 986: 33-40.

Ali SMF. 1973. Influence of cations on aluminium toxicity in wheat (*Triticum aestivum* Vill. Host). Ph. D. Thesis, Oregon State University, Corvalis.

Andrade LRM de, Ikeda M, Ishizuka J, De-Andrade LRM. 1996. Effect of nitrogen sources on aluminium toxicity in wheat varieties differing in tolerance to aluminium. Soil Sci Plant Nutr. 42(3): 651-657.

Aniol A. 1991. Genetics of acid tolerant plants. In: Plant Soil Interaction at Low pH. Kluwer Academic Publishers, The Netherlands. pp. 1007-1017.

Aniol AM. 1995. Physiological aspects of aluminum tolerance associated with the long arm chromosome *2D* of the Wheat (*Triticum aestivum* L.) genome. Theor Appl Genet. 91: 510-516.

Arihara A, Kumagai R, Koyama H, Ojima K. 1991. Aluminium tolerance in carrot (*Daccus carota* L.) plants regenerated from selected cell cultures. Soils Sci Plant Nutr. 37: 699-705.

Badora A, Filipek T. 1998. An assessment of rehabilitation of strongly acidic sandy soils. In: Towards Sustainable Land Use. Furthering cooperation between people and institutions. Volume 1. Peoc of the Intrnl Soil Conservation Organisation, Bonn, Germany, 26-30 August 1996. Advances in Geoecology. No. 31: 681-688.

Baggie I, Zapata F, Sanginga N. 2002. Genotypic respose to aluminium toxicity of some rice. Soil, Water and Nutrient Mangement, June 2002. pp. 42-43.

Baligar VC, Wright RJ, Kinraide TB, Foy CD, Elgin JH Jr. 1988. Differential responses of forage legumes to aluminium.. J Plant Nutr. 11: 549-561.

Basu V, Godbold D, Taylor GJ. 1994. Aluminium resistance in *Triticum aestivum* associated with enhanced exudation of malate. J Plant Physiol. 144: 474-753.

Bennet RJ, Breen CM, Fey MV. 1985. Aluminium-induced changes in the morphology of the quiescent centre, proximal meristem and growth region of the root of *Zea mays*. S Afr Tydskr Planick. 51: 355-362.

Bennet RJ, Breen CM, Fey MV. 1987. The effects of aluminium on root cap function and root development in *Zea mays* L. Environ Exp Bot. 27: 91-104.

Bennet RJ, Breen CM, Fey MV. 1991. The aluminium signal: new dimensions of aluminum tolerance. Plant and Soil. 134: 153-166.

Blamey FPC, Edmeades DC, Wheeler DM. 1990. Role of cation –exchange capacity of differential aluminium tolerance of *Lotus* species. J Plant Nutr. 13: 729-744.

Bollard EG, Butler GW. 1966. Mineral nutrition of plants. Annu Rev Plant Physiol. 17: 77-112.

Borgonovi RA, Schaffert RE, Pitta GVE, Magnavaca R, Aliue SUMC. 1985. Aluminium tolerance in sorghum. In: Genetics Aspects of Plant Mineral Nutrition, Madison, USA. pp. 16-20.

Bouharmont J, Dekeyser A, Van Sint Jan V, Dogbe YS. 1991. Application of somaclonal variation and *in vitro* selection to rice improvement. In: Rice Genetics II, IRRI, Philippines. pp. 271-277.

Bradshaw AD. 1974. The evolutionary lesions of metal tolerance. Heredity. 33: 450.

Brondani C, Paiva E. 1996. RFLP analysis of aluminium toxicity tolerance on maize chromosome 2. Pesquisa Agropecuaria Brasileria. 31(8): 575-579.

Budikova S, Uzik M. 1999. Structural reactions of different crop plant roots to aluminium toxicity. Nove poznatky z genetiky a sl'achtenia pol'nohospodarskych rastlin. Zbornik z 5. odbborneho seminara, Vyskumny Ustav Rastlinnej Vyroby, Piest'any, 8 december 1999. pp. 155-159.

Bunta G. 2000. Resulta regarding genetic control of wheat tolerance to aluminium ion toxicity. Analele Institutului de Cercetari pertru Cereale si Plante Tehnice, Fundulea. 67: 27-38.

Cakmak I, Horst WJ. 1991. Effect of aluminium on lipid peroxidation, superoxide dismutase, catalase, and peroxidase activities in the root tips of soybean (*Glycine max*). Physiol Plant. 83: 463-468.

Camargo CE de O, Felicio JC, Ferreira-Filho AWP, De-Freitas JG. 1995. Durum wheat: tolerance of aluminium, manganese and iron toxicity in nutrient solutions. Braagantia. 54(2): 371-383.

Camargo CE de O. 1998. Genetic control aluminium toxicity tolerance in wheat in nutrient solutions. Bragantia. 57(2): 215-225.

Camargo CE de O. 1999. Yield component and tolerance to aluminium toxicity in hybrid populations of wheat: variance, heritability and correlation estimates. Ciencia e Agrotechnologia. 23(3): 500-509.

Campbell KAG Jr, Carter TE, Anderson JM. 1987. Screening for aluminium tolerance in soybeans: A comparison of cell culture, seedling and whole plant methods. In: 79th Annu Meet Amer Soc Agron, Madison, USA. p. 56.

Campbell KAG, Caster TE Jr, Anderson JM. 1989. Aluminium tolerance in soybean callus culture screening methods. Soybean Genet Newsl. 16: 191-193.

Campbell TA, Xia ZL, Jacson PR, Baligar VC. 1994. Diallel analysis of tolerance to aluminium in alfalfa. Euphytica. 72: 157-162.

Castillo AE, Carbonell RM, Vazquez S. 1999. Influence of organic amendment on the exchangeable aluminium of an Oxisol. Ciencia del Suelo. 17(1): 58-59.

Chen M, Chen YH, Shen ZG, Shen QR. 2002. Amelioration of aluminium toxicity with pig manure in acid red soil. Plant Nutr Fer Sci. 8(2): 176.

Chowdhury B, Mandal AB, Bandopadhyaya AK. 1998. Development of Al-toxicity tolerant rice through *in vitro* screening at cellular level. Asia Pacific J Mol Biol Biotechnol. 6: 61-67.

Chowdhury B, Sheeja TE, Mandal AB. 2001. Modulation in isozyme profiles in relation to aluminium toxicity tolerance in rice. J Genet. 55: 111-118.

Ciamporova M, Baluska F. 2000. Divers responses of root cell structure to aluminium stress. Plant Soil. 226(1): 113-116.

Clark RB, Brown JC. 1974. Differential phosphorus uptake by phosphorus stressed corn inbreds. Crop Sci. 14: 5-8.

Clark RB, Pier HA, Knudsen D, Maranville JW. 1981. Effect of trace element deficiencies and excesses on mineral nutrition in sorghum. J Plant Nutr. 3: 357-374.

Clark RB. 1973. Twentieth Annual Corn an Sorghum Research Conference. Am Seed Trade Assoc, Washington DC. pp. 144-160.

Clarkson DT, Sanderson J. 1969. The uptake of a polyvalent cation and its distribution in root apices of *Allium cepa*. Tracer and autoradio-graphic studies. Planta. 89: 136-154.

Cocker KM, Evans DE, Hodson MJ. 1998. The amelioration of aluminium toxicity by silicon in wheat (*Triticum aestivum* L.): malate exudation as evidence for an *in planta* mechanism. Planta. 204(3): 318-323.

Conner AJ, Meredith CP. 1985a. Simulating the mineral environmental of aluminium toxic soil in plant cell culture. J Exp Bot. 36: 870-880.

Conner, AJ, Meredith, CP. 1985b. Strategies for the selection and characterization of aluminium-resistant variants from cell cultures of *Nicotiana plumbaginifolia*. Planta. 166: 466-473.

Conner, AJ, Meredith, CP. 1985c. Large scale selection of aluminium-resistant mutants from plant cell culture: expression and inheritance in seedlings. Theor Appl Genet. 71: 159-165.

Conway G, Toenniessen G. 1999. Feeding the world in the twenty-first century. Nature (London). 402: 55-58.

Costa de Macedo C, Kinet JM, Van Sint Jan V. 1997. Effects of duration and intensity of aluminium stress on growth parameter in 4 rice genotypes differing in Al-sensitivity. J Plant Nutr. 20(1): 181-193.

Coulombe EJ, van Scoyoe SW. 1990. *In vitro* selection methods for aluminium tolerance in alfalfa. Plant Breed Abstr. 60: 58-59.

Cuenca G, Herrera R, Medina E. 1990. Aluminium tolerance in trees of tropical cloud forest. Plant Soil. 125: 169-175.

Czembor HJ, Wiewiora M. 2001. The inheritance of tolerance to aluminium toxicity in chosen spring wheat (*Triticum aestivum* L.) cultivars. Biuletyn Institutu Hodowli I Aklimatyzacji Roslin. No. 220. pp. 45-51.

Czembor HJ. 2003. The inheritance of tolerance to aluminium toxicity in some spring wheat (*Triticum aestivum* L.) cultivars. Biuletyn Institutu Hodowli I Aklimatyzacji Roslin. No. 230. pp. 103-107.

Darko E, Ambrus H, Stefanovits Banyai E, Fodor J, Bakos F, Barnabas B. 2004. Plant Sci. 166(3): 583-591.

Datta SK. 2002. Recent development in transgenic for abiotic stress tolerance in rice. JIRCAS Working Report. pp. 43-53.

Deak M, Horvath GV, Davletova S, Torok K, Sass L, Vass I, Barna B, Kiraly Z, Dudits D. 1999. Plants ectopically expressing the iron-binding protein ferratin are tolerant to oxidative damage and pathogen. Nature Biotechnol. 17: 196-197.

Delhaize E, Ryan PR, Randal PJ. 1993. Aluminium tolerance in wheat (*Triticum aestivum* L.). II. Aluminium-stimulated expression of malic acid from root apices. Plant Physiol. 103: 695-702.

Delhaize E, Ryan PR. 1995. Aluminium toxicity and tolerance in plants. Plant Physiol. 107: 315-321.

Dobermann A, Fairhurst T. 2000. Nutrient disorder and nutrient management. In: Handbook Series. Potash Phosphate Institute of Canada and International Rice Research Institute, Philippines. pp. 191.

Dornelles ALC, Carvalho FIF de, Federizzi LC, Sereno MJC de M, Handel CL, Mittelmann A. 1997. Somaclonal variation for aluminium toxicity tolerance and gibberellic sensitivity in wheat. Pesquisa Agropecuaria Brasileira. 32(2): 193-200.

Durieux RP, Jackson WA, Kamprath EJ, Moll RH. 1993. Inhibition of nitrate uptake by aluminium in maize. Plant Soil. 151: 97-104.

Eckert MI, Viegas J, Osorio EA, Silva JB da, Brauner JL, Da Dilva JB. 1996. Aluminium toxicity in barley (*Hordium valgare* L.) root tips. Brazilian J Genet. 19(3): 429-433.

Endo T, Morishima H. 1983. Current status of isozyme research in individual plant species. 6. Rice. In: Isozyme in plant genetics and breeding, Part B, Tanskley SD and Orton TJ (eds). Elsivior, Amsterdam. pp. 129-146.

Ezaki B, Gardner RC, Ezaki Y, Matsumoto H. 2000. Expression of aluminium-induce gene in transgenic *Arabidopsis* plants can ameliorate aluminium stress and/or oxidative stress. Plant Physiol. 122: 657-665.

Ezaki B, Tsugita S, Matsumoto H. 1996. Expression of a moderately anionic peroxidase is induced by aluminium treatment in aluminium ion stress. Physiol Plant. 96: 21-28.

Fageria NK, Zimmerman FJP. 1979. Screening rice varieties for resistance to aluminium toxicity. Pesquisad Agropecuaria Bras. 14: 141-147.

Farina MPW, Mendes P, Gevers HO, Channon P. 1982. Differential tolerance to soil acidity among several S African maize genotypes. Crop Prod. 11: 133-134.

Ferreira R de P, Cruz CD, Sediyama CS, Pinheiro BS, de P Ferreira R. 1999. Inhertance of aluminium toxicity tolerance on rice based on diallelic analysis. Pesquisa Agropecuaria Brasileria. 34(4): 615-621.

Ferreira R de P, Sediyama CS, Cruz CD, Freire MS. 1997. Inheritance of tolerance of aluminium toxicity in rice based on analyses of means and variances. Pesquisa Agropecuaria Brasileria. 32(5): 509-515.

Fleming AL, Schwartz JW, Foy CD. 1974. Aluminium toxicity in plants. Agron J. 66: 715-719.

Foy CD, Fleming AL, Armiger WJ. 1969. Aluminium tolerance in soybean varieties in relation to calcium nutrition. Agron J. 61: 505-511.

Foy CD, Fleming AL. 1982. Aluminium tolerance of two wheat cultivars related to nitrate reductase activities. J Plant Nutr. 5: 1313-1333.

Foy CD, Lee EH, Wilding SB. 1987. Differential aluminium toleraces of two barley cultivars related to organic acids in their roots. J Palnt Nutrition. 10: 1089-1101.

Foy CD, Scott BJ, Fisher JA. 1988. Genetic differences in plant tolerance to manganese toxicity. In: Manganese in Soils and Plants, Graham RD, Hannm RJ, Uren NC (eds). Kluwer Academic Publishers, Dordrecht, The Netherlands. pp. 293-307.

Foy CD. 1974a. Manganese and plants. In: Manganese. Nat Acad Sci Nat Res Council, Washington DC. pp. 51-76.

Foy CD. 1974b. Effect of aluminium on plant growth. In: The Plant Root and its Environment, Carson EW (ed). Charlottesville, Univ Press, Virginia. pp. 601-642.

Foy CD, Chaney RL, White MC. 1978. The physiology of metal toxicity in plant. Annu Rev. Plant Physiol. 29: 511-566.

Foy CD. 1984. Physiological effect of hydrogen, aluminium and manganese toxicities. In: Acid in Soil Acidity and Liming, 2nd edition, Adams F (ed). American Society of Agroniomy, Madison, WI. pp. 57-97.

Foy CD. 1988. Plant adaptation to acid, aluminium toxic soils. Commun Soil Sci Plant Anal. 19: 959-987.

Foy CD. 1996. Tolerance of barley cultivars to acid, aluminium-toxic subsoil related to mineral element concentrations of their shoots. J Plant Nutr. 19: 1361-1380.

Frageria NK, Barbosa-Fillo MP. 1983. Upland rice varieties reactions to aluminium toxicity on a oxisol in central Brazil. Int Rice Res Newsl. 8: 18-19.

Freire JRJ. 1976. Inoculation of soybean. In: Exploiting the Legum-*Rhizobium* Symbiosis in Tropical Agriculture, Vincent JM, Wintney AS, Bose J (eds). Col Tropical Agric Misc Pub, Deptt Agron Annu Soil Sci, University of Hawaii, Honolulu. pp. 145.

Frye SU, Anderson IC. 1978. Effects of aluminium on elongation and DNA synthesis of wheat roots (Abstr). In: Agron Abstr, Madison, USA, ASA. pp. 152-153.

Furlani RR, Clark RB. 1981. Screening sorghum for aluminium tolerance in nutrient solution. Agron J. 73: 587-594.

Gartside DW, McNeilly T. 1974. The potential evaluation of heavy metal tolerance in plants. II. Copper tolerance in normal populations of different plant species. Heredity. 32: 335.

Grabski S, Schindler M. 1995. Aluminium indices rigor within the action net work of soybean cells. Plant Physiol. 108: 897-901.

Gregorio GB, Senadhira D, Mendoza RD, Manigbas NL, Roxas JP, Guerta CQ. 2002. Progress in breeding for salinity tolerance and associated abiotic stresses in rice. Field Crops Res. 76: 91-101.

Gu MH, Li XF. 2002. Alleviation of aluminium toxicity to rice by silicon and its mechanisms. Plant Nutr Fer Sci. 8(3): 360-366.

Gunse B, Garzon T, Bercelo J. 2003. Study of aluminium toxicity by vital staining profiles in four cultivars of *Phaselus vulgaris* L. J Plant Physiol. 160(2): 1447-1450.

Hairiah K, Adawiyah R, Widyaningsih J. 1996. Amelioration of aluminium toxicity with organic matter: selection of organic matter based on its total cation concentration. Agrivita (Special Issue: Biological Management of Soil Fertility for Sustainable Agriculture on an Ultisol). 19(4): 158-164.

Hanson WD, Kamprath EJ. 1979. Selection for aluminium tolerance in soybean based on seedling-root growth. Agron J. 71: 581-586.

Hao LN, Lfuo HT. 1989. Effects of aluminium on physiological functions of rice seedlings. Acta Bot Sinica. 31: 847-853.

Haug A, Caldwell CR. 1985. Aluminium toxicity in plants: The role of root plasma membrane and calmodulin. In: Frontiers of Membrane Research in Agriculture, St Jhon JB, Berlin E, Jackson E, Ottawa PC (eds). Rowman and Allanheld. pp. 359-381.

Haug A, Shi B, VitorelloV. 1994. Aluminium interaction with phosphoinsitide-associated signal transduction. Arch Toxicol. 68: 1-7.

Haug A. 1984. Molecular aspects of aluminium toxicity. CRC Crit Rev Plant Sci. 1: 345-373.

Henning SJ. 1975. Aluminium toxicity in the primary meristem of wheat roots. Ph.D. thesis, Oregon State Univ, Corvallis, Oregon. USA.

Howeler RH, Cadavid LF. 1976. Screening of rice cultivars for tolerance to Al-toxicity in nutrient solution as compared with a field screening method. Agron J. 68: 551-555.

Hu HQ, Huang QY, Li XY, Xu FL. 1995. The diminishing effect of phosphate rock application on aluminium toxicity in acid soil. Scientia Agricultura Sinica. 28(2): 51-57.

Huang CY, Shen B. 2003. Amelioration of aluminum toxicity to barley by silicon application. Plant Nutr Fer Sci. 9(!): 98-101.

Huang JW, Grunes DL, Kochian LV. 1992a. Aluminium effects on the kinetics of calcium uptake into cells of the wheat root apex: Quantification of calcium fluxes using a calcium-selective vibrating microelectrode. Planta. 188: 414-421.

Huang JW, Grunes DL, Kochian LV. 1993. Aluminium effect on calcium ($^{45}Ca^{2+}$) translocation in aluminium-tolerant and aluminimum-sensitive wheat (*Triticum aestivum* L.) cultivars. Plant Physiol. 102: 85-93.

Huang JW, Grunes DL, Kochian LV. 1995. Aluminium and calcium transport interaction in intact roots plasmalemma vesicle from Al-sensitive and tolerant wheat cultivars. Plant Soil. 171: 131-135.

Huang JW, Pellet DM, Papanik LA, Kochian LV. 1996. Aluminium interaction with voltage dependant calcium treatment in plasmalemma membrane vesicle isolated from root of aluminium sensitive and resistant wheat cultivars. Plant Physiol. 110: 561-569.

Huang JW, Shaff JE, Grunes DL, Kochian LV. 1992b. Aluminium effects on calcium fluxes at the root apex of aluminium-tolerant and aluminium-sensitive wheat cultivars. Plant Physiol. 98: 230-237.

IRRI. 1977. Annual Report. International Rice Research Institute, Manila, Philippines. pp. 548.

IRRI. 1990. Program Report. International Rice Research Institute, Manila, Philippines.

Jaffe W, Rojar M. 1994. Abiotic stress and biotechnology in Latin America. Biotech Development Monit. 18: 6-7.

Jariel DM, Wallace SU, Jones US, Samonta HP. 1990. Aluminium tolerance in maize: Minimizing the pH changes in screening solutions. J Plant Nutr. 13: 995-1016.

Jones DL, Kochian LV. 1995. Aluminium inhibition of the inositol 1, 4, 5-triphosphate signal transduction pathway in wheat roots: A role of aluminium toxicity. Plant Cell. 7: 1913-1922.

Kerridge PC, Dawson MD, Moore DP. 1971. Separation of degrees of aluminium tolerance in wheat. Agron J. 63: 586-591.

Khatiwada SP, Senadhira D, Carpena AL, Zeigler RS, Fernandez PG. 1996. Variability and genetics of tolerance for aluminium toxicity in rice (*Oryza sativa* L.). Theor Appl Genet. 93: 738-744.

Kidd PS, Lugany M, Poschenrieder C, Gunse B, Barcelo J. 2001. The role of root exudate in aluminium resistance and silicon-induced amelioration of aluminium toxicity in three varieties of maize (*Zea mays* L.). J Exp Bot. 52(359): 1339-1352.

Kinraide TB. 1988. Proton extrusion by wheat roots exhibiting severe aluminium toxicity symptoms. Plant Physiol. 88: 418-423.

Klimashevskii EL, Markova Yu A, Zyabkina SM,Zirenki GK, Zolotukin TE, Pavolva SE. 1976. Aluminium absorption and localization in tissues of different pea varieties. Fiziol Biokhim Kul't Rust. 8: 396-401.

Kochian LV. 1995. Cellular mechanisms of aluminium toxicity and resistance in plants. Ann Rev Plant Physiol Plant Mol Biol. 46: 237-260.

Konzak CF, Polle E, Kittrick JA. 1976. Screen several crops for aluminium tolerance. In: Plant Adaptation to Mineral Stress in Problem Soils, Write MJ, Ferrari AS (eds). Cornell Univ Agric Exp Stn, Ithaca, NY. pp. 311-327.

Krizek DT, Foy CD, Mirecki RM. 1997. Influence of aluminium stress on shoot and root growth of contrasting genotypes of Coleus. J Plant Nutr. 20: 1045-1060.

Lafever HN, Campbell LG, Foy CD. 1977. Differential response of wheat cultivars to aluminium. Agron J. 69: 563-568.

Lafever HN. 1981. Genetic differences in plant response to soil nutrient stress. J Plant Nutr. 4: 89-109.

Lazof DB, Goldsmith JG, Rufty TW, Linton RW. 1994a. Rapid uptake of aluminium into cells of intact soybean root tips. Plant Physiol. 106: 1107-1114.

Lazof DB, Goldsmith JG, Rufty TW, Linton RW. 1996. The early entry of Al to cell of intact soybean roots. Plant Physiol. 112: 1289-1300.

Lazof DB, Rincon M, Rufty TW, MacKnow CM, Carter TE. 1994b. The preparation of cryosection from plant tissue: An alternative method for secondary ion mass spectrometry studies of nutrient traces and trace metals. J Plant Microsc. 176: 99-109.

Lazof DB, Rincon M, Rufty TW, MacKown CM, Carter TE. 1994c. Aluminium absorption in morphological regions of intact soybean root and associated effect on $^{15}NO_3$- influx. Plant Soil. 164: 291-297.

Lenoble ME, Blevins DG, Miles RJ. 1996b. Prevention of Aluminium toxicity with supplemental boron. II. Stimulation of root growth in acidic, high-aluminum subsoil. Plant Cell Environ. 19(10): 1143-1148.

Lenoble ME, Blevins DG, Sharp RE, Cumbie BG. 1996a. Prevention of Aluminium toxicity with supplemental boron. I. Maintenance of root elongation and cellular structure. Plant Cell Environ. 19(10): 1132-1142.

Ligon WS, Pierre WH. 1932. Soluble aluminium studies. II. Minimum concentration found to be to corn, sorghum and barley in culture solution. Soil Sci. 34: 307-321.

Linderberg S. 1990. Aluminium interaction with K^+ ($^{86}Rb^+$) and $^{45}Ca^{2+}$ fluxes in three cultivars of sugar beet (*Beta valgaris*). Physiol Plant. 79: 275-282.

Lu-Ning, Hou-Tian. 1989. Effects of aluminium on physiological functions of rice seedlings. Acta Bot Sinica. 31:847-853.

Luo MC, Dvorak J. 1996. Molecular mapping of an aluminium tolerant locus on chromosome *4D* of Chinese Spring wheat. Euphytica. 91: 31-35.

Lutz JA Jr, Hawkins GW, Genter CF. 1971. Differential responses of corn inbreds and single crosses to certain properties of an acid soil. Agron J. 63: 303-305.

Ma JF, Hiradate S, Nomoto K, Iwashita T, Matsumoto H. 1997. Internal; detoxification mechanism of Al in *Hydragena*. Plant Physiol. 113: 1033-1039.

Ma JF, Shen RF, Zhao ZQ, Wissuwa M, Takeuchi Y, Ebitan T, Yano M. 2002. Response of rice to Al stress and identification of quantitative trait loci for Al tolerance. Plant Cell Physiol. 43(2): 652-659.

Ma JF, Zheng SJ, Li XF, Takeda K, Matsumoto H. 1997. A rapid hydrophonic screening for aluminium tolerance in barley. Plant and Soil. 191: 133-137.

MacDiarmid CW, Gardner RC. 1996. Al toxicity in Yeast. A role for Mg? Plant Physiol. 112: 1101-1109.

MacDonald TL, Humpreys WG, Martin LB. 1987. Promotion of tubulin assembly by aluminium ion *in vitro*. Science. 236: 183-186.

Mackay AD, Caradus JR, Wewala S. 1991. Aluminium tolerance of forage species. In: Plant-soil Interactionsat Lower pH, Wright RJ, Baligar VC, Murrman RP (eds). Kluwer Academic Publishers, Dordrecht. pp. 25-30.

Mandal AB, Basu AK, Roy B, Sheeja TE, Roy T. 2004. Genetic management for increased tolerance to aluminium and iron toxicities in rice- A review. Indian j Biotechnol. 3(3): 359-368.

Martin RB. 1988. Bioinorganic chemistry of aluminium. In: Metal Ion in Biological system: Aluminium and its Role in Biology, Vol. 24, Siegel H, Siegel A (eds). Marcel Dekker, New Yourk. pp. 1-57.

Matsumoto H, Hirasawa E. 1976. Localization of aluminum intea leaves. Plant Physiol. 125: 169-175.

Matsumoto H, Morimura S, Takahashi E. 1977. Binding of aluminium to DNA of DNP (deoxyribonucleoprotein) in pea root nuclei. Plant Cell Physiol. 18: 987-993.

Meredith CP. 1987a. Response of cultured tomato cells to aluminium. Plant Sci. 12: 17-24.

Meredith CP. 1987b. Selection and characterization of aluminium-resistant variants from tomato cell cultures. Plant Sci Lett. 12: 25-34.

Minocha R, Minocha SC, Long SL, Shortle WC. 1992. Effects of aluminium on DNA synthesis, cellular polyamines, polyamine biosynthetic enzymes and inorganic ions in cell suspension cultures of a woody plant, *Catharanthus roseus*. Physiol Plant. 85: 417-424.

Miyasaka SC, Kochian LV. 1989. Mechanisms of aluminium tolerance in wheat. Plant Physiol. 91: 1188-1196.

Mohanty S, Das AB, Das P, Mohanti P. 2004. Effect of a low dose of aluminium on mitotic and meotic activity, 4C DNA content, and pollen sterility in rice, *Oryza sativa* L. cv. Lalat. Ecotoxicology Environl Safty. 59(1): 70-75.

Moon DH, Ottoboni LMM, Souza AP, Sibov ST, Gaspar M, Arruda P. 1997. Somaclonal variation induced aluminium-sensitive mutant from an aluminium-inbred maize tolerant line. Plant Cell Reports. 16(10): 686-691.

Morikawa CK, Saigusa M. 2002. Si amelioration of Al toxicity in barley (*Hordium vulgare* L.) grown in two Andosols. Plant Soil. 240(1): 161-168.

Morimura S, Takahashi E, Matsumoto H. 1978. Z Pflanzenphysiol. 88: 395-401.

Mugwira LM, Patel SV. 1977. Root zone pH changes and ion uptake imbalances in triticle, wheat, and rye. Agron J. 69: 719-722.

Nawrot M, Szarejko I, Maluszynski M. 2001. Barley mutants with increased tolerance to aluminium toxicity. Euphytica. 120(3): 345-356.

Neogy M, Datta JK, Mukherji S, Roy AK. 2001. Effects of Aluminium on pigment content, Hill activity and seed yield in mungbean. Indian J Plant Physiol. 6(4): 381-385.

New PB, Kennedy IR. 1989. Regional distribution and pH sensitivity of *Agospirilum* associated with wheat roots in eastern Australia. Microbial Ecol. 17: 299-309.

Nguyen BD, Brar DS, Bui BC, Nguyen TV, Pham LN, Nguyen HT. 2003. Identification and mapping of the QTL for aluminium tolerance introgressed from the aluminium tolerance FURIPOGON Grif., into indica rice (*Oryza sativa* L.). Theor Appl Gnet. 106: 583-593.

Nguyen VT, Nguyen BD, Sarkarung S, Martinez C, Paterson AH, Nguyen HT. 2002. Mapping of genes controlling aluminium tolerance in rice: comparison of different backgrounds. Mol Genet Genomics. 267: 772-780.

Nichol BE, Oliveira LA, Glass ADM, Siddiqi MY. 1993. The effect of aluminium on the influx of calcium, potassium, ammonium nitrate and phosphate in an aluminium-sensitive cultivar of barley (*Hordium vulgare* L.). Plant Physiol. 101: 1263-1266.

Nichol BE, Oliveira LA. 1995. Effects of aluminium on the growth and distribution of calcium in root of an aluminium-sensitive cultivar of barley (*Hordium vulgare* L.) Can J Bot. 73: 1849-1858.

Nicol BE, Oliveira L, Class ADM, Siddiqui MV. 1991. The effects of short-and long-term aluminium treatment on potassium fluxes in roots of an aluminium-sensitive cultivar of barley. In: Plant Soil Interaction at Low pH, Wright RJ, Baligar VC, Murrmann RP (eds). Kluwer Academic Publishers, Dordrecht. pp. 741-746.

Ohki K. 1987. Aluminium stress on sorghum growth and nutrient relationships. Plant and Soil. 98: 195-202.

Ono K, Yamamoto Y, Hachiya A, Matsumoto H. 1995. Synergistic inhibition of growth by aluminium and iron of tobacco (*Nicotiana tabaccum* L.) cells in suspension culture. Plant Physiol. 36: 115-125.

Pan A, Yang M, Tie F, Li L, Chen Z, Ru B. 1994. Expression of mouse metallothionein-I gene confers cadmium resistance in transgenic tobacco plants. Mol Biol. 24: 341-351.

Parrot WA, Button JH. 1990. Aluminium tolerance in alfalfa as expressed in tissue culture. Crop Sci. 30: 387-389.

Pellet DM, Grumes DL, Kochian LV. 1995. Organic acid exudation as an aluminum-tolerance mechanism in maize *(Zea maize L.)* Planta. 196: 788-795.

Petolino JF, Collins GP. 1983. *In vitro* approaches to acid soil stress tolerance in tobacco. In: 75th Annu Meet Amer Soc Agron, Madison, USA. pp. 75.

Pintro J, Barloy J, Fallavier P. 1995a. Aluminium toxicity in corn plants in a low ionic strength nutrient solution. I. Discrimination of two corn cultivars. Revista Brasileira-de-Fisiologia Vegeta. 7(2): 121-128.

Pintro J, Barloy J, Fallavier P. 1995b. Aluminium toxicity in corn plants in a low ionic strength nutrient solution. II. Distribution of Al in the principal root tip zone. Revista Brasileira-de-Fisiologia Vegeta. 7(2): 129-134.

Plucinska GL, Zeigler H. 1995. The effect of aluminium on adenylate levels in Scots pine roots. Acta Physiol Plant. 17: 225-232.

Plucinska GL, Zeigler H. 1996. Changes in ATP levels in Scots pine needles during aluminium stress. Photosynthetica. 32: 141-144.

Prakash CS. 2000. Can genetically engineered crops feed a hungry world? San Francisco Chronicle, March 30, 2000.

Prathap S, Ramarao G, Reddy KB. 2002. Response of groundnut genotypes to aluminium toxicity in solution culture. India J Plant Physiol. 7(4): 396-400.

Prestes AM, Konzak CF, Hendrix JW. 1975. An improved seedling culture method for screening wheat for tolerance to toxic levels of aluminium. Agron Abdtr. 67: 60.

Ramaskevicene A, Kupcinskiene E, Sliesaravicius A. 2000. Aluminium sensivity among spring barley (*Hordium vulgare* L.) cultivars grown in Lithuania. Sodininkyste ir Darzininkyste. 19: 3(2) 419-428.

Reid DA. 1970. Genetic control of reaction to aluminium in winter barley. In: Proc 2nd Intr Barley Genet Symp, Washington State University Press. pp. 409-413.

Reid DA. 1971. Genetic control of reaction to aluminium in winter barley. In: Barley Genetics II. Proc 2nd Int Barley Genetics Synp, Nilan RA (ed.). Pullman, Washington State Univ Press. pp. 409-413.

Rengel Z. 1990. Competative Al^{3+} inhibition of net Mg^{2+} uptake by intact *Lolium multiflorum* roots II. Plant age effects. Plant Physiol. 93: 1261-1267.

Rhue RD, Grogan CO, Stockmeyer EW, Evert HL. 1978. Genetic coltrol of aluminium tolerance in corn. Crop Sci. 1063-1067.

Rhue RD, Grogan CO. 1977. Screening cotton for aluminium tolerance using different Ca and Mg concentrations. Agron J. 69: 755-760.

Rice WA. 1982. Performance of *Rhizobium meliloti* strain selected for low pH tolerance. Canadian J Palnt Sci. 62(4): 1941-1948.

Ried CR, Anderson JA. 1996. Linkage of RFLP markers to a aluminium tolerance gene in wheat. Crop Sci. 26: 905-909.

Rincon M, Gonalez RA. 1992. Aluminium partitioning in intact root tips of aluminium tolerant and sensitive wheat (*Triticum aestivum* L.) cultivars: Reduced aluminium accumulation and rapid translocation mechanisms operate in aluminium tolerant wheat roots. Plant Physiol. 99: 1021-1028.

Robinson DL, Rengel Z. 1991. Aluminium influences on magnesium uptake and on grass tendency potential of ryegrass. Curr Trop Plant Biochem Physiol. 10: 107-116.

Rorison IH. 1958. The effect of aluminium on legume nutrition. In: Nutrition of the Legumes, Hallsworth EC (ed). Scientific Publishers, Butterworths, London. pp. 43-58.

Roy AK, Sharma A, Talukder G. 1988. Some aspects of aluminium toxicity in plants. The Bot Rev. 54: 145-178.

Roy B, Mandal AB. 2005. Towards development of Al-toxicity tolerant lines in indica rice by exploiting somaclonal variation. Euphytica. 145(3): 221-227.

Roy B. 2002. Genetic studies on seed components and standardization of genetic transformation in rice. Ph. D. Thesis, Bidhan Chandra Krishi Viswavidyalaya, Mohanpur, Nadia, West Bengal, India. pp. 187-196.

Rufty TW, MacKnow CT, Lazof DB, Carter TE. 1995. Effects of aluminium on nitrate uptake and assimilation. Plant Cell Environ. 18: 1325-1331.

Ruftyikiri G, Ceclerck S, Dufy JE, Delvaux B. 2000. Arbuscular mychorrizal fungi might alleviate aluminium toxicity in banana plants. New Phytologist. 148(2): 343-352.

Ryan PR, Delhaize E, Randall PJ. 1995b. Malate efflux from root apices and tolerance to aluminium are highly correlated in wheat. Aust J Plant Physiol. 22: 531-536.

Ryan PR, Delhaize E, Randall PJ. 1995a. Characterization of Al-stimulated efflux of malate from the apices of Al-tolerant wheat roots. Planta. 196: 103-110.

Ryan PR, Dimoso JM Kochian LV. 1993. Aluminium toxicity in root: An investigation of spatial sensitivity and the role of the root cap. J Exp Biol. 44: 437-446.

Samac DA, Tesfaye M. 2003. Plant improvement for tolerance to aluminium in acid soils- a review. Plant Cell Tiss Organ Cult. 75: 189-207.

Schmohl N, Horst WJ, Romheld V. 2001. Cell-wall composition modulates aluminium toxicity. In: Plant Nutrition Food Security and Sustainability of Agro-ecosystems through Basic and Applied Research- Fourteenth Intrnl Plant Nutrition Colloquium, Hannover, Germany. pp. 262-263.

Scholl L van, Keltjens WG, Hoffland E, Breem N van. 2005. Effect of ectomycorryzal colonization on the uptake of Ca, Mg and Al by *Pinus sylvestris* under aluminium toxicity. Forest Ecology and Mangement. 215(1/3): 352-360.

Schutzendubel A, Polle A. 2002. Plant response to abiotic stress: Heavy metal-induced oxidative stress and protection by mycorrhization. J Exp Bot. 53: 1351-1365.

Seigel B, Haug A. 1983. Calmodulin-dependent formation of membrane potential in barley root plasma membrane vesicle: A biochemical model of aluminium toxicity in plants. Physiol Plant. 59: 285-291.

Severi A. 1997a. Aluminium toxicity in *Lemna minor* L.: Effect of citrate and kinetin. Environ Exp Bot. 37: 53-61.

Severi M. 1997b. Aluminium toxicity in *Lemna minor* L. Atti Soc Nat e Mat di Modena. 122: 95-108.

Sibi M. 1990. Genetic bases of variation from *in vitro* tissue culture. In: Somaclonal variation in crop improvement, Bajaj Y.P.S. (Ed), Springer–Verlag, Berlin, Heidelberg. pp. 271-287.

Sivaguru M, Paliwal K. 1993. Differential aluminium tolerance in some tropical rice cultivars. J Plant Nutr. 16: 1705-1716.

Sledge MK, Bouton JH, Dall'Agnoll M, Parrot WA, Kochert G. 2002. Identification and confirmation of aluminium tolerance QTL in diploid *Medicago sativa* subsp. *Coerulea*. Crop Sci. 42: 1121-1128.

Slootmaker IAJ. 1974. Tolerance to high soil acidity in wheat related species, rye and triticle. Euphytica. 23: 505-513.

Snowden KC, Gardner RC. 1993. Five genes induced by aluminium in wheat (*Triticum aestivum* L.) roots. Plant Physiol. 103: 855-861.

Snowden KC, Richards KD, Gardner RC. 1995. Aluniminium-induced genes. Induction by toxic metals, low calcium, and wounding and pattern of expression in root tips. Plant Physiol. 107: 341-348.

Sousa CAN, De sousa CAN. 1998. Classification of Brazilian wheat cultivars for aluminium toxicity in acid soils. Plant Breed. 117(3): 217-221.

Spehar CR. 1994. Aluminium tolerance in soybean genotypes in short-term experiments. Euphytica. 76: 73-80.

Stolen O, Anderson S. 1978. Inheritance of tolerance to low soil pH in barley. Hereditas. 88: 101-105.

Strid H. 1996. Effect of root zone temperature on aluminium toxicity in two cultivars of spring wheat with different resistance to aluminium. Physiologia Plantarum. 97(1): 5-12.

Sun Z, Sun L, Shu L. 1991. Utilization of somaclonal variation in rice breeding. In: Biotechnology in Agriculture and Forestry, Rice, Bajaj YPS (ed.), Springer-Verlag, Berlin, Heidelberg. 14: 328-446.

Taghian AS, El-Enany AE. 1996. Genotypic differences and alteration of protein patterns of tomato explants under aluminium stress *in vitro*. Assiut J Agril Sci. 27(4): 164-178.

Takeda K, Kariuda M, Itoi H. 1985a. Bluing of sepals colour of *Hydrangea macrophylla*. Phytochem. 24: 2251-2254.

Takeda K, Kubota R, Yogioka C. 1985b. Co-pigments in the bluing of sepal colour of *Hydrangea macrophylla*. Phytochem. 24: 1207-1209.

Tang VH, Houben V, Mbouti CN, Dufey JE. 1993b. Early diagnosis of resistance of rice (*Oryza sativa* L. cultivars) to aluminium toxicity. Agronomie. 13(9): 853-860.

Tang VH, Parijs B- van, Van Parijs B, Mensvoort MEF-van. 1993a. Effect of fluoride on aluminium toxicity in rice. In: Selected Papers of the Ho Chi Minh City Symp on Acid Sulphate Soils, Dent DL (ed.), Ho Chi Minh City, Viet Nam, March 1992. pp. 261-264.

Taylor GJ. 1991. Current views of the aluminium stress response: The physiological basis of tolerance. Curr Top Plant Biochem Physiol. 10: 57-93.

Taylor GJ. 1995. Over coming barriers to understanding the cellular basis of aluminum resistance. Plant Soil. 171: 89-103.

Thorton FC, Davey CB. 1983. Response of the clove-*Rhizobium* symbiosis to soil acidity and *Rhizobium* strain. Agron J. 75: 557-560.

Tonniessen GH, Khush GS. 1991. Prospect for the future in biotechnology and agriculture. In: Rice Biotechnology, Vol. 6, Khush GS, Toennessen GH (eds). CAB International/IRRI, Wallington, UK/Manila, Philippines. pp. 309-313.

Tulmann -Neto A, Camarago CE de O, Alves MC, Santos RR-dos, Freitas JG de, De Freitas JG, Dos Santos RR. 1995. Mutation breeding in wheat cultivar IAC-24 for resistance to disease. Pesquisa Agropecuaria Brasileira. 30(4): 497-504.

Van Sint Jan S, de Macedo CC, Kinel JM, Bouharmont J. 1997. Selection of Al-resistant plants from a sensitive rice cultivar, using somaclonal variation, *in vitro* and hydrophonic cultures. Euphytica. 97: 303-310.

Vannavanich D, Vannavanich Danai. 2003. Alleviation of aluminium toxicity in soil by using lime gypsum and organic matter. In: Proc of 41st Kasetsart University Annual Conf, 3-7 February, 2003. Subject: Plant and Agricultural Extension and Communication. pp. 42-51.

Vierstra R, Haug A. 1978. The effect of aluminium on the physical properties of membrane lipids in *Thermoplasma acidophilum*. Biochem Biophys Res Commun. 84: 138-143.

Vitorello VA, Haug A. 1996. Short term aluminium uptake by tobacco cells: Growth dependence and evidence for internationalization in a discrete peripheral region. Physiol Plant. 97: 536-544.

Wagatsuma T, Ishikawa S, Obata H, Tawaraya K, Katohad S. 1995. Plasma membrane of younger and outer cells in the primary specific site for aluminium toxicity in roots. Plant Soil. 171(1): 105-112.

Wagatsuma T, Kaneko M, Hyasaka Y. 1987. Destruction process of plant root cells by aluminium. Soil Sci Plant Nutr. 33: 161-175.

Wan UH, Nian H, Yan XL. 2001. Some indices of tolerance of soybean to aluminium and low phosphorus and their interaction. Plant Nutr Fer Sci. 7(2): 199-204.

Wendell RR, Ritchey KD. 1996. High-calcium flue gas desuifurization products reduce aluminium toxicity in wheat varieties differing tolerance to aluminium toxicity in an Appalachian soil. J Environ Quality. 25(6): 1401-1410.

Wersuhn G, Kalta T, Genapp R, Reinke G, Schulz D. 1994. Problem posed by *in vitro* selection for aluminium tolerance when using cultivated plant cells. Plant Physiol. 143: 92-95.

Wheeler DM, Edmeades DC, Christie RA, Gardner R. 1992. Comparison of techniques for determining the effect of aluminium on the growth and inheritance of aluminium tolerance in wheat. Plant Sci. 146: 1-8.

Wheeler DM. 1994. Effects of growth period, plant age and change in solution aluminium concentrations on aluminium toxicity in wheat. Plant Soil. 166(1): 21-30.

Wild H, Bradshaw AD. 1977. The evolutionary effects of metaliferous and other anomalous soils in South Central Africa. Evaluation. 31: 383.

Woolhouse HW.1969. Differences in the properties of the acid phosphatases of plant roots and their significancein the evolution of edaphic ecotypes, *In*: Ecological aspects of the mineral nutrition of plants, Rorison I.H. (Ed). Blackwell Publishers, Edinburg, Oxford, pp. 357–380.

Wu P, Liao CY, Hu B, Yi KK, Jin WZ, Ni JJ, He C. 2000. QTLs and epistasis for aluminium tolerance in rice (*Oryza sativa* L.). Theor Appl Genet. 100: 1295-1303.

Xue Y, Jiang L, Ma JF, Zhai HQ, Wan JM. 2004. Identification of quantitative traits loci for Al toxicity tolerance in rice (*Oryza sativa* L.). Rice Genet Newsl. 20: 60-62.

Xue Y, Jiang L, Zhang WW, Wang CM, Ma JF, Wan JM, Zhai HQ, Yoshimura A. 2005. Mapping of quantitative trait loci associated with aluminium tolerance in rice (*Oryza sativa* L.) using recombinant inbred lines. Acta Agronomica Sinica. 31(5): 560-564.

Yamamoto Y, Rikiishi S, Chang YC, Ono K, Kasai M, Masumoto H. 1994. Quantitative estimation of aluminium toxicity in cultured tobacco cells: Correlation between aluminium uptake and growth inhibition. Plant Cell Physiol. 35: 575-583.

Yamaoto Y. Masamoto K, Rikishii S, Hachiya A, Yamaguchi Y, Matsumoto H. 1996. Aluminium tolerance acquired during phosphate starvation in cultured tobacco cells. Plant Physiol. 112: 217-227.

Yoshida S, Forno DA, Cock JH, Gomez KA. 1976. Routine procedure for growing rice plants in culture solution. In: Laboratory Manual for Physiological Studies of Rice. Los Banos (Philippines): International Rice Research Institute, pp. 61-66.

Zhang WH, Rengel Z, Kuo J, Yan G. 1999. Aluminium effects on pollen germination and tube growth of *Chamelaucium uncinatum*: A comparisonwith other Ca^{2+} antagonists. Ann Bot. 84: 559–564.

Zhang X, Humphries A, Auricht G. 2007. Genetic variability and inheritance of aluminium tolerance as indicated by long root regrowth in lucerne (*Medicago sativa* L.). Euphytica. 157: 177-184.

9

Iron Toxicity Tolerance

9.1. INTRODUCTION

Iron (Fe) is important component of a plant's energy-power house chlorophyll. It also interacts with all the major enzymes and proteins at higher concentration in the soil, and it is toxic to plants. Fe-toxicity creates a range of nutrient disorders and deficiencies of P, K, Ca, Mg, Mn and Zn in plants. These elements were reported to play decisive roles in manifesting toxicity symptoms in rice (Sahu, 1968; Ottow *et al.*, 1983; Yamauchi, 1989; Sahrawat *et al.*, 1996). Excess absorption and translocation of Fe in the rice plants led to toxicity, which has been a major limiting factor in wetland rice. Average yield loss due to Fe-toxicity accounts to 50% and in fact ranges from 10-100% (Brigit *et al.*, 1993).

Fe-toxicity had been reported as one of the major soil constraint of lowland acid soils, inland valley swamps, costal swamps and irrigated lowlands in utisols and oxisols. Fe is also abundantly found in heavy soils (Tanaka and Yamaguchi, 1973). In the humid forest and moist Savana zones of Africa interflow of ferrous ion occurs from upper slops (Fig. 9.1). More than 50% lowland rice is being affected with Fe-toxicity in Siera Leone, Liberia, Guinea, Nigeria, Ivory Coast, and Senegal. It was also reported in Sri Lanka, Vietnam, Malaysia, India (Kerala, Orissa, West Bengal, and Andaman and Nicobar Islands), Brazil, Colombia, Madagascar and Indonesia (Kalimanta and

Sumatra), Philippinese (Ponnaperuma, 1978; Sahrawat and Singh, 1995). Young acid sulphate soil in the coastal areas also show rampant symptoms and low lying valleys of Andaman and Nicobar Islands constitute a vast area constrained with Fe-toxicity, which is seriously limiting rice production (Mandal and Roy, 2000). In the People's Republic of China gleyic paddy soils are estimated to cover 7.5-8.0 m ha, where ferrous iron toxicity in the soil reduce rice yield by 10-20% depending on the intensity of toxicity and tolerance of the rice cultivar (Li *et al.*, 1991, Wan *et al.*, 2003a).

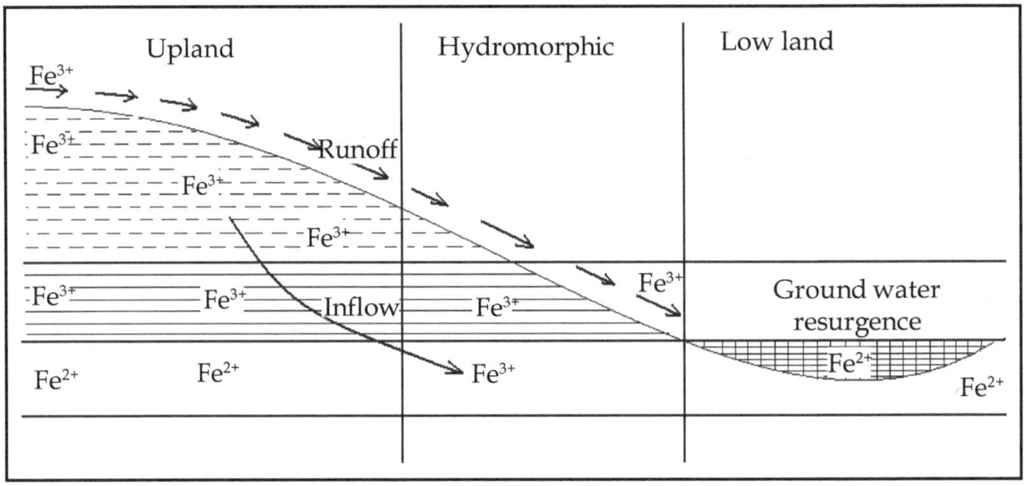

Fig. 9.1. Cross section of toposequence showing movement of iron and reducing process.

9.2. SYMPTOMS

9.2.1. Soil Characteristics

Fe-toxicity is generally found in acid soil (over a wide range of soil pH, 4.0-7.0), which contains moderate to high amounts of organic matter and reactive iron. Fe-toxicity occurs in young acid sulfate soils, poorly drained, coalluvial and alluvial sandy soils, in valleys receiving interflow water from adjacent acid highlands, and in alluvial or coalluvial clayey acid soils, and acid peat soils. Soil feature that are also associated with Fe-toxicity, aside from low pH, are low cation exchange capacity, low base status, low levels of potassium, phosphate and zinc, low supply of easily reducible manganese, acute deficiency of exchangeable basses especially of potassium (Detruck, 1994), but well provided with P and Zn.

Redish ferric iron on the soil surface becoming polarized under the process of reduction at close-up of ground water resurgence zone, where the slope and stationary water meet. Flooded field with high iron concentration also turns reddish, a typical symptom of iron toxic soil.

9.2.2. Plant Symptoms of Fe-toxicity

1. Typical symptom of iron toxicity in rice is *'bronzing'* of leaves (Becker and Asch, 2005), it is small brown spots, which starts appearing at the leaf tips. These spread, merge and result in reddish colour leaves.
2. Leaves of affected plants are narrow, but often remain green.
3. Leaf tips become orange-yellow (Baruah *et al.*, 1996) and dry up in some varieties of rice. Leaves appear purple-brown if Fe-toxicity is severe.
4. Leaves of some varieties may roll.
5. Roots are stunted, coarse, and reddish-brown or dark brown (a coating of iron oxide reduce root surface and decrease capacity to absorb soil nutrients). Freshly uprooted rice hills often have poor root systems with many black roots. Many dead roots of crop plants are found in Fe-toxic soil.
6. Plants are stunted and tiller poorly.
7. If Fe-toxicity occurs at late stage of crop growth, repetitively growth is not severely affected, but the grain yield is reduced because of sterility.
8. With increase in Fe-toxicity, plant height, number of ear-bearing tillers per plant, panicle length and number of ear-bearing tillers per plant, panicle length and grain yield get decreased, whereas vegetative period gets extended with increased spikelet sterility (Majumder *et al.*, 1995). Abraham and Pandey (1989) also reported that Fe-toxicity reduce plant height, panicle length, number of productive tillers and grain yield in rice.
9. In soybean, Fe-toxicity symptoms became apparent from the third trifoliate leaf upward. The affected and unaffected leaves of soybean variety Bossier contained 702 and 290 ppm Fe, respectively.

A specific physiological disorder, *bronze speckle*, was continuously induced in *Tagetes erecta* (L.) cv. Frist Lady and Voyager.

9.3. INJURY MECHANISMS

The cause of Fe-toxicity is technically complex and hence difficult to describe concisely. In rice it is caused by its accumulation in the leaves in toxic concentrations. Further, it is associated with high concentration of ferrous ion (Fe^{2+}) in the soil solution, mobilized *in situ*. It is specially concentrated in the lowland soils of the region. Many upland slopes also have high concentrations of Fe. But, just because of Fe is present in the soil does not necessarily mean that there will be a toxicity problem. In the upland and on the slopes, Fe is typically in ferric form (Fe^{3+}). This form is non-soluble, therefore, not available to plants. Ferric iron is harmless to rice, and Fe-toxicity is not a problem on the upper slopes. In lowlands, Fe^{3+} reduced to Fe^{2+}, which is soluble in water and available for uptake by rice plants (Fig. 9.1). In a typical iron toxic valley, ferric iron is washed down into the valley lowland, either by seepage trough the soil, else by run off and erosion of the upper slopes in the valley bottom. Either *en route* or else upon arrival in the valley bottom, the ferric iron encounters waterlogged conditions, and become ferric iron.

Rice is grown in low to medium-elevation lands with P and K deficient soil and adjacent to leached uplands often suffer from Fe-toxicity associated with interflow of water from the uplands. The mechanisms by which interflow exacerbates the toxicity is uncertain, but it appears to involve dilution of plant nutrients and upsetting of the plants ability to exclude the toxic Fe, rather than the inflow of large amounts of dissolved Fe (van Breemen and Moormann, 1978).

9.3.1. Inhibition of Nutrient Absorption

Sometime, presence in the soil of excessive quantities of iron inhibits nutrient absorption by the plants and leads to severe nutritional imbalances, which manifest themselves in the typical symptoms. The intake phosphorus and potassium is specially inhibited, and it is the relative lack of these two primary nutrients (in conjunction with the relative abundance of nitrogen), which causes many of the symptoms.

Preponderance influence of native non-applied elements, like Fe, Ca, Mn, and Zn in laterite and organic '*kole*' soils, and Ca, S, and Mn in alkali black soils, result in excessive absorption and accumulation of Fe that severely changes the cation-anion balance. This facilitated substantial chlorophyll degradation that led to low yield in rice, especially in West coast of India (Brigit and Potty, 2002). Excessive Fe uptake by plants results in increased polyphenol oxidase activity, leading to the production of oxidized phenols, the cause of leaf *bronzing*.

9.3.2. Oxidative Damage

The appearance of Fe-toxicity symptoms in rice involves an excessive uptake of Fe^{2+} by the rice roots and its acropetal translocation into the leaves where an elevated production of toxic oxygen radicals can damage cell structural components and impair physiological processes (Beker and Asch, 2005). Oxygen radicals are highly phototoxic and responsible for protein degradation and peroxidation of membrane lipids (Fang *et al.*, 2001).

Auto-oxidation of redox active metals, such as Fe^{2+}, or Cu^{+} results in O_2^- formation and subsequently in H_2O_2 and OH production via Feriton-type reactions. Cellular injury by this type of mechanisms is well documented for Fe (Halliwell and Guttridge, 1986; Imlay *et al.*, 1988).

$$Fe^{3+} + O_2^- \rightarrow Fe^{2+} + O_2$$

$$Fe^{2+} + H_2O_2 \rightarrow Fe^{3+} + OH^{\cdot} + OH^-$$

The net result is therefore,

$$O_2^- + H_2O_2 \rightarrow O_2 + OH^{\cdot} + OH^-$$

Reaction of ROS with lipids leads to highly damaging reaction- lipid peroxidation. Singlet oxygen reacts with unsaturated fatty acid and forms lipid hydroperoxides that breakdown to several products of lipid peroxidation. Lipid peroxidation causes degredation of cellular membranes (Devasagayam and Kamal, 2002) and affects cellular functions that lead to cell death.

9.3.3. Inhibition of Photosynthesis

Fe-toxicity also effects photosynthesis and other biochemical process. The green leaves of the susceptible variety become thinner at the higher green leaf concentration of iron, whereas those of the tolerant variety did not. This leaf thinning may result from or in reduced photosynthesis and therefore reduced plant growth.

9.4. TOLERANCE MECHANISMS

Nowadays, plant breeding becomes increasing complex; breeder has to exactly know how resistance or tolerance to stress works. Finding out the mechanism behind a variety's resistance is the first step in working-out which combination of plants should produce high levels of resistance by blending their mechanisms. Tolerance of rice plants to higher concentrations of Fe^{2+} is governed by age, nutritional status of the plant and chemical environment of the root zone (Ponnaperuma, 1972). Therefore, knowledge of the tolerance mechanism is important. The possible tolerance mechanisms towards Fe-toxicity have been briefly described in this section.

9.4.1. Rhyzosphere Oxidation

Fe-toxicity tolerant cultivars display superior performance under Fe-toxic environment partly due to avoidance and partly due to Fe-toxicity tolerance. Fe-stress avoidance because of Fe^{2+} oxidation in the rhyzosphere and the participation of Fe^{3+} hydroxide in the rhyzosphere by the healthy roots. Consequently, such plants may show a crust of insoluble ferric iron in the root zone and the reddish brown coating on roots, which prevents excessive Fe^{2+} uptake.

In strongly reduced soils containing very large amounts of Fe, however there may be insufficient O_2 at the root surface to oxidize Fe^{2+}. In such cases, Fe uptake is excessive and roots appear black because of presence of sulfide. Root oxidation power includes the exertion of O_2 transported from the roots and oxidation mediated by enzymes such as peroxidase or catalase. An inadequate supply of nutrients (K, Si, P, Ca and Mg) and excessive amounts of toxic substances (H_2S) reduce root oxidation power. Rice varieties differ in their ability to reduce O_2 from root to oxidize Fe^{2+} in the rhyzosphere and protect the plants from Fe-toxicity.

9.4.2. Sap Exudation

Fe-toxicity tolerant lines exudates less sap and the sap contains higher concentrations of Fe than those from susceptible lines (Singh, 1977).

9.4.3. Physiological Avoidance

The Fe-tolerant rice cultivars absorbed less Fe or transported less from roots to leaves, indicating presence of a physiological avoidance mechanism (Audebert and Sahrawat, 2000). Further they reported that this mechanism can be enhanced through application of P, K and Zn.

9.4.4. Compartmentalization of Iron

The total uptake of Fe did not differ significantly among the cultivars (Anonymous, 2002). However, there were significant differences in the partitioning among the plants organ. The tolerant varieties partitioned more Fe into its stem and dead leaves, leaving the green leaves significantly free of Fe than those of either the susceptible or the very susceptible varieties.

9.4.5. Enzymatic Oxidation of Iron

Enzymatic oxidation of Fe was found to be the principle mechanism for higher oxidation activity of rice roots (Armstrong, 1969). Glycolic acid pathway present in the rice root system was reported to be chiefly responsible for the root oxidation power in addition to catalase and peroxidase enzymes (Mitsui *et al*, 1962). *Oryza rufipogon* and *O. sativa* cv. TRC229-F41 with higher glycolic acid oxidase and leaf nitrate reductase activities were found to be more tolerant to the adverse effects of higher concentration of Fe (Nath and Borah, 1999).

Application of Fe down regulated the glycolic acid oxidase activities both in root and shoot tissues irrespective of genotypes (Nath and Borah, 1999). This might be attributed to the fact that with the increased levels of Fe amount of the available Fe increased under submergence (Pathak *et al.*, 1978). The increased levels of Fe^{2+} enhance the excessive intake by the rice roots, which might disturb some of the normal metabolic process, including the disruption of glucolic acid pathway. Upon addition of Fe to the growth media leaf nitrate reductase activity was found decreased (Nath and Borah, 1999). The higher uptake of plant reduces the protein synthesis in leaf (Baba *et al.*, 1964) and subsequently inhibits or suppresses the normal nitrate reductase activity. The *in vitro* nitrate reductase activity was found to be lower at the panicle initiation stage than active tillering stage (Deka and Borah, 1993).

Findings of Hu *et al.* (1997, 1999) indicated that Fe stress significantly increased the activities of ascorbate peroxidase, dehydroascorbate reductase [glutathione dehydrogenase (ascorbate)] and glutathione reductase. Activities in the tolerant lines were higher than those in sensitive lines, and significantly negatively correlated with a relative decrease in shoot dry weight. Thus, they concluded that the ascorbate-specific H_2O_2-scavenging system may play an important role in the detoxification of Fe-toxicity in rice.

9.5. GENOTYPIC DIFFERENCE IN IRON TOXICITY TOLERANCE

Rice cultivars differ in tolerance to Fe-toxicity and in general, the upland cultivars are more sensitive to Fe-toxicity than are the lowland cultivars. There are useful landraces, which tends to have reasonable levels of Fe-toxicity tolerance. The rice cultivars viz., Pokkali, IR50, IR2151-190-3, IR2153-6-1,

IR20, IR29, and IR32 (IRRI, 1974, 1975, 1977) were found to be the source of tolerance to Fe-toxicity. The genotypes CL Selecao 307, SR 661-500-B2-1 and RS724-500-B1-29G-1 were classified as resistant and Blucbelle, Dawan, Caloro, CL Selecao 27 and CL Selecao Central1 were classified as tolerant (Machado *et al.*, 1988). The rice cultivars INCA, BP-IRGA410, GA3914, GA3916, CNA5189 and CNA796019 were found to be Fe-toxicity tolerant (Soares and Mattos, 1987). Fe-toxicity genotypes of different crop species have been listed in Table 9.1.

Table 9.1. Fe-toxicity tolerant crop genotypes

Crop/species	Genotype	Reference
Rice *Oryza sativa*	Pokkali, IR50, IR2151-190-3, IR2153-96-1, IR20, IR29, IR32	IRRI, 1974, 1975, 1977
	Suakoko 8, Gossi 27	Abifarin, 1986
	INCA, BP-IRGA410, GA3914, GA3916, CNA5189, CNA796019	Soares and Mattos, 1987
	CL Selecao 307, RS661-500-B2-1, RS724-500-B1-29G-1, Blucbelle, Dawn, Caloro, CL Selecao 287, CL Selecao Central1	Machado *et al.*, 1988
	RCPL-87-7	Abraham and Pandey, 1989
	IR 9764-45-2	Bode *et al.*, 1996
	Chakau, Amubi, Budwan local, Sheeta Bhog, Ramsail, Ngoba, Black Jonga, Meetmuri, Digio, Sendurijoba, Nania, Chipti, Chinadhan, Sirsoria, Thaothabi, Laiphoe, Phouren, Kalma, Framdhan, Dharaimukhi	Majumder *et al.*, 1995
	EMBRAPA 39 (Agrisul)	Terres *et al.*, 1996
	Pankaj, TTB 10-14	Baruah *et al.*, 1996
	At69-2, At76-1, Bw267-3, Bw85, Bw295-5, At69-5	Chandrasiri *et al.*, 1995
	BR5331-93-2-7-2, BR5881-93-2-8-3, IFR53647-3B-9-1, Madhukar	IRSSTN, 1998
	Oryza rufipogon, *O. sativa* cv. TRC-229-F-41	Nath and Borah, 1999
	IR8192-200, IR9764-45, Kantic Putin, Mashuri	Doebermann and Fairhurst, 2000
	CK4	Sahrawat *et al.*, 2001
	ROK24	Baggie and Bah, 2001
	WITA-1, WITA-2, WITA-3	Anonymous, 2002
	CK4, CG14 (*O. glaberima*)	Sahrawat and Sika, 2002

Contd...

Table 9.1. Contd...

	Mashuri	Nayak *et al.*, 2004
	Burdwan local, Seeta Bhog, Black jeera, Meetmuri, Digo, Sindurioba, Chipti, Chinadhan, Framdhan, Charamukhi, Pokkali, Damodar, Vantana, Kurunthakuruka, Matla, Hamilton, Cisadane, Intan , Gauri	Mandal *et al.*, 2004
	Phalguni	Thampatti *et al.*, 2005
Maize *Zea mays*	Oh40B	Odurukwe and Maynard, 1969
Wheat *Triticum* sp.	Siete Cerros	Camargo, 1985
Soybean *Glycin max*	Lee T203	Brown and Jones, 1977a and b

Fe-toxicity tolerance ability also differs among the types (Sub-species) of rice. At the reproductive stage Korean *japonica* varieties are tolerant, the *indica* varieties in Thailand is moderately affected, and the Korean *indica* × *japonica* hybrids differ in their response, among them, Hangancharlbeyo, Manseogbyeo and Yeongpunbyeo are tolerant (Kang and Oh, 1986).

9.6. GENETICS OF IRON TOXICITY TOLERANCE

Very little works have been done on the genetic inheritance of Fe-toxicity tolerance in crop plants. Inheritance of Fe-toxicity tolerance in rice was found to be simple. Two different genes were reported to govern Fe-toxicity tolerance in rice genotypes Suakoko 8 and Gossi 27. Suakoko 8 habours a dominant gene, while Gossi 27 posses a recessive gene for tolerance (Abifarin, 1986). Camargo (1985) reported partial dominance of Fe-toxicity tolerance in wheat, with high broad sense heritability. Gorsline *et al.* (1964) found that the additive gene action was more important in maize. Nipah *et al.* (1999) found that the most genes responsible for tolerance of Fe-toxicity were dominant, but epistatic effects were also observed.

9.7. SCREENING FOR Fe-TOXICITY TOLERANCE

Germplasm can be screened as described in the previous Chapter (Section 8.8). Rice may be evaluated for Fe-toxicity tolerance in between seedling or transplanting and booting stages on 1-9 scale as outlined by IRRI (1988) and the detailed has been presented in Table 9.2. Lines scoring 0-4 can be classified as tolerant and those scoring 5-9 as susceptible.

Table 9.2. Scoring of Fe-toxicity tolerance in rice (IRRI, 1988)

Scale	Expression
1	Growth and tillering nearly normal
2	Growth and tillering nearly normal, reddish brown spots or orange discolourization on older leaves
3-4	Growth and tillering nearly normal, older leaves reddish brown, purple or orange yellow
5-6	Growth and tillering retarded, most leaves dicoloured or dead
7-8	Growth and tillering ceases, most leaves dicoloured or dead
9	Almost all plants dead or drying

The time of occurrence, severity of symptoms and yield responses vary widely among soil types, years, seasons and genotypes. Cultivars resistant to one system may fail when transferred to another. Thus, targeting of varietal improvement requires selection tools improving our understanding of the resistance mechanisms and strategies of crop in the presence of excess Fe. A phototron study was done by Asch *et al.* (2005) to screen rice seedlings resistant to excess Fe based on individual plant subjected to varying levels of Fe (0-300 ppm) for 1-5 days exposure. Scoring was done at 28 days old seedlings. They observed that in most cases, leaf-bronzing scores were highly correlated with tissue Fe concentration. This screening method allows selecting genotypes with low-bronzing score as resistant to Fe-toxicity, and additional analyses of the tissue Fe concentration of those can identify the general adaptation strategy to be utilized in breeding programmes.

9.8. CROP MANAGEMENT

In the recent-past Fe-toxicity has emerged as a serious nutrient disorder of lowland rice. A range of agronomic management interventions have been advocated to reduce the Fe^{2+} concentration in the soil or foster the plants' ability to cope with the excess Fe in either soil or the plant. Keeping knowledge about the build-up Fe-toxic conditions, some field level management may help alleviate the problem of Fe-toxicity.

9.8.1. Improvement of Nutritional Status

Excessive Fe^{2+} levels notably influenced the morphological parameters and concentrations of K, P, and N in the plants. Although, the Fe-toxicity would not be eliminated entirely by improving the nutritional status of the rice growing lands, however application of the other nutrients, especially N, P, K and Zn was found to reduce Fe-toxicity and increase the yield of lowland rice cultivars considerably (Sahrawat *et al.*, 1996; Sahrawat *et al.*, 2000; Audebert and Sahrawat, 2000). Use of appropriate nutrient could reduce the intensity of Fe toxicity and bring about sustainable yield increases in Fe-toxic sulfate soils (Thampatti *et al.*, 2005). High K supply ameliorated the negative effects of higher Fe^{2+} (Li *et al.*, 2001). Proper K supply reduced the translocation of Fe^{2+} from roots to shoots, especially to the upper leaves, and to the improved the plant nutritional status. They also found that the P was antagonistic to Fe^{2+}, and significantly increased the K supply also. Balanced use fertilizers (NPK or NPK + lime) to avoid nutrient stress (Doebermann and Fairhurst, 2000). Do not apply excessive amounts of Fe and organic matter where drainage is poor. Use of urea (less acidifying) is suggested instead of ammonium sulfate (more acidifying) fertilizer. The application of NPK fertilizer and Zn is recommended. Application of additional K, P and Mg fertilizer improve the crop stand in Fe-toxic soils. Reclamation can be carried out by means of liming, which raise the pH in acid soils. Application of manganese oxide (MnO_2) at a rate of 50-100 kg/ha may help correct Fe-toxicity by maintaining the correct Fe/Mn ratio in the plant. Whereas, Dobermann and Fairhurst (2000) suggested that the incorporation of about 100-200 kg MnO_2 per ha in the topsoil decrease Fe^{3+} reduction. Sahu *et al.* (2001) reported that K application at the rate of 132 kg/ha improved yield of rice in Fe-toxic soil. The response of K application was greater in wet season varieties than in dry season varieties.

The efficacies of various amendments for Fe-toxicity amelioration were studied by Nayak *et al.* (2004). The treatments were application of lime, fly ash (20 and 10 t/ha), K (66 kg/ha) Zn (10 and 5 kg/ha) and foliar spray of $MnSO_4$ (0.6%). The symptoms of Fe-toxicity decreased upon application of different amendments. The application of Zn and lime at higher doses resulted in minimum toxicity symptoms. Devi *et al.* (1996) reported that application 600 kg lime/ha as single topdressing resulted in significantly higher grain and straw yield in rice.

Fe chlorosis often persists in upland rice (Asewar *et al.*, 2000). However, rice cultivars display differential tolerance towards Fe-toxicity. In general, upland cultivars are more sensitive to the Fe-toxicity than the lowland cultivars (Anonymous, 1973). Fe concentration was found to be higher at shallow submergence depths in the same field. Therefore, farmers' are generally advised to cultivate deep rooted varieties in shallow submergence depth (Tran *et al.*, 2001).

9.8.2. Physical Barrier to Minimize Soil Erosion

It is better to block the movement of iron from the uplands flush fields to wash out excess Fe, and try to minimize plant uptake of Fe or find varieties tolerant to high concentrations of Fe in the soil. Use of legumes in the uplands to diversify rice-based cropping systems, it improve soil fertility, reduce soil erosion and alleviate weed pressure. They also help in Fe-toxicity problem by stabilizing the upland soil, so that they are not washed down the slope in the first rain (Anonymous, 2002).

9.8.3. Washout Excess Fe

Water in the field becomes increasingly depleted in oxygen the longer it stays there. Circulation of water within the field to keep it aerated, and thereby helps to keep iron in the ferric form. Sometimes, regular flash of fresh water through the field would tend to wash the Fe rather than have it accumulate (Anonymous, 2002). Use of intermittent irrigation and avoidance of continuous flooding on poorly drained soils containing a large concentration of Fe and organic matter reduces the toxicity.

9.8.4. Dry Tillage

Carry out dry tillage after the rice harvest to enhance Fe oxidation during the fallow period. This reduces Fe^{2+} accumulation during the subsequent flooding period. Mid-season drainage remove accumulated of Fe^{2+}. At the mid-tillering stage of rice, draining the field and keeping it flood free for 7-10 days improve oxygen supply during tillering. Delay planting to avoid the peak Fe^{2+} concentration period in the flood prone area is suggested in Fe-toxic soils.

9.9. BIOTECHNOLOGY IN Fe-TOXICITY TOLERANCE

The soils influenced by human activities a range of different problems have been reported. Increase emissions of metals are dangerous because they may get into food chain with risk for human health (Angelone and Bini, 1992). For the re-cultivation of degraded lands and the reclamation of industrial sites, stress tolerant plants are the demand of the days. Biotechnological efforts are underway to improve plants tolerant to abiotic stresses (Salt *et al.*, 1995).

9.9.1. Somaclonal Variation and *in vitro* Screening

Selection imposed to cells under *in vitro* culture acts as directly on cellular phenotypes. Consequently, the characters of interest must be a well-defined cellular manifestation, which must express in culture cells. Many agronomical characters are too complex residing at the levels of many functions. In fact, if the sought cellular phenotype is known, its isolation is possible by adoption of appropriate selection strategy (Fig. 9.2). Comprehensive understanding of the condition of which the crop deems to be grown is essential to stimulate appropriately *in vitro* selection pressure. However, the selected tolerant calli from *in vitro* condition need not necessarily produce tolerant lines. Reaction norms towards different stress were found to be diverse at cellular and *in vitro* levels.

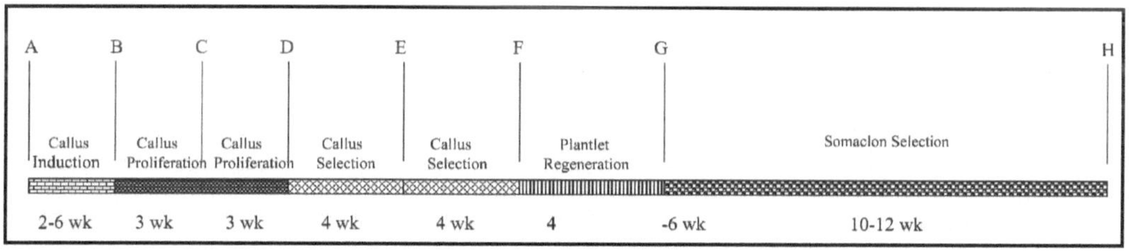

Fig. 9.2. Generalized diagramatic representation of *in vitro* selection procedure for Fe-toxicity tolerance. (A) Explant inoculation on Callus Induction Medium, (B) First sub-culture on Callus Maintenance Medium, (C) Second sub-culture on Callus Maintenance Medium, (D) First sub-culture on stressed Callus Maintenance Medium, (E) Second sub-culture on stressed Callus Maintenance Medium, F) Transferred to stress free medium, (G) *In vivo* selection, (H) Selected somaclones

Fe in the culture medium at toxic levels reduce the plantlet regeneration and it was found to be inversely proportional to the increment of metal concentration (Mandal and Roy, 2000; Roy and Mandal, 2005). More Fe concentration in the culture medium was found to create necrosis (Fig. 9.3c) and metal toxicity that renders total loss of plantlet regeneration at higher concentrations (Mandal and Roy, 2003) and it was reported that concentration of 400 ppm was highly toxic, which killed the cells (Table 9.3). Probably at low pH (~5.0) abundance of Fe cation causes toxicity at cellular level by increasing dedifferentiation of calli into plantlets. Subsequently, Mandal and Roy (2003) developed putative plantlets tolerant to Fe toxicity were hardened and transplanted under glass-house condition. Those plants could flower and mature normally. However, high spikelet sterility was observed in the plants in comparison to the control. This character was found to be dramatically affected in the somaclones. Tonelli (1990) and van Sint Jan *et al.* (1997) also reported reduced fertility in rice plants regulated from the calli after a period of *in vitro* culture. This offers ample scope to develop plants tolerant to Fe-toxicity, which bears immense importance to harness Fe-constrained soils. Pathirana *et al.* (2002) also developed *in vitro* protocol for selecting rice mutants tolerant to Fe-toxicity. They used MS medium (Murashige and Skooge, 1962) for callus induction and supplemented with 2,4-D (dichlorophenoxyacetic acid), CPA (Cholorophenoxyacetic acid), BA (6-bezylaminopurine) and different doses of Fe.

Table 9.3. *In vitro* screening of mature seed derived calli for increased Fe-tolerance (stress was induced in alternate subculture) in rice

Variety	FeNa EDTA (ppm)	Seed germination (%)	Callus induction (%)	Callus health (1-6)*		Plantlet regeneration		Total No. of plant developed
				After 1st selection	After 2nd selection	Somatic embryogenesis	Rhizogenesis	
Taichung Sen Yu	Control	77.16 bcd	82.40 ab	2	2	30.00 c	12.50	41
	50	81.15 bcd	81.59 ab	3	3	10.52 fg	36.84	6
	100	80.00 bcd	68.75 cde	3	4	18.75 de	18.75	14
	200	67.16 efg	60.24 ef	3	4	0.00 i	26.66	-
	400	74.00 def	27.02 i	6	6	0.00 i	0.00	-
	Mean	**75.89**	**64.00**	-	-	**11.85**	-	-
IR 72	Control	86.36 abc	74.21 bcd	2	2	53.84 a	15.38	29
	50	80.18 bcd	72.29 cde	3	3	18.52 de	7.41	17
	100	86.90 ab	77.22 bc	3	4	26.08 c	23.91	28
	200	87.02 ab	41.14 g	4	4	16.21 de	16.21	14
	400	75.71 cdef	43.39 g	4	4	16.66 de	0.00	4
	Mean	**83.23**	**61.64**	-	-	**26.26**	-	-
IR 2269-424-298-18	Control	77.53 bcde	79.92 abc	2	2	40.00 b	6.66	22
	50	64.55 fgh	88.23 a	2	3	28.57 c	28.57	11
	100	56.77 h	84.09 ab	3	3	14.63 ef	9.75	13
	200	61.00 gh	75.41 bcd	3	4	21.42 d	14.28	10
	400	58.50 h	55.55 f	4	5	0.00 i	0.00	-
	Mean	**63.67**	**76.64**	-	-	**20.92**	-	-
C 14-8	Control	93.00 a	81.18 ab	2	2	11.25 fg	12.24	30
	50	87.16 ab	75.12 bcd	2	3	8.11 gh	19.51	24
	100	81.09 bcd	75.13 bcd	2	3	4.82h	18.91	11
	200	72.66 defg	63.33 de	3	4	2.89 i	20.00	7
	400	67.00 efgh	55.97 f	3	5	0.00 i	0.00	-
	Mean	**80.18**	**70.14**	-	-	**5.41**	-	-

****1**-Excellent; **2**-Good; **3**-Medium; **4**-Weak; **5&6**-Poor

Means having same letter in column are not significantly different by Duncan's Multiple Range Test at P= 0.05

Fig. 9.3. Development of plantlets tolerant to Fe-toxicity in rice through *in vitro* screening of Taichung Sen Yu (Roy, 2002; Mandal and Roy, 2003). **a)** An embryogenic callus-mass; **b)** Embryogenic callus showing green spots and initial redifferntiation on regeneration medium; **c)** Calli showing necrosis under high concentration of Fe; Rhizogenesis on regeneration medium; **e)** A few regenerants from selected calli; **f)** *In vitro* selected plants at *in vivo* condition. *(For colour version of this figure see page 539)*

Toxicity tolerance induced *in vitro* mutants might have separate mechanism and would be worthy to use as *de novo* source of variation. *In votro* culture and exploitation of somaclonal variation deem to be useful in developing genotypes tolerant to Fe-toxicity. Further more, this can be combined for progeny testing and backcrossing so as to introgress new tolerant gene(s) in well adopted background. Biotechnological approaches may not to be able to substitute conventional breeding; however, would be supplementary in maintaining such characters in future.

Table 9.4. Alteration in isozyme profiles in rice callus of var. C 14-8 in response to Fe-toxicity (stress induced in the callus induction medium)

Enzyme	Band	Rf. value	Band intensity at different Fe-stress (ppm)*					Stress response
			C	50	100	200	400	
Esterase	1	0.59	A	D	D	A	A	New band with dark intensity appeared at low levels of Fe-stress and disappeared at high levels of stress.
	2	0.61	M	D	D	A	A	Degraded at higher concentrations (200 and 400) of Fe.
	3	0.63	A	L	L	A	A	New band appeared at low levels of stress, but disappeared at high levels.
	4	0.66	A	Vl	Vl	L	M	New band appeared in response to stress and intensity increased with increased stress levels.
	5	0.69	A	Vl	Vl	Vl	Vl	New band appeared at all stress levels, seems to be stress specific.
	6	0.75	D	D	D	D	D	Monomorphic, no change in band intensities across stress levels.
	7	0.78	D	D	D	D	D	Monomorphic, no change in band intensities across stress levels.
Peroxi-dase	1	0.26	L	L	L	L	L	Monomorphic, no change in band intensities across varied stress levels.
	2	0.42	L	L	L	L	L	Monomorphic, no change in band intensities across varied stress levels except at 100 ppm of Fe.
	3	0.66	A	L	Vl	L	L	New band appeared in response to stress with varied intensity.
	4	0.75	Vl	M	L	M	L	Monomorphic, no trend discernible in band intensities under varied stress levels.
	5	0.80	A	L	Vl	L	Vl	New band appeared, showed null relation to stress gradient.

Contd....

Table 9.4. Contd....

G6PDH	1	0.04	D	D	D	D	D	Monomorphic, no change in band intensities across varied stress levels.
	2	0.26	Vl	L	Vl	Vl	Vl	Monomor phic, however displayed more intense band than control at 50 ppm og Fe - stress.
	3	0.45	L	D	M	D	D	Monomorphic, however displayed more intense band than control.
	4	0.59	L	D	M	D	D	Monomorphic, however displayed more intense band than control.
MDH	1	0.55	D	D	D	D	D	Monomorphic, no change in band intensities across varied stress levels.
	2	0.59	L	M	L	L	L	Monomophic, intensity decreased at higher concentration of Fe -toxicity.
	3	0.62	L	D	L	D	A	Intensity varied across stress gradient, however, degraded at hi gh concentration level of Fe -stress.
	4	0.66	L	L	L	A	A	Degraded at higher concentrations (200 and 400 ppm) of Fe stress.
	5	0.72	D	D	D	D	D	Monomorphic, null relationship with increased stress levels.
LDH	1	0.01	A	M	M	M	M	New band (medium intensity) app eared in response to Fe stress of stress, seems to be specific
	2	0.16	L	M	M	M	M	Monomorphic, intensity increased in comparison to control, and remained uniform across stress levels.
	3	0.27	A	L	L	L	L	New band appeared in response to stress environment, seems to be specific.
	4	0.34	L	M	L	M	M	Monomorphic, band intensity decreased at 100 ppm of Fe.
	5	0.76	A	L	A	L	A	New band appeared with light intensity at 50 and 200 ppm of Fe.
ADH	1	0.19	A	L	L	L	L	Light new band appeared under Fe stress.
	2	0.52	A	D	M	D	D	New band appeared, however, no relationship in band intensities across increased stress levels discernible.

* A: Absent, D: Dark, M: Medium, L: Light, Vl: Very Light, C: Control

Table 9.5. Alteration in isozyme profiles in rice callus of var. IR 72 in response to Fe-toxicity (stress induced in the callus induction medium)

Enzyme	Band	Rf. value	Band intensity at different Fe -stress (ppm)*					Stress response
			C	50	100	200	400	
Esterase	1	0.59	D	A	D	A	A	Disappeared at 50, 200, and 400 ppm, howeve r retained at 100 ppm of Fe -toxicity.
	2	0.61	M	A	M	A	A	Disappeared at 50, 200, and 400 ppm, however retained at 100 ppm of Fe -toxicity.
	3	0.63	L	L	L	L	L	Monomorphic, no change in band intensity across increased Fe levels.
	4	0.66	L	L	L	L	L	Monomorphic , no change in band intensity across increased Fe levels.
	5	0.69	D	D	D	D	D	Monomorphic, no change in band intensity across increased Fe levels.
	6	0.75	M	M	M	M	M	Monomorphic, no change in band intensity across increased Fe levels.
Peroxi-dase	1	0.26	L	L	L	L	L	Monomorphic, no change in band intensity across the increased Fe levels.
	2	0.42	L	L	L	L	L	Monomorphic, no change in band intensity across the increased Fe levels.
	3	0.66	L	Vl	Vl	Vl	Vl	Dcreased under stress environment, but intensity remained un altered across the Fe -stress.
	4	0.75	M	L	L	L	L	Decreased under intensity as compared to control but remained unchanged across stress levels.
	5	0.80	L	Vl	Vl	Vl	Vl	Decreased in intensity as compared to control, but remained unchanged across stress levels .
G6PD H	1	0.04	D	D	D	D	D	Monomorphic, no change in band intensity across the increased stress levels.
	2	0.26	L	L	L	L	L	Monomorphic, intensity remained unaltered in increased stress levels.
	3	0.45	M	M	M	D	L	Monomorphic, intensity varied across the stress gradient.
	4	0.59	D	D	M	D	D	Monomorphic, intensity varied across the stress gradient.
MDH	1	0.55	D	D	D	D	D	Monomorphic, intensity remained unchanged under increased stress levels.
	2	0.62	D	A	A	A	A	Degraded under stress environment.
	3	0.72	M	M	M	M	M	Monomorphic, intensity remained unaltered under increased stress levels.

Contd....

Table 9.5. Contd....

LDH	1	0.01	L	L	L	L	VI	Monomorphic, intensity decreased substantially at 400 -ppm stress level.
	2	0.16	L	L	L	L	VI	Monomorphic, intensity decreased substantially at 400 -ppm Fe con centrations.
	3	0.27	VI	VI	A	A	A	Degraded at higher levels of Fe stress.
	4	0.34	L	VI	VI	L	VI	Monomorphic, band intensity varied across stress levels.
	5	0.45	VI	A	VI	VI	A	The band disappeared at 100 and 200 ppm of stress environment.
	6	0.76	VI	VI	A	A	A	Degraded in response to stress environment.
ADH	1	0.19	VI	A	A	VI	L	The band disappeared at 50 and 100 ppm of Fe and intensity increased at 400 ppm of Fe stress.
	2	0.52	M	L	M	L	A	Band intensity varied across the stress gradient and finally degraded a t 400 -ppm of Fe-toxicity.

* A: Absent, D: Dark, M: Medium, L: Light, Vl: Very Light, C: Control

9.9.2. Isozyme Modulation for Fe-toxicity Tolerance

Isozyme, the multiple molecular forms of a single enzyme with identical substrate specificity (Merkert and Moller, 1959) were found to vary in their patterns in intact plants as a function of various physiological and developmental statuses. They are recognized as suitable biochemical markers under physiological stress conditions (Maheswaran and Rangaswamy, 1990). Banding pattern expressing differential intensity indicates the status of an enzyme activity affected by stress and this stress which can be simultaneously used for identification of tolerant lines.

Substantial variation was observed in the banding pattern of isozymes (Mandal and Roy, 2003, Roy and Mandal, 2005) in response to Fe-toxicity in rice var. IR72 and C14-8 (Table 9.4 and 9.5). Various bands showed a gradual increase or decrease in activity with increased Fe levels. Novel bands appeared and disappeared at certain stress levels perhaps due to cleavage of indigenous bands or homo and/or hetero polymer formation or owing to *de novo* synthesis of unique isozyme form newly activated domines of the genome in response to various levels of Fe-toxicity. The major changes in the zymogram are detailed in the Table 9.4. and 9.5. Finally, it could be concluded that isozyme can be used as biochemical markers in breeding programmes for development of Fe-toxicity tolerant plants, which will reduce the time taken in conventional screening. Further, the accuracy of genetic selection will be greatly enhanced because plants are screened at the gene product level with reduce involvement of environmental factors, which effects the qualitative characters.

9.9.3. Molecular Marker Assisted Selection

Molecular markers linked to Fe and Al was mapped (Wu *et al.*, 1998). Rapid and reliable techniques were developed at International Rice Research Institute, Philippines to detect tolerance to Fe-toxicity. Wu *et al.* (1998) mapped the QTLs by using a double haploid population, which revealed that Fe-toxicity tolerance as measured by the *leaf bronzing index* (LBI) was associated with two loci of chromosome 1. The first locus was found to be associated with Fe-induced enzyme glutathione-*S*-reductase involve in ascorbate-specific H_2O_2 scavenging system.

Wan *et al.* (2003 a, b) used 96 backcross inbred lines (BILs) of rice derived from Nippobare (*japonica*) × Kasalath (*indica*) for molecular mapping. Fe-toxicity tolerance was measured by LBI, stem dry weight, tiller number and root dry weight after 4 weeks of solution culture. They identified a total of 8 QTLs associated with ferrous iron tolerance. QTL was detected for LBI on chromosome 1. It explained 20.5% of the variation for LBI observed among BILs. Of the 2 QTLs mapped for stem dry weight, one on chromosome 1 coincided with the QTL for LBI. Three putative QTLs were detected for root dry weight on chromosome 1 and 3. The QTL on chromosome 1 coincided with the QTL for LBI, while the other QTL on chromosome 3 coincided with QTL for stem dry weight on chromosome 3. Of the 2 QTL mapped for tiller number located on chromosome 1, one QTL located at the region of C955-C885 on chromosome 1 also coincided with QTL for LBI. In another study, Wan *et al.* (2003c) detected four QTLs on chromosome 1 and 3, respectively. One QTL controlling leaf bronzing index, stem dry weight, tiller number and root dry weight was located at the region of C955-C885 on chromosome 1. This QTL may be important to ferrous iron toxicity tolerance in rice. Another QTL for shoot dry weight and root dry weight was located at the region of C25-C515 on chromosome 3. Their findings revealed that there is a linkage between a QTL detected under Fe^{2+} stress condition for stem and root dry weight and a QTL detected under phosphorus-deficiency condition for dry weight on chromosome 3. Further, Wan *et al.* (2004) detected a total of 3 QTLs on chromosome 3 using mapping population derived from a cross between Kinmaze (japonica) and DV85. Two QTLs controlling LBI was located at the region of X279-C25 and X144-X362, and their contribution to the total variation were 17.38 and 22.07%, respectively. One QTL for plant height was located at the region of R1468A-R1468B, with 23.18% contribution rate. Another QTL for LBI located at the region of X144-X362 linked with QTL for chlorophyll content which was located at the region of C136-C944 on a rice function map.

Molecular marker loci associated with variations in index values and in relative decrease in shoot dry weight (RDSDW), and gene loci for tolerance were detected by Wu *et al.* (1997) using 175 markers mapped on all 12 chromosomes by single marker loci and interval mapping. Two gene loci were identified as flanked by RG345 and RG381 and linked to RG810, respectively, on chromosome 1 for index value and RDSDW. These explained 32 and 13% of the total variation in the index values, and 15 and 21% in the RDSDW in the population, respectively. Comparing the two marker genotypic class means the have suggested that the tolerant alleles were from Azucena at the first locus on chromosome 1 and the locus on chromosome 8, and at the second locus on chromosome 1 from IR64.

QTL for iron accumulation and related mineral contents in each plant were analyzed with composite interval mapping (Shimizu *et al.*, 2005). QTL mapping for the Fe, P, and Mg content in shoots were compared in the maps of F_3 and F_8. The QTLs for Fe content in the shoots varied in three types of nutritional conditions, but consistently indicated two overlapping regions on chromosome 3 and 4. Some of the QTLs were indicative of iron excluding the power of the root, which was expressed under reduced P content.

9.9.4. Transgenic Development

The transgenic development study in case of Fe-toxicity tolerance is very scanty. Deak *et al.* (1999) developed transgenic rice with *ferritin* gene to enhance high iron storage and they suggested that the enhanced Fe storage ability can reduce reactive oxygen species. *Ferritin* gene has been also transferred into rice (Khalekuzzaman *et al.*, 2006), which enhanced iron level in the endosperm.

9.10. REFERENCES

Abifarin AO. 1986. Inheritance of tolerance to iron toxicity in two rice cultivar. In: Inter Rice Genet Symp (Rice Genetics I), International Rice Research Institute, Philippines. pp. 423-427.

Abraham MJ, Pandey DK, 1989. Performance of selected varieties and advanced generation genotypes in rainfed low land and iron-toxic soils. Inter Rice Res Inst Newsl. 14: 16.

Angelone M, Bini C. 1992. Trace elements concentrations in soils and plants of western Europe. In: Biochemistry of Trace Metals, Adriano DC (ed). Boca Raton, FL; Lewis Publishers. pp. 19-60.

Anonymous. 1973. Annual Report. International Rice Research Institute, Manila, Philippines. pp. 100-104.

Anonymous. 2002. Painting the rice red: Iron toxicity in lowlands. WARDA Annual Report 2001-02. pp. 29-37.

Armstrong W. 1969. Rhyzosphere in rice : An analysis of inter-varietal difference in oxygen flux from the root. Physiol Plant. 59: 285-291.

Asch F, Becker M, Kpongor DS. 2005. A quick and efficient screen for resistance to iron toxicity in lowland rice. J Plant Nutrition Soil Sci. 168(6): 764-773.

Asewar BV, Dahiphate VV, Chauhan GV, Katare NB, Sontakke JS. 2000. Effect of ferrous sulfate on grain yield of Basmati rice. J Maharashtra Agric Univ. 25: 209-210.

Audebert A, Sahrawat KL. 2000. Mechanism of iron toxicity tolerance in lowland rice. J Plant Nutr. 23: 11-12.

Baba I, Inada K, Jajima K. 1964. Mineral nutrition and occurrence of physiological disease. In: Proc Plant Nutr Symp, International Rice Research Institute, Philippines. pp. 173-195.

Baggie I, Bah AR. 2001. Low cost management of iron toxicity in farmers' fields in Sierra Leone. International Rice Research Notes. 26(1): 35-36.

Baruah KK, Bharat N, Nath B. 1996. Changes in growth, ion and metabolism of rice (*Oryza sativa* L.) seedlings at excess iron in growth medium. Indian J Plant Physiol. 1(2): 114-118.

Baruah KK, Nath BC, Gogoi N. 1996. Physiological and biochemical traits of rice (*Oryza sativa* L.) genotypes associated with tolerance of iron toxicity. In: Plant nutrition-food security and sustainability of agro-ecosystems through basic and applied research. Fourteen International Plant Nutrition Colloquium, Horst WJ, Schenk MK, Burkert A, Claassen N, Flessa H, Frommer WB, Goldbach H, Olfs HW, Romheld V (eds.). Hannover, Germany. pp. 476-477.

Beker M, Asch F. 2005. Iron toxicity in rice- conditions and management concepts. J Plant Nutr Soil Sci. 168(4): 558-573.

Bode K, Doring O, Luthje S, Bottger M. 1995. Induction of iron toxicity symptoms in rice (*Oryza sativa* L.). Mitteilungen aaus dem Institut fur Allgemeine Botanika Hamburg. 25: 35-43.

Brigit TK, Potty NN, Makutty KC, Anil Kumar K. 1993. Anionic relation to iron in rice culture in lateritic soil. In: Proc 5th Kerala Sci Congr, Kottayam, Kerala, India. pp. 28-30.

Brigit TK, Potty NN. 2002. A new production technology for rice in iron-toxic laterite soils. SAIC Newsl. 12: 10.

Brown JC, Jones WE. 1977a. Manganese and iron toxicities dependent of soybean variety. Commu Soil Sci Plant Anal. 8: 1-15.

Brown JC, Jones WE. 1977b. Fitting plants nutritionally to the soils. Agron J. 69: 399-404.

Camargo CE de O. 1985. Melhoramento do trigo XI. Estudo da tolerancia a toxicidade de ferro. Bragantia. 44: 87-96.

Chandrasiri PAN, Pathirana R, Probert ME. 1995. Use of improved tolerant cultivars to increase rice production on acid histolsols of southern Sri Lanka. In: Plant Soil Interaction at Low pH: Principles and Management, Proc Third Intrl Symp, Brisbane, Queensland, Australia, 12-16 September, 1993. pp. 407-411.

Deak M, Horvath GV, Davletova S, Torok K, Sass L, Vass I, Barna B, Kiraly Z, Duditd D. 1999. Plants ectopically expressing the iron binding-protein ferritin are tolerant to oxidative damages and pathogens. Nature Biotechnol. 17: 192-196.

Deka SC, Borah, RC. 1993. Effect of nitrogenous fertilizer on N availability in soil and N assimilation in rainfed low land rice. Oryza. 30: 219-225.

Detruck, P. 1994. Iron toxicity to rainfed lowland rice in Sri Lanka. Leuven (Belgium) KULFTB.

Devasagayam TPA, Kamal JP. 2002. Biological significance of singlet oxygen. Indian J Exp Biol. 40: 680-692.

Devi KMD, Gopi CS, Santhakumari G, Prabhakaran PV. 1996. Effect of water management and lime on iron toxicity and yield of paddy. J Trop Agri. 34(1): 44-47.

Dobermann A, Fairhurst T. 2000. Nutrient disorder and management. In: Handbook Series (Potash & Phosphate Institute of Canada and IRRI, Philippines). p. 191.

Fang WC, Wang JW, Lin CC, Kao CH. 2001. Iron induction of lipid peroxidation and effects on antioxidative enzyme activities in rice leaves. Plant Growth Regulation. 35(1): 75-80.

Gorsline GW, Thomas WI, Baker DE. 1964. Inheritance of P, K, Mg, Cu, B, Zn, Mn, Al and Fe concentrations by corn (*Zea mays* L.) leaves and grains. Crop Sci. 4: 207-210.

Halliwell B, Gutteridge JMC. 1986. Iron and free radical reactions: two aspects of antoxidant protection. Trend in Biochemical Sci. 11: 375.

Hu B, Wu P, Liao CY, Jing WZ, Zhu JM, Hu B. 1999. Effect of ascorbate-specific $H_2O_2^-$ scavenging on tolerance to iron toxicity in rice. Acta Phytophysiologica Sinica. 25(1): 43-48.

Hu B, Zhu JM, Wu YR, Luo AC, Wu P. 1997. Effect of peroxidase on tolerance to ferrous iron toxicity in rice. J Zhejiang Agril University. 23(5): 557-560.

Imlay JA, Chin SM, Linn S. 1988. Toxic DNA damage by hydrogen peroxidase through the Fenton reaction *in vitro* and *in vivo*. Science. 240: 640-642.

IRRI. 1974. Rice II: Annual Report for 1973. International Rice Research Institute, Los Banos, Philippines.

IRRI. 1975. Rice II: Annual Report for 1974. International Rice Research Institute, Los Banos, Philippines. pp. 113-129.

IRRI. 1977. Rice III: Annual Report for 1976. International Rice Research Institute, Los Banos, Philippines.

IRRI. 1988. Standard Evaluation System for Rice. International Rice Research Institute, Manila, Philippines. pp. 54

IRSSTN. 1998. Final Report of the 21st International Soil Stress Tolerance Nursery (IRSSTN), International Rice Research Institute, Philippines.

Kang YS, Oh YJ. 1986. Studies on irrigated rice cultivation in Ghana, West Africa. Res Rep Rural Dev Adms, Korea Republic, Crops. 28: 68-78.

Khalekuzzaman M, Datta K, Oliva N, Alam MF, Jorder OI, Datta SK. 2006. Stable integration, expression and inheritance of ferritin gene in transgenic elite *indica* cultivar BR29 with enhanced iron level in the endosperm. Indian J Biotechnol. 5(1): 26-31.

Li DM, Tang JJ, Zhou JL, Li SQ. 1991. The eco-physiological mechanism of rice tolerance for gleyic soil stress and the breeding of varieties tolerance for soil-related stress. Rice Rev Abstr. 10: 1-4.

Li H, Yang XE, Luo AC. 2001. Ameliorating effect of potassium on iron toxicity in hybrid rice. J Plant Nutr. 24(12): 1849-1860.

Machado MO, Nachtigall GR, Terres AL, Gomezes A da S. 1988. The identification of rice genotypes tolerant of iron toxicity. In: Anis 17th Reuniao da cultura do Arroz Irrigado, Pelotas, RS de 26 a 30 de Setembro de 1988. Pelotas, Brazil, Centro de Pesquisa Agropecuaria de Terras Baixas de Clima Temperado. pp. 178-183.

Maheswaran M, S Rangaswamy R. 1990. Peroxidase isozyme system: A potential marker in *Oryza* cultivar *in vitro*. Oryza. 27: 129-132.

Majumder ND, Mandal AB, Ram T, Singh S, Ansari MM. 1995. Improvement of crop productivity in Bay Islands- Approaches and achievements. Research Bulletin-10, Central Agricultural Research Institute, Port Blair. pp. 132.

Mandal AB, Majumder ND, Chowdhury B, Roy B, Elanchezhian R. 2004. Genetic improvement of rice in Andaman and Nicobar Islands. In: Genetic improvement of rice of India, Sharma SD, Prasad Rao U (eds.). Today and Tomorrow's Printer and Publ (India), New Delhi. pp. 1161-1184.

Mandal AB, Roy B. 2000. Development of Fe tolerant rice somaclones through *in vitro* screening. In: 4th International Rice Genet Symp, International Rice Research Institute, Philippines. p. 356.

Mandal AB, Roy B. 2003. *In vitro* selection of mature seed derived calli for increased tolerance towards Fe-toxicity in rice and their isozyme profiling. J Genet Breed. 57: 325-340.

Merkert CL, Moller F., 1959. Multiple forms of enzymes: tissue ontogenic and species specific patterns. Proc Nat Acad Sci, USA. 45: 753-763.

Mitsui S, Kumazawa K, Varaki J, Hirala H, Ishizuka K. 1962. Dynamic aspect of NPK uptake and oxygen secretion in relation to the metabolic pathway with in the plant roots. Soil Sci Plant Nutr. 8(2): 25-30.

Murashige T, Skoog F. 1962. A revised medium for rapid growth and bioassays with tobacco tissue culture. Plant Physiol. 15: 473-497.

Nath T, Borah RC. 1999. Effect of application of iron on glycolic acid oxidase and nitrate reductase in rice. Oryza. 36: 167-168.

Nayak SC, Sahu SK, Mishra GC, Sandha B. 2004. Comparison of different amendments for alleviating iron toxicity in rice. International Rice Research Notes. 29(1): 51-53.

Nipah JO, Safo-Kantanka O, Jones MP, Singh BN. 1999. Genetics of tolerance for iron toxicity in rice. Intrnl Rice Res Notes. 24(1): 11.

Odurukwe SO, Mynard DN. 1969. Mechanism of the differential response of wf9 and Oh40B seedlings to iron nutrition. Agron J. 61: 694-697.

Ottow JCG, Benkiser G, Watanabe I, Santigo S. 1983. Multiple nutritional soil stress as the prerequisite for iron toxicity of wetland rice (*Oryza sativa* L.). Trop Agric (Trinidad). 60: 102-106.

Pathak AN, Singh RK, Singh RS. 1978. Effect of Fe and Mn interaction in yield, chemical composition and their uptake in crop (1): Direct and residual effects in rice crop. Fert News. 24: 35-40.

Pathirana R, Wijithawarna WA, Jagoda K, Ranawaka AL, de-Klerk GJM. 2002. Selection of rice for toxicity tolerance through irradiated caryopsis culture. Plant Cell Tiss Organ Cult. 70(1): 83-90.

Ponnaperuma FN. 1972. The chemistry of submerged soils. Adv Agron. 24: 29-96.

Ponnaperuma NN. 1978. Toxic rice soils. In: Plant Nutrition, Ferguson AR, Bieleshi RL, Ferguson IB (eds.), Vol. II. 8th International Colloquium on Plant Analysis and Fertilizer Problem, Aukland, New Zealand.

Roy B, Manal AB. 2005. Increased Fe-toxicity tolerance in rice calli and modulation in isozyme profiles. Indian J Biotechnol. 4(1): 65-71.

Roy B. 2002. Genetic studies on seed components and standardization of genetic transformation in rice. Ph. D. Thesis, Bidhan Chandra Krishi Viswavidyalaya, Mohanpur, Nadia, West Bengal, India. pp.197-223.

Sahrawat KL, Diatta S, Singh BN. 2000. Reducing iron toxicity in lowland rice trough an integrated use of tolerant genotypes and plant nutrient management. Oryza. 25: 209-210.

Sahrawat KL, Diatta S, Singh BN. 2001. Nutrient application reduces iron toxicity in lowland rice in West Africa. International Rice Research Notes. 26(2): 51-52.

Sahrawat KL, Mubah CK, Diatta S, Deleaux RD, Patrick Jr WH, Singh BN, Jones MP. 1996. The role of tolerant genotypes and plant nutrients in the management of iron toxicity in lowland rice. J Agric Sci (Cambridge). 126: 143-149.

Sahrawat KL, Sika M. 2002. Comparative tolerance of *Oryza sativa* and *O. glaberima* rice cultivars for iron toxicity in West Africa. Intrl Rice Res Notes. 27(2): 30-31.

Sahrawat KL, Singh BN. 1995. Management of iron toxic soil for lowland rice cultivation in West Africa. Proc 3rd African Soil Sci Soc Conf. pp. 617-624.

Sahu BN. 1968. Bronzing disease of rice in Orissa as influenced by soil type and manuring and its control. J Indian Soil Sci. 16: 41-45.

Sahu SK, Sadhana B, Dev G. 2001. Relationship between applied potassium and iron toxicity in rice. Inter Rice Res Newsl. 26(20: 52-53.

Salt DE, Blaylock M, Kumar NPBA, Dashenkov V, Ensley BD, Chet I, Rahkin I. 1995. Phyto-remediation: A novel strategy for the removal of toxic metals from the environment using plants. Biotechnology. 13: 468-474.

Shimizu A, Guerta CQ, Gregorio GB, Kawasaki S, Ikehashi H. 2005. QTLs for nutritional contents of rice seedlings (*Oryza sativa* L.) in solution cultures and its implication to tolerance to iron-toxicity. Plant Soil. 275(1/2): 57-66.

Singh KK. 1977. Exudate delivery and iron and manganese transport determining different responses of navy bean lines (*Phaseolus vulgaris* L.) to iron-induced manganese chlorosis. Abstr. P 48: In: Seminar on recent advances in plant science, Kalyani.

Soares SF, Mattos T. 1987. Performance of cultivar and lines irrigated rice in Epirito Santo. Lavoura Arrozeira. 40(376): 22-25.

Tanaka A, Yamaguchi K. 1973. A role on the nutritional status of the rice plant in Italy, Portugal and Spain. Siol Sci Plant Nuttr. 19: 161-171.

Terres AL, Machado MO, Ribeiro AS, Martins JF, Fagundes PR, Magalhaes Jr AM de, Galli J, Pinto JJ de O, De Magalhaes Jr AM. 1996. EMBRAPA 38-Ligeirinho and EMBRAPA 39-Agrisul: new cultivars of irrigated rice released by the Temparate Climate Centre in Pelotas. Lavoura Arrozeira. 49: (425): 9-12.

Thampatti KCM, Cherian S, Iyer MS. 2005. Mapping iron toxicity in acid sulfate rice soils by integrating genetic tolerance and nutrition. International Rice Research Notes. 30(1): 37-39.

Tonelli C. 1990. Somaclonal variation in cereals. In: Biotechnology in Agriculture and Forestry, Somaclonal Variation in Crop Inmprovement, Vol. II. Springer-Verlag, Berlin, Heidelberg. pp. 271-287.

Tran KT, Huynh TTH, Nilsson SI. 2001. Rice soil interaction in Vietnamese acid sulfate soils: Impact of submergence depth on soil solution chemistry and yield. Soil Use Manage. 17: 67-76.

van Breemen N, Moormann FR. 1978. Iron toxic soils. In: Soils and Rice. International Rice Research Institute, Philippines. pp. 781-800.

van Sint Jan V, de Macedo CC, Kinel JM, Bouharmont J. 1997. Selection of Al-resistant plants from a sensitive rice cultivar, using somaclonal variation, *in vitro* and hydrophonic cultures. Euphytica. 97: 303-310.

Wan JL, Zhai HQ, Wan JM, Ikehashi H. 2003a. Detection and analysis of QTLs for ferrous iron toxicity tolerance in rice (*Oryza sativa* L.). Rice Genetics Newsl. 19

Wan JL, Zhai HQ, Wan JM, Ikehashi H. 2003b. Detection and analysis of QTLs for ferrous iron toxicity tolerance in rice *Oryza sativa* L. Euphytica. 131(2): 201-206.

Wan JL, Zhai HQ, Wan JM, Ikehashi H. 2003c. Detection and analysis of QTLs for ferrous iron toxicity tolerance in rice (*Oryza sativa* L.). Euphytica. 131(2): 201-206.

Wan JL, Zhai HQ, Wan JM, Yoshimura A. 2004. Detection and analysis of QTLs associated with resistance to ferrous iron toxicity in rice (*Oryza sativa* L.), using recombinant inbred lines. Acta Agronomica Sinica. 30(4): 329-333.

Wu P, Hu B, Liao CY, Zhu JM, Yu YR, Senadhira D, Paterson A. 1998. Characterization of tissue tolerance to iron by molecular markers in different lines of rice. Plant Soil. 203: 217-226.

Wu P, Luo A, Zhu J, Yang J, Huang N, Sendhira D. 1997. Molecular markers linked to genes underlying seedling tolerance for ferrous iron toxicity. Plant and Soil. 196(2): 317-320.

Yamauchi M. 1989. Rice bronzing in Nigeria caused by nutrient imbalances and its control by potassium sulphate application. Plant Sci. 117: 275-286.

10

Other Metal Toxicity Tolerance

10.1. CADMIUM

10.1.1. Introduction

Cadmium (Cd) is a strongly phytotoxic heavy metal in an increasing environmental problem worldwide. It is one of the most dangerous metal due to its high mobility and the small concentration at which its effects on the plants being to appear. It is released into the environment by the power stations, heating systems, metal-working industries or urban traffic. It is recognized as an extremely significant pollutant due to its high toxicity and large solubility in water (Pinto *et al.*, 2004). Soil solutions which have a Cd concentration ranging from 0.32 to 1.00 mM can be regarded as polluted to a moderate level (Sanita di Toppi and Garbrielli, 1999).

Cd toxicity is highest in acidic environment and decreased as the soil pH increased (Przybulewska, 2004). Availability of Cd to plants is regulated by pH, redox potential and other physicochemical parameters. The effect of Cd salt on the growth of seedlings was weaker in loamy soil and stronger in sandy soil. The root system was more sensitive to Cd ions than the cotyledons.

10.1.2. Symptoms of Cd-toxicity in Plants

1. *Leaf symptoms:* Red-brown discolourization of leaf veins, petioles and stem. The leaves cup and roll downward with puckering of lamina (Woltz and Chambliss, 1981). Curling of tender leaves of *Ceratophyllum demersum* L. (Kumar and Prasad, 2004).
2. *Chlorosis and necrosis:* Interveinal chlorosis and necrosis in tomato. Cd-toxicity induces necrotic patches on the base and sheath parts of the oldest leaves of wheat cultivars (Koleli *et al.*, 2004). Leaf chlorosis followed by necrosis in Kiwifruit vines (Sotiropoulos *et al.*, 2006).
3. *Reduced plant growth:* Severe reduction in growth and development. Young pea plant exhibited retarded growth and low productivity (Kamenova-Jouhimenko, 2003). Barley seedlings grew shorter in proportion to increased contamination (Przybulewska, 2004). Reduces root and shoot length and dry weight in wheat seedlings (Naquib *et al.*, 1986). Spinach crop suffering from severe Cd toxicity had small roots and narrow yellowish leaves (Adhikari *et al.*, 2005). Higher levels of Cd inhibited the growth and biomass of the crop.
4. *Reduction in yield:* Decrease in yield of rice under high Cd environment (Adhikari *et al.*, 2006). At 40 mg/kg soil, yield of spinach was reduced to 38% of control plants (Adhikari *et al.*, 2005).

10.1.3. Injury Mechanisms

The sensitivity of plants to heavy metals depends on an interrelated network of physiological and molecular mechanisms that includes uptake and accumulation of metals through binding top extra-cellular exudates and cell wall, complextion of ions inside the cell by various substances, such as, organic acids, amino acids, ferritins, phytochilatins, and metalothioneins.

10.1.3.1. Effect on metabolism

Cd has been shown to affect various aspects of metabolism in different plant systems (Chen and Kao, 1995; Shah and Dubey, 1997; Vanaja *et al.*, 2000). Cd ions inhibit the activities of RuBP carboxylase in barley (Stiborova *et al.*, 1986; Stiborova, 1988; De Filippis and Ziegler, 1993) and PEP carboxylase in maize. Inhibition in the activity of RuBP carboxylase in ascribed to the formation of mercaptides with the enzyme's thiol-groups (Stiborova, 1988). Cd treatment has been shown to reduce ATPase activity of the plasma membrane fraction of wheat and sunflower roots (Fodor *et al.*, 1995). The inhibition of root Fe (III) reductase induced by Cd led to Fe (II) efficiency, and it seriously affected photosynthesis (Alcantara *et al.*, 1994). Poschendieder *et al.* (1989) reported a significant increase of ABA content after 120 hours exposure to Cd. The activity of glutamate dehydrogenase, glutamic synthetase and nitrate contents are the most sensitive in chlorophyll synthesis are also under Cd toxicity. Enzyme involved under Cd-toxicity (Rani *et al.*, 1987).

10.1.3.2. Decrease nitrogen fixation in leguminaceae

Nitrogen fixation and primary ammonia assimilation decreased in nodules of soybean plants during Cd treatments (Balestrasse *et al.*, 2003).

10.1.3.3. Induction of free radicals

Cd stimulates free radical production by imposing oxidative stress (Foyer *et al.*, 1997), resulting cell damage. Cd toxicity involved in either in inducing oxygen free radicals production, or by decreasing enzymatic and non-enzymatic antioxidants (Somashekaraiah *et al.*, 1992; Stohs and Bagchi, 1995; Gallego *et al.*, 1996; Sandalio *et al.*, 2001; Balestrasse *et al.*, 2001; Fornazier *et al.*, 2002; Cho and Seo, 2004; Balestarasse *et al.*, 2004).

Contrasting results obtained by Demirevska-Kepova *et al.* (2006) in barley (*Hordium vulgare* L. cv. 'Obzor') seedlings and they suggested that only minor effects via oxidative damage due to Cd-toxicity.

10.1.3.4. Inhibition of photosynthesis

Cd ions inhibit chlorophyll biosynthesis in plant affecting the synthesis of 5-amino laevutinic acid (Stobart *et al.*, 1985), oligomeric forms of light harvesting complexe II (Krupka *et al.*, 1987). Hassan *et al.* (2005) reported significant decrease in chlorophyll content resulting reduction in photosynthesis.

10.1.3.5. Cell damage

Heavy metal injury mechanisms including the oxidation and cross-linking of protein thiols, inhibition of key membrane proteins, or changes in the composition and fluidity of membrane lipids (Meharg, 1993). Mukherjee and Sharma (1988) found that Cd ions reduced cell division, but increased chromosomal aberrations in the root tip cells of garlic. Nouairi *et al.* (2006) reported drastically decrease in total lipid content in Cd-treated *Brassica napus* leaves and it also showed higher peroxidation.

10.1.3.6. *Alter uptake of nutrients*

Cd can alter the uptake of minerals by plant through its effects on the availability of minerals from the soil, or through a reduction in the population of soil microbs (Morreno *et al.*, 1999). It inhibits Ca and Fe in cucumbers (Burzynski, 1987a), caused reflux of K at higher concentrations. In general, Cd has to shown to interfere with the uptake, transport and use of several elements (Ca, Mg, P and K) and water by plants (Das *et al.*, 1997). Metwally *et al.* (2005) also observed that toxic Cd levels inhibit uptake nutrient elements, such as P, K, S, Ca, Zn, Mn, and B by plant in an organ- and genotype specific manner. Cd also reduced the absorption of nitrate and its transport from root to shoots (Hernendez *et al.*, 1996).

Cd ions are likely entre into the cells through cation transporters with specific metal specificity (Clemens, 2001). Cd and Zn have been found to be co-accumulated in aerial parts of *Arabidopsis halleri* (Bert *et al.*, 2003). Most of the transporters thought to be involved in the uptake of micronutrients are in the *ZIP* (*ZRT, IRT*-like protein) and *Nramp* (natural resistance-associated macrophase protein) family (Williams *et al.*, 2000). *ZNT1* transporter (Lasat *et al.*, 2000) has low affinity Cd transport. IRT1, an iron transporter has a broad substrate range and possibly Cd (Korshunova *et al.*, 1999; Clemens, 2001). *AtNramp3* involves in iron metal uptake, showed Cd^{2+} transport activity (Thomine *et al.*, 2000).

10.1.3.7. *Reduced enzymes activities*

Toxic levels of Cd in *Pisum sativum* exhibited low expression of ribulose-bisphosphate carboxylase, phosphenolpyruvate carboxylase, malic enzyme, adenosinetriphosphate, peroxidase, glutathione reductase, ascorbate peroxidase, superoxide dismutase (Chein *et al.*, 2001; Kamenova-Jauhimenko *et al.*, 2003; Metwally *et al.*, 2005). Cd induced a significant decrease in nitrate content and inhibition of nitrate reductase, glutamine and ferredoxin-glutamate synthase (Chaffei *et al.*, 2004).

10.1.4. TOLERANCE MECHANISMS

Plant has developed some functions for Cd^{2+} tolerance, which include cell wall binding, chelation with phytochelations, compartmentation of Cd^{2+} in vacuole and enrichment in leaf trichomes. Cd may be immobilized by the cell wall of roots (Nishizono *et al.*, 1989) and extra-cellular carbohydrates, like mucilase, callose (Verkleij and Schat, 1990; Wanger, 1993). In roots and leaves of bush bean, Cd ions seem to be mostly bound by pectic sites and hytidyl groups of the cell wall (Leita *et al.*, 1996).

10.1.4.1. Compartmentation of Cd^{2+} in vacuole

Plant cell can be used to remove heavy metals by accumulating Cd in vacuole. This cost effective eco-friendly technology is known as 'phytoremedation' (Salt *et al.*, 1995a). The plant species that accumulate high concentrations of heavy metals in shoot-offer one option for the phytoremediation of metal-contaminted soils. Song *et al.* (2003) observed a correlation between high expression levels of *YCF1* and resistance of the corresponding yeast to Cd. *YCF1*, which detoxifies Cd by transporting it into vacuoles, enhancing Cd tolerance in plants. Vacular compartmentalization prevents the free circulation of Cd ions in the cytosol and forces them into a limited area (Sanita di Toppi and Gabrielli, 1999; Hall, 2002). At every level, concentrations and affinities of chelating molecules, as well as the presence and selectivity of transport activities, affect metal accumulation rates (Clemens *et al.*, 2002). Nouairi *et al.* (2006) found that the tolerance exhibited by *Brassica juncea* to Cd toxicity may be attributed to the enhanced lipid synthesis leading to a better compartmentalization of Cd ions as Cd-phytochelatin complexes in the cell vacuoles or during vesicle formation.

10.1.4.2. Permeability of Cell Membrane

The cell walls of the endodermal cell layer act as a barrier for apoplastic diffusion into the vascular system. Cd enters first the roots, subsequently it can reach the zylem through an apoplastic and/or a symplastic pathway (Salt *et al.*, 1995a), complexed by several ligands, such as organic acids and/or phytochelatins (Senden *et al.*, 1992, 1994; Salt *et al.*, 1995b). In general, Cd ions are mainly retained in the roots, and only small amounts are transported to the shoots (Cataldo *et al.*, 1983). The content decreases in the order as: roots > stems > leaves > fruits > seeds (Blum, 1997).

10.1.4.3. Over production of organic acids

Plants response to elevated Cd levels involved overproduction of organic acids in maize roots as a resistance mechanism to alleviate Cd toxicity. Hang *et al.*, (2006) reported that acetic and malic acids increased the uptake of Cd by maize (*Zea mats* L. cv. TY2) roots and enhanced Cd accumulation in shoots under hydroponic condition. Concentration depends on net Cd influx in the presence and absence of organic acids could be resolve into linear and saturable components. Cd(II)-organic acid complexes associated with the root zone, could decompose and liberate Cd^{2+} was buffered by the presence of Cd(II)-organic acid complexes.

10.1.4.4. Chelation

One recurrent general mechanism for heavy metal detoxification in plant and other organisms is the chelation of the metal by ligand and, in some cases, the subsequent compartmentalization of the ligands-metal complex. Several of metal-binding ligands have now been recognized in plants and include organic acids, amino acids, peptides, and polypeptides. The two best characterized heavy metal-binding polypeptides involve in chelation and sequesteration of Cd includes metalothioneins and phytochelatins.

Phytochelations: Phytochelatins (PCs) are heavy metal binding peptides involved in heavy metal tolerance and sequestration (Steffens, 1990). PCs comprises a family of peptides with the general structure (g-Glu-Cys)$_n$ -Gly (Rauser, 1995). They were shown to be induced by heavy metals such as Cd (Zenk, 1996; Howden *et al.*, 1995). PCs are enzymatically synthesized, cysteine rich peptides.

Glutathione (GSH) is the direct precursor of phytochelatins. PCs synthesis from GSH is catalysed by a transpeptidase, named phytochelatin synthase (EC 2.3.2.15), which is a constitutive enzyme requiring post-translational activation by heavy metals (Grill *et al.*, 1989; Klapheck *et al.*, 1995). PC has been shown to be activated only in the presence of heavy metal ions. The capacity to synthesis PCs is supposed to present in all higher plants (Gekeler *et al.*, 1989).

10.1.4.5. Salicylic acid

Salicylic acid (SA) plays a key role in plant disease resistance and hypersensitivie cell death but is also implicated in hardening responses to abiotic stresses. Metwally *et al.* (2003) reported increased free SA content in barley under Cd toxicity. SA treatment strongly or completely suppressed the Cd-induced up-regulation of the antioxidant enzyme activities.

10.1.4.6. Microorganisms in tolerance of Cd-toxicity

Mycorrhizas proved to protect plants against to metal toxicity (Marschner, 1995; Jentshcke and Godbold, 2000). Detoxification of Cd in *Poxillus involotus* is done by accumulation of Cd in the vacuole (Blaudez *et al.*, 2000).

10.1.5. CROP MANAGEMENT FOR Cd-TOXICITY TOLERANCE

Cd has reverse effect on nitrogen fixation efficiency. Number of nodules and yield decreased with increased concentration of Cd. Singh *et al.* (1989) suggested application of organic manure in Cd toxic soils, which significantly improve yield of wheat. High clay content, higher CEC and presence of $CaCO_3$ were found to immobilize Cd in the soil. Kamenova-Jouhimenko *et al.* (2003) reported that addition of humic acid in the nutrient medium resulted in partial restoration of the normal physiological activity in the plant, indicating the potential use of humic acid against Cd-toxicity. Application of Zn (300 mmol/L) or ascorbic acid (250 mg/L) also alleviate Cd-toxicity in barley (Wu and Zhang, 2002).

10.1.6. BREEDING FOR Cd-TOXICITY TOLERANCE

Development of genotypes less sensitive to Cd-toxicity is the feasible solution for Cd-contaminated soils. The existence of genetic difference in heavy metal uptake and accumulation, and tolerant to metal toxicity have been identified in many crops (Zhang *et al.*, 2000; Arao and Ae, 2003; Liu *et al.*, 2003). In rice, a wide difference exists among genoptypes in their ability to Cd accumulation in grains (Arao and Ae, 2003; Liu *et al.*, 2003). Development of low Cd accumulating cultivars have been also undertaken in sunflower (Li *et al.*, 1997; Penner *et al.*, 1995).

10.1.7. TRANSGENIC DEVELOPMENT

Transgenic development is an alternative approach to improve Cd tolerance and its accumulating capacity in crop plants. There are few successful reports on genetically engineered crop plant tolerant to Cd-toxicity (Zhu *et al.*, 1999a; Creissen *et al.*, 1999). Over expression of *Escherichia coli gshI* gene in Indian mustard (*Brassica juncia*) increased γ-glutamylcysteine synthetase (g-ECS) activity leading increased Cd tolerance and accumulation (Zhu *et al.*, 1999b). Transgenic tobacco (*Nicotiana tabaccum*, var Wisconsin 38) bearing the transgene coding for the polyhistidine cluster, combined with yeast metallothione (HiCUP) showed increasing resistance to the stress response induced by cadmium (Pavlikova *et al.*, 2004). The Cd content of the aerial biomass of transgenic tobacco increased by 45-75% compared with control. Another transgenic tobacco developed by Antosiewicz and Henning (2004) carrying cDNA *LTC1*, a nonspecific transporter for Ca^{2+}, Cd^{2+}, Na^{+} and K^{+}, which displayed substantially higher level of tolerance to Cd and accumulated less Cd in roots.

Plant tolerate to heavy metals through sequestration with cysteine-rich peptides, phytochelatins. Cysteine is synthesized by cyteine synthase (CS, E.C. 4.2.99.8) from hydrogen sulfide and O-acetylsrine. Transgenic *Nicotiana tabaccum* var. Xanthi transformed with *RCS1*, a cytosolic cysteine synthase gene of rice showed higher cyteine synthase activity and Cd toxicity tolerance than the wild type (Harada *et al.*, 2001). Their result suggested that introduction of a cysteine synthase gene into tobacco plants resulted in high level production of sulfur-containing compounds that detoxify the Cd.

Saccharomyces cerevisiae protein *YCF1*, a member of the ATP-binding cassette pumps Cd^{2+} conjugated to glutathione into vacuoles. Song *et al.* (2003) found improved Cd (II) resistance, enhanced Cd accumulation into vacuoles and elevated Cd content in transgenic plants, which are desirable characteristics for phytoremedition.

10.2. LEAD TOXICITY

10.2.1. Introduction

Lead is a potential pollutant of soil and air, and it get easily absorbed and accumulated in different plant parts. It is not an essential element of plants. Pb contamination results from natural weathering, mining, smelting activities, paints, gasoline, explosive, disposal of municipal wastes and processed food preservatives (Fig. 10.1). Yang *et al.* (2000) cited that the soil contamination with Pb is not likely to decrease in the near future. The main source of Pb contamination are fumes of automobiles, chimneys of factories using Pb, effluents from the storage battery, industry, mining and smelting of Pb ores, metal plating and finishing operations, fertilizers, pesticides and additives in pigments and gasoline (Eick *et al.*, 1999). Lead compounds used as agricultural chemicals, like Pb arsenate also contaminate soils.

Pb affected soils contain Pb in the range of 400-800 mg kg^{-1} of soil; it may reach even 1000 mg kg^{-1} of soil in the industrial area (Angelon and Bini, 1992). Uptake of Pb by plants depends on the soil pH, particle size and cation exchange capacity, root surface area, root exudation mycrrhization, and rate of transpiration and other physiochemical parameters. The absorption of Pb in soil increases with increasing pH between 3.0 and 8.5 (Lee *et al.*, 1998). Roots have the ability to take up a significant quantity of Pb, which is simultaneously translocated to the other parts of the plants. A part of Pb also absorbed by leaves from the contaminated atmosphere.

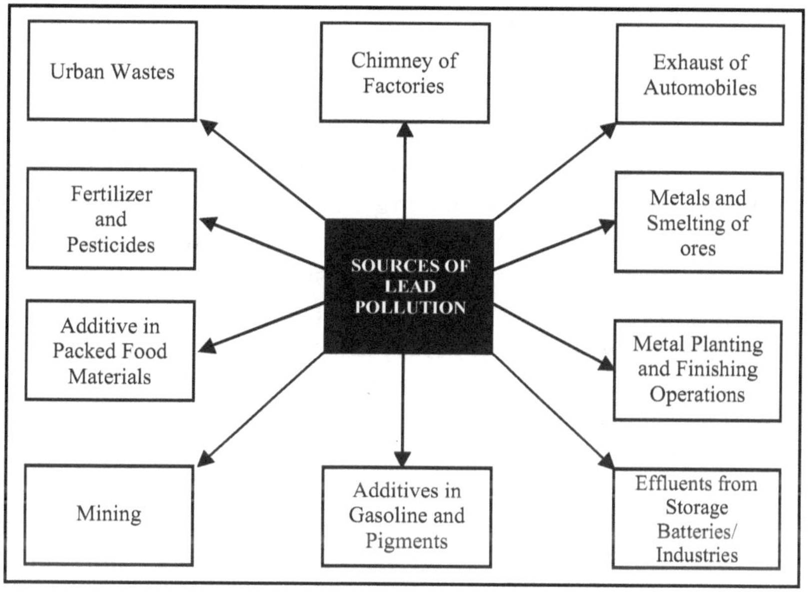

Fig. 10.1. Sources of lead pollution in air and soil

10.2.2. Symptoms of Pb-toxicity in Crop Plants

1. Reduced leaf-blade area (Iqbal and Mostaq, 1987)
2. Stunted growth of effected plants (Burton *et al.*, 1984)
3. *Leaf chlorosis:* The green tissues are more susceptible to Pb-toxicity than etiolated tissues. Chlorophyll content of barley leaves decreased with added Pb (Kacabova and Natr, 1986). Reduced chlorophyll content in Pb-toxic environment was also observed by Phetsombat *et al.* (2006) in *Salvinia cuculata*. Subsequently, it reduces photosynthesis.
4. High concentrations of Pb (> 0.1 mM Pb) in cabbage cv. Pride of India depressed plant growth; young leaves developed interveinal chlorosis along the margins; chlorosis gradually intensified; the leaves thickened; and head size reduced markedly (Chatterjee *et al.*, 2006).
5. Pb-toxicity inhibits root growth (Burton *et al.*, 1984). Toxic amounts of Pb accumulate in roots, which turn black. Phetsombat *et al.* (2006) reported higher content of Pb in roots of *Salvinia cuculata* than leaves suggesting that the metals were bound to the root cells and were partially transported to the leaves.
6. Pb inhibits elongation of etiolated wheat segments, and etiolated green sunflower hypocotyls (Burzynski and Jacob, 1983)

10.2.3. Injury Mechanisms

The toxic concentrations of Pb specific to plant species generally 10 ppm severly restricts elongation of primary roots and 25 ppm completely stops it (Burzynski, 1987b). Plant growth and development was reduced at 5 mg Pb/l, and the plants were killed at 500 mg Pb/l. Pb upsets mineral nutrition, water balance, affects membrane structure and permeability, and changes hormonal status. It inhibits apparent photosynthesis, photorespiration, CO_2 uptake, stomatal opening and transpiration (Poskuta *et al*, 1987). The plastic and elastic extensibility of wheat coleoptile cells walls was reduced and the hydration of sunflower hypocotyls segments was decreased.

10.2.3.1. Reduction in germination and seedling vigour

Pb-toxicity reduces germination percent and seedling vigour. Mishra and Choudhari (1998) reported decreased germination index, root/shoot length, tolerance index and dry mass of roots and shoots. Pb is more toxic to plants during the early stages of plant growth. High concentrations of Pb (1 mM) caused 14-30% decrease in germination and reduced seedling growth by 13-45% (Verma and Dubey, 2003). Pb reduces elongation of etiolated wheat segments, green pea epicotyl fragments and etiolated green sunflower hypocotyls (Burzynski and Jacob, 1983). In *Lupinus* reduced the membrane of germinating seeds and caused shortening of hypocotyls as well as roots (Wozny *et al.*, 1982).

10.2.3.2. Inhibition of root growth

Pb-toxicity reduces root growth and its branching pattern (Breckle, 1991). Lateral roots are more sensitive than the primary roots. Obroucheva *et al.* (1998) observed strong inhibition of primary root growth and a shorter branching zone with more compact lateral roots occupying a position much closer to the root tip of maize.

10.2.3.3. Mitotic irregularities

Lead reduced the mitotic index in the root tip cells of maize (Radecki *et al.*, 1989). Pb inhibits cell division in root tips (Eun *et al.*, 2000). Wierzbicka (1994) observed irregularities and chromosome stickiness under Pb-toxicity. The interphase cells possessed micronuclei, irregularly shaped nuclei, and nuclei with decomposed nuclear material under Pb treatment. It perturbs the aligment of microtubules in a concentration depended manner beginning at 10 mM (Yang *et al.*, 2000) and destroys microtubules of the mitotic spindle. Synthesis of DNA, RNA and protein were greatly reduced in the embryo axis and with increasing concentration of Pb (Maitra and Mukherji, 1977).

10.2.3.4. Membrane injury

Membrane injury due to Pb-toxicity causes leakage of solutes from the cells (Malikowski *et al.*, 2002). Pb-toxicity reduces the protein content of tissues and cause significant alterations in lipid composition (Przymusinski *et al.*, 1991; Stefanov *et al.*, 1995). Stefanov *et al.* (1993) observed substantial changes in the level of glycolipids, which are associated with membrane permeability in chloroplast.

10.2.3.5. Inhibition of enzyme activities

The activities of various enzymes were found to be inhibited by Pb-toxicity in the soil (Table 10.1). It also reduces the activity of catalase, peroxidase, acid phosphatase and ribonuclease in leaves (Chatterjee *et al.*, 2006). The inhibition of enzymes by Pb-toxicity results from the interaction of Pb with -SH group of enzymes (Levina, 1972). Pb interacts with free -SH groups that are present at the active site of the enzyme and essential for enzyme activity. Pb-toxicity also inhibits enzyme's activity by blocking -COOH groups.

Pb-toxicity inhibits the activities of RuBP carboxylase in barley and PEP carboxylase in maize. PEP carboxylase is more sensitive and was inhibited at 5 ppm Pb. Sung and Kwon (1980) reported decreased ATP content in the seedlings of kidney bean and buckwheat. Pb was found to be highly effective in inhibiting ATP synthetase/ATPase (Tu Shu and Brouillette, 1987). The ð-amino laevuline decarboxylase, the key enzyme of chlorophyll biosynthesis, is inhibited Pb-toxicity (Prasad and Prasad, 1987). Ribulose-bisphosphate carboxylase/oxygenase (Vallee and Ulmer, 1972) and lactate dehydrogenase were found to be inhibited at low concentrations (5 mM) of Pb. Mukherji and Maitra (1976) observed decreased activities of protease and -amylase in rice endosperm. A decrease in activity of glutamate dehydrogenase and a low content of nitrate have in malate dehydrogenase activity (Lee *et al.*, 1976). Pb treatment also enhances the activities of some stress-enzymes (Table 10.1). Lee *et al.* (1976) reported increased activity of acid phosphatase, δ-amylase peroxidase in soybean leaves. An increase in the activity of the RNA hydrolyzing enzyme ribonuclease and of protease has been observed in submerged aquatic angiospermic plants growing in presence of Pb acetate

Table 10.1: Effect of Pb on activities of enzymes of different metabolic processes.

Metabolic process	Enzyme	Plant species	Effect of Pb	Reference
Chlorophyll synthesis	δ-Aminoaevulinate	*Pennisetum typhoidenum*	Inhibition	Prassa and Prassad, 1987
CO_2 fixation	Ribose-1,5-bis-phosphate	*Avena sativa*	Inhibition	Moustakas *et al.*, 1994
	Phosphoenol pyruvate carboxylase	*Zea mays*	Inhibition	Vojtechova and Leblova, 1991
Calvin cycle	Glyceraldehyde-3-phosphate dehydrogenase	*Spinach oleracea*	Inhibition	Vallee and Ulmer, 1972
	Ribulose-5-phosphate kinase	*Spinach oleracea*	Inhibition	Vallee and Ulmer, 1972
Pentose phosphate pathway	Glucose-6-phosphate dehydrogenase	*Spinach oleracea*	Inhibition	Vallee and Ulmer, 1972
Nucleolytic enzymes	Decarboxylase	*Hydrilla verticillata*	Enhancement	Jana and Choudhary, 1982
	Ribonuclease	*Hydrilla verticillata*	Enhancement	Jana and Choudhary, 1982
Protein hydrolysis	Protease	*Hydrilla verticillata*	Enhancement	Jana and Choudhary, 1982
Phosphohydrolase	Alkaline phosphatase	*Hydrilla verticillata*	Enhancement	Jana and Choudhary, 1982
	Acid phosphatase	*Glycine max*	Enhancement	Lee *et al.*, 1976
Energy generation	ATP synthetase	*Zea mays*	Inhibition	Tu Shu and Brouillette, 1987
	ATPase	*Zea mays*	Inhibition	Tu Shu and Brouillette, 1987
Antioxidative metabolism	Catalase	*Oryza sativa*	Inhibition	Verma and Dubey, 2003
	Guaiacol peroxidase	*Glycine max*	Enhancement	Lee *et al.*, 1976
	Ascorbate oxidase	*Phaseolus aureus*	Enhancement	Rashid and Mukherji, 1991
	Ascorbate peroxidase	*Oryza sativa*	Enhancement	Verma and Dubey, 2003
	Glutathion reductase	*Oryza sativa*	Enhancement	Verma and Dubey, 2003
	Superoxide dismutase	*Oryza sativa*	Enhancement	Verma and Dubey, 2003

(Jana and Choudhari, 1982). Verma and Dubey (2003) reported increased activities of superoxide dismutase, guaiacol peroxidase, ascorbate peroxidase and glutathione reductase in roots and leaves of rice.

10.2.3.6. Inhibition of photosynthesis

Plants grown under Pb-toxicity showed reduced photosynthetic rate and this is the results of distorted chloroplast ultra-structure, restrained synthesis of chlorophyll, plastoquinone and carotenoids, obstructed electron transport, inhibited activities of Calvin cycle enzymes and reduced CO_2 aviability due to stomatal closure (Pinchasov *et al.*, 2006). Pb-toxicity also changes the lipid composition of thylakoid membrane (Stefanove *et al.*, 1995). Pb inhibits chlorophyll synthesis by causing impaired uptake of essential elements like Mg, and Fe by plants (Burzynski, 1987b). It damages the photosynthetic apparatus due to its affinity for protein N- and S-ligands (Ahmed and Tajmir-Raihi, 1993). An enhancement of chlorophyll degradation occurs in Pb treated plants due to increased chlorophyll activity (Drazkiewicz, 1994). Chlorophyll *b* is more affected than *a* by Pb-toxicity (Vodnik *et al.*, 1999). Pb-toxicity also inhibits electron transport (Rashid *et al.*, 1994).

10.2.3.7. Inhibition of nutrient uptake

Pb blocks the entry of cations (K^+, Ca, Mg, Mn, Zn, Cu, Fe^{3+}) and anions (NO_3^-) in the roots. The inhibition of root growth after exposure to Pb may be due to a decrease in cell division or cell elongation (Haussling *et al.*, 1988). Walker *et al.* (1997) reported decreased uptake of K^+, Ca, Mg, Fe and NO_3^- in *Cucumis sativus* seedlings and Ca, Mg, K and P in *Zea mays*. Phosphorus content was found to be negatively correlated with soil Pb (Paivoke, 2002). Pb-toxicity significantly reduces the root nitrogen content by lowering nitrate reductase activity and disturb nitrogen metabolism (Burzynski and Gabrowski, 1984; Paivoke, 2002).

10.2.3.8. Decline in transpiration and water content in tissues

Pb-toxicity declines transpiration rate and water content in plant tissues. Pb lowers the compounds that are associated with maintaining cell turgor and cell wall plasticity and thus lower the water potential within the cell. Disordered respiration and oxidative photorespiration observed under Pb-toxicity may also cause disarry in the plant water regime. Pb increases the content of ABA and induces stomatal closure (Bazzaz *et al.*, 1974), and it may be due to inhibition of an energy system or due to alterations of K^+ fluxes through membrane.

10.2.3.9. Decrease in respiration

Toxic concentration of Pb decrease respiration rate (Reese and Roberts, 1985). It affects the flow of electrons via the electron transport system (Bazzaz and Govindjee, 1974). The inhibitory effect of Pb at higher concentrations appears to be due to uncoupling of oxidative phosphorylation (Miller *et al.*, 1973).

10.2.3.10. Oxidative damage

The oxidative stress induced by Pb in plants appears to be an indirect effect of Pb toxicity leading to production of ROS (reactive oxygen species), enhancing pro-oxidant systems and affecting iron-mediated processes. ROS produced during membrane-linked electron transport. Catalase enzyme decomposes H_2O_2 to water and molecular oxygen (O_2). Its activity decline under Pb stress environment (Verma and Dubey, 2003). Such a decrease appears to be due to decline in enzyme synthesis or a change in the assembly of enzyme subunits.

10.2.3.11. Reduction in biomass production and yield

Lead toxicity significantly reduced the plant growth and biomass productivity in *Salvinia cuculata* (Phetsombat *et al.*, 2006). Chatterjee *et al.* (2006) also reported significant reduction in biomass production of cabbage under high concentrations of Pb.

10.2.4. TOLERANCE MECHANISMS

Heavy metals are toxic to higher plants by causing oxidative stress, displacing other essential metals in plants pigments or enzymes, leading to disruption of functions of theses molecules and of many metabolic processes, and finally reducing growth and yield (Seregin and Ivanov, 2001; Zhang *et al.*, 2002; Wang *et al.*, 2003). To overcome the heavy metal toxicity problem in agricultural soil, development of genetically tolerant genotypes of crop species are very essential. Therefore, the knowledge of the tolerance mechanisms is important. The two basic principles of Pb-toxicity tolerance are exclusion and accumulation of Pb. Other tolerance mechanisms are avoidance, detoxification and biochemical tolerance.

10.2.4.1. Exclusion of Pb

Tolerant genotypes of plants under Pb-toxicity soils, maintain optimum concentrations of Pb by excluding it from plant parts. Ye *et al.* (1997) reported that the metal exclusion ability of *Typha latifolia* appeared to be related to its oxygen transport capability and radial oxygen loss from the roots resulting in a greater ability to modify the rhizosphere. Tolerant rice varieties up-regulate the synthesis and secretion of oxalate, a compound that precipitate Pb, thereby reducing its uptake (Yang *et al.*, 2000).

10.2.4.2. Binding of Pb

At the root surface, Pb binds to carboxyl groups of mucilage uronic acids and this restricts metal uptake into the root and establishes an important barrier protecting the root system (Rudakova *et al.*, 1988). Some of the bound metals are released when mucilage is biodegraded (Morel *et al.*, 1986).

10.2.4.3. Accumulation and discharge of Pb

Another mechanism of Pb detoxification is sequestering it in the vacuole in the form of complexes. The content of Pb in various plant organs tends to decrease in the following order:

Root > leaves > stem > inflorescence > seeds

In onion (*Allium cepa*) plants, absorbed Pb is localized in root (Michalak and Wierzbika, 1998). Leaves differ in their abilities to accumulate Pb depending on age. Maximum Pb content is found in senescence of leaves and least in young leaves (Godzik, 1993). Within the cells, the major part of Pb is sequestered in the vacuoles. Pinocytosis is observed in leaf cells of many plants treated with Pb salt solution. Formation of such vacuoles is important for the sequestering of excess metal ions, as these vacuoles could protect the cells contents from toxic effects of Pb (Silverberg, 1975). Sometimes, a Pb particle accumulates within the cell wall towards its periphery. The deposition of Pb particles occurs possibly trough the action of pinocytotic vesicles (Ksiazek *et al.*, 1984).

10.2.4.4. Phytochelatins

Plants under metal toxic environment synthesize low molecular weight polypeptides, called phytochilatins (PCs). Binding of Pb metal with PCs leads to sequestration of Pb ions in plants. Synthesis mechanisms of photochelatins have been detailed in section 10.1.4.4.

10.2.4.5. Osmoprotectants

Proline and polyamines involve in defense mechanisms against Pb-toxicity. It has been shown that free proline acts as an osmoprotectant (Paleg *et al.*, 1984).

10.2.4.6. Oxidative enzymes

The reactive oxygen species (ROSs) generated under Pb stress is being catalized by different oxidative enzymes. Pb stress increased peroxidase activity in soybean and rice seedlings (Lee *et al.*, 1976; Verma and Dubey, 2003). Increase in the activity of (super oxide dismutase) SOD in response of Pb appears to be due to *de novo* synthesis of enzymatic protein. Pb-stressed rice seedlings showed increased activities of ascorbate peroxidase (APX) and glutathione reductase (GR).

10.2.4.7. High light intensity

The electron transport rates and coupling activity in the chloroplasts, aldehyde contents, rate of photosynthesis and respiration play important roles in detoxification of lead in plant. Romanowska *et al.* (2006) reported that that the ATP content in the Pb-treated leaves increased to a greater extent in high light intensity than low light intensity grown plants *Pisum sativum*. The presence of Pb ions was found inhibit ATP synthesis activity only in chloroplast from low light intensity grown plants. Low light intensity during growth also lowered PSI electron transport rates and Pb^{2+} induced changes in photochemical activity of this photosystem were visible only in the chloroplast isolated from the plants grown under low light intensity. Their results demonstrated that leaves from plants grown under high light intensity were more resistant to lead toxicity. The light condition during growth might play a role in regulation of photosynthetic and respiratory energy conservation in heavy metal stressed plants by increasing the flexibility of the stoichiometry of ATP to ADP production.

10.2.4.8. Microorganisms

Soil microorganisms may affect heavy metal aviability by the process of bioabsorption, bioaccumulation and solubilization. Ectomychrrhiza influenced uptake, transport and toxicity of Pb in Norway sprouce plants (Marschner *et al.*, 1996).

10.3. NICKEL TOXICITY TOLERANCE

Nickel (Ni) is required for healthy plant growth and it is required at very low concentration (Brown *et al.*, 1987). It forms the active metallo-centre of the hexameric enzyme urease (Genedas *et al.*, 1999). Ni is not toxic at low concentrations, but it becomes toxic at a level of 100 ppm or higher.

10.3.1. Injury Symptoms

1. At higher concentration of Ni inhibits plant growth (Rao and Sresty, 2000; Wang *et al.*, 2001).
2. Ni-toxicity reduces the root growth in rice (Lin and Kao, 2005). Their findings suggested that cell-wall stiffening and lignification are the processes that are enhanced by $NiSO_4$ to permit growth reduction of rice roots.
3. Under Ni-toxicity, oats (*Avena byzantina*) give rise to chlorotic bands (Anderson *et al.*, 1979).

10.3.2. Injury Mechanism

Phytotoxicity of Ni varies within plant species and cultivars as well as with the concentration of Ni in the rooting medium. Ni is rapidly taken by the plant root system and research with different plant species has shown that Ni is able to inhibit a large number of plant enzymes such as those of the Calvin Cycle and chlorophyll biosynthesis (van Assche and Clijsters, 1990). It is also able to inhibit enzymes related to nitrogen metabolism (Boussama *et al.*, 1999). Ni was shown to accumulate mainly in the root system, with little being translocated to the shoots (Cardoso *et al.*, 2005). Thus, toxic levels of Ni inhibit plants root growth. Independent of rooting region and tissue, Ni content in the protoplast exceeded that in the cell walls. Ni penetrates the endodermal barrier and accumulated in the endodermis and pericycle to the highest concentration (Seregin *et al.*, 2003). Thus, accumulation of Ni in the pericycle and suppress the cell division, which restrict the root growth.

Cardoso *et al.* (2005) reported that glutathione reductase (GR) activity responded to the stress induced by Ni *Crotolaria juncea*. Their result suggested that in *C. juncea* GR participates in the detoxification of Ni-induced reactive oxygen species via the glutathione-ascorbate cycle. GR also involves in antioxidative defense mechanism under Cd-induce oxidative stress (Pereira *et al.,* 2002).

10.4. MOLYBDENUM TOXICITY TOLERANCE

10.4.1. Introduction

Molybdenum (Mo) is an essential element in higher plants. Most crops require less than 1 ppm of Mo. Normal plant tissues usually contain between 0.8 and 5 ppm, some plant may contain up to 15 ppm. Soil contains about 0.25-5.0 ppm total Mo. Molybdenum is a major factor for crop growth acidic and poorly drained soils. In acidic soils, Fe and Al hydroxides strongly hold Mo; Ca compounds in calcareous soils do not. Thus, Mo availability increases with increase in soil pH. Soils derived from shale and granite tend to have higher levels of Mo. Molybdenum availability also generally higher in young volcanic soils and in soils high in organic matters.

Mo is an essential component of two major enzymes involved in N-metabolism. Nitrogen fixation by symbiotic N-fixing bacteria requires Mo, and the reduction of NO_3^- anion by the enzyme nitrate reductase requires Mo (Welch, 1995). The usual carriers of Mo are sodium or ammonium molybdate. Mo also acts as an essential element for the development of reproductive parts of plant (Shkolnik, 1984).

Mo does not exist naturally in pure metallic form. Plants take up Mo as $MoO_4^=$ ion. The major fertilizer sources of Mo have been given in Table 10.2. Uptake Mo at high concentrations induces physiological disorders and changes in metabolic pathways in plants.

Table 10.2. Fertilizer sources of molybdenum

Source	Formula	Mo percent
Ammonium molybdate	$(NH_4)_6Mo_7O_{24}.2H_2O$	54
Molybdenum trioxide	MoO_3	66
Sodium molybdate	$Na_2MoO_4.2H_2O$	39

10.4.2. Injury Symptoms

1. Marginal leaf scorch and abscission as found in typical salt damage.
2. Yellowing or browning of leaves.
3. Reduced tillering.

4. In green house, tomato leaves turned golden-yellow and cauliflower seedlings turned purple.

Uptake of Mo at high concentrations induces physiological disorders and changes in metabolic pathways in plants (Warner and Klinhofs, 1992). It also interacts with other mineral elements in plant nutrition.

Molybdenum correlated with increased peroxidase and catalase activity in different cultivars of rice (Routh and Das, 2002). They have suggested that this method could be employed for quick screening of rice cultivars for Mo tolerance in breeding programme. Rout and Das (2002) based on root tolerance index and shoot tolerance index classified Annapurna, Kusuma and Deepa as tolerant rice varieties.

10.5. COPPER TOXICITY

Copper (Cu) is an essential element for plant growth and development, and it is important for various biochemical processes. But, at high concentrations, it is toxic to plants, which interferes many physiological processes.

Cu at toxic levels causes cell membrane damage by binding membrane proteins and lipid peroxidation (De Vos *et al.*, 1992). At higher concentrations of Cu induces free radical formation particularly H_2O_2 (Gallego *et al.*, 1996; Chen and Kao, 1999; Chen *et al.*, 2000). Hydrogen peroxide (H_2O_2) is a necessary substance for cell wall stiffening process, which is considered to be one of the mechanisms resulting in growth inhibition. Cellular damage caused by free radicals might be reduced or prevented by antioxidant, namely ascobate peroxidase, catalase, glutathione reductase, peroxidase and superoxide dismutase.

10.6. CROP MANAGEMENT UNDER HIGH METAL-TOXICITY

10.6.1. Soil Reclamation

The labile fraction of heavy metals in soils is the most important for toxicity for plants and microorganisms. Thus, it is crucial to reduce this fraction in contaminated soils to decrease the negative effect of heavy metals. This can be done through soil reclamation involving *phytostabilization*. It is the ability of plant by which it reduces the intrinsic hazards posed by the contaminated soils, and thus reduces bioaviability in the soil. Usman *et al.* (2006) reported that the labile fraction was reduced due to addition of Na-bentonite and Ca-bentonite for Zn, Cd, Cu, Ni and Pb. Further, the shoot heavy metal concentrations with the exception of Zn were reduced below the phytotoxicity range. The addition of phosphate fertilizers (notably Novaphos) strongly reduced the bioavailability of Pb for wheat plants. Among the heavy metal immobilizing compounds, Na-bentonite and Ca-bentonite have the most promising potential to reduce the bioavailability for the heavy metals.

10.6.2. Biological Agents in Metal Toxicity Tolerance

10.6.2.1. Metal Absorbing Plants

Plant posses homeostatic cellular mechanisms to regulate the concentration of metal ions inside the cell to minimize the potential damage that could result from the exposure to nonessential metal ions. Pytoremediation is the most feasible and eco-friendly method to be applied in the Pb contaminated soils. There are many plants which can absorb heavy metals from the environment without affecting their normal growth and this is known as *phytoextraction*. The phytoextraction involves in accumulation of Pb in the plant parts. It is a remediation that uses plants to remove heavy metals from soil. The success of a phytoextraction process depends on adequate plant yield (aerial parts) and high metal concentrations in plant shoots.

The Pb accumulation in the roots is significantly higher than in shoots (Cunningham *et al.*, 1995; Verma and Dubey, 2003). Plant species with higher shoot/root Pb concentrations are more efficient in Pb translocation. Reeves and Brooks (1983) reported 130-8200 mg Pb/kg shoot dry weight in *Thlaspi rotundifolium*. Some varieties of *Brassica juncia* have been shown to accumulates as high as 1.5% Pb in the shoot when grown in nutrient solution containing 760 mM Pb (Kumar *et al.*, 1995). It has been shown that corn also accumulated high amount of Pb in their shoot (Huang and Cunningham, 1996). Tung and Temple (1996) reported a strong deposition of Pb (0.66 mM) occurs in roots. Malkowski *et al.* (2002) also observed 138430 mg of Pb per kg of dry weight in root tips when corn plants were treated with 10^{-3} M Pb.

Singh *et al.* (2006) studied the adsorption of heavy metals (Cd, Pb, Hg, Cr, Fe, Co, Ni and Cu) by three bryophytes in polluted zones of Delhi. They found that *Riccia* sp. (thalloid frowth form) was able to absorb considerable amount of heavy metals. *Physcomitrium* sp. and *Barbuls* sp. (erect growth form) were low growing and form compact wide turfs on the ground with all parts of their gametophores remaining in contact with the substratum. They concluded that the mosses have better toxicity tolerance than liverworts. Mahmoud *et al.* (2005) reported sun flower (*Helianthus annus*) has high extraction potential for removal of Pb and Zn from polluted soils.

10.6.2.2. Mychorryza in metal toxicity tolerance

Mychorryzal symbiosis reduces the transportation of heavy metals from root to shoot by immobilizing heavy metals in the mychorryza. Arbuscular mychorryzal fungus *Glomus mosseae* inoculated plants of *Vicia faba* showed significantly increased growth and tolerance to toxicity induced by heavy metals compared with uninoculated plants (Zhang *et al.*, 2006). *Glomus mosseae* treatment decreased oxidative stress by intricating antioxidative systems such as peroxidase and non-enzymatic systems including soluble protein. Zhang *et al.* (2006) reported that DNA damage due to oxidative stress in plants was decreased by the *Glomus mosseae* treatment.

10.6.2.3. Green sea weed in metal toxicity tolerance

Cadmium toxicity and problems with regards to tolerance, physiological processes and ecological significance with green sea weeds was studied by Rafia *et al.* (2006). The sea weed *Codium iyengarii*, which contains highest amount of essential mineral ions (Ca, Fe, K, Na and Mg) used to control the toxicity of Cd in bean plant. The result indicated that the sea weed increased the tolerance ability of bean plant under higher concentration of Cd. Healthy growth of plants were recorded up to 100 ppm of Cd concentration which improved up taking of mineral ions from soil into the roots and shoots. It was also observed that sea weed acts as catalyst starter in the soil irrigated with contaminated water. *Codium iyengarii* improves the morphology and physiology processes of plant and help in the population of soil bacteria in heavy metal contaminated environment.

10.7. REFERENCES

Adhikari T, Biswas AK, Saha JK, Ajay. 2005. Cadmium toxicity in spinach with or without spent wash in a Vertisol. Comm Soil Sci Plant Analysi. 36(11/12): 1499-1511.

Adhikari T, Tel-Or E, Libal Y, Shenker M. 2006. Effect of cadmium and iron on rice (*Oryza sativa* L.) plant in chelator-buffered nutrient solution. J Plant Nutr. 29(11): 1919-1940.

Ahmed A, Tajmir-Riahi HA. 1993. Interaction of toxic metal ions Cd^{2+}, Hg^{2+} and Pb with light-harvesting proteins of chloroplast thylakoid membranes. An FTIR spectroscopic study. J Inorg Biochem. 50: 235-243.

Alcantara E, Romera FJ, Cañete M, De La Guardia MD. 1994. Effects of heavy metals on both induction and function of root Fe (III) reductase in Fe-deficient cucumber (*Cucumis sativus* L.) plants. J Exp Bot. 45:1893-1898.

Anderson AJ, Meyer DR, Mayer FK. 1979. Effects of the environment on the symptom pattern of nickel toxicity in the oat plant. Annals Bot. 43: 271-283.

Angelone M, Bini C. 1992. Trace elements concentrations in soils and plants of western Europe. In: Adriano DC (ed.), Biogeochemistry of Trace Metals. Lewis Publishers, Boca Raton, London. pp.19-60.

Antosiewicz DM, Henning J. 2004. Overexpression of LTC1 in tobacco enhances the protective action of calcium against cadmium toxicity. Environl Pollution. 129(2): 237-245.

Arao T, Ae N. 2003. Genotypic variation in cadmium levels of rice grain. Soil Sci Plant Nutr. 49(4): 473-479.

Balestrasse KB, Benavides MP, Gallego SM, Tomaro ML. 2003. Effect on cadmium stress on nitrogen metabolism in nodules and roots of soybean plants. Func Plant Biol. 30:57-64.

Balestrasse KB, Gallego SM, Tomaro ML. 2004. Cadmium-induced senescence in nodules of soybean (*Glycine max* L.) plants. Plant Soil. 262:373-381.

Balestrasse KB, Gardey L, Gallego SM, Tomaro ML. 2001. Response of antioxidant defence system in soybean nodules and roots subjected to cadmium stress. Aust J Plant Physiol. 28: 497-504.

Bazzaz FA, Rolfe GL, Windle P. 1974. Differing sensitivity of corn and soybean photosynthesis and transpiration to lead contamination. J Environ Qual. 3:156-158.

Bazzaz MB, Govindjee. 1974. Effect of lead chloride on chloroplast reactions. Environ. Lett. 6: 175-191.

Bert V, Meerts P, Saumitou-Laprade P, Salis P, Gruber W, Verbruggen N. 2003. Genetic basis of Cd tolerance and hyperaccumulation in *Arabidopsis halleri*. Plant Soil. 249: 9-18

Blaudez D, Botton B, Chalot M. 2000. Cadmium uptake and subcellular compartmentation in the ectomycorrhizal fungus *Paxillus involutus*. Microbiology. 146: 1109-1117.

Blum WH .1997. Cadmium uptake by higher plants. In: Proceedings of extended abstracts from the Fourth International Conference on the Biogeochemistry of Trace Elements, pp.109-110, Berkeley, USA. University of California.

Boussama N, Quariti O, Ghorbal MH. 1999. Changes in growth and nitrogen assimilation in barley seedlings under cadmium stress. J Plant Nutr. 22: 731-752.

Breckle SW. 1991. Growth under stress. Heavy metals. In: Waisel Y, Eshel A, Kafkafi U (eds.), Plant Roots: The Hidden Half. Marcel Dekker Inc., New York, USA. pp. 351-373.

Brown PH, Welch RM, Cary EE. 1987. Nickle: A micronutrient essential for high plants. Plant Physiol. 85: 801-803.

Burton KW, Morgan E, Roig A. 1984. The influence of heavy metals on the growth of sitka-spruce in South Wales forests. II green house experiments. Plant Soil. 78: 271-282.

Burzynski M, Grabowski A. 1984. Influence of lead on nitrate uptake and reduction in cucumber seedlings. Acta Soc Bot Pol. 53:77-86.

Burzynski M, Jacob M. 1983. Influence of lead on auxin-induced cell elongation. Acta Societatis Botanicorum Poloniae. 52: 231-239.

Burzynski M. 1987a. The influence of lead and cadmium on the absorption and distribution of potassium, calcium, magnesium and iron in cucumber seedlings. Acta Physiol Plant. 9: 229-238.

Burzynski M. 1987b. The uptake and transpiration of water and accumulation of lead by plants growing on lead chloride solutions. Acta Societatis Botanicorum Poloniae. 56: 271-280.

Cardoso PF, Gratao PL, Gomes-Junior RA, Medici LO, Azevedo RA. 2005. Response of *Crotolaria juncea* to nickel exposure. Braz J Plant Physiol. 17(2): 267-272.

Cataldo DA, Garland TR, Wildung RE. 1983. Cadmium uptake kinetics in intact soybean plants. Plant Physiol. 73: 844-848.

Chaffei C, Pageau K, Suzuki A, Gouia H, Ghorbel MH, Masclaux-Daubresse C. 2004. Cadmium toxicity induced changes in nitrogen management in *Lycopersicon esculentum* leading to a metabolic safeguard through an aminoacid storage strategy. Plant Cell Physiol. 45(11): 1681-1693.

Chatterjee C, Sinha P, Dube BK, Parul S. 2006. Lead toxicity induced damages in growth and metabolism of cabbage. Indian J Hort. 63(4): 393-396.

Chen LM, Kao CH. 1999. Effect of excess copper on rice leaves: evidence for involment of lipid peroxidation. Bot Bull Acad Sin. 40: 283-287.

Chen LM, Lin CC, Kao CH. 2000. Copper toxicity in rice seedling: Changes in antioxidative enzyme activities, H_2O_2 levels, and cell wall peroxidase activity in root. Bot Bull Acad Sin. 41: 99-103.

Chen SL, Kao CH. 1995. Cd induced changes in proline levels and peroxidase activity in the root of rice seedlings. Plant Growth Regulation. 17: 67-71.

Chien HF, Wang JW, Lin CC, Kao CH. 2001. Cadmium toxicity of rice leaves is mediated through lipid peroxidation. Plant Growth Regulation. 33(3): 205-213.

Cho U, Seo N. 2004. Oxidative stress in *Arabidopsis thaliana* exposed to cadmium is due to hydrogen peroxide accumulation. Plant Sci. 168: 113-120.

Clemens S, Palmgreen MG, Kramer U. 2002. A long way ahead: understanding and engineering plant metal accumulation. Trends Plant Sci. 7: 309-315.

Clemens S. 2001. Molecular mechanisms of plant metal tolerance and homeostasis. Planta 212: 475-486.

Cressen G, Firmin J, Fryer M, Kular B, Leyland N, Reynolds H, Pastari G, Wellburn F, Baker N, Wellburn A, Mullineause P. 1999. Elevated glutathione biosynthetic capacity in the chloroplasts of transgenic tobacco plants paradoxically causes increased oxidative stress. Plant Cell. 11: 1277-1291.

Cunningham SD, Berti WR, Huang JW. 1995. Phytoremediation of contaminated soils. Trends Biotechnol. 13: 393-397.

Das P, Samantaray S, Rout GR. 1997. Studies on cadmium toxicity in plants: a review. Environ Pollution. 98: 29-36.

De Filippis LF, Ziegler H .1993. Effect of sublethal concentrations of zinc, cadmium and mercury on the photosynthetic carbon reduction cycle of *Euglena*. J Plant Physiol. 142:167-172.

De Vos CHR, Vonk MJ, Schat H. 1992. Glutathione depletion due to copper induced phytochelatin synthesis cause oxidative stress in *Silene cucbalus*. Plant Physiol. 98: 853-858.

Demirevska-Kepova K, Simova-Stoilova L, Stoyanova ZP, Feller U. 2006. Cadmium stress in barley: growth, leaf pigment, and protein composition and detoxification of reactive oxygen species. J Plant Nutrn. 29930; 451-468.

Drazkiewicz M. 1994. Chlorophyll-occurrence, functions, mechanism of action, effects of internal and external factors. Photosynthetica. 30: 321-331.

Eick MJ, Peak JD, Brady PV, Pesek JD. 1999. Kinetics of lead adsorption and desorption on goethite: Residence time effect. Soil Sci. 164: 28–39.

Eun SO, Youn HS, Lee Y. 2000. Lead disturbs microtubule organization in the root meristem of *Zea mays*. Physiol Plant. 110: 357-365.

Fodor A, Szabó-Nagy A, Erdei L. 1995. The effects of cadmium on the fluidity and H^+-ATPase activity of plasma membrane from sunflower and wheat roots. J Plant Physiol. 14: 787–792.

Fornazier RF, Ferreira RR, Vitória AP, Molina SMG, Lea PJ, Azevedo RA. 2002. Effects of cadmium on antioxidant enzyme activities in sugar cane. Biol Plant. 45: 91-97.

Foyer CH, Lopez-Delgado H, Dat JF, Scott IM. 1997. Hydrogen peroxidase- and glutathione-associated mechanisms of acclamatory stress tolerance and signaling. Physiol Plant. 100: 241-254.

Gallego SM, Benavides MP, Tomaro ML. 1996. Effect of heavy metal ion excess on sunflower leaves: evidence for involvement of oxidative stress. Plant Sci. 121: 151-159.

Gekeler W, Grill E, Winnacker EL, Zenk MH .1989. Survey of the plant kingdom for the ability to bind heavy metals through phytochelatins. Zeitschrift Naturforsch. 44: 361-369.

Genendas J, Polacco JC, Freermuth SK, Sattelacher B. 1999. Significance of nickel for plant growth and metabolism. J Plant Nutr SAoil Sci. 162: 241-256.

Godzik B. 1993. Heavy metal contents in plants from zinc dumps and reference area. Pol Bot Stud. 5: 113-132.

Grill E, Loffler S, Winnacker EL, Zenk, M H. 1989. Phytochelatins, the heavy-metal-binding peptides of plants, are synthesized from glutathione by a specific ?-glutamylcysteine dipeptidyl transpeptidase (phytochelatin synthase). Proc Natl Acad Sci, USA 86: 6838–6842.

Hall JL. 2002. Cellular mechanisms for heavy metal detoxification and tolerance. J Exp Bot. 53: 1-11.

Hang F, Shang XQ, Zhang SZ, Bei W, Owens G. 2006. Enhanced cadmium accumulation in maize roots- the impact of organic acids. Plant and Soil. 289(1/2): 355-368.

Harada E, Choi YE, Tsuchisaka A, Obata H, Sano H, Choi YE. 2001. Transgenic tobacco plants expressing a rice synthase gene are tolerant to toxic levels of cadmium. J Plant Physiol. 158(5): 655-661.

Hassan MJ, Shao GS, Zhang GP. 2005. Influence of cadmium toxicity on growth and antioxidant enzyme activity in rice cultivars with different grain cadmium accumulation. J Plant Nutr. 28(7): 1259-1270.

Haussling M, Jorns CA, Lehmbecker G, Hecht-Buchholz C, Marschner H. 1988. Ion and water uptake in relation to root development of Norway spruce (*Picea abies* (L.) Karst). J Plant Physiol. 133: 486-491.

Hernández LE, Cárpena-Ruiz R, Garate A. 1996. Alterations in the mineral nutrition of pea seedlings exposed to cadmium. J Plant Nutr. 19: 1581–1598.

Howden R, Anderson CR, Goldsbrough PB, Covvett CS. 1995. A cadmium-sensitive, glutathione-deficient mutant of *Arabidopsis thaliana*. Plant Physiol. 107: 1067-1073.

Huang JW, Cunningham SD. 1996. Lead Phytoextraction: species variation in lead uptake and translocation. New Phytol. 134: 75-84.

Iqbal J, Mushtaq S. 1987. Effect of lead on germination, early seedling growth, soluble protein and acid phosphatase content in *Zea mays*. Pak J Sci Ind Res. 30: 853-856.

Jana S, Choudhari MA. 1982. Senescence in submerged aquatic angiosperms: effects of heavy metals. New Phytol. 90: 477- 484.

Jentshcke G, Godbold DL. 2000. Metal toxicity and ectomycorrhizas. Physiol Plant. 109: 107-116.

Kacabova P, Natr L. 1986. Effect of lead on growth characteristics and chlorophyll content in barley seedlings. Photosynthetica. 20: 411-417.

Kamenova-Jouhimenko S, Georgieva V, Hristov H, Markovska Y. 2003. Protective action of humic acids against cadmium toxicity in plants (*Pisum sativum*). Rasteniev' dni Nauki. 40(3): 283-287.

Klapheck S, Schlunz S, Bergmann L. 1995. Synthesis of phytochelatins and homo-phytochelatins in *Pisum sativum* L. Plant Physiol. 107: 515-521.

Koleli N, Eker S, Cakmak I. 2004. Effect of Zn fertilization on cadmium toxicity in durum and bread wheat grown in Zn deficient soil. Environl Pollution. 131(3): 453-459.

Korshunova YO, Eide D, Clark WG, Guerinot ML, Prakasi HB. 1999. The IRT1 protein from *Arabidopsis thaliana* is a metal transporter with a broad substrate range. Plant Mol Biol. 40: 37-44.

Krupka Z, Zkorzynska E, Maksymiec W, Basznski T. 1987. Effect of cadmium treatment on the photosynthetic apparatus and its phytochemical activities in green radish seedlings. Photosynthetica. 21: 156-164.

Ksiazek M, Wozny A, Mlodzianowski F. 1984. Effect of $Pb(NO_3)_2$ on poplar tissue culture and the ultrastructural localization of lead in culture cells. For Ecol Manag. 8: 95-105.

Kumar GP, Prasad MNV. 2004. Cadmium toxicity to *Ceratphyllum demersum* L.: Morphological symptoms, membrane damage, and ion leakage. Bulletin of Environmental Contamination and Toxicology. 72(5): 1038-1045.

Kumar NPBA, DushenkovV, Motto H, Raskin I. 1995. Phytoextraction: the use of plants to remove heavy metals from soils. Environ Sci Technol. 29: 1232-1238.

Lasat MM, Pence NS, Garvin DF, Ebbs SD, Kochian LV. 2000. Molecular physiology of zinc transport in the Zn hyperaccumulator *Thlaspi caerulescens*. J Exp Bot. 51: 71–79.

Lee KC, Cunningham BA, Poulsen GM, Liang JM, Moore RB. 1976. Effects of cadmium on respiration rate and activities of several enzymes in soybean seedlings. Physiol Plant. 36: 4-6.

Lee S-Z, Chang L, Yang H-H, Chen C-M, Liu M-C. 1998. Absorption characteristics of lead onto soils. J Haz Mat. 63:37-49.

Leita L, De Nobili M, Cesco S, Mondini C. 1996. Analysis of intercellular cadmium forms in roots and leaves of bush bean. J Plant Nutr. 19: 527–533.

Levina EN. 1972. Obshchaya tosikologiya metallov (General metal toxicology). Leningrad, Meditsyna.

Li YM, Chaney RL, Schneiter AA, Miller JF, Elias EM, Hammond JJ. 1997. Screening for low grain cadmium wheat and flax. Euphytica. 94(1): 23-30.

Lin YC, Kao CH. 2005. Nickle toxicity of rice seedlings: Cell wall peroxidase, lignin, and $NiSO_4$-inhibited root growth. Crop, Environ & Bioinformatics. 2: 131-136.

Liu JG, Liang JS, Li KQ, Zhang ZJ, Yu BY, Lu XL, Yang JC, Zhu QS. 2003. Correlation between cadmium and minearal nutrients in absorption and accumulation in various genotypes of rice under cadmium stress. Chemosphere. 52(9): 1467-1473.

Mahmoud S, Hossain S, Hajabbasi MA. 2005. Lead and zinc extraction potential of two common crop plants, *Helianthus annuus* and *Brassica nupus*. Water, Air Soil and Pollution. 167(1/4): 59-71.

Maitra P, Mukherji S. 1977. Effect of lead on nucleic acid and protein contents of rice seedlings and its interaction with IAA and GA_3 in different plant systems. Ind J Exp Biol. 17: 29-31.

Malkowski E, Kita A, Galas W, Karez W, Michael K. 2002. Lead distribution in corn seedlings (*Zea mays* L.) and its effect on growth and the concentration of potassium and calcium. Plant Growth Regul. 37: 69-76.

Marschner H. 1995. Mineral nutrition of higher plants, 2nd edn. Academic Press, London.

Marschner P, Godbold DL, Jutschhe G. 1996. Dynamics of lead accumulation in mycorrhizal and non-mycorrhizal Norway spruce (*Picea abies* (L.) karst.). Plant Soil. 178: 239-245.

Meharg AA. 1993. The role of plasmalemma in metal tolerance in angiosperms. Physiol Plant. 88: 191-198.

Metwally A, Finkemeier I, Georgi M, Dietz KJ. 2003. Salicylic acid alleviates the cadmium toxicity in barley seedlings. Plant Physiol. 132(1): 272-281.

Metwally A, Safronova VI, Belinove AA, Diez KJ. 2005. Genotype variation of the response to cadmium toxicity in *Pisum sativum* L. J Exp Bot. 56(409): 1167-178.

Michalak E, Wierzbicka M. 1998. Differences in lead tolerance between *Allium cepa* plants developing from seeds and bulbs. Plant Soil. 199: 251-260.

Miller RJ, Biuell JE, Koeppe DE, 1973. The effect of cadmium on electron and energy transfer reactions in corn mitochondria. Physiol Plant. 28: 166-171.

Mishra A, Choudhari MA. 1998. Amelioration of lead and mercury effects on germination and rice seedling growth by antioxidants. Biol Plant. 41: 469-473.

Morel JL, Mench M, Guckert A. 1986. A measurement of Pb, Cu, Cd binding with mucilage exudates from maize (*Zea mays* L.) roots. Biol Fertil Soils. 2: 29-34.

Moreno JL, Hernandez T, Garcia C. 1999. Effects of a cadmium-containing sewage sludge compost on dynamics of organic matter and microbial activity in an arid soils. Biol Fert Soils. 28: 230-237.

Moustakas M, Lanaras T, Symeonidis L, Karataglis S. 1994. Growth and some photosynthetic characteristics of field grown *Avena sativa* under copper and lead stress. Photosynthetica. 30: 389-396.

Mukherjee A, Sharma A. 1988. Effects of cadmium and selection on cell division and chromosomal aberration in *Allium sativum*L. Water, Air and Soil Pollution. 37: 433-438.

Mukherji S, Maitra P. 1976. Toxic effects of lead on growth and metabolism of germinating rice (*Oryza sativa* L.) root tip cells. Ind J Exp Biol. 14: 519-521.

Naquib MI, Hamed AA, Al-Waheel SA. 1986. Effect of cadmium on growth criteria of some crop plants. Egyptian J Bot. 25: 1-12.

Nishizono H, Kubota K, Suzuki S, Ishii F. 1989. Accumulation of heavy metals in cell walls of *Polygonum cuspidatum* roots from metalliferous habitats. Plant Cell Physiol. 30: 595-598.

Nouairi I, Ammar WB, Youssef NB, Miled DDB, Gjorbal MH, Zarrouk M. 2006. Variation in membrane lipid metabolism in *Brassica juncea* and *Brassica napus* leaves as a response to cadmium exposure. J Agron. 5(2): 299-307.

Obroucheva NV, Bystrova EI, Ivanov VB, Anupova OV, Seregin IV. 1998. Root growth responses to lead in young maize seedlings. Plant Soil. 200: 55-61.

Paivoke AEA. 2002. Soil lead alters phytase activity and mineral nutrient balance of *Pisum sativum*. Environ Exp Bot. 48: 61-73.

Paleg LG, Stewart GR, Bradbeer JW. 1984. Proline and glycine betaine influence on protein solvation. Plant Physiol. 75:974-978.

Pavlicova D, Macek T, Mackova M, Szakova J, Balik J. 2004. Cadmium tolerance and accumulation in transgenic tobacco plants with a yeast metallothionein combined with a polyhistidin tail. Intrnl Biodeterioration and Biodegradation. 54(2/3): 233-237.

Penner GA, Clarke J, Bezte LJ, Leise D. 1995. Identification of RAPD markers linked to a gene governing cadmium uptake in durum wheat. Genome. 38: 543-547.

Pereira GJG, Molina SMG, Lea PJ, Azevedo RA. 2002. Activity of antioxidant enzymes in response to cadmium in *Crotolaria juncea*. Plant Sci. 239: 123-132.

Phetsombat S, Kruatrachue M, Pokethitiyook P, Upatham S. 2006. Toxicity and bioaccumulation of cadmium and lead in *Salvinia cuculata*. L Environl Biol. 27(4): 645-652.

Pinchasov Y, Berner T, Dubinsky Z. 2006. The effect of lead on photosynthesis, as determined by photoacoustics in *Synechococcus leopoliensis* (Cyanobacteria). Water, Air and Soil Pollution. 175(1/4): 117-125.

Pinto AP, Mota AM, de Varennes A, Pinto FC. 2004. Influence of organic matter on the uptake of cadmium, zinc, copper and iron by sorghum plants. Sci Tot Environ. 326: 239-247.

Poschendieder C, Gunse B, Barcelo J. 1989. Influence of cadmium on water relations, stomatal resistance, and abscisic acid content in expanding bean leaves. Plant Physiol. 90: 1365-1371.

Poskuta JW, Parys E, Romanowska E. 1987. The effect of Pb on the gaseous exchange and photosynthetic carbon metabolism of pea sedlings. Acta Societas Botanicorum Polonial. 56: 127-137.

Prassad DDK, Prassad ARK. 1987. Altered δ-aminolaevulinic acid metabolism by lead and mercury in germinating seedlings of Bajra (*Pennisetum typhoideum*). J Plant Physiol. 127: 241-249.

Przybulewska K. 2004. Influence of the concentration of calcium on growth and development of barley seedlings depending upon soil reactions. J Elementology. 9(2): 157-163.

Przymusinski R, Spychala M, Gwozdz EA. 1991. Inorganic lead changes growth polypeptide pattern of lupin roots. Biochem Physiol Pflan. 187: 51-57.

Radecki J, Banaszkiewicz T, Klasa A. 1989. The effect of different leaf compounds on mitotic activity of maize root tip cells. Acta Physiol Plant. 11: 125-130.

Rafia A, Aliya H, Tanveer K, Rukhasna T, Fahim U. 2006. Effect of micronutrients of *Codium iyengarii* on metal toxicity in bean plants. J Biol Sci. 6(1): 173-177.

Rani SMV, Muthuchelian K, Paliwal K. 1987. Differntial toxicity of Cu^{2+} and Cd^{2+} on chlorophyll biosynthesis and nitrate reductase activity in *Phaseolus mungo* L. Annals Plant Physiol. 1: 126-135.

Rao KVM, Sresty TVS. 2000. Antioxidative parameters in the seedling of pigeon pea (*Cajanus cajan* (L.) Mullsourgh) in response to Zn and Ni stresses. Plant Sci. 157: 113-128.

Rashid A, Camm EL, Ekramoddoullah KM. 1994. Molecular mechanism of action of Pb and Zn^{2+} on water oxidizing complex of photosystem II. FEBS Lett. 350: 296-298.

Rashid P, Mukherji S. 1991. Changes in catalase and ascorbic oxidase activities in response to lead nitrate treatments in mungbean. Ind J Plant Physiol. 34: 143-146.

Rauser WE. 1995. Phytochelatins and related peptides: structure, biosynthesis, and function. Plant Physiol. 109: 1141–1149.

Reese RN, Roberts LW. 1985. Effects of cadmium on whole cell and mitochondrial respiration in tobacco cell suspension cultures (*Nicotiana tobacum* L. var. Xanthi). J Plant Physiol. 120: 123-130.

Reeves RD, Brooks RR. 1983. European species of *Thlaspi* L. (Cruciferae) as indicators of nickel and zinc. J Geochem Explor. 18: 275–283.

Romanowska E, Wroblewska B, Drozak A, Siedlecka M. 2006. High light intesisty protects photosynthetic apparatus of pea plants against exposure to lead. Plant Physiol Biochem. 44(4/5): 387-394.

Rout GR, Das P. 2002. Rapid hydroponic screening for molybdenum tolerance in rice through morphological and biochemical analysis. Rostlinna Vyroba. 48(11): 505-512.

Rudakova EV, Karakis KD, Sidorshina ET. 1988. The role of plant cell walls in the uptake and accumulation of metal ions. Fiziol Biochim Kult Rast. 20: 3-12.

Salt DE, Blaylock M, Kumar NPBA, Dushenkov V, Ensley BD, Chet I, Raskin I. 1995b. Phytoremediation: a novel strategy for the removal of toxic metals from the environment using plants. Biotechnol. 13: 468–474.

Salt DE, Prince RC, Pickering IJ, Raskin I. 1995a. Mechanisms of cadmium mobility and accumulation in Indian mustard. Plant Physiol. 109: 427-433.

Sandalio LM, Dalurzo HC, Gomez M, Romero-Puertas MC, del Río LA. 2001. Cadmium-induced changes in the growth and oxidative metabolism of pea plants. J Exp Bot. 52: 2115-2126.

Sanitá di Toppi L, Gabbrielli R. 1999. Response to cadmium in higher plants. Environ Exp Bot. 41: 105-130.

Senden MHMN, Van der Meer AJGM, Verburg TG, Wolterbeek HT. 1994. Effects of cadmium on the behaviour of citric acid in isolated tomato xylem cell walls. J Exp Bot. 45: 597–606.

Senden MHMN, Van Paassen FJM, Van der Meer AJGM, Wolterbeek HT. 1992. Cadmium-citric acid xylem cell wall interactions in tomato plants. Plant Cell Environ. 15: 71–79.

Seregin IV, Ivanov VB. 2001. Physiological aspects of cadmium and lead toxicity effects on higher plants. Russ J Plant Physiol. 48(4): 523-544.

Seregin IV, Kozhevinkova AD, Kazyumina EM, Ivanov VB. 2003. Nickle toxicity and distribution in maize roots. Russian J Plant Physiol. 50(5): 711-717.

Shah K, Dubey RS. 1997. Effect of cadmium on proteins, amino acids and protease, aminopeptidase in rice seedlings. Plant Physiol Biochem. 24: 89-95.

Shkolnik MY. 1984. Molybdenum. In: Trace Elements in Plants. Vol. 6, Shikolnik MY (ed.). Elsevier Sci Publ, The Netherlands. pp. 195-231.

Silverberg BA. 1975. Ultrastructural localization of lead in *Stigeoclonium tenue* (Chlorophyseae Ulotrichales) as demonstrated by cytochemical and X-ray microanalysis. Phycologia 14: 265-274.

Singh L, Uniyal PL, Sharma KR, Santosh V, Preeti V. 2006. Study on the metal absorption by bryophytes in Delhi Region. Biochem Cellular Archives. 6(1): 129-133.

Singh SP, Takker PN, Nayyar VK. 1989. Effect of cadmium on wheat as influenced by line and manure and its toxic level in the plant and soil. Int J Environ Studies. 33: 59-66.

Somashekaraiah BV, Padmaja K, Prasad ARK. 1992. Phytotoxicity of cadmium ions on germinating seedlings of mung bean (*Phaseolus vulgaris*): involvement of lipid peroxides in chlorophyll degradation. Physiol Plant. 85: 85–89.

Song WY, Sohn EJ, Martino E, Lee YJ, Yang YY, Jasinski M, Forestier C, Hwang I, Lee Y. 2003. Engineering tolerance accumulation of lead and cadmium in transgenic plants. Nature Biotechnol. 21: 914-919.

Sotiropoulos TE, Therios IN, Dimassi KN. 2006. Seasonal accumulation and distribution of nutrient elements in fruit of kiwifruit vines affected by boron toxicity. Australian J Exp Agri. 46(12): 1639-1644.

Staffens JC. 1990. The heavy metal-binding peptides of plants. Annu Rev Plant Physiol Plant Mol Biol. 41: 553-575.

Stefanov K, Popova I, Kamburova E, Pancheva T, Kimenov G, Kuleva L, Popov S. 1993. Lipid and sterol changes in *Zea mays* caused by lead ions. Phytochemistry. 33: 47-51.

Stefanov K, Seizova K, Popova I, Petkov VL, Kimenov G, Popov S. 1995. Effects of lead ions on the phospholipid composition in leaves of *Zea mays* and *Phaseolus vulgaris*. J Plant Physiol. 147: 243-246.

Stiborova M, Doubarovova M, Leblova S. 1986. A comparative study of heavy metal ions in ribulose 1,5-biphosphate carboxylase and phosphoenol pyruvate carboxylase. Biochem Physiol Pflanzen. 181: 373-379.

Stiborova M. 1988. Cd^{2+} ions affect the quaternary structure of ribose-1,5-biphosphate carboxylase from barley leaves. Biochem Physiol Pflanzen. 183: 371-378.

Stobart AK, Griffiths WT, Ameen-Bukari I, Shearwood RP. 1985. The effect of Cd^{2+} on the biosynthesis of chlorophyll in leaves of barley. Physiol Plant. 63: 293-398.

Stohs SJ, Bagchi D. 1995. Oxidative mechanisms in the toxicity of metal ions. Free Rad Biol Med. 18: 321-336

Sung MW, Kwon BK. 1980. Effect of Pb surplus and P deficiency on ATP content of plant leaves. Korean J Bot. 23(2): 37-44.

Thomine S, Wang R, Ward JM, Crawford NM, Schroeder JI. 2000. Cadmium and iron transport by members of a plant metal transporter family in *Arabidopsis* with homology to Nramp genes. Proc Natl Acad Sci, USA. 97: 4991-4996.

Tu Shu I, Brouillette JN. 1987. Metal ion inhibition of corn root plasmamembrane ATPase. Phytochemistry. 26: 65-69.

Tung G, Temple PJ. 1996. Uptake and localization of lead in corn (*Zea mays* L.) seedlings: a study by histochemical and electron microscopy. Sci Total Environ. 188: 71-85.

Usman ARA, Kuzyakov Y, Stahr K. 2006. Remediation of a soil contaminated with heavy metals by immobilizing compounds. J Plant Nutr Soil Sci. 169(2): 205-212.

Vallee BL, Ulmer DD. 1972. Biochemical effects of mercury, cadmium and lead. Annu Rev Biochem. 41: 91-128.

Van Assche F, Clijsters H. 1990. Effects of metal on enzyme activity in plants. Plant Cell Environ. 13: 195-206.

Vanaja M, Charyulu NVN, Rao KVN. 2000. Effect of cadmium on carbohydrate, nucleic acid, amino acid and phenolic content in *Stgeoclonium tenue* Kutz. Indian J Plant Physiol. 5(3): 253-256.

Verkleij JAC, Schat H. 1990. Mechanisms of metal tolerance in higher plants. In: Shaw J. (eds.), Heavy metal tolerance in plants: evolutionary aspects. CRC Press, Boca Raton. pp.179-193.

Verma S, Dubey RS. 2003. Lead toxicity induces lipid peroxidation and alters the activities of antioxidant enzymes in growing rice plants. Plant Sci. 164: 645-655.

Vodnik D, Jentschke G, Fritz E, Gogala N, Godbold DL. 1999. Root-applied cytokinin reduces lead uptake and affects its distribution in Norway spruce seedlings. Physiol Plant. 106: 75-81.

Vojtechova M, Leblova S. 1991. Uptake of lead and cadmium by maize seedlings and the effect of heavy metals on the activity of phosphoenolpyruvate carboxylase isolated from maize. Biol Plant. 33: 386-394.

Wagner GJ. 1993. Accumulation of cadmium in crop plants and its consequences to human health. Adv Agron. 51: 173-212.

Walker WM, Miller JE, Hassett JJ. 1997. Effect of lead and cadmium upon the calcium, magnesium, potassium and phosphorus concentration in young corn plants. Soil Sci. 124: 145-151.

Wang CX, Mo Z, Wang H, Wang ZJ, Cao ZH. 2003. The transportation, time-dependant distribution of heavy metals in paddy crop. Chemosphere. 50 (6): 717-723.

Wang HH. Kang J, Zeng FH, Jiang MY. 2001. Effect of nickel at high concentration of enzymes of rice seedlings. Acta Agron Sin. 247: 953-957.

Warnner RL, Kleinhofs A. 1992. Genetics and molecular biology of nitrate metabolism in higher plants. Physiol Plant. 85: 245-252.

Welch RM. 1995. Micronutrient nutrition of plants. Crit Rev Plant Sci. 14: 49-82.

Wierzbicka M. 1994. Resumption of mitotic activity in *Allium cepa* root tips during treatment with lead salts. Environ Exp Bot. 34: 173-180.

Williams LE, Pittman JK, Hall JL. 2000. Emerging mechanisms for heavy metal transport in plants. Biochim Biophys Acta. 1465: 104–126.

Woltz SS, Chambliss CG. 1981. Cadmium toxicity in some vegetables and test plant species. Proc Florida State Soc Pub. 92: 110-112.

Wozny A, Zatorska B, Mlodzianowski F. 1982. Influence of lead on the development of lupin seedlings and ultrastructural localization of this metal in the roots. Acta Soc Bot Pol. 51: 345-351.

Wu FB, Zhang GP. 2002. Alleviation of cadmium–toxicity by application of zinc and ascorbic acid in barley. J Plant Nutr. 25(12): 2745-2761.

Yang Y-Y, Jung J-Y, Song W-Y, Suh HS, Lee Y. 2000. Identification of rice varieties with high tolerance or sensitivity to lead and characterization of the mechanism of tolerance. Plant Physiol. 124: 1019-1026.

Ye ZH, Baker AJM, Wong MH, Willis AJ. 1997. Zinc, lead and cadmium tolerance, uptake and accumulation in populations of *Typha latifolia* L. New Phytol. 136: 469-480.

Zenk MH. 1996. Heavy metal detoxification in higher plants: a review. Gene. 197: 21-30.

Zhang GP, Fukami M, Sekimoto H. 2002. Influence of cadmium on mineral concentrations and yield components in wheat genotypes differing in Cd tolerance at seedling stage. Field Crops Res. 77(2-3): 93-98.

Zhang GP, Fukami M, Sekimoto H. 2000. Genotypic differences in effects of cadmium on growth and nutrient compositions in wheat. J Plant Nutr. 23(9): 1337-1350.

Zhang XH, Lin AJ, Chen BD, Wang YS, Smith SE, Smith FA. 2006. Effects of *Glomus mosseae* on the toxicity of heavy metals to *Vicia faba*. J Environl Sci. 18(4): 721-726.

Zhu YL, Pilon-Smits EAH, Jounin L, Terry T. 1999b. Over expression of glutathione synthetase in *Brassica juncia* enhances cadmium accumulation and tolerance. Plant Physiol. 199: 73-79.

Zhu YL, Pilon-Smits EAH, Tarun AS, Weber SU, Jounin L, Terry T. 1999a. Cadmium tolerance and accumulation in Indian mustard is enhanced by overexpression of g-glutamycysteine synthetase. Plant Physiol. 121: 1169-1177.

Unit-VI
NON METAL TOXICITY TOLERANCE

11

Non Metal Toxicity Tolerance

11.1. BORON TOXICITY

11.1.1. Introduction

Boron (B) is essential micro-nutrient. It is necessary for cell wall formation, membrane integrity, calcium uptake and carbohydrate synthesis and transportation. B affects many functions in plants, which includes flowering, pollen germination, fruiting, cellular activities (division, differentiation, maturation, respiration etc.), water relationships and the movement of hormones. It is not translocated and leached easily from soils.

B is also essential for many plant functions, such as, maintaining a balance between sugar and starch , the translocation of sugar and carbohydrates, it is important in pollination (germination of pollen grains and growth of pollen tubes) and seed production. Deficiencies cause sterility by killing terminal buds. It is necessary for normal cell division, nitrogen metabolism, and protein synthesis, cell wall formation. It plays an important role in the proper function of cells for the proper control of internal water balance.

Boron becomes toxic to plants at higher concentrations in the soil. Most of the crop plants affected at 8-16 kg of B /ha or 1 mM of $B(OH)_3$ in the soil. Excess concentrations of B in the soil result in a significant yield reduction. Cartwright *et al.* (1984) reported 17% reduction in yield of barley. High concentrations of B tend to occur where marine sediments (high in B) have influence soil fractions. B salts are slightly soluble; they are leached out of the root zone in higher rainfall areas, but lower rainfall areas, or on land where impermeable clay layers at depth prevent leaching, B concentration can be high. B may occurs naturally in the soil or in ground water, or be added to the soil from mining, fertilizer, or irrigation water. It also can be traced directly to water from certain wells, or indirectly to land application of drainage water, and in soils with high B availability. High B concentrations are most frequently with soils formed in parent material of marine origin and deep seated fault system (Paull *et al.*, 1991a). The primary sources of B in most soils are tourmaline and the volatile emanations of volcanoes (Chesworth, 1991). In igneous, metamorphic, sedimentary rocks, B occurs as borosilicate, which are resistant to weathering. The other major sources of B have been presented in Fig. 11.1.

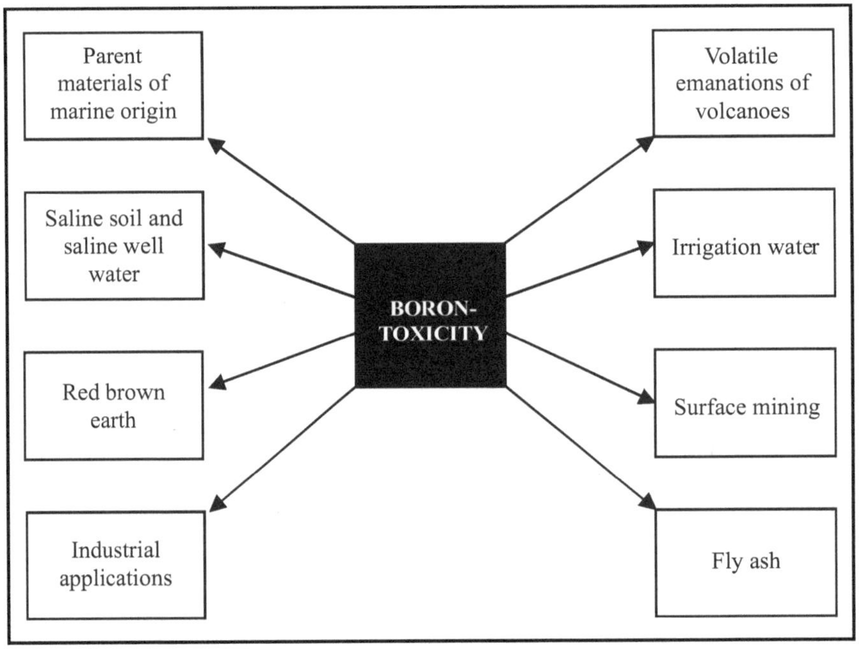

Fig. 11.1. Sources of B, which increase B in soil causing toxicity

B-toxicity can occur in arid and semi arid area such as the dry lands of South Autralia (Cartwright *et al.*, 1984, 1986), the Middle East (Ravikovitch *et al.*, 1961), the West Coast of Malaysia (Shorrcks, 1964), valleys along the southern coast of Peru (Masson, 1967), the Aden's foothills in northern Chile (Caceres *et al.*, 1992), solonchaks and solonetz soils of USSR (Zyrin and Zborishchuk, 1975), ferralsols of India (Takkar, 1982), rendzinas in Israel (Ravikovitch *et al.*, 1961) and Turkey (Sillanpoa, 1982; Nable *et al.*, 1997; Kalayci *et al.*, 1998; Genzin *et al.*, 2002; Avci and Akar, 2005).

11.1.2. B-toxicity Symptoms in Plants

There are great variations in expression of B-toxicity symptoms in plants. The appearance of B-toxicity symptoms is therefore determined solely by the patterns of B distribution in the plants. The critical plant level for toxicity can ranges from 10 to 50 ppm in sensitive crops.

1. In general, B-toxicity will be exhibited as leaf tips/edge burn in old leaves followed by stem death. Because of low translocation of B in the plants, symptoms generally appears on older leaves as it accumulate in the leaf margin or leaf chlorosis, browning of leaf tips (with accumulation at the end of the transpiration stream), which quickly followed by the death of effected tissues or defoliation (Bennett, 1993; Bergmann, 1992). In grapevine, symptoms of chlorosis and necrosis appear at the leaf margins, cause reduce in leaf size and internodes distance (Yermiyahu *et al.*, 2006).
2. Leaf cupping has been suggested by Loomis and Durst (1992), and it results from inhibition of cell wall expansion, through the disturbance of cell wall crosslinks.
3. Brownish spots prominent at panicle initiation stage. The chlorotic/ necrotic patches are greatly elevated B concentrations compared with the surrounding leaf tissues.
4. Wheat cultivation in B-toxic area shows chlorosis, necrosis and yield reduction.
5. In soybeans, the leaves may have a rust-like appearance.
6. Some species of crop plants in which B is phloem mobile (e.g. *Prunus, Malus, Pyrus*) and accumulations in developing sinks rather than at the end of the transpiration stream. Here the symptoms of toxicity are gummy rusts, internal fruit necrosis, bark necrosis which appears due to death of the cambial tissues, and stem die back (Brown and Hu, 1996), gum exudation in leaf axils and buds (El-Motaium *et al.*, 1994). In this species B will accumulate in the meristimatic region or fruit, but not in the mature leaves, of species in which sugar alcohol are present. And, B-toxicity will likely appear in apical and meristematic tissues while the typical symptomps of B-toxicity such as leaf tips/edge burn will not be observed.
7. Reduced root growth as compared to shoot growth. The resistant wheat genotypes and longer root axes and more lateral roots than the sensitive or moderately sensitive genotypes (Huang and Graham, 1990).

8. Reduction in plant hight and number of internodes under B-toxicity (30-40 mg B of soil).
9. Fruit cracks in pear (Chee *et al.*, 1986).

11.1.3. Injury Mechanisms

B-toxicity decrease CO_2 fixation, inhibit ureide metabolism in the leaves of nodulated soybean (Lukaszewski *et al.*, 1992) and complexing of rhibonucleotides which causes metabolic disturbances (Loomis and Dust, 1992).

B-toxicity symptoms reflect the patterns of B-accumulation, which is characterized by burning of tips and margin of older leaves. The accumulation of B depends on the phloem mobility of B in plants. And the phloem mobility varies among the plant species. The variation in phloem B mobility occurs as a consequence of the presence of polyols. In the polyol producing species, 'classic' B-toxicity symptoms were absent and B-toxicity was expressed as meristematic *die-back* and an accumulation of B in apical tissues. The unusual symptoms of B-toxicity are the result of the high phloem mobility of B in many plant species (Brown and Hu, 1996), which occurs as a result formation and phloem transport of a B-sugar alcohol compel (Hu *et al.*, 1997). As sugar alcohol, viz., sorbitol has very high affinity to bind B (Makkee *et al.*, 1985), and as the represent primary products of photosynthesis, it was predicted that the B-sugar alcohol complexes will be ready transported to active sinks such as apical tissues (Brown and Hu, 1996; Hu *et al.*, 1997). Thus, B will be accumulated in the meristematic region or fruit, but not in mature leaves of plant species in which sugar alcohol are present.

11.1.4. Tolerant Mechanisms

To develop genetically high tolerant ability of crop plants to B-toxicity, it is essential to know the injury and tolerance mechanisms. Differences in root uptake, root-to-shoot transport and shoot accumulation of B should play a decisive role in differential expression of B-tolerance among genotypes. Sometimes, reduced uptake of B by roots the play an important role in B-toxicity tolerance. B-exclusion, detoxification at cellular level also involve in tolerance of crop plants.

11.1.4.1. Reduced Accumulation of B in Root and Shoot

The tolerant species reduced accumulation in both root and shoot (Paull *et al.*, 1992a, b). Paull *et al.* (1991b) reported significantly higher B-concentration in susceptible Chinese Spring wheat as compared to amphidiploid var. Halberd, and it was due to reduced accumulation of B in shoots. However for most of the genotypes of wheat , the shoot concentrations of B were not related to the B-toxicity-induced decreases in shoot growth (Mahalakshmi *et al.*, 1995; Torun *et al.*, 2003; Torun *et al.*, 2006).

11.1.4.2. Exclusion of Excess Boron

Sensitivity to B-toxicity in barley and wheat is governed by the ability of cultivars to exclude B (Chippa and Lal, 1990; Paull, 1992a, b), thus they maintain lower B in the shoot. Eaton and Blair (1935) reported higher concentrations of B in leves of *Helianthus tuberosus* (Jerusalem artichoke) than the more B-tolerant *Helianthus annus* (Sunflower); *Citrus limonin* (leomon) than its B-tolerant relative *Severina buxifolia* (Chinese box orange). *Lycopersicon esculantum* is more susceptible to B-toxicity and accumulates more B in its shoots than its wild relative *Lycopersicon cheesmanii* (Toledo and Spurr, 1984).

Secretion of B-chelating compounds in rhizosphere associated with mucilage, which reduce the amount of B around the root. Thus it restricts B to enter into plant system. Inactivation of cell wall or cytoplasm of root cells is another exclusion mechanism. In B-toxic soils, it adsorbed by the cell wall, which prevents entry of B (Torun *et al.*, 2006).

11.1.4.3. Altered Distribution of Accumulated Boron

Accumulated B is distributed at the cellular tissues or organ level. The difference in phloem mobility of B amongst species results in B being accumulated either in leaf margin or in fruits, cambial tissue and stem (Brown and Hu, 1996). B accumulated in the older leaves of barley and caused senescence (Riley, 1987). The phloem immobility keeps B away from important metabolic sites, retaining it in leaf margins. Compartmentation of B in vacuoles could be a more plausible explanation for B tolerance (Torun *et al.*, 2006).

11.1.5. Genetic Variation for B-toxicity

Existing genetic variation is promising and can be exploited in breeding programmes aiming at development of B-toxicity tolerant genotypes.

Germplasm collection of several crop species have been screened for response to B-toxicity to determine the extent of variation within the species and to identify accessions that could be used in breeding more tolerant varieties (Table 11.1). A large portion of existing genotypes of different crop species are susceptible to B-toxicity. Proper screening of genotypes of a particular crop may result in isolation of tolerant ones. Genetic variation is present at both the inter- and intra-species levels. Few examples of inter-species variation in response to B-toxicity has bee presented in Table 11.2. A wide range of intra-species variation also reported in many crops, viz., *Triticum aestivum* (Chaterjee *et al.*, 1980; Paull *et al.*, 1988a), *Triticum turgidum* var. *durum* (Yau *et al.*, 1995; Torun *et al.*, 2006), *Hordium vulgare* (Nable, 1988), *Oryza sativa* (Cayton, 1985; Ponnamperuma *et al.*, 1979), *Pisum sativum* (Bagheri *et al.*, 1992), *Citrus* spp. (Chapman and Vanselow, 1955), *Carya illinoensis* (Picchioni and Miyamoto, 1991), *Fragaria ananssa* (Blatt, 1976). Genotypes with higher tolerance to B-toxicity may be used in breeding programmes to develop a new and more B-toxicity tolerance cultivar for B-toxic soils.

Table 11.1. List of crop species in which genetic variation has been studied and screened for B-toxicity tolerance

Crop	Species	Number of genotypes tested	Reference
Wheat	*Triticum aestivum*	1576	Moody *et al.*, 1988
	Triticum turgidum var. *durum*	150	Cartwright *et al.*, 1987
		1600	Ralph, 1992
		19	Yau *et al.*, 1995
		300	Jamjod, 1996
		70	Torun *et al.*, 2006
Barley	*Hordium vulgare*	350	Jenkin, 1993
Pea	*Pisum sativum*	135	Paull *et al.*, 1992b
		617	Bagheri *et al.*, 1994
Medics	*Medicago* sp.	681	Paull *et al.*, 1992b

Torun *et al.* (2006) reported a substantial range of genotypic tolerance to B-toxicity in soil between 70 durum wheat genotypes. Among the genotypes that they have tested, Sabil-1 and Stn'S' were most tolerant, while genotypes Jabiru-4 and Gerbrach-1 were classified as the most sensitive to B-toxicity.

It has been found that genotypes having higher tolerance to B-toxicity had accordingly lower concentration of B in shoot (Nable, 1988; Paull *et al.*, 1992a, b). Whereas, differences in susceptibility to B-toxicity in soil did not correlate with leaf or shoot concentration of B (Yau *et al.*, 1995; Torun *et al.*, 2006).

Table 11.2: Classification for B-toxicity tolerant crop plants based on B in irrigation water

Tolerance level and concentration	Common name	Botanical name
Tolerant (2-4 mg B-L in irrigation water)	Carrot	*Daucus carrota*
	Alfalfa	*Medicago sativa*
	Sugar beat	*Beta vulgaris*
	Coprosma	*Coprosma kirkii*
	Pomegranate	*Punica granatum*
		Olea skylark
		Maytens boaria
Semi-tolerant (1-2 mg B L-1 in irrigation water)	Oat	*Avena sativa*
	Maize	*Zea myas*
	Potato	*Solanum tuberosum*
Sensitive (0.3 mg B in irrigation water)	Avacado	*Persicum americana*
	Apple	*Malus domestica*
	Bean	*Galium odororatum*

11.1.6. Genetics of Boron Toxicity

Soil B-toxicity tolerance is genetically correlated (Paull *et al.*, 1988b) and this has been investigate4d for several species to enable efficient strategies for breeding of B-toxicity tolerant cultivars, viz., bread wheat (Paull *et al.*, 1991a; Punchana *et al.*, 2004), durum wheat (Jamjod, 1996; Torun *et al.*, 2006), barley (Jenkin, 1993), field bean (Bagheri *et al.*, 1996) etc. B-toxicity tolerance of bread wheat is partially controlled by dominant nuclear genes (Paull *et al.*, 1991a). They observed monogenic segregation in F_2 and F_3 for combinations WI*MMC ´ Kenya Farmer, Warigal ´ WI*MMC and Halberd ´ Warigal.

When Chinese Spring substitution lines carrying chromosomes from the highly sensitive variety Kenya Farmer were grown in 75 mg B/kg, results indicated that chromosome 4B has a major effect on B response, but modifying genes appear to be present in other chromosomes (Paull *et al.*, 1988b). Genes, viz., *Bo1, Bo2,* and *Bo3* have been identified and are being incorporated into breeding lines (Ralph, 1992). Paull *et al.* (1992a) also reported that the response of wheat to high B supply is under the control of several major additive genes, one of which is located in chromosome 4A. Excess B-stress tolerance is controlled by factors predominantly on chromosome 5R of Imperial Rye 5S' (*Aegiolops sharonnensis*, Manyowa and Miller, 1991).

B-toxicity tolerance is conferred by major additive genes; therefore, backcrossing is an appropriate breeding method to transfer tolerance ability into well adaptive cultivars. Transgressive segregation occurred among the progeny derived from crosses the tolerant and sensitive genotypes. Thus, it is possible to develop lines with a greater level of B-toxicity tolerance than that available in a germplasm collection.

11.1.7. Breeding for Tolerance to B-toxicity

Conventional methods of removing B from soil, such as leaching or increasing pH by liming, may be unsuitable for B-contaminated regions for a variety of reasons, such as low rainfall, lack of water or high lime content of the soil. Furthermore, such techniques are often expensive to apply, requiring chemical additives and regular maintenance. They can also result in the generation of hazardous byproducts, such as chemical sludge, potentially leading to high disposal costs. Therefore, it is suggested that either developing B-tolerant plants of using B-accumulating plants, which may provide a more effective and cheaper solution. Breeding for B-toxicity tolerance may involves the association between high levels of B and the cultivation of tolerant varieties (Rathjen and Pederson, 1986), and the performance of near isogenic lines (NIL) under normal and high levels of B (Moody *et al.*, 1993; Campbell *et al.*, 1994, 1995).

Several sets of NILs of wheat have been developed through backcrossing; mostly transferring the *Bo1* allele from Halberd and yield advantage of 5-10% had been achieved in Southern Australia (Moody *et al.*, 1993; Campbell *et al.*, 1994). Rathjen *et al.* (1987) demonstrated that three years of selection at five sites from a broadly based population of breeding lines resulted in disruptive selection that could be attributed to the B-status of the site of selection.

Success of breeding for B-toxicity tolerance is greatly aided by the genotype within the species, subsequently application of appropriate screening methods. Screening under controlled environments in the glasshouse and measurement of tissue B concentration has been an approach that is successfully utilized in several breeding programmes (Bagheri *et al.*, 1996; Paull *et al.*, 1990). Screening diverse populations under field condition with B-toxicity soils is the most appropriate method of screening for this (Paull *et al.*, 1990). Determination of root growth pattern and effects of excess B on root growth on filter paper also may be used as screening method (Chantaschume *et al.*, 1995).

The degree of dominance expressed by an F_1 hybrid varies according to the B treatment (Paull *et al.*, 1991a; Jenkin, 1993). Thus, the most appropriate B treatment for screening segregating populations will depend on the genetic composition of the test population. A high treatment would be appropriate for identifying homogenous tolerant plant in an F_2 population than for identifying heterogygous plants during backcrossing where an excessive high treatment would result in the heterogygous plants being phenotypically similar to the more sensitive parent.

The variation in B-toxicity tolerance in lentil (*L. culinaris*) was screened by Yau (1999). He suggested symptom score (0-9 scale; 0 = no symptom and 9 = symptoms on all leaves) as a selection criteria for B-toxicity tolerance at seedling stage. ILL5993 was found as most tolerant genotype.

In wheat, several chromosomal regions and DNA markers have been identified to use in molecular-assisted selection for B-toxicity tolerance (Jeferies *et al.*, 2000).

11.1.8. Management of B-toxic Soils for Crop Production

B-accumulation in crop plant could be reduced by using alternative cultivars with lower accumulation of by improving agronomic practices, such as water and fertilizer management, which lower the availability of excess B in the rhizosphere.

11.1.8.1. Leaching with Excess Irrigation Water

Since B is a mobile element in the soil, excess B from over application can be corrected over-time with leaching. A commonly used method of reclaiming high B soils is to extensively leach with B free irrigation water (Leyshon and Jame, 1993). Applying water in excess, of a plant water requirement is generally referred to as the leaching fraction (Hoffman, 1990). Longer intervals of good quality irrigation water and/or intermittent ponding are necessary to remove B from the root zone (Shennan *et al.*, 1995). Leaching with excess irrigation water is not a permanent solution for B-toxic soils. B can be regenerated through the mineralization of B-from soil organic matter, or through weathering processes of soil organic matter, or through weathering processes of soil minerals (Peryea *et al.*, 1985).

11.1.8.2. Soil Amendment

Application of lime and other sources of soluble Ca have been shown to be effective in reducing B-toxicity. Liming increases soil pH solution and thus promotes the adsorption of B from soil solution may provide a short term solution. Addition of gypsum improves water infiltration and converts readily soluble Na-metaborate to less soluble Ca-metaborate (Bhumbla and Ckhabra, 1982). Heavy applications of $Ca(H_2PO_4)_2$ also lower the plant available B in acid soils. Sulphuric acid decreases soil pH, which increases the concentration of water soluble B in the soils (Prather, 1977).

Additional N-application can be of benefit. Under some circumstances, B-toxicity in plant may be reduced by application of Zn in the soil or as foliar spray (Graham *et al.*, 1986; Swietlik, 1995).

11.1.5.3. B-clean up by Plants in Toxic Soils

B-accumulating plants may be provides a more effective and cheaper solution of clean up soils contaminated with excess B. *Gypsophila sphaerocephala*, a plant that usually grows on dry slops and lime stone rocks, can survive in B-contaminated soils. It accumulates B in its tissues. It was also suggested that, once it has been harvested, the plant could be transported to sites with very low levels of B, helping to add the trace elements to the soil in regions where it is B-deficient.

11.1.5.4. Growing of Tolerant Varieties

Growing of tolerant genotypes is the best and eco-friendly approach of management B-toxic soil. Some of the B-toxicity tolerant plant genotypes of few crops have been presented in Table 11.3. Several years of field study showed that growing of tolerant plants on B-toxic condition maintained the level of soluble soil B to a non-toxic level for the planted species. Some potential plant species that can be grown on high B-toxic soils are saltbush (Watson *et al.*, 1994), milkvetch (Parker *et al.*, 1991), barley (Nable *et al.* 1990), wheat (Paull *et al.*, 1988a), Indian mustard (Banuelos *et al.*, 1993), tall fescue (Banuelos *et al.*, 1995).

Natural hyper-accumulators, such as *Gypsophila sphaerocephala*, tend to grow slowly. But, genetic engineering could be used to create fast-growing plants that carry B- tolerance genes to improve their ability to tolerate, accumulate B and it is known as *'cleaning by genes'*. This techniques could be particularly useful in developing countries, because the cost of growing plants are minimal compared to those of soil removal and replacement.

Adaptive characters for boron-toxicity tolerance for screening against boron-toxicity tolerance of wheat have been presented in Table 11.4.

Table 11.3. B-toxicity tolerant genotypes in crop plants

Crop/species	Genotypes	Reference
Triticum turgidum var. *durum*	Sabil-1, Stn'S' Halbred, G61450	Torun *et al.*, 2006 Ralph, 1992; Paull *et al.*, 1992b
Triticum aestivum	Bonza	Punchna *et al.*, 2004
Pisum sativum	Early Dun, Partridge, Dun, Derrimut, Dundale, Alma, Maitland	Paull *et al.*, 1992b; Bagheri *et al.*, 1992
L culinaris	ILL5883	Yau *et al.*, 1999

Table 11.4. Adaptive traits for boron-toxicity tolerance of bread wheat.

Crop/species	Trait	Reference
Triticum aestivum	1. Root and shoot length 2. Number of tiller per plant 3. Leaf chlorosis and necrosis 4. Dry matter production 5. Days to effective tillering	Punchana *et all.*, 2004 Punchana *et all.*, 2004 Punchana *et all.*, 2004 Paull *et al.*, 1988a Paull *et al.*, 1988a

11.2. ARSENIC TOXICITY

11.2.1. Introduction

Arsenic (As) is ubiquitous in the environment and classified as metalloid, which can exist both in solid and liquid states. It occurs in several forms, often in compounds with the other chemical elements. All of its compounds as well as pure state are poisonous to both plants and animals. There is no evident that As is essential for plant. Sever As contamination in soil may cause a variety of problems such as loss of vegetation, ground water contamination and As toxicity in plants, animals and human beings. Agricultural crops, namely rice, vegetables and cereals are vurnable to As contamination from contaminated irrigation water. Agriculture soils of many areas of Bangladesh have been found to be contains high levels of As.

11.2.2. Sources of Arsenic

Arsenic contamination takes place from various sources such as rocks and minerals like arsenopyrites, realgar, aerosenolyte. Anthropogenically, it also entres into the environment from pesticides, such as calcium arsenolate

[$Ca_3(AsO_4)$], defoliants [arsenicacid (H_3AsO_4)], sodium arsenate (Na_3As_4), sodium arsenite ($NaAsO_2$), and dimethyl arsenic acid- $(Ch_3)_2AsO_2H$], fertilizers, wood preservatives, smelting wastes of copper, lead, zinc, gold and silver, burning of fossil fuels, agriculture and silviculture products etc. (Smith *et al.*, 1998). Arsenic residues can accumulate to very high levels in agricultural area where arsenic pesticides or defoliants are repeatedly used. It is apparent that agricultural uses can cause surface soil accumulation of 600 mg/l or more. Arsenate, arsenite methyl arsenic acid and dimethyl arsenic acid are usually found in water. Arsenate is predominant As species in aerobic soils, whereas arsenite dominates under anaerobic conditions (Smith *et al.*, 1998). It is well known that As associations with Fe and Al control arsenic behaviour in the soil.

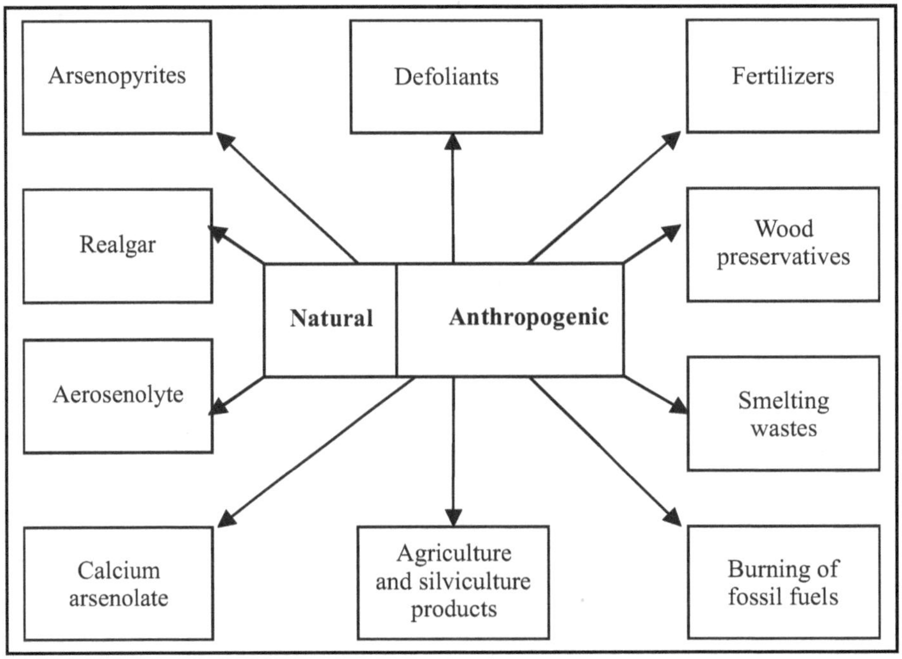

Fig. 11.2. Sources of arsenic

The permissible contamination in fruits and vegetables is 2.6 mg/kg of fresh weight. The allowable concentration in agricultural soils is 20 mg/kg of soil. Whereas, 5 ppm As in soil was found to be toxic to sensitive crop plants. The mean As toxicity threshold for plant is 40 and 200 mg/kg in sandy and clay soils, respectively. Dimethyl arsenic acid, a herbicide, apparently was more readily translocated to the shoot than the inorganic arsenicals or MMA (Monomethyl arsenic acid), thus, was phytotoxic. Arsenite has been considered at least twice as phytotoxic as arsenate either foliar or root applied. The range for cereal crops is highly variable. The yields of barley and ryegrass were significantly reduced by addition of 50 mg/kg of soil. In rice the critical level in tops of the plant ranges from 20 to 100 mg/l of As, and in roots 1000 mg/l. Wet land rice is known to be very susceptible to arsenic toxicity as compared to upland rice, since As^{3-} would be more prevalent under reducing conditions. Barley seedlings are sensitive at 20 ppm of As in the soil.

Solubility and subsequently the availability are greatly influence by the different forms of As (arsenicals) as well as plant species. The As in a soils may exist as arsenite [As(III)], arsenate [As(V)], monomethyl arsenic acid (MMA) and dimethyl arsenic acid (DMA). The presence of other ions also affected As availability and phytotoxicity. The availability of arsenicals to *Spartina alterniflora* L. followed the order: DMA < MMA < As(V) < As(III) . Whereas, Marin *et al.* (1992) reported sequence of As availability to rice (*Oryza sativa* L.): DMA < As(V) < MMA < As(III). DMA is readily translocated to the plant shoot, but As(III), As(V) and MMA accumulated primarily in the roots upon uptake (Marin *et al.*, 1992). Both MMA and DMA had a great upward translocation than arsenite and arsenate in *Lycopersicon esculantus* Mill. (Burlo *et al.*, 1999).

The availability of As to plants may also vary depending upon soil pH, nature of minerals constituting the soil and other competing ions present in soil solution. The plant availability of arsenic may be high at high soil pH due to presence of soluble arsenic in soil solution. The soil rich in oxidic materials exhibit low arsenic bioavailability. The increased availability of As in sandy soils where the added phosphorous may displace some of the bound As into the soil solution.

Irrigation water contributes a major role to contaminate vegetable crops, viz. potato, tomato, brinjal, okra, bitter gourd, chilli, cabbage, Indian spinach, amaranthaus, red amaramthaus, katua data, China shak and cauliflower. The aesenic content of these vegetables was found to be higher when irrigated with As contaminated water than those graown with As free water. The trend of As accumulation in leafy vegetables is higher and lower in fruit vegetables and it was noted that As uptake by different crop vegetables is variable.

11.2.3. Arsenic Toxicity Symptoms in Plants

The external symptoms of As toxicity vary with plant species. Followings are some examples of As toxicity symptoms in crop plants.

1. Arsenic toxicity induced chlorosis symptoms in the youngest leaves of rice seedlings by decreasing chlorophyll content (Shaibur *et al.*, 2006). Leaf numbe and width of leaves also decrease with increase concentrations of As.
2. Tomato plants grown in soils with high As concentraions (100-130 mg As/kg) show leaf dieback from the tip and poor fruit set.
3. Fruit plants grown on replanted orchard sites commonly show retarded growth due to As toxicity.
4. Rice grown in former cotton producing soils that had a history of repeated monomethyl As acid applications show straight head disease or spikelet sterility under flooded conditions.
5. Biomass production and yield of crops reduce significantly at elevated arsenic concentrations.

11.2.4. Injury Mechanisms

11.2.4.1. Biochemical Injury of As in Plants

The major biochemical actions of As are coagulation of protein, complexion with coenzymes and uncouplexing of phosphorelation. Generally As toxicity results from inhibition by trivalent As of enzymes containing sulphydryl. Arsenate acts as a phosphate analogue and can disrupt phosphate metabolism, whereas arsenite reacts with sulfhydryl groups of enzymes and tissue proteins, leading to inhibition of cellular function and death (Meharg and Hartley-Whitaker, 2002). In most cases, the enzyme activity can be restored by adding an excess of monothiol such as glutathione, suggesting that the inhibition is due to a reversible reaction of arsenic with single sulphydryl group in the enzyme molecule. The availability of As(III) to inhibit ATP production results in the cessation of organ functions rapidly specially in the event of acute arsenic poisoning. In contrast, As(V) induces its toxicity in uncoupling mitochondrial oxidative phosphorylation by acting as inorganic phosphate.

Mishra and Dubey (2006) reported that *in situ* As^{+3} treatment caused a marked inhibition in activities ribonuclease (RNase, EC 3.4.16.5), protease and leucine aminopeptidase (LAP, EC 3.4.11.1.) in rice (*Oryza sativa* L.) cvs. Malviya-36 and Pant-12. Isoform pattern of RNase extracted from As^{+3}-exposed seedlings showed a significant alteration compared to its pattern in unexposed seedlings. The findings suggest that As exposure impairs hydrolysis of RNA and proteins in rice seedlings due to inhibition of RNase and protease activities and that proline accumulation under As^{+3} toxicity appears to serve as enzyme protectant.

11.2.4.2. Reduction in Yield and Biomass Productivity

Arsenic at toxic concentrations severely affects crop productivity. In rice yield and shoot dry weight decreased significantly under As-toxic condition. Whereas, decrease in root dry weight was comparatively less than the shoot (Shaibur *et al.*, 2006). This indicated that the shoot was more sensitive to As than the root in rice.

11.2.5. Tolerance Mechanisms

11.2.5.1. Reduced As uptake

Arsenate resistance has been identified in a range of plant species. Tolerant plant showed a decreased uptake of arsenate because of suppression of the light-affinity phosphate uptake system (Meharg and Macnair, 1991, '92; Meharg and Hartley-Whitaker, 2002). The low uptake of As from soil may be of several other resons, such as restricted uptake by plant roots or limited translocation of As from roots to shoots. Chaturvedi (2006) reported that arsenite uptake was higher than arsenate. Asrsenic extraction by plant increased with increasing soil arsenic concentration. The author found that the mustard variety, DHR 9504 was more tolerant to arsenic than Varuna.

11.2.5.2. Hyper Accumulation

Hyper accumulation of As is another feature of As-toxicity tolerance by plants. The brake fern (*Pteris vittata*) has been found to accumulate As from soil (Ma *et al.*, 2001; Tu and Ma, 2002; Wang *et al.*, 2002) and it is highly tolerant of As toxicity. It has the ability to accumulate upto 22630 mg As/kg in the frond dry weight (Wang *et al.*, 2002). *Pityrogramma calomelanos* (Francesconi *et al.*, 2002), *Pteris cretica*, *Pteris longifolia* and *Pteris umbrosa* (Zhao *et al.*, 2002) were also identified as As hyperaccumulators.

As-hyperaccumulating plants may be as used successfully in remediating arsenic-contaminated sites. Laddler brake furn has potential to produce large plant biomass, thus it is a suitable soil remediation plant for As-toxic soils.

11.2.6. Reclamation

A wide range of technologies available for the removal of As from water. The most commonly used technologies include oxidation, co-precipitation and adsorption onto coagulated flocs, lime treatment, adsorption onto sorptive media, ion exchange resin and membrane techniques (Shen, 1973; Cheng *et al.*, 1994; Hering *et al.*, 1997; Kartinene and Martin, 1995; Joshi and Chaudhari, 1996). Many of the As removal technologies have been discussed in details in AWWA reference book (Pontious, 1990).

11.3. REFERENCES

Avci M, Akar T. 2005. Severity and spatial distribution of boron toxicity in barley cultivated areas of Central Anatolia and Transitional Zones. Turk J Agric For. 29: 377-382.

Bagheri A, Paull JG, Rathjen AJ, Ali SM, Moody DB. 1992. Genetic variation in the response of pea (*Pisum sativum* L.) to high concentrations of boron. Plant Soil. 146: 261-269.

Bagheri A, Paull JG, Rathjen AJ. 1994. the response of *Pisum sativum* L. germplasm to high concentrations of soil boron. Euphytica. 44: 9-17.

Bagheri A, Paull JG, Rathjen AJ. 1996. Genetics of tolerance to high concentrations of soil boron in peas (*Pisum sativum* L.). Euphytica. 87: 69-75.

Baltt CR. 1976. Phosphorus and boron interactions on growth of strawberries. HortScience. 11: 597-599.

Banuelos GS, Akohoue S, Zambrzushki S, Mead R. 1993. Trace elements composition of different plant species used for remediation of boron-laden soils. In: Plant Nutrition form Genetic Engineering to Field Practice, Barrow NJ (ed.). Kluwer Academic Publishers, Dordrecht, The Netherlands. pp. 425-428.

Banuelos GS, Mackey B, Wu L, Zambrzuski S, Akohouse S. 1995. Bioextraction of soil boron by tall fescue. Ecotoxicology Environ Safty. 31: 110-116.

Bennett WF. 1993. Nutrient deficiencies and Toxicities in Crop Plants. APS Press, St Paul, MN, USA.

Bergmann W. 1992. Color Atlas: Nutritional Disorders of Plants. Gustav Fisher, New York. pp. 204-239.

Bhumbla DR, Ckhabra R. 1982. Chemistry of sodic soils in review of soil research in India. In: Trans 12th Intr Cong Soil Sci, New Delhi, India. p. 169.

Brown PH, Hu H. 1996. Phloem mobility of boron is species dependent: Evidence for phloem mobility in sorbitol-rich species. Ann Bot. 77: 497-505.

Burlo F, Gijarro I, Carbonell-Barrachina AA, Valero D, Martinez-Sanchez F. 1999. Arcenic species: effects on and accumulation by tomato plants. J Agric Food Chem. 47: 1247-1253.

Caceres L, Grutter D, Contreras N. 1992. Water recycline in arid zones: Chilean case. Ambio. 21: 138-144.

Campbell TA, Moody DB, Jerreries SP, Cartwright B, Rathjen AJ. 1995. Grain yield evaluation of near isogenic lines for boron tolerance. In: Pro 8th Intr Wheat Genet Symp, Beijing, Li ZS, Xin Zy (eds.). pp. 1021-1027.

Campbell TA, Rathjen AJ, Jefferies SP. 1994. Breeding wheat (*Triticum aestivum* L.) for tolerance to boron toxicity. In: Proc 7th Assembly Wheat Breeding Soc of Australia, Paull JG, Dundas IS, Shepherd KW, Hollamby GJ (eds.). Wheat Breeding Society of Australia. pp. 111-114.

Cartwright B, Rathjen AJ, Sparrow DHB, Paull JG, Zarcinas BA. 1987. Boron tolerance in Australian varieties of wheat and barley. In: Genetic Aspects of Plant Mineral Nutrition, Gabelman HW, Loughsnan BC. Martinus Nijhoff, Dordrecht (eds.). The Netherlands. pp. 139-151.

Cartwright B, Zarcinas BA, Mayfield AH. 1984. Toxic concentrations of B in red-brown earth at Goldstone, South Australia. Aust J Soil Res. 22: 261-272.

Cartwright B, Zarcinas BA, Spouncer LA. 1986. Boron toxicity in South Australian barley crops. Aust J Agric Res. 37: 351-359.

Cayton MTC. 1985. Boron toxicity in rice. IRRI Research Paper Series. p. 113.

Chantachume Y, Smith D, Hollamby GJ, Paull JG, Rathjen AJ. 1995. Screening for boron tolerance in wheat (*T. aestivum*) by solution culture in filter paper. Plant and Soil. 177: 249-254.

Chapman HD, Vanselow AP. 1955. Boron deficiency and excess. Calif Citrograph. 41: 31-34.

Chatterjee BW, Chatterjee M, Das NR. 1980. Note on the differences in the response of wheat varieties to boron. Indian J Agric Sci. 50: 796-

Chaturvedi I. 2006. Effect of arsenic concentrations and forms on growth and arsenic uptake and accumulation by Indian mustard (*Brassica juncea* L.) genotypes. J Central European Agri. 7(1): 31-39.

Cheng CR, Liang S, Wang HC, Beuhler MD. 1994. Enhanced coagulation for arsenic removal. J American Water Works Assoc. 86(9): 79-90.

Chesworth W. 1991. Geochemistry of micronutrients. In: Micronutrients in Agriculture, Mortvedt JJ, Cox FR, Shuman LM, Welch RM (eds.). Soil Science Society of America, Madison WI, USA. pp. 1-30.

Chippa BR, Lal P. 1990. A comparative study on the effect of soil B on yield, yield attributets and nutrient uptake by susceptible and tolerant varieties of wheat. An Edafol Agrobiol. 48: 489-498.

Chee JS, Lee JC, Kim SB, Moon JY. 1986. Studies on the causes of shoot die-back in pear tree (*Pyrus serotina* Rehder). J Korean Soc Hort Sci. 27: 149-156.

El-Montaium R, Hu H, Brown PH. 1994. The relative tolerance of six *Prunus* root stocks to boron and salinity. J American Soc Hort Sci. 119: 1169-1175.

Eaton FM, Blair GY. 1935. Accumulation of boron by reciprocally grafted plants. Plant Physiol. 10: 411-424.

Franceconi K, Visoottiviseth P, Sridockhan W, Goessler W. 2002. Arsenic species in an arsenic hyperaccumulating fern, *Pityrogramma calomelanos*: a potential hyperaccumulator of arsenic contaminated soils. Sci Total Environ. 284: 27-35.

Genzgin S, Dursun N, Hamurcu M, Harmakaya M, Onder M, Sade B, Topal A, Ciftci N, Acar B, Babaoglu M. 2002. Determination of boron content of soils in Central-Anatolia Cultivated Lands and its relationship between soil and water characteristcs. In: Boron in plants and Animal Nutrition, Goldbah HE (ed.). Kluwer Academic Pub, New York. pp. 391-400.

Graham RD, Welch RM, Grunes DL, Cary EE, Norvell WA. 1986. Effect of zinc deficiency on the accumulation of boron and other mineral nutrients in barley. Soil Sci Soc Am J. 51: 652-287.

Hering JG, Chen P, Wilkie JA, Elimelech M. 1997. Arsenic removal from drinking water during coagulation. J Env Eng, ASCE. 123(8): 800-807.

Hoffman GJ. 1990. Leaching fraction and root zone salinitycontrol. In: Agricultural Salinity Assessment and Management, Tanji KK (ed.). Amer Soc of Civil Eng, New York. pp. 238-261.

Hu H, Penn SG, Lebrilla CB, Brown PH. 1997. Isolation and characterization of soluble B-complex in higher plants. Plant Physiol. 113: 649-655.

Huang CY, Graham RD. 1990. Resistance of wheat genotypes to boron toxicity is expressed at the cellular level. Plant Soil. 126: 295-300.

Jamjod M. 1996. Genetics of boron tolerance in durum wheat. Ph.D. Thesis, The University of Adelaide, South Australia.

Jefferies SP, Palotta MA, Paull JG. 2000. Mapping and variation of chromosome regions conferring boron toxicity tolerance in wheat (*Triticum aestivum*). Theor Appl Genet. 101: 767-777.

Jenkin MJ. 1993. The genetics of boron tolerance in barley. Ph.D. Thesis, The University of Adelaide, South Australia.

Joshi A, Chaudhury M. 1996. Removal of arsenic from ground water by iron-oxide-coated sand. J Env Eng, ASCE. 122(8): 769-771.

Kalayci M, Alkan A, Cakmak I, Bayramoglu O, Yilmaz A, Aydin M, Ozbek V, Ekiz H, Ozberisoy F. 1998. Studies on differential response of wheat cultivars to boron toxicity. Euphytica. 100: 123-129.

Kartinene EO, Martin CJ. 1995. An overview of arsenic removal process. J Desalination. 103: 79-88.

Leyshon AJ, Jame YM. 1993. Boron toxicity and irrigation management. In: Boron and its Role in Crop Production, Gupta (ed.). CRC Press, Boca Raton, FL, USA. pp. 207-226.

Loomis WD, Durst RW. 1992. Chemistry and biology of boron. BioFactors. 3: 229-239.

Lukaszewski KM, Blevins DG, Randal DD. 1992. Asparagine and boric acid cause allantoate accumulation in soybean leaves by inhibiting manganese-dependant allantoate amidohydrolase. Plant Physiol. 99: 1670-1676.

Ma LQ, Komar KM, Tu C, Zhang WH, Cai Y, Kenneley ED. 2001. A Fern that hyperaccumulates arsenic: a hardy, versatile, fast-growing plants helps to remove arsenic from contaminated soils. Nature. 409: 579.

Mahalakshmi V, Yau SK, Ryan J, Peacock JM. 1995. Boron toxicity in barley (*Hordium vulgare* L.) seedling in relation to soil temperature. Plant Soil. 177: 151-156.

Makkee M, Kieboom APG, Bekkum H. 1985. Studies on borate esters III. Borate esters of D-manitol, D-glucose and D-fructose in water. Rec des Chim Pays-Bas. 104: 230-235.

Manyowa NM, Miller TE. 1991. The genetics of tolerance to high mineral concentrations in the tribe *Triticeae*- A review and update. Euphytica. 57: 175-185.

Marin AR, Masscheleyn PH, Patrick WH. 1992. The influence of chemical from and concentration of arsenic on rice growth and tissue arsenic concentration. Plant Soil. 139: 175-183.

Masson L. 1967. Alaunos problemas relacianados con la calinidad in los vales de la costa sur. Officeing Nacional de Evaluacion di Recursos Nationals. Lima, Peru.

Meharg AA, Hartley-Whitaker J. 2002. Arsenic uptake and metabolism in arsenic resistant and non-resistant plant species. New Phytol. 154: 29-43.

Meharg AA, Macnair MR. 1991. The mechanisms of arsenic tolerance in *Deschampsia cespitosa* (L.) Brauv and *Agrostis capillaries* L. New Phytol. 119: 291-297.

Meharg AA, Macnair MR. 1992. Suppression of the high-afficnity phosphate uptake system: a mechanism of arsenate tolerance in *Holcus lanatus* L. J Exp Bot. 43: 519-524.

Mishra S, Dubey RS. 2006. Inhibition of ribonuclease and protease activities in arsenic exposed rice seedlings: role proline as enzyme protectant. J Plant Physiol. 163(3): 927-936.

Moody B, Rathjen AJ, Cartwright B, Paull JG, Lewis J. 1988. Genetic diversity and geographical distribution of tolerance to high levels of soil boron. In: Proc 7th Intr Wheat Genet Symp, Miller TE, Koebner RMD (eds.), Cambridge, UK, 13-19 July, 1988, Inst Plant Sci Res. pp. 859-865.

Moody B, Rathjen AJ, Cartwright B. 1993. Yield evaluation of a gene for boron tolerance using back-cross derived lines. In: Genetic Aspect of Plant Mineral Nutrition, Randall PJ, Delhaize E, Richards RA, Munns R (eds.). Kluwer Academic Publishers, Dordrecht, The Netherlands. pp. 363-366.

Nable RO, Banuelos GS, Paull JG. 1997. Boron Toxicity. Plant Soil. 198: 111-115.

Nable Ro, Cartwright B, Lance RCM. 1990. Genotypic differences in boron accumulation in barley: relative susceptibility to boron deficiency and toxicity. In: Genetics Aspects of Plant Mineral nutrition, Bassam NEl (ed.). Kluwer Academic Publishers, Dordrecht, The Netherlands. pp. 243-251.

Nable RO. 1988. Resistance to boron toxicity amongst several barley and wheat cultivars: a prelimary examination of the resistance mechanism. Plant Soil. 112: 45-57.

Parker DR, Page AL, Thomas DN. 1991. Salinity and boron tolerances of candidate plants for the removal of selenium from soils. J Environ Qual. 20: 157-164.

Paull JG, Nable RO, Rathjen AJ. 1992a. Physiological genetic control of the tolerance of wheat to high concentrations of boron and implecation of plant breeding. Plant Soil. 146: 251-260.

Paull JG, Rathjen AJ, Cartwright B, Nable RO. 1990. Selection parameters assessing the tolerance of wheat to high concentration of boron. In: Genetics aspects of Plant Mineral Nutrition, El Bassam N (ed.). Kluwer Academic Publishers, Dordrecht, The Netherlands. pp. 361-369.

Paull JB, Rathjen AJ, Cartwright B. 1988b. Genetic control of tolerance to high concentrations of soil boron in wheat. In: Proc 7th Intr Wheat Genet Symp, Miller TE, Koebner RMD (eds.), Cambridge, UK, 13-19 July, 1988, Inst Plant Sci Res. pp. 871-877.

Paull JG, Nable RO, Lake AWH, Materne MA, Rathjen AJ. 1992b. Response of annual medics (*Medica* spp.) and field peas (*Pisum sativum*) to high concentration of boron: genetic variation and the mechanism of tolerance. Aust J Agric Res. 43: 203-213.

Paull JG, Rathjen AJ, Cartwright B. 1988a. Response of wheat and barley genotypes to toxic concentrations of soil boron. Euphytica. 39: 137-144.

Paull JG, Rathjen AJ, Cartwright B. 1991a. Major gene control of tolerance of bread wheat (*Triticum aestivum* L.) to high concentrations of soil boron. Euphytica. 55: 271-282.

Paull JG, Rathjen AJ, Cartwright B. 1991b. Tolerance to high concentration of boron for the amphidiploid of *Triticum aestivum* × *Agropyron elongtum*. Plant Soil. 133: 297-299.

Peryea FJ, Bingham FT, Rhoades JD. 1985. Mechanisms for boron regeneration. Soil Sci Soc Am J. 49: 840-843.

Picchioni GA, Miyamoto S. 1991. Growth and boron uptake of five pecan cultivar seedlings. HortScience. 26: 386-388.

Ponnaperuma FN, Lantin RS, Cayton MTC. 1979. Boron toxicity in rice soils. Intr Rice Res Newl. 4: 8.

Pontious FW. 1990. Water quality treatment: A handbook of Quality Water Supplies. American Water Works Association, McGraw Hill, New York.

Prather RJ. 1977. Sulphuric acid as an amendment for reclaiming soils high in boron. Soil Sci Soc Am Proc. 41: 1098-1101.

Punchana S, Jamjod S, Rerkasen B. 2004. Response to boron toxicity efficient and inefficient wheat genotypes. In: Proc of the 4th Crop Sci Cong, 26 Sep-01 Oct, 2004, Brisbane, Australia.

Ralph W. 1992. Boron problem in the wheat belt. Rural Research No. 153. pp. 4-8.

Rathjen AJ, Cartwright B, Paull JG, Moody BD, Lewis J. 1987. Breeding for tolerance for mineral toxicities in Australian cereals with special reference to boron. In: Priorities in Soil/Plant Relation Research for Plant production, Searle PGE, Davey BG (eds.). School of Crop Sciences, The University of Sydney. pp. 111-130.

Rathjen AJ, Pederson DG. 1986. Selecting for improved grain yields in variable environments. In: Plant Breeding Symp. DSIR, Williams TA, Wratt GS (eds.). Agronomy Soc of New Zealand, Special Publication No. 5.

Ravicovitch S, Margolin M, Navrath J. 1961. Miceroelements in soils of Israel. Soil Sci. 92: 85-89.

Riley MM. 1987. Boron toxicity in barley. J Plant Nutr. 10: 2109-2115.

Shaibur MR, Kitajima N, Sugawara R, Kondo T, Huq SMI, Kawai S. 2006. Physiological and mineralogical properties of arsenic-induced chlorosis in rice seedlings grown hydrphonically. Soil Sci Plant Nutr. 52(6): 691-700.

Shen YS. 1973. Study of arsenic removal from drinking water. J American Water Works Assoc. 65(8): 543-548.

Shennan C, Grattan SR, May DM, Hillhouse RG, Schachtman DP, Wander M, Robert B, Tafoya S, Burau RG, McNeish C, Zelinski L. 1995. Feasibility of cyclic reuse of saline drainage in a tomato cotton rotation. J Environ Qual. 24: 476-486.

Shorrocks VM. 1964. Boron toxicity in *Hevea brasiilensis*. Nature. 204: 599-600.

Sillanpoa M. 1992. Micronutrients and the nutrient status of soils: A global study. FAO Soils Bulletin 48. Rome: Food and Agriculture Organization of the United Nations.

Smith E, Naidu R, Alston AM. 1998. Arsenic in the soil environment: A review. Adv Agron. 64: 149-195.

Swietlik D. 1995. Interaction between zinc deficiency and boron toxicity on growth and mineral nutrition of sour orange seedlings. J Plant Nutr. 18: 1191-1207.

Takkar PN. 1982. Micronutrients: forms, content, distribution in profile, indices of availability and soil test methods. In: 12th Intr Soil Soc Cong, Part I, New Delhi. p. 136.

Torun B, Kalayci M, Ozturk L, Torun A, Aydin M, Cakmak I. 2003. Differences in shoot boron concentrations, leaf symptoms, and yield of Turkish barley cultivars grown on boron-toxic soil in field. J Plant Nutr. 26: 1735-1747.

Toledo J, Spurr J. 1984. Plant growth and ion uptake by *Lycopersicon esculentum* and *L. cheesmanii* f. minor. Turriaba. 34: 111-115.

Torun AA, Yazici A, Erdem H, Cakmak I. 2006. Genotypic variation in 70 Durum wheat genotypes. Turk J Agric For. 30: 49-58.

Tu c, Ma LQ. 2002. Effects of arsenic concentrations and forms on arsenic uptake by the hyperaccumulator ladder brake. J Environl Quality. 31: 641-647.

Wang J, Zhao FJ, Meharg AA, Roab A, Feldmann J, MC Grath P. 2002. Mechanisms of arsenic hyperaccumulation in *Pteris vittata*. Uptake kinetics, interactions with phosphate and arsenic speciation. Plant Physiol. 130: 1552-1561.

Watson MC, Banuelos GS, O'Leary JW, Riley JJ. 1994. Trace element composition of *Antriplex* grown with saline drainage water. Agri Ecosys Environ. 48: 157-167.

Yau SK, Nachit MM, Ryan J, Hamblin J. 1995. Phenotypic variation in boron toxicity tolerance at seedling stage in durum wheat (*Triticum durum*). Euphytica. 83: 185-191.

Yau SK. 1999. Boron toxicity in lentil: Yield loss and variation between contrasting lines. LENS Newsl. 26: 14-17.

Yermiyahu U, Ben-Gal A, Sarig P. 2006. Boron toxicity in grapevine. HortScience. 41(7): 1698-1703.

Zhao FJ, Dunham SJ, McGrath SP. 2002. Arsenic hyperaccumulation by different fern species. New Physiol. 156: 27-31.

Zyrin NG, Zborishchuk JN. 1975. Boron in the ploughed layer of soil of the European part of the USSR. Pochvovedensise. 5: 44.

Unit-VII
ATMOSPHERIC STRESSES

12

Air Pollution

12.1. INTRODUCTION

Air is polluted when it contains enough unhealthy particles and gases to harm human beings, animals, plants and even objects such as building and statues. Air pollutant may present as solid, liquid or gas form.

Industrial revaluation in the 19th and 20th centuries leads faster increase in concentration of air pollutants in the atmosphere. The most important air pollutants that damage the plants' growth and development, subsequently the economic yields of crop plants are ozone (O_3), nitrous oxides (NO_2 and NO), carbon monoxide (CO), ammonia (NH_4), peroxiacetyl nitrate (PAN), sulphur dioxide (SO_2), hydrogen fluorides (HF), particulate matters (cement dust, magnesium-lime dust, carbon shoot etc.), smoke etc. Burning of hydrocarbons in motor vehicle engines gives rise to CO_2, CO, SO_2, and NO in varying proportions and ethylene (C_2H_4) as well as other hydrocarbons. Industrial plants release SO_2, H_2S, NO_2 and HF into atmosphere. Hydrogen peroxide (H_2O_2) is another potentially injurious molecule, can form by the reaction between O_3 and naturally release volatiles (terpenes) from forest trees.

Air pollutant injure plant foliage, significantly alter growth and yield and are known to change quality of the marketable products depending up on concentration and duration of exposure. Plant damage caused by air pollution is usually most sever during warm, clear, calm, humid weather, when pressure is high, as these conditions can cause an air inversion. During air inversion, warm air above the earths' surface traps cooler air at ground level, allowing pollutants to accumulate.

Initially injury takes place at the biochemical level and progressing to the ultrastructural level and finally cellular level, which expresses as visible symptoms. Air pollution injury may be visible to foliage as necrotic lesion (dead tissue), or it can develop slowly as a yellowing or chlorosis of the leaf. There may be reduction in growth of various portions of a plant. Plants may be killed outright, but they usually do not succumb until they have suffered recurrent injury. A general view of air pollution has been presented in Fig. 12.1. Individual pollutant and its effects on plants have been discussed in detail in this chapter.

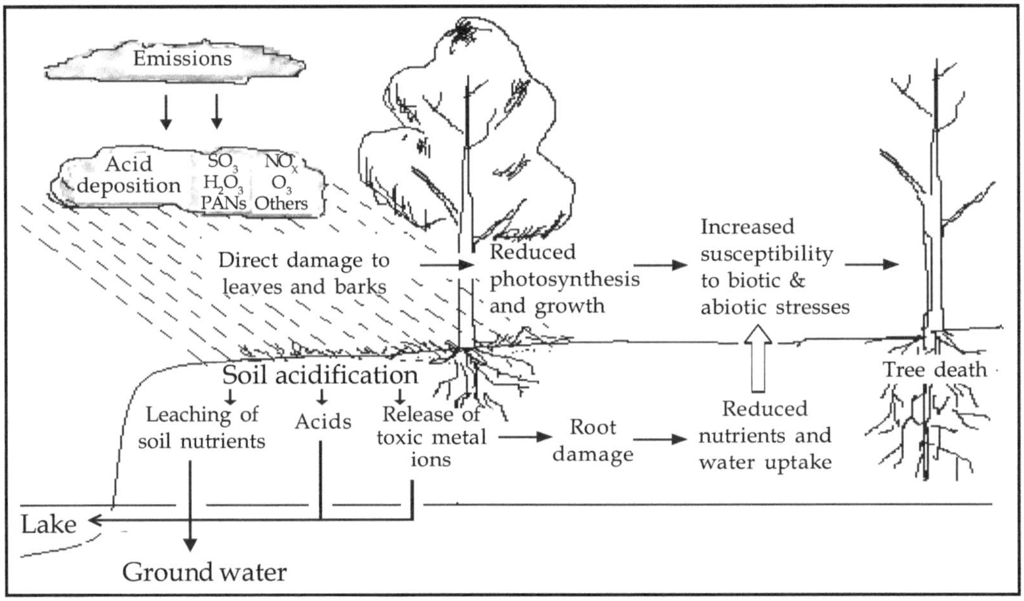

Fig. 12.1. A generalized view of acid rain and its effect

12.2. OZONE

It is a common air pollutant in the atmosphere. The concentration at the earth's surface is around 0.02-0.03 ppm. Whereas it may be 10 times higher in urban atmosphere, again, O_3 concentration is higher during day time as compared to night. At a concentration of 0.1 ppm can reduce photosynthesis by 50%. It is the main oxidant component of photochemical smog and cause leaf damage in many plant species (Kangasjarvi *et al.*, 1994), contributing substantially to crop loss and forest decline (Preston and Tingey, 1988).

It is formed by electrical discharge during thunder storms, the action of sunlight on oxygen, gasses liberated by combustion engine and as a by-product of photochemical reactions. Sometimes it is resulted from stratospheric ozone intrusion.

12.2.1. Injury Symptoms

1. *Leaf symptoms:* Ozone toxicity characteristically occurs on the upper surface of the affected leaves and appears as a flecking, bronzing or bleaching of leaf tissues. Fully mature leaves are more susceptible than younger and old leaves (Simini *et al.*, 1989), interveinal chlorosis leading to bleaching of foliage and necrosis. It decrease leaves and subsequently leads to immature senescence (Leadley *et al.*, 1990) in soybean. In tomato, symptoms appear on upper surface of leaves as small, irregular shaped spots that range in colour from dark brown to black or light tan to white (Heck *et al.*, 1965).
2. *Inhibition of growth:* Reduced morphological growth of *Penicum miliaceum* plant (Agrawal *et al.*, 1985). Reduced root growth than shoot growth (Velissariou *et al.*, 1992).
3. *Yield losses:* The yield losses in cotton due to ozone are directly related with the degree of determinacy (Temple, 1990). Early maturing white bean (*Phaseolus vulgaris*) cultivars tends to suffer reduced yield and earlier senescence (Hucl and Beversdorf, 1982). Ozone-toxicity can reduce soybean yields by an estimated 5-20% (Foy *et al.*, 1995). Heagle (1989) has been reported reduction of yield of soybean, maize and winter wheat by 5, 10, and 16%, respectively.

12.2.2. Injury Mechanisms

Critical ozone level must be determined to assess ozone damage on plant. Crop plants can be injured when exposed to high concentrations of various air pollutants. Injury ranges from visible marking on the foliage to reduced growth and yield, to permanent death of the plant. The development and severity of the injury depends not only the concentrations, but also the plant species and the stage of development (Iriti *et al.*, 2006) as well as the environmental factor conducive to a build up of the precondition of the plant which make it either susceptible or resistance to injury. Pollution injury mostly associated with interference with photosynthesis, respiration, lipid and protein biosynthesis, disorganization of cellular membranes, cell wall, mesophyll and nuclear breakdown.

Susceptibility to O_3 injury is influenced by many environmental and plant growth factors. High relative humidity optimum soil-nitrogen levels and water availability increase susceptibility. The susceptibility of O_3 injury also depends on leaf maturity stage. The youngest leaves of broad leafed-plants are resistance to O_3 injury. Significant differences also have been observed between species toward O_3 exposure. Based on genotypes screened and combining different sensitivity criteria (vegetative growth, reproductive growth, exposure-growth response relationships) *Malva sylvestris* has been regarded as the most sensitive species (Bender *et al.*, 2005). Some of the injury mechanisms of O_3 in plants have been brieft below.

12.2.2.1. Age of plants or plant parts

The sensitivity of O_3-toxocity in plants not only depends on concentration of O_3, but also the age of plants or plant parts and season. Muntifering *et al.* (2006) reported that younger leaves of alfalfa (*Medicago sativa*) were more tolerant than the older leaves.

12.2.2.2. Reduced photosynthesis

Ozone reduces the activity of RuBPC in rice. The concentration of RuBPCase in the leaves of *Medicago sativa* decreased (Pell and Pearson, 1983). It inhibits electron transport in photosynthesis in *Spinacia oleracea* (Coulson and Heath, 1974). Lee (1991a) also noted that O_3-induced stress blocks photosynthetic electron transport between PS II and PS I. It affects the chlorophyll content of effected plants. Knudson *et al.* (1977) found a high correlation between chlorophyll loss and visible necrosis in *Phaselus vulgaris* under O_3 stress condition. They also observed decreased ratio of chlorophyll *a* and *b* with increasing O_3 injury. Ariyaphanphitak *et al.* (2006) reported the most severe damage of O_3 in photosynthetic components, namely chlorophyll and carotinoid content, and rate of net photosynthesis. Thus, O_3 reduces the photosynthetic efficiency (Ormrod *et al.*, 1981).

12.2.2.3. Ozone disturbs carbohydrate levels

Ozone alters the allocation pattern of carbohydrate in plant and it effects on the allocation on leaves, stem, flowers/fruits and seeds (Bender *et al.*, 2006; Grantz *et al.*, 2006). Ozone disturbs sugar and carbohydrate levels in leave; it can either decrease or increase (Tingey *et al.*, 1973). However, starch content does not change. It reduces soluble sugar content in Pinto beans (Bender *et al.*, 1991).

12.2.2.4. Injury to cell structure and cellular organelles

Ozone is highly reactive, it binds to plasma membranes. It destroys the semi-permeability of the cellular membranes by reacting with the carbon-carbon double bonds of unsaturated fatty acid in cell membranes (Rich, 1964; Goldstein and Balchum, 1968). Ozone leaks guard cells of onion, thereby closing the stomata in tolerant plants of onion. Whereas, susceptible plants allow O_3 pass into the substomatal cavity causing damage to the surrounding tissues (Engle and Gableman, 1966). Cell wall extension is retarded by inhibiting cellulose synthesis and subsequently cell wall formation (Ordin, 1965). The inhibition is mainly due to oxidizing effect of O_3 of the -SH groups of enzymes.

Electron microscopy showed swelling of thylakoids in the chloroplast due to O_3 injury, which was followed by swelling of golgibodies, endoplasmic reticulum and nuclear envelopes (Miyake *et al.*, 1984). They also reported reduced internal space of mitrocondrial cristae. Finally, SO_2 toxicity leads to deformed chloroplast.

12.2.2.5. Effect on amino acid and protein synthesis

Ozone markedly reduced protein content (Ting and Mukherji, 1971), suggesting either breakdown of existing protein or no synthesis of new protein. In contrast, Tingey *et al.* (1973) found an increase in biosynthesis of amino acids and proteins under O_3 stressed environment.

12.2.2.6. Oxidative damage

Ozone reacts with O_2 and produce reactive oxygen species, including hydrogen peroxide (H_2O_2), superoxide (O_2^-), single oxygen ($^1O_2^·$) and hydrogen radical ($^·OH$). These denature proteins and damage nucleic acids, and cause lipid peroxidation, which breaks down lipids in membranes (for detail see chapter 7). Plant sensitivity to O_3 could be due to a deficiency in the antioxidant pools (Iriti *et al.*, 2006).

Kubo *et al.* (1995) investigated the effect of O_3 on antioxidant enzymes in *Arabidopsis thaliana*. It increased the activity of ascorbate peroxidase and guaiacol peroxidase in leaves, but had little effect on the activities of superoxide dismutase, catalase, monodehydroascorbate reductase.

12.2.2.7. Damage to waxy coating

Chronic exposure of leaves and needles to O_3 (similar damage also caused by some other air pollutants, namely SO_2, NO_x, PAN) can break down the waxy coating that helps prevent excessive water loss and damage from diseases, pests, drought and frost.

12.2.3. Tolerance Mechanisms

Plants differ in their ability of tolerance to air pollutants and generally influenced by the ability of plant organs/organelles to convert the dissolve pollutant into relatively less toxic forms. However, the ability of tolerance depends on the concentrations and the duration of exposure of pollutants, subsequently the ability alters when the plant genetic ability to withstand the stress. Different crop plants respond differently to the air pollutants. Plant resistance to stress is a consequence of two possible mechanisms, stress avoidance or stress tolerance (Levitt, 1972).

Pollution stress avoidance: Plants exclude partially or completely the environmental stress, therefore avoids the specific strain induced by the stress. Plant may alter both the external and internal concentrations of air pollutants affecting its absorption rates. The morphological and physiological features of foliage, like leaf pubescence, surface water and reactive cellular compounds may provide effective sites for removal of pollutant molecules from atmosphere, thereby reducing the external concentration (Fowler and Unsworth, 1974).

Avoiding ozone-induced damage depends on the ability of different genotypes to reduce O_3 uptake through stomatal closure and on the capacity non-photochemical quenching scavenge reactive oxygen (Crous *et al.*, 2006). The tolerant plant closes stomata avoiding intake of air pollutants. May studies associated the levels of leaf susceptibility with the openness of the stomata. Runeckles and Rosen (1977) reported variation in the extent of O_3-induced leaf strain with O_3^- pretreated plants of *Phaseolus vulgaris* is a consequence of differences in the ability of the stability of the stomata to close following acute exposure levels.

Leaf cuticle acts as an effective barrier to air pollutants to prevent into the leaf interior. Thus, variation in cuticular thickness during different phases of leaf development leads in differ tolerance ability of leaf. It has been reported that fully expanded leaves are more susceptible to air pollutants than the fully matured and older leaves.

Pollution stress tolerance: Plants may experience internally the stress but not the debilitating effect of the strain. The capacity to tolerate potentially toxic pollutant derivatives is a continuous unaffected despite the presence of harmful pollutant molecules. However, the capacity to assimilate the plants ability to metabolize the pollutant to less harmful, or remove them by deposition or translocation to sinks within the leaf or plant organs.

12.2.3.1. Water-stress reduces ozone injury

Water-stressed plants are more tolerant of O_3 exposure than are unstressed plants (Tingey and Hogsett, 1985). The probable reasons for this tolerance are stomatal closure, which reduces O_3 uptake and biochemical or anatomical changes within the leaves. Engle and Gabelman (1966) reported that resistant genotypes of onion close stomata when exposed to O_3, thus preventing the entry of gas into the leaf tissue. Ozone resistance in *Pelargonium hortorum* (Bonte *et al.*, 1975) and potassium deficient tomatoes (Leone, 1976) results in part from higher stomatal resistance which lessens the amount of ozone entering the leaf.

12.2.3.2. Antioxidants

Experimental evidences indicated that antioxidant defenses systems existing in plant tissues may function to protect cellular components from deleterious effects of photochemical oxidants through endogenous and exogenous controls. Ozone-toxicity tolerance was associated with the presence of kaempferol glycosides, a naturally occurring antioxidant (Foy *et al.*, 1995). Soybean lines containing no kaempferol glycosides were among the most sensitive to O_3 stress.

Ascorbic acid is the most abundant antioxidant in plants and serves as a major contributor to the cell redox state. Exposure to environmental O_3 can cause significant damage to plants by imposing condition of oxidative stress. Lee (1991b) reported that superior O_3 tolerance in the Hood soybean (O_3^- toxicity tolerant) cultivars were associated with a greater increase in endogenous levels of ascorbic acid compared to the O_3^- susceptible cultivar Hark. Ascorbic acid may scavenge free radicals and thereby protect cells from injury by O_3 or other oxygen-radical products. Increasing the levels of ascorbic acid through enhanced it's recycling would also limit the deleterious effects of environmental oxidative stress (Chen and Gallie, 2005). Dehydroascorbic reductase-overexpressing plants had a lower oxidative load, lower levels of chlorophyll, and a level of photosynthetic activity.

12.3. SULPHUR DIOXIDE

Sulfur dioxide is the most widespread air pollutant that can cause severe damage to vegetation. Sulfur is necessary for general metabolism of plant because it is a major component of amino acid, proteins and some vitamins. In healthy leaves sulphur content ranges from 500-14000 ppm by dry weight (0.5-14.0 mg/g dry weight) depending upon species. If the concentration of SO_2 increases beyond a certain critical level that may vary with species. It can result in the general disruption of photosynthesis, respiration and other fundamental cellular process. Injury becomes irreversible, leading to death, as concentration and time of exposure increase further.

Sulfur dioxide is formed during the coal burning operations, especially those providing electric power and space heating. Sulfur dioxide emission can also result from burning of petroleum and the smelting of sulphur containing ores and act directly as an air pollutant. It combines with water (air vapour) to form sulfuric acid tiny particles. These tiny particles come to the earth surface with rains, and this is known as acid rain.

$$SO_2 + 2H_2O \rightarrow H_2SO_4 + O_2$$

Rain is slightly acidic, with pH close to 5.6, because the CO_2 dissolve in it produces the weak acid, H_2CO_3. Dissolution of NO_x and SO_2 in water droplets in atmosphere causes the pH of the rain to decrease to 3-4. Most wet acid deposition forms when NO_x and SO_2 are converted to nitric acid (HNO_3) and sulfuric acid (H_2SO_4) through oxidation and dissolution. Dilute acidic solution can remove mineral nutrients from leaves, depending on the age of the leaf and the integrity of the cuticle and therefore are not strongly buffered; such addition of acid can be harmful to plants. Furthermore, the added acids can result the release of aluminium ions from soil minerals, causing Al-toxicity.

Acid rain as augmented by acid snow, acid mist, acid fog, and dry deposition of acid gases and particles damage ecosystem. It can lower pH of soil and natural water bodies, cause mineral leaching, damage vegetation and deplete fish populations in some lakes and streams.

12.3.1. Injury Symptoms

1. The symptoms appear as a yellowing or chlorosis of the leaf and occasionally as a bronzing on the under surface of the leaves.
2. Above the threshold concentrations of SO_2 (approximately 0.1 µl/L) elicits the acute injury syndrome of interveinal necrotic lesions. The

symptoms appear as two-sided (bifacial) lesions that usually occur between the veins and occasionally along the margins of the leaves. The colour of the necrotic area may vary from a light tan or near white to an orange-red or brown depending on the time of year, the plant species attacked and weather condition.

3. *Plant growth and yield:* Decreased morphological growth of *Panicum miliaccum* plants (Agrawal *et al.*, 1985). Buckenham *et al.* (1982) had reported decreased straw and grain yield of barley. Soybean plants are more susceptible at reproductive phase (Pandey and Rao, 1977). Root dry weight of wheat cultivars decreased significantly under higher SO_2 environment (Gould and Mannsfield, 1989). Jain and Chauhan (1990) observed reduction in pollen fertility in *Capsicum annum*.
4. In small grained crops, *tip die-back* is a common symptom.

12.3.2. Injury Mechanisms

Injurious effects result when SO_2 is taken up in excess of the capacity of tissue to incorporate sulphur into the normal metabolic activities.

12.3.2.1. Decrease in photosynthesis

Diffusion through stomata, SO_2 dissolves in aqueous solutions leading in formation of $SO_3^{=}$ (suphite) and HSO_3^{-}, which are phytotoxic to many biochemical physiological process (Malhotra and Khan, 1984). In such transformation cellular pH would also be influenced by the generation of protons. Higher concentrations or longer exposure of plants in SO_2 significantly reduced photosynthesis in rice (Matsuoka, 1978) and *Vicia faba* (Ormrod *et al.*, 1981). SO_2 exposure significantly depressed total chlorophyll concentration in *Eucaliptus crebra* and *E. moluccana* (Murray, 1984).

12.3.2.2. Inhibition of ATP synthesis

It also attacks electron transport and photophosphorelation reactions. $SO_3^{=}$ inhibits formation of ATP (Ballantyne, 1973; Malhotra and Hocking, 1975) causing reduction of electron transport components and uncoupling, which resulted in subsequent decreased ATP synthesis. $SO_3^{=}$ inhibits the PEPCase involved in C_4 pathway of photosynthesis.

12.3.2.3. Decrease in total amino acid

Several enzymes involve in amino acid metabolism have been shown to be affected by SO_2 (Pahlich *et al.*, 1972). They have observed that fumigation with SO_2 affected glutamate dehydrogenase activity in *Pisum sativum* by stimulating reductive amination and inhibiting oxidative deamination. Increased SO_2 in the atmosphere caused a decrease in total amino acids and amines, asparagine, and glutathione. This decrease could attribute to a breakdown of the existing proteins and to reduced *de novo* synthesis.

A decline in linoleic acid content by SO_2 stimulated free-radical peroxidation reactions has been observed by Khan and Malhotra (1977) in pine. Higher concentrations of SO_2 also lead formation of malonaldehyde (Peiser and Yang, 1979). Decrease in lipid content following SO_2 exposure may be brought about by reduced synthesis, increased lipase activity, peroxidation of fatty acid chains, or a combination of all.

12.3.2.4. Effect on oxidative enzymes

Sulfur dioxides effect oxidative enzymes in *Arabidopsis thaliana* (Kubo *et al.*, 1995). They observed increased activities of ascorbate peroxidase and guaiacol peroxidase in leaves, but had little effect on the activities of superoxide dismutase, catalase, monohydroascorbate reductase, dehydro-ascorbate reductase or glutathione reductase.

12.3.2.5. Imbalance in soluble sugar and reducing sugar

An increase in amount of soluble sugar, and increased content of reducing sugar and decreased content of non-reducing sugar have been noted by Malhotra and Sarkar (1979) in pine under high SO_2 environment. The increase may be due to breakdown of polysaccharides rich in reducing sugar. The total carbohydrate and nitrogen of seeds of chickpea decreased significantly (Kumar and Singh, 1987).

12.3.2.6. Injury to cell structure and cellular organelles

Electron microscopy has been used to demonstrate the extent of ultra-structural injury caused by air pollutants. At aqueous concentration of 100-500 ppm SO_2 caused swelling of thylakoid and disintegrated other intra-chloroplast membrane, resulting in the formation of small vesicles (Malhotra, 1975). Chloroplast structural injury was more pronounced in old tissues than in younger and more metabolically active tissues. Sulfur dioxide injury first appears as swelling of the stomata and deformation of the chloroplast and swelling of thylakoid appeared latter (Miyake *et al.*, 1984).

12.3.3. Tolerance Mechanisms

12.3.3.1. Protein and enzyme accumulation

Anbazhagan *et al.* (1988) noted that the tolerant genotypes of rice accumulated the high concentration of proline in response SO_2 fumigation. The total and reduced-glutathione and glutathione reductase activity increased more rapidly in *Pisum sativum* under SO_2 pollution (Madamanchi and Alscher, 1991). They also reported increased activities of superoxide dismutase.

12.3.3.2. Emission of H_2S

Under SO_2 exposure, plant absorbs it rapidly through stomata. The SO_2 dissolves in tissue water and ionizes to HSO_3^- or SO_3^{2-}. It has been shown in several plant species that most of SO_2 absorbed is oxidized to SO_4^{2-} rapidly (Garsed and Read, 1977a, b). The susceptibility of leaves to injury by SO_2 changes systematically during development. In cucurbitaceae, as in many other species, young leaves are resistant to injury from acute exposure to SO_2 and mature leaves are sensitive. The difference is because young cucurbit leaves actually absorb SO_2 at a high rate than mature leaves. Sekiya *et al.* (1982) observed emission of volatile sulfur (H_2S) on exposure of cucumber (*Cucumis sativus* L.) plants to SO_2 at injurious concentrations. Young leaves emitted H_2S many times more rapidly than do mature leaves. Their findings suggested that a high capability for the reduction of SO_2 to H_2S and emission of the H_2S is a part of the biochemical basis of the resistance of young leaves to SO_2.

12.3.3.3. Sulfurdioxide and water stress

Plant subjected to water stress favours root growth instead of shoot growth and responds to water stress by potential stomatal closure which will reduce uptake air pollutants. Qifu and Murray (1991) reported reduction in growth induced by SO_2 in well watered plants, but usually in not water-stressed plants, indicating a protective function of soil water stress in the response of plants to SO_2. This could be caused by reduced SO_2 uptake in water-stressed plants, as the well-watered plants showed much higher leaf sulfur concentration than did the water-stressed plants at the same SO_2 concentration.

12.4. AMMONIA

Accidents in storage, transportation or application of anhydrous and aqua-ammonia fertilizers usually release huge quantities of NH_4 into the atmosphere for brief periods of time and cause severe injury to vegetation in the immediate vicinity. Wetlands are also a prominent source of NH_4. Ammonia emission from wetland depends on the air and soil temperature, soil redox potential, and standing water depth (Yang *et al.*, 2006).

12.4.1. Injury Symptoms

1. Symptom appears as irregular, bleached, bifacial, necrotic lesions.
2. Grasses often show reddish, interveinal necrotic streaking or dark upper surface discoloriztion. Flowers, fruits and woody tissues usually are not affected.

12.5. FLUORIDE

Fluorides are discharged into the atmosphere from combustion of coal, the production of bricks, tiles, enamel frit, ceramics, and glass, the manufacture of aluminium and steel, and the production of hydrofluoric acid, phosphate chemicals and fertilizers (Stein, 1971). It is an extremely toxic ion, near sources of fluoride air pollution, vegetation gets destroyed. Gaseous fluoride enters the leaf through stomata then it dissolve in the water permeating the cell wall. The natural flow of water in a leaf towards the site of greatest evaporation, which is the margins and tip, so it shows visible injury. Young still expanding leaves are most sensitive. The translocation of fluoride throughout the plant structure damaging leaves, blossoms and fruits is much more pronounced (Garber. 1967).

12.5.1. Injury Symptoms

Air born fluoride can damage either the foliage or the fruit of a wide range of plants, and the amount of fluoride necessary for toxicity depends on the species.

1. Visible symptoms appear to the tip of the monocotyledonous leaves, known as *tip burn* (Treshow and Pack, 1970). The injury starts as a gray or light green water-soaked lesion, which turns tan to reddish brown. Gradually necrotic areas increase in size, spreading downward. The

necrotic area is sharply delineated from the healthy portion of the leaf blade by a narrow band of clorotic tissue sometimes streaks with red as in some varieties of sorghum.

2. In broad-leafed species, the young developing leaves and occasionally petals, the translocation of fluoride to the margins and tips leads to a distorted shape. This occurs because cells in the mid-parts of the leaves have low fluid and expand normally, but those on the margins are slower-growth so the leaf buckles and distorts, becoming cupped and concave or convoluted like a Savoy cabbage.
3. In some broad leaf species the dead, necrotic areas are pale white to tan, in others they are brown and they may be black (*Populus* spp.) or have reddish tinges. Characteristically, there is a dark brown margin along the basal part of the necrotic area. This line of demarcation is very useful in identifying multiple exposures.
4. Peach get soft structure disease or *suture red spot* in which the region of the peach along the season near the tip ripens before the rest of the fruit. The ripening of the tissue considerably precedes that of the normal fruit and is often accompanied by splitting of the flesh along the suture (Benson, 1959; MacLean *et al.*, 1984).
5. Fluoride causes *snub nose* or *shrivel tip* in cherries (Treshow and Pack, 1970).
6. Dead dry pieces of leaf may become brittle and fall off, giving leaf a tatterd appearance, such as in Chinese apricot, Italian prune and many species of *Populus*.
7. In pine (*Pinus ponderosa*) needle exhibit a lightening in colour, which turns light brown to reddish-brown at the tip and progress basipentally along the needle.
8. Fluoride induced distortion of strawberry fruits (*Fragaria* sp.) was caused by lack of fertilization of the some of the ovules, which are responsible for hormonal-induced swelling of the fruit (Bonte, 1982).
9. In corn (*Zea mays*), and other grass species, symptoms begin as scattered chlorotic flecks at the tips and upper margins of middle-aged leaves. On progress, the flecking becomes more intense and extends downward, especially along the margin.

12.5.2. Tolerant Mechanism

The survival of grapevine was related to portioning of fluoride in leaf boundaries and extremities with plant tendency to delimit the new necrotic tissues by a narrow dark violet borderline, leading to the development of concentric halos of necrotic zones, and giving a blade a mosaic aspect (Abdallah *et al*., 2006). This response occurs concomitantly with the ability of healthy leaf areas to preserve plant photosynthesis capacity, as long as necrosis did not exceed 10-20% of the leaf surface. Fluoride also temporarily closes the stomates restricting the entry of fluoride. Fluoride accumulation accompanied by calcium leading to formation of CaF_2, which involve in detoxifying and trapping of fluoride.

12.6. OTHER GREENHOUSE GASES

12.6.1. Carbon Dioxide

Carbon dioxide (CO_2) is one of the major air pollutants in the atmosphere. The atmospheric concentration of CO_2 in 2005 exceeds by far the natural range over the last 65000 years (180-300 ppm) as determined from ice core. In 1950, levels of CO_2 were around 270 ppm, by 2005 these levels were at 379 ppm (IPCC, 2007). Carbon dioxide growth rate (1.9 ppm per year) was larger during the last 10 years (1995-2005), than it has been since the beginning of continuous direct atmospheric measurement (1960-2005, average- 1.4 ppm per year) although there is year-to-year variability in growth rates. According to the congressional Office of Technology Assessment (OAT, 1993), 70-90% of the CO_2 added to the atmosphere is due to the burning of fossil fuels, and the rest is from deforestation. Increase of CO_2 concentration in the atmosphere at this rate is estimated that the concentrations of CO_2 will double pre-industrial concentration by a least 2100 (IPCC, 2001).

Globally, fossil fuel use results each year about 6 billion metric tons of C, of which USA alone accounts for about 23% of global carbon emission. Industrial countries account for 65% of CO_2 emissions with the United States and Soviet Union responsible for 50%. Less developed countries account, with 80% of world's population, are responsible for 35% of CO_2 emissions but may contribute 50% by 2020. With destruction and burning of forest also released a huge amount of CO_2 into the atmosphere. Carbon dioxide emissions are increasing by 4% per year. About half of the CO_2 emissions resulting from human activities are absorbed by natural '*sink*' on land and the oceans. Ocean water contains about 60 times more CO_2 than the atmosphere. If the equilibrium is disturbed by externally increasing the concentration of CO_2 in the air, then the oceans would absorb more and more CO_2. If the ocean can no longer keep the pace, then more CO_2 will remain into the atmosphere causing increase in temperature of the atmosphere. The increasing atmospheric temperature will warm up the ocean water, which will further reduce the ability of ocean water to absorb CO_2. Natural disaster may happen, and ocean water may releases the locked CO_2 due to some natural catastrophe, such as earth quack or any disturbance in the ocean. It was observed that the plant's ability to absorb CO_2 emissions due to human activities has decreased considerably. About 50 years ago, for every tonne of CO_2 emitted, 600 kg were removed by natural sinks. In 2006, only 550 kg were removed per tonne and that amount is falling.

12.6.1.1. Effect of CO_2 on plants

Elevated CO_2 is known to stimulate leaf level photosynthesis and to reduce leaf transpiration. This may result in alter biomass production of agricultural plants and subsequent secondary feedback effects on ecosystem properties, such as water relations, carbon turnover and soil biology (Weigel *et al.*, 2006).

12.6.2. Carbon Monoxide

Carbon monoxide (CO) is a colourless, odourless poisonous gas produced by incomplete oxidation. Motor vehicle is one of the major man-made sources of CO production. It remains in the atmosphere for 1-2 months. It is removed by oxidation to form carbon dioxide, absorption by some plants and micro-organisms and rain.

12.6.3. Chlorofluorocarbons

Chlorofluorocarbons (CFCs) are greenhouse gases that contribute global warming. Aerosol accounts about 25% of global CFC use. Other sources of CFCs are spray cans, discarded or leaking refrigeration and air conditioning equipments, and the burning plastic foam products. Depending upon the type, CFCs stay in the atmosphere from 22-111 years. Chlorofluorocarbons move up to the atmosphere gradually over several decades. It lowers the average concentration of ozone in the atmosphere. Under high energy UV-radiation, they break down and release chlorine atoms, which speed up the breakdown of ozone into oxygen.

12.6.4. Methane

Methane is an important greenhouse gas in the atmosphere. Methane remains in the atmosphere for a short time (about 12 years) as compared to CO_2 (with 50-100 years). Its atmospheric concentration is 1.675 ppm, and every year it increases about 1%. Its relative green house effect (considering CO_2 as standard) is 25 and contribution in greenhouse effect is about 12%. Vast water sources, such as dams, release substantial quantities of methane, especially early in their lifetime, due to rotting vegetation beneath the water line. Other sources of methane are solid wastes, livestock, hard coal, rice field and pipeline leakage.

12.7. PARTICULATE MATTERS

There are solid or liquid particles that may suspended in the air, which is referred as particulate matters. These are very small air borne particles less than 10 micron in diameter. It includes dirt, smoke, soot, plant spores, pollen grains, bacteria and salts and liquid droplets directly emitted into the air by sources such as factories, power plants, cars, construction activities, fire and natural wind blown dusts. Particulate matters can be usefully classified by size. Large particles settle out of the air quickly while smaller particles may remain suspended for days or months. Rainfall is an important mechanism for removing particles from the air.

Major sources of particulate matter are burning fuels, such as wood in woodstoves, and fire places, or diesel in motor vehicles, crushing or grinding, dust from unpaved roads and construction sites, and from industrial processes.

12.7.1. Injury Symptoms

Cement dust may cause chlorosis and death of leaf tissue by the combination of thick crust and alkaline toxicity produced in wet weather.

12.7.2. Injury Mechanisms

Particulate matters deposition on leaf-surface inhibit normal respiration and photosynthesis. Cement respiration and photosynthesis. Dust pollution is of localized importance in sandy and loose soil areas. It affects the plant near roads, quarries, cement works, and other industrial places. It inhibits sunlight inception to leaves, blocks stomata leading reduction of CO_2 conductance, simultaneously interfering photosynthesis. Accumulation of alkaline dust on soil may increase soil pH. Usual dust has no injurious effect on plants, but it may reduce pesticide efficiency, inhibition of respiration and photosynthesis.

12.8. OXIDES OF NITROGEN

The main oxides of nitrogen present in the atmosphere are nitric oxide (NO), nitrogen dioxide (NO_2) and nitrous oxide (N_2O). Nitrous oxides occurs in much smaller quantities than the other two, but is of interest as it is a powerful greenhouse gas and thus contributes to global warming. The global atmospheric nitrous oxide concentration increased from a pre-industrial value of about 270 pph to 319 ppb in 2005.

The major source of NO_x is fuel combustion, such as automobiles, and industries. Oxides of nitrogen form in the air when fuel is burnt at high temperatures. This is mostly in the form of nitric oxide with usually less than 10% as nitrogen dioxide. Once emitted, nitric oxides combine with O_2 to form NO_2, especially in warm sunny days. The other sources of NO_x are bacterial action in soil, forest fires, volcanic actions and natural lighting.

Agricultural fields are significant sources of anthropogenic atmospheric nitrous oxide (N_2O). The mean N_2O emission from Japanese paddy fields is 0.36 kg N/ha for the cropping season (Akiyama *et al.*, 2006).

12.8.1. Injury Symptoms

1. NO_2 can suppress plant growth
2. Increased concentrations of nitrite nitrogen decreased dry matters yield, total acidity, the concentration of nitrogen, phosphorus and potassium in tomato plants (Philipps and Cornforth, 1970).
3. It increased chlorophyll of leaves and lignification of roots (Philipps and Cornforth, 1970).
4. Nitrous oxides (NO_x) in cells and give rise to nitrite ions (NO^{2-}), which are toxic at high concentrations that enter into nitrogen metabolism as if they had been absorbed through the roots.

12.9. PEROXYACETYL NITRATE

Peroxyacetyl nitrate (PAN) is one of the most abundant of the gaseous organic matters. PAN is toxic to plants and also acts as a reservoir for the transport NO_2 in the troposphere (Teklemariam and Sparks, 2004). PAN reduced fresh weight and photosynthesis pigment content; and increased ion leakage in *Petunia hybrida* (Oka *et al.*, 2004). Morphological observations reveled that ion leakage started concurrently with the start of plasmolysis-like symptoms at the mesophyll cells of injured leaves.

12.10. VOLATILE ORGANIC COMPOUNDS

Volatile organic compounds (VOCs) are organic gases and vapour that evaporate into the air easily. At room temperature vapour and readily escaped from volatile liquid chemicals. VOCs include gasoline, industrial chemicals such as benzene, nail polish remover, barbecue fluid, and solvents such as toluene, xylene and perchloroethylene (particularly dry-cleaning solvents), paints, glues and other products used at home such as inks, aerosol spray products. VOCs are also released from burning of fuel, such as wood, coal, natural gas. Vehicle emissions are important source of VOCs. These compounds react with nitrogen oxides, sunlight and that to form ozone.

12.11. SMOG

Photochemical smog is the product of chemical reaction driven by sunlight and involving NOx of urban and industrial origin and volatile organic compounds from either vegetation or human activities. Simply, it is chemical mixture of gases that form a brownish-yellow haze primarily over urban areas. Components of smog include ground level O_3, NO_x, volatile organic compounds, SO_2, acidic aerosols, gases, and particulate matters. Smog is most prevalent in the summer months, where there is the most sunlight and temperatures are the highest. The air born pollutants which make up 90% of all smog found in urban areas is ground level O_3. When stagnant air masses linger over urban areas, the pollutants are held in place for longer periods of time. Sunlight interacts with NO_x and VOCs, transforming them into ground level O_3. The O_3 remains in the lower atmosphere until weather system flush out a given area and dissipate them. Photochemical smog is also appearing in regions of the tropics and subtropics where Savanna grasses are periodically burned.

Smog is common in cities, such as London, Chicago and Pittsburgh. When these cities burn large amount of coal and heavy oil without control of the output, large-scale problems were witnessed. It had taken 4000 life during 1952 in London.

12.11.1. Injury Symptoms

1. The unique symptoms are silvering and bronzing of the lower leaf surfaces and more sever cases necrosis (Middleton *et al.*, 1950).
2. A decrease in total growth of plants without visual damage was also noticed (Koritz and Went, 1952).

12.12. FACTORS AFFECTING POLLUTANT'S TOLERANCE ABILITY

Several environmental and organizational factors affect the tolerance ability of crop plants against air pollutants have been presented in Table 12.1.

Table 12.1. Factors (environmental and organizational) affecting the susceptibility of foliage to air pollution stress (Taylor Jr, 1978)

Source of variation to pollutants	Possible physiological site of mechanism	Stress avoidance	Tress tolerance	
			Strain avoidance	Strain tolerance
1. Climatic				
i) Light intensity	$R_B + R_S + R_M$	√	√	√
ii) Wind velocity	$R_B + R_S$	√	×	×
iii) Time of day	$R_B + R_S + R_M$	√	√	×
2. Atmospheric				
i) Pollutant dosage	C_E	√	×	×
ii) CO_2 concentration	$R_S + R_M$	√	×	×
iii) Humidity	$R_S + C_E$	√	×	×
iv) Pre-exposure	$R_S + R_M + C_E$	√	√	×
v) Synergism	$R_S + R_M$	√	×	×
3. Edapic				
i) Water stress	$R_B + R_S + R_M$	√	√	×
ii) Macro and micronutrient availability	BTL	×	√	×
4. Biotic				
i) Fungus	$R_S + R_M$	√	√	×
ii) Virus	$R_S + R_M$	√	√	×
5. Organismal				
i) Interspecific	Variable	√	√	√
ii) Intraspecific	Variable	√	√	√
iii) Developmental	Variable	√	√	√
iv) Foliar	Variable	√	√	√

R_B: Boundary layer resistance, R_S: Stomatal resistance, R_M: Mesophyll resistance, C_E: External pollutant concentration, BTL: Biochemical threshold level tolerance

12.13. VARIABILITY FOR AIR POLLUTANTS TOLERANCE

Existing variability is most important than the creation of variability for crop improvement programme to select/screen/isolate desirable tolerant genotype(s) of crop plants against different air pollutant. Intraspecies-genetic variations in susceptibility to air pollutants have been screened for many crops (Table 12.2). Cultivars sensitive to one pollutant may be sensitive or tolerant to another pollutant. Feder *et al.* (1969) studied the response of 15 petunia cultivars to O_3, SO_2, PAN, NO_3 and irradiated auto exhaust. They reported that some cultivars were sensitive to only one of the pollutants, while the other was sensitive to several pollutants. Significant variations in leaf injury were noted by Youngner and Nudge (1980) among the species and cultivars of turgrass (*Lolium* sp.).

Table 12.2. Pollution tolerant genotypes of crop species

Crop	Tolerant genotype	Pollutant	Reference
Rice *Oryza sativa*	GR_3, TKM 9	SO_2	Anbazhagan *et al.*, 1988
Petunia *Petunia hybrida*	White Cascade	O_3, SO_2, NO_2, PAN	Feder *et al.*, 1969
Bean *Phaselous vulgare*	Berner, Narda, BBL274, Horticultural	O_3	Tonneijck, 1983; Heck *et al.*, 1988
Soybean *Glycin max*	Essex	O_3	Heagle and Letchnorth, 1982
	Mukden, Lee	O_3	Foy *et al.*, 1995
Tomato *Lycopercicon esculentum*	Green Mountain, Irish Cobbler, Belrus, Superior	O_3	Clarke *et al.*, 1990
	Naebyongjangsu	O_3	Ku *et al.*, 1989
Wheat *Triticum* spp.	Abc, Arthur71	O_3	Kress *et al.*, 1985
	Grana, Beta	SO_2	Bialobok, 1990
	Sonalika, HI 10-77, K 8565, K 7410, Huw206	SO_2	Pavgi *et al.*, 1991

Different plant species may vary considerably in there sensitivity to SO_2 (Table 12.3). This variation occurs because of the differences in geographical location, climate, stage of growth and maturity.

Table 12.3. Tolerant and susceptible crop species to air pollutants

Scientific name	Common name	Resistant to air pollutants			
		O_3	SO_2	NH_4	Fluoride
Abies balsamea	Fir, Balsam	T	–	–	–
Abies concolor	Fir, white	T	–	–	–
Acer negundo	Boxelder	M	M	–	–
Acer plantanoides	Maple	T	–	–	–
Acer rubrum	Maple		M	–	–
Acer saccharum	Maple	T	T	–	–
Acer saccharinum	Maple	–	T	–	–
Alium cepa	Onion	S	S	T	–
Amelanchier	Serviceberry	–	S	–	–
Apium graveolence	Celery	–	T	–	T
Avena sativa	Oats	–	S	–	–
Asparagus cooperi	Asparagus	–	T	–	T
Beta vulgaris	Beet	–	–	T	
Betula pendula	Birch	T	–	–	–
Betula spp.	Birch		S	–	–
Brassica oleraceae capitata	Cabbage	–	T	–	T
Brassica oleraceae botrytis	Cauliflower	–	–	–	T
Catalpa spp.	Catalpa	S	–	–	–
Cercis canadensis	Redbud	M	–	–	–
Cornus florida	Dogwood	T	–		–
Cucurbita melamosperma	Squash	–	S	–	T
Cucumis sativus	Cucumber	T	S	T	–
Cucurbita pepo	Pumkin	–	S	–	–
Daucus carrota	Carrot	–	–	T	T
Fagopyrum esculentum	Buckwheat	–	S	–	–
Fraxinus americana	Ash, white	S	–	–	–
Fraxinus pennsylvanica	Ash, green	S	S	–	–
Ginkgo biloba	Ginkgo	T	T	–	–
Gladiolus grandiflorus	Gladiolus	–	S	–	S
Gledistsia triacanthos	Honeylocust	S	–	–	–
Glycin max	Soybean	–	S	S	–
Hordium vulgare	Barley	–	S	S	S (young)
Ilex spp.	Holly	T	–	–	–
Jagulans regia	Walnut	S	–	–	–

Contd...

Table 12.3. Contd...

Jagulans nigra	Walnut	T	–	–	–
Juniperus spp.	Juniper	T	–	–	–
Latuca sativa	Lettuce	S	–	–	–
Liquidambar styracifolia	Gum, Sweet	M	–	–	–
Liriodendron tulipifera	Poplar	S	–	–	–
Malus spp.	Crabapple	S	–	–	–
Malus sylvestris	Apple	–	–	S	–
Medicago sativa	Alfalfa	–	S	T	T
Nicotiana tubccum	Tobacco	S	S	–	T
Nyssa sylvatica	Gum, black	T	–	–	
Phaseolus	Snap bean	–	–	S	T
Picea abies	Spruce	T	–	–	–
Picea pungens	Spruce	T	T	–	–
Pinus echinata	Pine	M	–	–	–
Pinus nigra	Pine	S	M	–	–
Pinus resinosa	Pine	T	–	–	–
Pinus strobus	Pine	M	S	–	–
Pinus sylvestris	Pine	M		–	–
Piper nigrum	Pepper	–		–	T
Pisum sativum	Pea	–	S	–	T
Pinus virginiana	Pine	S	–	–	–
Pinus toeds	Pine	S	–	–	–
Plantanus occidentalis	Sycamore	S	–	–	–
Populus deltoids	Cottonwood	M	–	–	–
Populus nigra	Poplar	–	S	–	–
Prunus ameniaca	Apricot	T	–	–	S
Prunus communis	Pear	T	–	–	T
Prunus domestica	Plum	–	–	–	T
Prunus persica	Peach	–	–	T	–
Pseudotsuga menziesii	Fir, Douglas	T	–	–	–
Quercus alba	Oak	S	M	–	–
Quercus coccinea	Oak	M	–	–	–
Quercus palustris	Oak	S	T	–	–
Quercus robur	Oak	T	–	–	–

Contd...

Table 12.3. Contd...

Quercus rubra	Oat	T	T	–	–
Quercus velutina	Oat	M	–	–	–
Raphanus sativa	Radish	S	S	S	–
Rubus sp.	Raspberry	–	–	S	
Salix nigra	Willow	S	–	–	–
Salix babylonica	Willow	S	–	–	–
Solanum esculentum	Tomato	S	–	T	–
Solanum melongena	Brinjal		–	T	T
Solanum tuberosum	Potato	S	S	T	T
Sorbus aucuparia	Mountain ash	S	M	–	–
Spinacia olericeae	Spinach	S	S	–	–
Syringa spp.	Lilac	M	M	–	–
Taxus spp.	Yew	T		–	–
Thuja spp.	Arborvitae	T	T	–	–
Zelkova serrata	Zelkova	S	–	–	–
Tilia americana	Liden	T	M	–	–
Tilia cordata	Linden	T	T	–	–
Trifolium sp.	Clover	–	S	S	–
Triticum aestivum	Wheat	–	–	–	T
Tulipa hybrida	Tulips	–	S	–	S
Ulmus parviflolia	Elm	M	S	–	–
Ulmus americanum	Elm	M	–	–	–
Vitis sp.	Grape	S	S	–	S
Vaccinium corymbosum	Blue berry	–	–	–	S
Zea mays	Corn/Maize	S (sweet corn)	–	T	S (sweet corn)

T: Tolerant, **M:** Moderately tolerant, **S:** Susceptible, **-:** Information not available

12.14. SCREENING METHODS

12.14.1. Screening Methods for Sulfur Dioxide Tolerance

Screening against air pollutants, a controlled environment such as greenhouse is essential. Air pollution can be originated both inside and outside of the greenhouse. Sulfur dioxide is generated by burning high-sulfur coal or fuel oil in the greenhouse or directly the pollutants in the greenhouse from industrial facilities such as refineries, chemical plants, and power generating plants.

Sekiya *et al.* (1982) used Plesiglas chamber to find out resistance mechanism of Cucurbitaceae against SO_2. The chamber had an air stirrer built in, and was illuminated cool-white fluorescent lamps (0.8 mw cm-2). The plant in plastic pot was placed inside the chamber along with a beaker containing mixture, which would generate SO_2 upon acidification. After sealing the chamber, lactic acid was added to the beakers by Telfon tubing. The mixture after acidification contained in 30 ml: $KHSO_4$ (60 m mol), Na_2CO_3 (40 m mol), 6.7% (v/v) ethanol, and 12% (v/v) lactic acid.

Agrawal *et al.* (1985) used polyethelene chamber to find out the effect of SO_2 on plants. Sulfur dioxide generated by bubling air through 1% aqueous solution, which was ionized under turbulence producing SO_2. The gas concentration was monitored by sucking a known volume of air pollutant mixture and absorbing it in sodium tetrachloromercurate solution, which in turn was colorimetrically assayed for the SO_2 concentration.

12.14.2. Screening Methods for Ozone Tolerance

Agrawal *et al.* (1985) studied effect of O_3 on *Penicum miliaccum* in polyethelene chambers. The O_3 was generated by passing charcoal-filtered air across a UV lamp. Ozone concentrations was continuously monitored with a Mast ozone meter and were controlled by regulating the voltage applied to the UV lamp.

A chlorophyll fluorescence induction assay could be used to screen crop genotypes against O_3 injury. This approach became possible with the development of methods to ensure fluorescence induction from the illuminated surface of dense leaf material (Schreiber and Vidaver, 1974; Schreiber *et al.*, 1975). Nowadays, many companies make chlorophyll fluorescence meter, which can be directly used to detect air pollutants injury-tolerance ability of plants. It is easy and rapid way of studying O_3 effects in whole plants. It is conventionally applied in both in the laboratory and in field experiments. It is non-destructive nature and method is very simple.

12.15. TRANSGENIC DEVELOPMENT

The technology of transgenic development could be used for improvement of air pollutant tolerance in agriculturally important plant species. The findings of Nakajima *et al.* (2002) indicated that the introduction of antisense DNA for an ozone-inducible 1-amonocyclopropane-1-carboxylate (ACC) synthase (E.C. 4.4.1.14; LE-ACS6) can improve the O_3 tolerance of plants without reducing the gas absorption and productivity. They developed tobacco (*Nicotiana tabaccum*) with an antisense DNA for ACC synthase, which was O_3^- toxicity tolerance.

12.16. REFERENCES

Abdallah FB, Elloumi N, Mezghani I, Boukhris, Garrec JP. 2006. Response of a local grapevine to fluoride pollution. Canadian J Bot. 84(3): 393-399.

Agrawal M, Nandi PK, Rao DN. 1985. Ozone and sulphur dioxide effects on *Panicum miliaceum* plants. Bulletin of the Torrey Botanical Club. 110(4): 135-141.

Akiyama, H, Yuan Y X, Yagi K. 2006. Estimations of emission factors for fertilizer-induced direct N_2O emissions from agricultural soils in Japan: summary of available data. Soil Sci Plant Nutr. 52(6): 774-787.

Anbazhagan M, Krishnamurthy R, Bhagawat KA. 1988. Proline: An energetic indicator of air pollution tolerance in rice cultivars. J Plant Physiol. 133: 122-123.

Ariaphanphitak W, Chidthaisong A, Sarobol E, Bashkin VN, Towprayoon S. 2006. Effects of elevated ozone concentrations on Thai Jasmine rice cultivars (*Oryza sativa* L.). Water, Air and Soil Pollution. 167(1/4) 179-200.

Ballantyne DJ. 1973. Sulphite inhibition of ATP formation in plant mitochondria. Phytochemistry. 12: 1207-1209.

Bender J, Bergmann E, Weigel HJ. 2006. Responses of biomass production and reproductive development to ozone exposure differ between European wild plant species. Water, Air and Soil pollution. 175(1/4): 253-267.

Bender J, Tingey DT, Jager HJ, Redecap KD, Clark CS. 1991. Physiological and biochemical responses of bush bean (*Phaselus vulgaris*) to ozone and drought stresses. J Plant Physiol. 137: 565-570.

Benson N. 1959. Fluoride injury or soft suture and splitting of peaches. Proc Amer Soc Hort Sci. 74: 184-198.

Bialobok S. 1990. Susceptibility of cereals to sulphur dioxide in the vegetative phase. Produkcja Roslinna. 107(4): 9-20.

Bonte J, de Cormis L, Louquet P. 1975. Influence d'une pollution par le dioxide de soufre sur le degree d'ouverture des stomates du *Pelargonium hortorum*. C R Hebd Seanc Acad Sci Ser D Seanc Nat. 280: 2377.

Bonte J. 1982. Effects of air pollutants on flowering and fruiting. In: Effects of Gaseous Pollutants in Agriculture and Horticulture, Unsworth MH, Omrod DP (eds.). Butterworth Scientific, London. pp. 207-223.

Buckenham AH, Parry MAJ, Whittingham CP. 1982. Effect of aerial pollutants on the growth and yield of spring barley. Annals Appl Biol. 100: 179-187.

Chen Z, Gallie DR. 2005. Increasing tolerance to ozone by elevating foliar ascorbic acid confers greater protection against ozone than increasing avoidance. Plant Physiol. 138(3): 1673-1689.

Clarke BB, Greenhalgh-Weidman B, Brennam EG. 1990. An assessment of the impact of ambient ozone on field-grown crops in New Jersey using the EDU method. I. White potato (*Solanum tuberosum*). Environ Pollution. 66: 351-360.

Coulson CL, Heath RL. 1974. Inhibition of photosynthetic capacity of isolated chloroplast by ozone. Plant Physiol. 53: 32-38.

Crous KY, Vandermeiren K, Ceulemans R. 2006. Physiological responses to cumulate ozone uptake in two white clover (*Trifolium repens* L. cv. Regal) clones with different ozone sensitivity. Envirnl Exp Bot. 58(1/3): 169-179.

Engle RL, Gableman WH. 1966. Inhertance and mechanism for resistance to ozone damage in onion *Allium cepa*. Proc Am Soc Hort Sci. 89: 423-430.

Feder WA, Fox FL, Heck WW, Campbell RJ. 1969. Varietal response of petunia to several air pollutants. Plant Dis Reotr. 53: 506-510.

Fowler D, Unsworth MH. 1974. Dry deposition of sulphur dioxide on wheat. Nature, London. pp. 249-389.

Foy CD, Lee H, Rowland R, Devine TE, Buzzell RI. 1995. Ozone tolerance related to flovanol glycoside genes in soybean. J Plant Nutr. 18: 637-647.

Garber K. 1967. Effects of Air Contamination. Grbruder Bontraeger, Berlin.

Garsed SG, Read DJ. 1977a. Sulphur dioxide metabolism in soybean, *Glycin max* var. Biloxi. II. Biochemical distribution of $^{33}SO_2$ products. New Physiol. 99: 583-592.

Garsed SG, Read DJ. 1977b. The uptake and metabolism of $^{33}SO_2$ in plant of differing sensitibility to sulphur dioxide. Environ Pollu. 13: 173-186.

Goldstein BD, Balchum OJ. 1968. Environment, Health Programme. 97th Annu Conf Califonia Med Assoc Meeting, San Francisco.

Gould RP, Mannsfield TA. 1989. The sensitivity of early century of wheat to air pollution. Environ Pollution. 56: 31-37.

Grantz DA, Gunn S, Vu HB. 2006. O_3 impacts on plant development: a meta-analysis of root/shoot allocation and growth. Plant Cell Envirn. 29(7): 1193-1209.

Heagle AS, Letchnorth MB. 1982. Relationship among injury growth and yield responses of soybean cultivars exposed to ozone at different light intensities. J Environ Quality. 11: 690-694.

Heagle AS. 1989. Ozone and crop yield. Annu Rev Phytopathol. 27: 397-423.

Heck WW, Dunning JA, Hindawi IJ. 1965. Interactions of environmental factors on the sensitivity of plants to air pollution. JAPCA. 15: 511-515.

Heck WW, Dunning JA, Reinert RA, Prieor SA, Rangappa M, Benepal PS. 1988. Differential responses of four bean cultivars to chronic doses of ozone. J Am Soc Hort Sci. 113: 46-51.

Hucl P, Beversdorf WD. 1982. The response of selected *Phaseolus vulgaris* L. to ozone under controlled fumigation and ambient field levels. Canadian J Plant Sci. 62: 561-569.

IPCC. 2001. Climate Change 2001: The Scientific Basis. In: Contribution of Working Group I to the Third Assesment Report of the Intergovernmental Panel on Climate Change, Horyhton JT, Ding Y, Griggs DJ, Noguer M, van der Linden PJ, Dai x, Maskell, Johnson CA (eds.). Cambridge University Press, Cambridge.

IPCC. 2007. Climate Change 2007: The Physical Science Basis- Summary for Policymakers. (http://www.ipcc.ch.).

Iriti M, Belli L, Nali C, Lorenzini G, Gerosa G, Faoro F. 2006. Ozone sensitivity of currant tomato (*Lycopersicon pimpinellifolium*), a potential bioindicator species. Environmental Pollution. 141(2): 275-282.

Jain PK, Chauhan SVS. 1990. Effect of SO_2 fumigation on flowering, pollen fertility and fruiting in *Capsicum annum* L. Plant Cell Incompatibility Newsl. No. 22: 15-17.

Kangasjarvi J, Talvinen J, Utriainnen M, Karjalainen R. 1994. Plant defense systems induced by ozone. Plant Cell Environ. 17: 783-794.

Khan AA, Malhotra SS. 1977. Effects of gaseous sulfur dioxide on pine needle glycolipids. Phytochemistry. 16: 539-543.

Knudson LL, Tibbits TW, Edwards GE. 1977. Measurement of ozone injury by determination of leaf chlorophyll concentration. Plant Physiol. 60: 606-608.

Kortiz HG, Went FW. 1952. The physiological actions of smog on plants. I. Initial growth and transpiration studies. Plant Physiol. 28: 50-62.

Kress LW, Miller JE, Smith HJ. 1985. Impact of ozone on winter wheat yield. Environ & Expt Bot. 25: 211-228.

Ku JH, Won DC, Kim TI. 1989. Studies of relative susceptibility of seven tomato cultivars to ozone. Korean Soc Hort Sci. 7: 44-45.

Kubo A, Saji H, Tanaka K, Kondo N. 1995. Expression of *Arbidopsis* cytosolic ascorbate peroxidasae gene in response to ozone or sulfur dioxide. Plant Mol Biol. 29: 479-489.

Kumar H, Singh J. 1987. Response of chickpea cultivars to sulphur and free amino acid content of pine seedlings. Plant Physiol. 47: 223-228.

Leadley PW, Reynold JF, Flagler R, Heagle AS. 1990. Radiation utilization efficiency and growth of soybean exposed to ozone: a comparative analysis. Agril & Forest Meteorology. 51: 293-308.

Lee EH. 1991a. Chlorophyll fluorescence as an indicator to detect differential tolerance of snap bean cultivars in response to O_3 stress. Taiwania. 36: 220-237.

Lee EH. 1991b. Plant resistance mechanism to air pollutants: rhythms in ascorbic acid production during growth under ozone stress. Chronobiol Int. 8(2): 93-102.

Leone IA. 1976. Response of potassium-deficient tomato plants. Contr Boyce Thompson Inst Plant Res. 20: 331.

Levitt J. 1972. Responses of plants to Environmental Stresses. Academic Press, New York.

MacLean DC, Weinstein LH, McCune DC, Schneider RE. 1984. Fluoride-induced suture red spot in 'Elberta' peach. Environ Exp Bot. 24: 353-367.

Madamanchi NR, Alscher RG. 1991. Metabolic basis of differences in sensitivity of two pea cultivars to sulphur dioxide. Plant Physiol. 97: 88-93.

Mahotra SS, Sarkar SK. 1979. Effect of sulphur dioxide on sulphur and free amino acid content of pine seedlings. Plant Physiol. 47: 223-228.

Malhotra SS, Hocking D. 1975. Biochemical and cytological effects of sulphur dioxide on plant metabolism. New Physiol. 76: 227-237.

Malhotra SS, Khan AA. 1984. Biochemical and physiological impacts of major pollutants. In: Air Pollution and Plant Life, Treshow M (ed.). Chichester, USA, John Wiley & Son. pp. 113-157.

Malhotra SS. 1975. Effects of suphur dioxide on biochemical activity and ultrasructural organization of pine needle chloroplasts. 76: 239-245.

Matsuoka Y. 1978. Injury of rice plants caused by sulfur dioxide and its mechanism. Japan Agric Res Quarterly. 12: 183-186.

Middleton JT, Kendrick JB Jr, Schwalm HW. 1950. Injury to herbaceous plants by smog or air pollution. Plant Dis Reporter. 34: 245-252.

Miyake H, Furukawa A, Totosuka T, Maeda E. 1984. Differential effects on ozone and sulphur dioxide on the fine structure of spinach leaf cells. New Physiol. 96(2): 215-228.

Muntifering RB, Manning WG, Lin JC, Robison GB. 2006. Short-term exposure to ozone altered the relative feed value of alfalfa cultivar. Environmental Pollution. 140(1): 1-3.

Murray F. 1984. Effect of sulphur dioxide on three *Eucalyptus* species. Australian J Bot. 32(2): 139-145.

Nakajima N, Itoh T, Takikawa S, Asai N, Tamaoki M, Aono M, Kubo A, Azumi Y, Kamada H, Saji H. 2002. Improvement in ozone tolerance of tobacco plants with an antisense

DNA for 1-aminocyclopropane-1-carboxylate synthase. Plant Cell Environ. 25: 727-735.

Oka E, Tagami Y, Oohashi T, Kondo N. 2004. A physiological and morphological study on the injury caused by exposure to the air pollutant, peroxyacetyl nitrate (PAN), based on the quantitative assessment of the injury. J Plant Res. 117(1): 27-36.

Ordin L. 1965. Effect of air pollutants on cell wall metabolism. Arch Environ Health. 10: 189-194.

Ormrod DP, Black VJ, Hnsworth MH. 1981. Depression of net photosynthesis in *Vicia faba* L. exposed to sulfur dioxide and ozone. Nature, UK. 291(5816): 585-586.

OTA (Office of Technology Assessment). 1993. Preparing for an Uncertain Climate, vol. I. OTA-O-567. US Government Printing Office, Washington D.C.

Pahlich E, Jager HJ, Steubing L. 1972. Effects of sulphite on metabolism isolated mesophyll cells from *Papaver somniferum*. Plant Physiol. 62: 210-214.

Pandey SN, Rao DN. 1977. Growth, productivity and yield of soybean plants under the influence of sulphur dioxide pollution. Acta Botanica Indica. 5(2, suppl): 34.

Pavgi S, Farroq M, Venilateshwar C, Beg MV. 1991. Physiology and biochemical effect of sulfur dioxide on wheat varieties. Environ & Ecol. 9: 760-765.

Peiser GD, Yang SF. 1979. Ethylene and ethane production from sulphur dioxide injured plants. Plant Physiol. 63: 142-145.

Pell EJ, Pearson NS. 1983. Ozone-induced reaction in quality of ribulose1-5-biophosphate carboxylase in alfalfa foliage. Plant Physiol. 73: 185-187.

Philipps RH, Cornforth IS. 1970. Factors affecting the toxicity of nitrite nitrogen to tomatoes. J Plant Soil. 33(1-3): 457-466.

Preston EM, Tingey DT. 1988. Assesment of Crop Loss from Air Pollution: The NCLAN Programme for Crop Loss Assesment, UK. pp. 45-62.

Qifu MA, Murray F. 1991. Responses of potato plants to sulphur dioxide, water stress and their combination. New Physiol. 118: 101-109.

Rich S. 1964. Ozone damage to plants. Annu Rev Phytopathol. 2: 253-260.

Runeckles VC, Rosen PM. 1977. Effective of ambient ozone pretreatment on transportation and susceptibility to ozone injury. Canadian J Bot. 55: 193.

Schreiber U, Groberman L, Vidaver W. 1975. Portable, solid- state fluorometer for the measurement of chlorophyll fluorescence induction in plants. Rev Sci Instrum. 46: 538-542.

Schreiber U, Vidaver W. 1974. Chlorophyll fluorescence induction in anaerobic *Scenedesmus abliquus*. Biochem Biophys Acta. 368: 97-112.

Sekiya J, Wilson LG, Filner P. 1982. Resistance to injury by sulfur dioxide. Plant Physiol. 70: 437-441.

Simini M, Simon JE, Reinert RA, Eason G. 1989. Identification of ozone-induced injury on field-grown muskmelons. HortScience. 24: 909-1012.

Stein L. 1971. Environmental sources and forms of fluoride, biologic effect of atmospheric pollutants- Fluorids. National Acad of Sci- National Research Council, Washington DC, NAS. pp. 5-28.

Taylor Jr GE. 1978. Plant and leaf resistance to gaseous air pollution stress. New Phytol. 80: 523-534.

Teklemariam TA, Sparks JP. 2004. Gaseous fluxes peroxyacetyl nitrate (PAN) in plant leaves. Plant, Cell and Environ. 27: 1149-1158.

Temple PJ. 1990. Growth form and yield responses of four cotton cultivars to ozone. Agron J. 82: 1045-1050.

Ting IP, Mukherji SK. 1971. Leaf ontogeny as a factor to susceptibility in ozone: Amiono acid and carbohydrate changes during expansion. American J Bot. 58: 497-504.

Tingey DT, Fites RC, Wickiloff C. 1973. Ozone alternation of nitrate reduction in soybean. Physiol Plant. 29: 33-38.

Tingey DT, Hogsett WE. 1985. Water stress reduces ozone injury via a stomatal mechanism. Plant Physiol. 77(4): 944-947.

Tonneijck AEG. 1983. Foliar injury responses of 24 bean cultivars (*Phaseolus vulgaris*) to various concentrations of ozone. Netherlands J Plant Physiol. 89: 99-104.

Treshow M, Pack MR. 1970. Fluoride, Reorganization of Air Pollution Injury to Vegetation: A Pictorial Atlas. JS Jacbson, AC Hill (eds). Pittsburgh, Air Poll Cont Assn. pp. D1-D17.

Velissariou D, Barnes JD, Davision AW. 1992. Has inadvertent selection by plant breeders affected the O_3 sensitivity of modern Greek cultivars of spring wheat? Agriculture, Ecosystem & Environ. 38: 79-89.

Wegel HJ, Pacholski A, Waloszczyk K, Fruhauf C, Manderscheid R, Anderson TH, Heinemeyer O, Kleikamp B, Helal M, Burkart S, Schrader S, Sticht C, Giesemann A. 2006. Effect of elevated atmospheric CO_2 concentration on barley, sugar beet and wheat in rotation: examples from the Braunschweig carbon project. Landbauforschung Volkenrode. 56(3/4): 101-115.

Young, JS, Lui JS, Wang JD, Yu JB, Sun ZG, Li ZH. 2006. Emission of CH_4 and N_2O from a wetland in Sanjiang plain. J Plant Eco. 30(3): 432-440.

Younger VB, Nudge FJ. 1980. Air pollution oxidant effects on cooling-season and warm-season turfgrass. Agron J. 72: 169-172.

13

Radiation Stress Tolerance

13.1. INTRODUCTION

Photosynthetically active radiation is a primary energy resource for terrestrial plants, necessary for the plant growth. Plants receives this photosynthetically active radiation from sunlight as a consequence, plants are also exposed to the ultraviolet (UV)-radiation. Seven percent of the electromagnetic radiation emitted by the sun is the UV range (200-400 nm). Though representing only a small portion of the total solar electromagnetic spectrum, UV-B has a disproportionately large photobiological effect. As it passes through the atmosphere, the flux transmitted is greatly reduced, and the composition of the UV-radiation is modified. This UV-radiation is mainly divided into three classes, namely UV-A, UV-B and UV-C. The UV-A region of the spectrum (wavelengths from 320 to 400 nm) are not attenuated by ozone, so their influence will be unaffected by ozone layer reduction. This portion of the UV radiation is less damaging and important for photomorphogenic signal in plant development (Bjorn, 1994). The wavelength of UV-B portion ranges from 280 to 320 nm. This region of UV is very energetic and effectively absorbed by ozone in the stratosphere. Thus, only a very small portion, approximately 4% is transmitted to the earths. Shortwave UV-C radiation (200-280 nm) is highly energetic and strongly absorbed by atmospheric oxygen and ozone. Thus, none of this sterilizing radiation is present in terrestrial sunlight.

In the past 50 years, the concentration of ozone has decreased by about 5%, mainly due to release of atmospheric pollutants (Pyle, 1996). Consequently, a large portion of UV-B spectrum reaches the earths surface with serious implications for all living organisms (Xiong and Day, 2001; Caldwell *et al.*, 2003; Paul and Gwynn-Jones, 2003).

The UV- portion of sunlight has received much attention, because irradiation from this special region will increase as the stratospheric ozone concentration decrease. Depletion of stratospheric ozone has increased solar UV-B radiation at high- and mid-latitudes (Madronich and de Gruijl, 1994) in both the Southern and Northern hemispheres (Seckmeyer *et al.*, 1994; McKenzie *et al.*, 1999). However, ozone destruction is about 50% more intense over the Souther hemisphere (Crutgen, 1992; Sechmeyer *et al.*, 1995).

13.2. RADIATION INDUCED PLANT SYMPTOMS

The UV-radiation is usually detrimental, particularly UV-B region. The visible symptoms of UV-radiation may be as follows:

1. The growth of many plant species is reduced by enhanced levels of UV-B radiation (Teramura and Murali, 1986), particularly herbaceous plants (Ballare *et al.*, 1996; Krizek *et al.*, 1998; Mazza *et al.*, 1999).
2. Decrease in plant height and leaf area (Correia *et al.*, 1998).
3. Reduced yield in soybean (Teramura *et al.*, 1990), wheat, rice (Teramura and Sullivan, 1991), sorghum, cotton and corn.

13.3. INJURY MECHANISMS

UV-radiation has many detrimental effects on plants. Elevated UV-B radiation has pleiotropic effects on plant development, morphology, and physiology (Table 13.1). UV-induced damages mainly includes tissue injury (Bornman *et al.*, 1986), induction of carotinoides and polyamines (Tevini and Teramura, 1989; Kramer *et al.*, 1991), damage to the photosynthetic apparatus (Kulandaivelu and Noorudeen, 1983), photoperiodism (Baskin and Iino, 1987), ATPase destruction (Imbrie and Murphy, 1984) and unscheduled DNA synthesis in pollen (Jackson, 1987). The damage caused by UV-radiation had been classified into two major groups:

1. *Heritable damage:* Damage to DNA, which is heritable and it is known as mutation.

2. *Non-heritable damage:* Damage to physiological processes, which are not heritable. For example, damage to leaf pigment, different metabolic process etc.

Table 13.1. Injury effects of UV-B radiation on plants

Particulars	Effects (molecular, biochemical, and physiological)	References
DNA	Formation of cyclobutane pyrimidine dimmers (CPDs) and 6-4 PPs (Pyrimidine (6-4) pyrimidinone dimmers)	Takahashi *et al.*, 2002; Britt and May, 2003
	Induction of repair mechanisms	Britt, 1995
	Stimulation of homologous recombination	Ries *et al.*, 2000a, b
Photosynthesis	Degradation of photosystem II, *D1* and *D2* proteins	Jordan, 1996; Mackerness *et al.*, 1997
	Damage to thylakoid membrane	
	Destruction of chlorophyll and carotinoides	Tevini and Teramura, 1989
Phytohormones	Photooxidation of indolacetamide	-
Membrane	Peroxidation of lipids	Britt, 1996
Secondary metabolism	Activation of phenylpropanoid biosynthetic pathway	Ulm and Nagy, 2005
	Accumulation of UV-protective pigments	Hoshimoto *et al.*, 1991
Stress response	Formation of ROS	Dai *et al.*, 1997
	Induction of superoxide dismutase, ascorbate peroxidase, and glutathione reductase accumulation of PR-1	Mazza *et al.*, 1999
Photomorphogenesis	Inhibition of hypocotyls elongation	Jenkins and Brown, 2007
	Cotyledon opening	-
	Morphology and anatomy	Searles *et al.*, 2001; Paul and Gwynn-Jones, 2003
	Alteration in the composition of epicuticular waxes	Correia *et al.*, 1998
	Reduction in leaf surface area	Murali and Teramura, 1985; Musil *et al.*, 2002
	Increased leaf thickness	Bornman and Vogelmann, 1991
	Shortened internode	-
	Branching and canopy structure	Barnes *et al.*, 1996

Contd...

Table 13.1. Contd...

	Influence of the whole plant, plant communities, and ecosystems	Ballare *et al.*, 1996; Krizek *et al.*, 1998; Mazza *et al.*, 1999
	Reduction in biomass production	Barnes *et al.*, 1996
	Reduction in crop yield	Teramura *et al.*, 1990; Teramura and Sullivan, 1991
	Altered competitive balance	-
	Altered flowering	-
	Reduced fertility	Jackson, 1987

13.3.1. Damage to DNA

DNA is particularly sensitive to UV-B radiation. UV radiation is readily absorbed by important macromolecules such as protein and nucleic acids. UV-induced DNA damage has two different effects on individuals-mutagenesis and toxicity. Absorption of UV-B causes phototransformations, resulting in the production of cyclobutane pyrimidine dimmers (CPD) and pyrimidine (6-4) pyrimidinone dimmers. DNA and RNA polymerases are not able to read these photoproducts, which misleading protein synthesis. Several investigators have measured CPD DNA damage directly in plants or in plant cell culture (Pang and Hays, 1991; Quaite *et al.*, 1992).

13.3.2. Change in Plant Morphology

It is a radiation-induced change in plant form. UV-B radition significantly reduced leaf area, ear length, mean relative growth rate, dry weight in maize (Correia *et al.*, 1998), thickening of leaves and culticular wax layer. UV-radition-induced photomorphogenic responses may change canopy structure, light interception and stand photosynthesis (Barnes *et al.*, 1996). Musil *et al.* (2002) reported significant alteration in specific leaf area (ratio of leaf dry mass) in *Colophospermum mopane* tree. The leaf area decreased due to increase in leaf thickness.

Alterations in leaf shape modify the area available for energy capture and may assest in reducing the harmful effects of incident UV-B (Jansen *et al.*, 1998). Significantly larger leaf shape indices, indicating more elongated leaves in *C. mopane* tree ecotype chota and the herbs *Glycine max* and *Vigna anguiculata* (Musil *et al.*, 2002). Conversly, smaller leaf shape indices, indicating broader leaves, were observed in the shrub *Barleria obtuse*. Alteration in leaf shape represent on of a suite of photomorphogenesis responses unrelated to changes in photosynthetic carbon assimilation and biomass production (Barnes *et al.*, 1996) that are especially sensitive to change in the wavelength most effected by stratospheric ozone reduction (Ensminger, 1993). The altered leaf shapes of species under elevated UV-B were significantly correlated with corresponding changes in their canopy area (Musil *et al.*, 2002).

13.3.3. Damage to Leaf Pigmentation and Inhibition of Photosynthesis

Under elevated UV-B radiation, plants displayed significantly altered leaf chlorophyll *a* and *b*, carotinoid and flavonoid concentrations. The decreased chlorophyll concentration, especially that of chlorophyll *b*, reported in *Barleria obtuse* and *Vigna anguculata* is a more common symptom of UV-B radiation stress (Musil *et al.*, 2002). Decrease chlorophyll is attributed to increased photo-degradation of chlorophyll (Strid and Porra, 1992) and lower rates of chlorophyll synthesis resulting from reduced expression of genes encoding chlorophyll-binding proteins (Strid *et al.*, 1994).

UV-B radiation damages photosynthetic apparatus in plant. Photosystem II is sensitive to UV-B radiation. The damage is assessed by measuring the increase in variable chlorophyll fluorescence, and an increase in fluorescence can be observed after dose of UV-B radiation in the physiologically relevant range (Tevini *et al.*, 1991a). UV-B degrades *D1* protein of Photosystem II and reduced pollen fertility (Caldwell *et al.*, 2003).

13.3.4. Damage to Plant Cells

UV-radiation inhibits elongation of epidermal cell in sunflower seedlings. This inhibition of cell elongation results from the photooxidation of indolacetic acid to 3-methylenoxidol, which inhibits hypocotyl elongation (Tevini *et al.*, 1989). UV-B radiation also damage cell plasma membrane ATPase of rose. The inactivation of ATPase results from singlet-oxygen mediated destruction of tryptophan residues in the ATPase protein (Imbrie and Murphy, 1984).

13.4. TOLERANT MECHANISMS

Plant must have some mechanisms to prevent UV-radiation to damage cells or mechanisms to repair the damage DNA. Thus, the sessile life style forces plant to adapt to changing environmental conditions.

13.4.1. Preventive Tolerant Mechanisms

The most common protective mechanism against potentially damaging irradiation is the biosynthesis of UV-absorbing compounds. These secondary metabolites are carotinoid (specially, echinenone and myxoxanthophyll), flavinoids, and hydroxycinamate esters, and UV-B absorbing phenolic compounds, accumulate in the vacuoles of the epidermal cells in response to UV-B irradiation and attenuate the penetration of the UV-B in deeper cell layers. The physiological relevance as UV-B sunscreens was confirmed by the UV-B hypersensitive phenotype of mutants devoid of these compounds on the one hand and the increased resistance to UV radiation of mutants with enhanced flavinoid on the other hand (Li *et al.*, 1993; Landry *et al.*, 1995; Bieza and Lois, 2001). The syntheses of these UV sunscreens are triggered by different UV photoreceptors.

13.4.1.1. Flavonoid and anthocyanins

UV-radiation activates the biosynthetic pathway of flavonoid. Flavonoid and/or anthocyanins are induced by UV-B exposure in plants (Beggs and Wellman, 1985). These molecules absorb UV-radiation, and they generally accumulate in the epidermis, where they could keep UV-radiation from reaching photosynthetic tissues (Hahlbrock and Scheel, 1989) and block transmittance of 95-99% of incoming UV-radiation. Induction of flavonoids in rye seedlings can prevent UV-B-induced damage to photosynthesis (Tevini *et al.*, 1991b). The needle flavonoid concentrations of Lonlolly pine increased under specific dose of UV-B radiation (Kossuth and Biggs, 1981; Sullivan and Teramura, 1988 and 1989). Rapid increase in flavonoid concentrations initially protect the plants from deleterious effects of UV-B radiation.

13.4.1.2. Carotenoid pigments

The photoprotective function of carotenoid is essential for photosynthetic organisms. UV-B radiation induced an increase in carotenoids synthesis especially, echinenone and myxoxanthophyll as a protective mechanism against UV radiation injury. In photosynthetic organisms, the protective role of carotenoids against high visible radiation is well known (Siefermann-Harms, 1987). The findings of Götz *et al.* (1999) indicated that carotenoids exert their protective function as antioxidants to inactivate UV-B-induced radicals in the photosynthetic membrane. Carotenoids are effective quenchers of triplet-state photosensitizers and protect against singlet oxygen and peroxy radicals (Krinsky, 1989).

Protection against UV-B radiation by carotenoids was demonstrated in several fungi and bacteria. Cyanobacteria produce a unique type of carotenoids, such as *keto*-carotinoids and glycosides (Hirschherg and Chamovitz, 1994). Xanthophylls and echinenone are other carotenoids predominant in *Nostoc commune* and also found in the outer membrane of *Synechocystis* as minor compound. Xanthophylls and echinenone may act as outer membrane-bound UV-B photoprotectors of *N. commune*. In UV protection, carotenoids provide a fast, active SOS response to counteract acute cell damage. Protective functions of carotenoids were also found in *Ustilago violacea* (Will *et al.*, 1984), *Neurospora crassa* (Blanc *et al.*, 1976), *Escherichia coli* (Sandmann *et al.*, 1998).

13.4.1.3. Mycosporine amino acids

Cyanobacteria withstand to UV-B radiation impairing similar protection mechanism of plants. They develop special class of compounds with absorption maxima between 310 and 360 nm as UV protectant. Mycosporine amino acids (MAA) are thought to fulfill a comparable purpose in lower organisms (Karentz *et al.*, 1991; Garcia-Pichel *et al.*, 1993). In the filamentous and heterocystous N_2-fixicing *Anabaene* sp., *Nostoc commune*, and *Scytonema* sp. posse shinorine, are presentative of mycosporine-like amino acids that accumulate in response to solar UV-B radiation, mostly during the daily light period (Sinha *et al.*, 2001). MAAs are water-soluble, substituted cyclohexanes which are linked to amino acids and iminoalcohols.

Scytonemin (a extracellular pigment of MAAs group), which has an *in vivo* absorption maxima 370 nm and is located in the cyanobacterial sheath, has been proposed to serve as UV-A sunscreen (Garcia-Pichel *et al.*, 1992). It is a yellow-brown, lipid-soluble dimeric pigment of terrestrial cyanobacteria with a structure based on indolic and phenolic subunit (Proteau *et al.*, 1993).

Scherer *et al.* (1988) also reported a UV-A/B absorbing pigment in *N. commune*. This pigment was structurally characterized by Bohm *et al.* (1995) as a water soluble oligosaccharide-MAA (OS-MAA) and it is located in the extracellular glycan sheath. A thicker sheath provides effective lengths for the absorption of radiation. The extracellular glycan with its UV-absorbing pigments is a passive UV screen against long-time exposure.

13.4.2. DNA Repair Mechanisms

To prevent mutation and/or cell death, UV-radiation-induced DNA damage must be repaired before DNA replication. Repair of UV-induced lesions may be of particular importance in plants. In response of UV radiation-induced DNA damage, most organisms have developed a complex set of repair mechanisms such as photoreactivation, excision, and recombination repair (Smith, 1989; Kornberg and Baker, 1992).

13.4.2.1. Photoreactivation

Cyclobutane pyrimidine dimmers (CPD) can be repaired by all the three methods, but other UV radiation-induce lesions can be repair only by excision of recombinational repair. It is a light-dependent enzymatic process using UV-A, and blue light to monomerize pyrimidine dimmers. Photolyase binds to the photoproducts and then uses light energy to initiate electron transfer to break the chemical bonds of the cyclobutane ring and restore integrity of the bases. Thus, CPD-type DNA damage is implicated in any UV radiation response that can be reversed by irradiation with 370-450 nm 'Photoreactivation' radiation (Beggs *et al.*, 1985; Cieminis *et al.*, 1987; Hashimoto *et al.*, 1991). Photoreactivation can be confirmed by measuring CDPs before and after 370-450 nm light treatment.

13.4.2.2. Nucleotide excision repair

Cyclobutane pyrimidine dimmers and cyclobutane can also be removed in the dark through nucleotide excision repair through endonucleolytic cleavage, release of the damage nucleotides, followed by strand resynthesis (Liu *et al.*, 2000). Excision repair cand be divided into three sages: nicking of the damage DNA near the site of the damage followed by removal of the multiple bases on the damage strand and resynthesis to fill the gap. This multistep process involving multi enzymes has been found to operate with only a low capacity in plant (Gallego *et al.*, 2000).

13.4.2.3. Recombination repair

In this DNA repair mechanism, DNA lesions are bypassed during DNA replication, and the resulting gaps are filled later using information from the sister duplex (Kornberg and Baker, 1992; Ries *et al.*, 2000a, b). Recombination repair in plant generally classified as dark repair process, it is stimulated by red but not by far-red exposure after UV-B treatment.

13.5. GENETICS VARIABILITY FOR RADIATION STRESS TOLERANCE

Plants have developed natural adaptations such as anatomical, morphological and biochemical changes which protect them from UV-B radiation. The extent of these natural adaptations may be related to geographic origin of the species. UV-B dose received at low latitude, high elevation sites (tropical mountains) can be nearly six-fold greater than the maximum dose received at arctic latitudes (Caldwell *et al.*, 1982). It has been postulated that species evolved from areas receives high levels of UV-B radiation would be more tolerant to UV-B. Thus, most plants from lower elevations are sensitive to UV-B, whereas plants from higher elevations, which receive highest levels of UV-B, are comparatively tolerant to UV-B radiation. Caldwell *et al.* (1982) found that arctic ecotype (variants from same species) of *Oxyria digyna* were consistently more sensitive to UV radiation than their counterparts collected from mountains of lower latitiudes. The effectiveness of UV-B radiation also depends on season, microclimate (Murali and Teramura, 1986) and soil fertility (Murali and Teramura, 1985).

Approximately two out of three species tested appeared to be sensitive and sensitivity differs among the cultivars of the same species (Teramura and Sullivan, 1991). Teramura and Murali (1986) reported that Essex cultivar of *Glycine max* (L.) sensitive to UV-B radiation, whereas Cultivar Williams was tolerant.

Attempts were also taken to study the effect of UV radiation on forest species. It has deleterious effect on tree growth and physiology (Kossuth and Biggs, 1981; Sullivan and Teramura, 1988 and 1989). Bogenrieder and Klein (1978) reported that the exclusion of naturally occurring UV-B radiation increased the growth of four broadleaf species. Studies showed that newly emerged loblolly pine seedlings were deleteriously affected by increased levels of UV-B radiation (Kossuth and Biggs, 1981; Sullivan and Teramura, 1988 and 1989). In another experiment, Teramura and Sullivan (1991) reported that only 8% of the species collected from the sea level to 500 m were tolerant to UV radiation, while tolerance increased markedly in species collected from higher elevation. All species collected from > 2000 m elevation were found to be tolerant of UV-B radiation, implying that plants of those high altitudes adapted to the high levels of UV radiations.

13.5. REFERENCES

Ballare CL, Scopel AL, Stapleton AE, Yanovsky MJ. 1996. Solar ultraviolet-B radiation affects seedling emergence, DNA integrity, plant morphology, growth rate, and attractiveness to herbivore insects in *Datura ferox*. Plant Physiol. 112: 161-170.

Barnes PW, Ballare CL, Caldwell MM. 1996. Photomorphogenic effects of UV-B radiation on plant: consequences for light competition. J Plant Physiol. 148: 15-20.

Baskin TI, Iino M. 1987. An action spectrum in the blue and ultraviolet for photoperiodism in alfalfa. Photchem Photobiol. 46: 127-136.

Beggs C, Stolzer-Jehle A, Wellman E. 1985. Isoflavonoid formation as an indicator of UV stress in bean (*Phaseolus vulgaris* L.) leaves. Plant Physiol. 79: 630-634.

Beggs C, Wellman E. 1985. Analysis of light controlled anthocyanin formation in coleoptiles of *Zea mays* L.: The role of UV-B, blue, red and far-red light. Photchem Photobiol. 41: 481-486.

Bieza K, Lois R. 2001. An *Arabidopsis* mutant tolerant to lethal ultraviolet-B levels shows constituvely elevated accumulation of flavonoids and other phenolics. Plant Physiol. 126: 1105-1115.

Bjorn LO. 1994. Introduction. In: Photomorphogenesis in Plants, Kendrick RE, Kronenberg (eds.), Edn 2. Kluwer Academic Publishers, Boston. pp. 3-25.

Blanc PL, Tuveson RW, Sargent ML. 1976. Inactivation of carotenoid-producing and albino strains of *Neurospora crssa* by visible light, black light and ultraviolet radiation. J Bacteriol. 125: 616-625.

Bogenrieder A, Klein R. 1978. Die abhangigketi der UV-empfindlichkeit von der lichtqualitat bei der aufzucht (*Lactuca sativa* L.). Angew Botanik. 52: 283-293.

Bohm GA, Pfleiderer W, Boger P, Scherer S. 1995. Structure of a novel oligosaccharide-mycosporine-amino acid ultraviolet A/B sunscreen pigment from the terrestorial cyanobacterium *Nostoc commune*. J Biol Chem. 270: 8536-8539.

Bornman JF, Evert RF, Mierzwa RJ, Bornman CH. 1986. Fine structural effects UV radiation of leaf tissue of *Beta vulgaries*. In: Stratospheric Ozone Reduction, Solar Ultraviolet Radiation and Plant Life, Worrest, Caldwell MM (eds.). Berlin, Heldelberg, Spring-Verlag. pp. 199-209.

Bornman JF, Vogelmann TC. 1991. Effect of UV-B radiation on leaf optical properties measured with fibre optics. J Expt Bot. 42: 547-554.

Britt AB, May GD. 2003. Re-engineering plant gene targeting. Trends Plant Sci. 8: 90-95.

Britt AB. 1995. Repair of DNA damage induced by ultraviolet radiation. Plant Physiol. 108: 891-896.

Britt AB. 1996. DNA damage and repair in plants. Annu Rev Plant Physiol Plant Mol Biol. 4: 75-100.

Caldwell MM, Ballare CL, Bornman JF, Flint SD, Bjorn LO, Teramura AH, Kulandaivelu G, Tevini M. 2003. Terrestorial ecosystems, increased solar ultraviolet radiation and interactions with other climatic change factors. Photochem Photobiol Sci. 2: 29-38.

Caldwell MM, Robberecht R, Nowak RS, Billings WD. 1982. Differential photosynthetic inhibition by ultraviolet radiation in species from the arctic-alpine life zone. Arctic and Alpine Res. 14: 195-202.

Cieminis KGK, Rancellene VM, Prijalgauskiene AJ, Tlunaltiene NV, Rudzianskalte AM, Jancys ZJ. 1987. Chromosome and DNA damage and their repair in higher plants irradiated with short-wave untravilet light. Mutat Res. 181: 9-16.

Correia CM, Areal ELV, Torres-Pereira MS, Torres-Pereira JMG. 1998. Interspecific variation in sensitivity to ultraviolet-B radiation in maize grown under field conditions. I. Growth and morphological aspects. Field Crop Res. 59(2): 81-89.

Crutzen PJ. 1992. Ultraviolet on the increase. Nature. 356: 104-105.

Dai Q, Yan B, Huang S, Liu X, Peng S, Miranda MLM, Chavez AQ, Vegara BS, Olszyk D. 1997. Response to oxidative stress defense systems in rice (*Oryza sativa*) leaves with supplemental UV-B radiation. Physiol Plant. 101: 301-308.

Ensminger PA. 1993. Control of development in plants and fungi by far-UV radiation. Physiologia Plantarum. 88: 501-508.

Gallego F, Fleck O, Li A, Wyrzykoswka J, Tinland B. 2000. AtRAD1, a plant homologue of human and yeast nucleotide excision repair endonucleases, is involved in dark repair of UV damages and recombination. Plant J. 21: 507-518.

Garcia-Pichel F, Sherry ND, Castenholz RW. 1992. Evidence of an ultraviolet sunscreen role of the extracellular pigment scytonemin in the terrestorial cyanobacterium *Chlorogloecopsis* sp. Photochem Photobiol. 56: 17-23.

Garcia-Pichel F, Wingard CE, Castenholz RW. 1993. Evidence regarding the UV sunscreen role of a mycosporine-like compound in the cyanobacterium *Gloeocapsa* sp. Appl Environ Microbiol. 59: 170-176.

Götz T, Windhövel U, Böger P, Sandmann G. 1999. Protection of photosynthesis against ultraviolet-B radiation by carotenoids in transformants of the Cyanobacterium *Synechococcus* PCC7942. Plant Physiol. 120(2): 599-604.

Hahlbrock K, Scheel D. 1989. Physiology and molecular biology of phenylpropanoid metabolism. Annu Rev Plant Physiol Plant Mol Biol. 40: 347-369.

Hirschhberg J, Chamovitz D. 1994. Carotinoids in cyanobactera. In: The Molecularbiology of Cyanobacteria, Bryant DA (ed.). Kluwer Academic Publishers, Dordrecht, The Netherlands. pp. 559-579.

Hashimoto T, Shichijo C, Yatsuhashi H. 1991. Ultraviolet action spectrum for the induction and inhibition of anthocyanin synthesis in broom soghum seedlings. J Photochem Photobiol. 40: 243-248.

Imbrie CW, Murphy TM. 1984. Mechanism of photoinactivation of plant plasma membrane ATPase. Photochem Photobiol. 40: 243-248.

Jackson JF. 1987. DNA repair in pollen, A review. Mutat Res. 181: 17-29.

Jansen MAK, Gaba V, Greenberg BM. 1998. Higher plants and UV-B radiation: balancing damage, repair and acclimation. Trend Plant Sci. 3: 131-135.

Jenkins GI, Brown BA. 2007. UV-B perception and signal transduction . In: Light and Plant Development, Vol. 30, Whitelam GC, Halliday KJ (eds.). Blackwell Publishing, Oxfor. pp. 155-182.

Jordan BR. 1996. The effects of UV-B radiation on plants: a molecular perspective. In: Advances in Botanical Research, Callow JA (ed.). Academic Press, Boca Raton, FL. pp. 97-162.

Karentz D, McEuen FS, Land MC, Dunlap WC. 1991.Survey of mycosporine-like amino acid compounds in Antarctic marine organisms: potential protection from ultraviolet exposure. Mar Biol. 108: 157-166.

Kornberg A, Baker TA. 1992. DNA replication. W.H. Freema, New York. pp. 771-791.

Kossuth SV, Biggs RH. 1981. Ultraviolet-B radiation effects early seedling growth of Pinaceae species. Can J For Res. 11: 243-248.

Kramer GF, Norman HL, Krizek DT, Mirecki RM. 1991. Influence of UV-B radiationon polyamine, lipid peroxidation and membrane lipid in cucumber. Phytochemistry.30: 2101-2108.

Krinsky NI. 1995. Antioxidant function of carotenoid. Free Radical Biol Med. 7: 617-635.

Krizek DT, Britz SJ, Mirecki RM. 1998. Inhibitory effects of ambient levels of solar UV-A and UV-B radiation on growth of cv. New Red Fire lettuce. Plant Physiol. 103: 1-7.

Kulandalvelu G, Noorudeen AM. 1983. Comparative study of the action of ultraviolet-C and ultraviolet-B on photosynthetic electron transport. Physiol Plant. 58: 389-394.

Landry L, Chapple C, Last R. 1995. *Arabidopsis* mutants lacking phenolic sunscreen exhibit enhanced ultraviolet-B injury and oxidative damage. Plant Physiol. 109: 1159-1166.

Li J, Ou-Lee T, Raba R, Amudson R, Last R. 1993. *Arabidopsis* flavonoid mutants are hypersensitive to UV-B irradiation. Plant Cell. 5: 171-179.

Liu Z, Hossain GS, Islas-Osuna MA, Mitchell DL, Mount DW. 2000. Repair of UV damage in plant by nucleotide excision repair: *Arabidopsis* UVH1 DNA repair gene is a homologue of *Saccharomyces cerevisiae* Radl. Plant J. 21: 519-528.

Mackerness SAH, Thomas B, Jordan BR. 1997. The effect of supplementary ultraviolet-B radiation on transcripts, translation and stability of chloroplast proteins and pigment formation in *Pisum sativum* L. J Exp Bot. 48: 759-768.

Madronich S, de Gruijl FR. 1994. Stratosphoric ozone depletion between 1979 and 1992. Implecations for biological active ultraviolet-B radiation and non-melanoma skin cancer incidence. Phytochemistry and Phytology. 59: 541-546.

Mazza CA, Battista D, Zima AM, Szwarcberg-Bracchitta M, Giordano CV, Acevedo A, Scopel AL, Ballare CL. 1999. The effects of solar UV-B radiation on the growth and yield of barley are accompanied by increased DNA damage and antioxidant responses. Plant Cell Environ. 22: 61-70.

McKenzie RL, Conner B, Bodeker GE. 1999. Increased summertime UV radiation in New Zealand in response to ozone loss. Science. 285: 1709-1711.

Murali NS, Teramura AH. 1985. Effects of ultraviolet radiation on soybean. VI. Influence of phosphorus nutrition on growth and flavinoid content. Physiol Plant. 63: 413-416.

Murali NS, Teramura AH. 1986. Effectiveness of UV-B radiation on the growth and physiology of field-grown soybean modified by water stress. Photochem Photobiol. 44: 215-220.

Musil CF, Chimphango SBM, Dakora FD. 2002. Effects of elevated ultraviolet-B radiation on native and cultivated plants of Southern Africa. Anna Bot. 90: 127-137.

Pang Q, Hays JB. 1991. UV-inducible and temperature sensitive photoreactivation of cyclobutane pyrimidine dimmers in *Arabidopsis thaliana*. Plant Physiol. 95: 536-543.

Paul ND, Gwynn-Jones D. 2003. Ecological roles of solar UV radiation: towards an integrated approach. Trends Ecol Evol. 18: 48-55.

Proteau PJ, Gerwick WH, Garcia-Pichel F, Castenholz R. 1993. The structure of scytonemin, an ultraviolet sunscreen pigment from the sheaths of cyanobacteria. Experimentia. 49: 825-829.

Pyle JA. 1996. Global ozone depletion: Obervation and theory. In: Plant and UV-B. Responses to Environmental Change, Lumbsden PJ (ed.). Cambridge University Press, Cambridge. pp. 3-12.

Quaite FE, Sutherland BM, Sutherland JC. 1992. Action spectrum for DNA damage in alfalfa lowers predicted impact of ozone depletion. Nature. 358: 576-578.

Ries G, Buchholz G, Frohnmeyer H, Hohn B. 2000a. UV-damage-mediated induction of homologous recombination in *Arabidopsis* is dependent on photosynthetically active radiation. Proc Natl Acad Sci USA. 97: 13425-13429.

Ries G, Heller W, Puchata H, Sandermann H, Seidlitz HK, Hohn B. 2000b. Elevated UV-B radiation reduces genome stability in plants. Nature. 406: 98-101.

Sandmann G, Kuhn S, Böger P. 1998. Evaluation of structurally different carotenoids in *Escherchia coli* transformant as protectants against UV-B radiation. Appl Environ Microbiol. 64: 1972-1974.

Scherer S, Chen TW, Boger P. 1988. A new UV-A/B protecting pigment in the terrestorial cyanobacterium *Nostoc commune*. Plant Physiol. 88: 1055-1057.

Searles PS, Flint SD, Caldwell MM. 2001. A meta analysis of plant field studies simulating stratospheric ozone depletion. Oecologia. 127: 1-10.

Seckmeyer G, Mayer B, Bernhard G, McEnzie RL, Johnston PV, Kotkamp M, Booth CR, Lucas T, Mestechkina T, Roy CR, Gies HP, Tomlinson D. 1995. Geographical differences in the UV measured by intercompared spectroradiometers. Geological Res Letters. 22: 1889-1892.

Seckmeyer G, Mayer B, Erb R, Bernhard G. 1994. UV-B in Germany higher in 1993 than in 1992. Geological Res Letters. 21: 577-580.

Sieferman-Harms D. 1987. The light-harvesting and protecting functions of carotinoids in photosynthetic membrane. Physiol Plant. 69: 561-568.

Sinha RP, Klisch M, Helbling EW, Hader DP. 2001. Induction of mycosporine-like amino acids (MAAs) in cyanobacteria by solar ultraviolet-B radiation. J Photochem Photobiol. 60: 129-135.

Smith KC. 1989. UV radiation effects: DNA repair and mutagenesis. In: The Science of Pysiology, Smith KC (ed.). Plenum Press, New York. pp. 111-134.

Strid A, Chow WS, Anderson JM. 1994. UV-B damage and protection at the molecular level in plants. Photosynthesis Res. 39:475-489.

Strid A, Porra RJ. 1992. Alteration in pigment content in leaves of *Pisum sativum* after exposure to supplementary UV-B. Plant Cell Physiol. 33: 1015-1023.

Sullivan JH, Teramura AH. 1988. Effects of ultraviolet irradiation on seedling growth in the pinaceae. Amer J Bot. 75: 225-230.

Sullivan JH, Teramura AH. 1989. Effects of ultraviolet radiation on loblolly pine. I. Growth, photosynthesis and pigment production in greenhouse-grown seedlings. Physiol Plant. 77: 202-207.

Takahashi S, Nakajima N, Saji H, Kondo N. 2002. Diurinal change of cucumber CPD photolase gene (*CsPHR*) expression and its physiological role in growth under UV-B irradiation. Plant Cell Physiol. 43: 342-349.

Teramura AH, Murali NS. 1986. Interspecific differences in growth and yield of soybean exposed to ultraviolet-B radiation under greenhouse and field conditions. Environ Expt Bot. 26: 89-95.

Teramura AH, Sulivan JH, Lydon J. 1990. Effect of solar UV-B radiation on soybean yield and seed quality: a six-year field study. Physiologia Plantarum. 80: 5-11.

Teramura AH, Sulivan JH. 1991. Potential effects of increased solar UV-B on global plant productivity. In: Photobiology, Riklis E (ed.). Plenum Press, New York. pp. 625-634.

Tevini M, Braun J, Fleser G. 1991a. The protective function of the epidermal layer of rye seedlings against ultraviolet-B radiation. Photochem Photobiol. 53: 329-333.

Tevini M, Braun J, Grusemann P, Ros J. 1989. UV-Wirkungen auf Nutzpflantzen. In: Akad Natursch Landschaftspfi. Laufen/Salzach, Germany. pp. 38-52.

Tevini M, Mark U, Fleser G, Salle M. 1991b. Effects of enhanced solar UV-B radiation on growth and function of selected crop plant seedlings. In: Photobiology, Riklis E (ed.). Plenum Press, New York. pp. 635-649.

Tevini M, Teramura AH. 1989. UV-B effects on terrestrial plants. Photochem Photobiol. 50: 479-487.

Ulm R, Nagy F. 2005. Signaling and gene regulation in response to ultraviolet light. Curr Opin Plant Biol. 8: 477-482.

Will OH, Newl NA, Reppe CR. 1984. The photosensitivity of pigmented and non-pigmented strains of *Ustilaga violacea*. Curr Microbiol. 10: 295-302.

Xion FS, Day TA. 2001. Effect of solar ultraviolet-B radiation during springtime ozone depletion on photosynthesis and biomass production of Antarctic vascular plant. Plant Physiol. 125: 738-751.

14

Climate Change

14.1. INTRODUCTION

The climate of a place is the average weather that is experiences over a period of time. It is the natural phenomenon that has occurred throughout the history of the earth. Climate change is strongly associated with higher temperature, altered precipitation, and higher levels of atmospheric CO_2 and other greenhouse gases. It is a very slow process; it takes a long time to settle in. Climate change in IPCC (2007) usage refers to any change in climate over time, whether due to natural variability or as a result of human activities. Over the last 150-200 years the changes to our climate are happening more quickly now than they have ever done before in the geological past. Human activities have altered natural climatic processes at a geographically rapid pace by booting atmospheric concentrations of several greenhouse gasses. Following are the changes taken place over years of industrial age (IPCC, 2007):

- Atmospheric CO_2 concentration from 280 to 360 ppmv (parts per million by volume).
- Escalation in methane gas concentration from 700 to 1700 ppmv.
- Growth in nitrous oxide concentration from 270 to 310 ppbv (parts per billion by volume).

- Increasing levels of ground-level ozone, CO, and other short-lived but potent greenhouse gases.
- Increased incidence of smog in urban areas.

Plants will have to adapt to new conditions more rapidly than they have ever had to do so (Ruddiman and Wright, 1987; Pielou, 1991; Peters and Lovejoy, 1992). Frequency and magnitude of climate change have varied substantially during and between glacial periods, and temperatures on both global and local scales have been both substantially warmer and colder than present day average.

Scientists have reported a number of substantial evidences revealing consistent with global change, such as:

- The National Academic and Space Administration (NASA), the National Ocenic Atmospheric Administration (NOAA), and the World Meteorological Organisation agreed that 1998 was the hottest year on record.
- Greenland ice sheet lost two cubic miles of mass per year during 1993-1998.
- Wider spread coral bleaching in 1998.

14.2. CAUSES OF CLIMATE CHANGE

Human activities to meet the luxurious and modern life-styles have increased greenhouse gases such as carbon dioxide, methane, water vapour, and nitrous oxide. These greenhouse gases are now for above their natural concentrations in atmosphere. Human activities that produce greenhouse gases are: industrial process, emission from power plants and vehicles, farming in the marshlands, dairy farming, explosion etc. The causes of climate change can be classified into two major categories- natural and man-made causes.

14.2.1. Natural Causes

23.2.1.1. Volcanoes

Volcano is a mountain or hill with an opening on top known as a crater. Hot melted rock (magma), gases, ash, and other materials from inside the earth mix together a few kilometers underground, rising up through cracks and weak spots in the mountain. On eruption, lava, rock fragments, hot vapour, and gases come out with a great force. Volcanic eruptions contain large quantities of SO_2, water vapour, dust, and ash and release these into the atmosphere. Huge volume of volcanic SO_2 may reach in the stratosphere. Sulfur oxide combines with water to form tiny droplets of sulphuric acid. These droplets are very small and many of them can stay aloft for several years. They are deficient reflectors of sunlight. The dust particles, ash and huge fiery clouds arise over the nearby areas block the incoming sunlight and screen the ground from some of the energy that it would ordinarily receive from the sun. Some of the major volcanic eruptions have been presented in Table 14.1. Eruption of Mt. Pinatoba (1991) in the Philippines Islands emitted huge volume of gases. This can reduce the amount of solar radiation reaching the earth's surface lowering temperatures of troposphere, and changing atmospheric circulation patterns. Eruption of Tambora volcano of Indonesia (1815) significantly changes the weather parameters in New England, Western Europe, USA and Canada.

Table 14.1. Major volcanic eruption of the world

Year (death approx.)	Volcano	Country	Causality
79 AD	Mt. Vesuvivus	Italy	16000
1568 BC	Kelut	Indonesia	10000
1792	Mt. Unzen	Japan	14500
1815	Tambora	Indonesia	10000
1883	Krakatoa	Indonesia	36000
1902	Mt Pelee	Martinique	28000
1980	Mt. St. Helens	USA	57
1982	El Chichon	Mexico	1880
1985	Nevado del Ruiz	Columbia	23000
1986	Lake Nyos	Cameroon	1700
1991	Mt. Pinatubo	Philippines	800

14.2.1.2. Tilting of earth axis

The earth is tilted at an angle of 23.5^0 to the perpendicular plane of its orbital path. Change in the tilt of the earth causes change of season from summer to winter. The earth axis always seems to point toward Pole Star. This axis moves very slowly, a little more than a half-degree each century. Around 2500 BC, the pole of the earth's axis was near the star Thuban (Alpha

Draconis). This gradual change in the direction of the earth's axis is also one of the causes for climate change.

14.2.1.3. Ocean currents

Global climate change is under way and intensity and distribution like *El-Nino* effects is predicted to further increase. *El-Nino* is defined as a 'large scale shift in water current and wind of the equatorial and tropical Pacific, resulting in extreme climate change characterized by excessive rains and strong winds in some areas and drought in others' (Tibig, 1995). Ocean covers about 71% of the earth surface, and it has a major influence on climate system. Ocean currents carry huge amount of heat and cold across the earth surface. The direction and speed of ocean current may change influencing the weather parameters of the regions. Much of heat that escapes from the ocean is in the form of water vapour, the most abundant greenhouse gas. Certain parts of the world are influenced by ocean currents more than others. Ocean current may have great impact on climate in long run. The coast of Peru and other adjoining regions are directly influenced by the Humbolt current that flows along the coast line of Peru.

14.2.2. Man-made Causes

The findings of the IPCC (2005) said that there was some human influence on climate change. Some of the man-made causes of climate change have been discussed here. Future climate change due to human influence could occur many times faster than any past episode of global climate change (IPCC, 1990, 1992; Schneider *et al.*, 1992).

14.2.2.1. Industrial revolution

During the renaissance- the revival of art and literature under the influence of classical styles in the 14th to16^{th} centuries followed by industrial revaluation in the 19th century lead a large scale use of fossil fuels. The land that covered with vegetation has been cleared to make way for houses. Natural resources are being used extensively for construction, industries, roads and other consumption purpose. All this has contributed to a rise in greenhouse gases in the atmosphere. Greenhouse gases include CO_2, water vapour, methane, nitrous oxides, chloroflorocarbons (CFCs) and ozone. (Details have been given in Chapter 12). The energy section is responsible about $3/4^{th}$ of the CO_2 emission, $1/5^{th}$ of the methane emission and large quantity of nitrous oxides.

Carbon dioxide is good transmitter of sunlight, but partially restricts infrared radiation going back from earth into space causing greenhouse effect that prevent drastic cooling of the earth during the night. Increasing the amount of CO_2 in the atmosphere reinforces the greenhouse effect and is expected to result in warming the earth's surface. Currently CO_2 is responsible for 57% of the global warming.

Methane is another important greenhouse gas in the atmosphere. It is emitted during the process of oil drilling, coal mining and also from the leaking gas pipelines. Oil exploration, production, refinement, transportation and storage also release methane. To lower the possibilities of explosions, most mines circulate underground air, thus venting large amounts of coal-associated methane. Incomplete fossil fuel combustion also may release small amounts of methane.

Global atmospheric concentration of methane has increased from a pre-industrial value of about 715 to 1732 ppb in the early 1990s and is 1774 ppb in 2005. The atmospheric concentration of methane in 2005 exceeds by far the natural range of the last 650000 years (320 to 790 ppb) as determined from ice cores (IPCC, 2007).

Nitrous oxides and CFCs (chlorofluorocarbans) are also increasing at an alarming rate. But, because CFCs destroy O_3, the net warming effect from CFCs is negligible (IPCC, 2001). Water vapour is another most important greenhouse gas. Rising temperatures increase evaporation, setting up a positive feed back loop for accelerating warming.

14.2.2.2. Agriculture

Agriculture activities play an important role in production of methane and release in the atmosphere. About $1/4^{th}$ of all methane emissions are said to come from domesticated animals, such as dairy cows, goats, pigs, buffaloes, camels, horses, and sheep. These animals produce during cud-chewing process and decomposition of dung and litters. Methane is also released from rice fields that are flooded during its growth periods (Fig. 14.1) by anaerobic decomposition of organic matters. When soil is covered with water it becomes anaerobic, under such conditions, methane producing bacteria and other organisms decompose organic matter in the soil to form methane. Methane escapes into the atmosphere in three different from the rice field- up to 80% travel through the roots, ebullition, where gas bubbles up to the water surface and diffusion which contributes small amount.

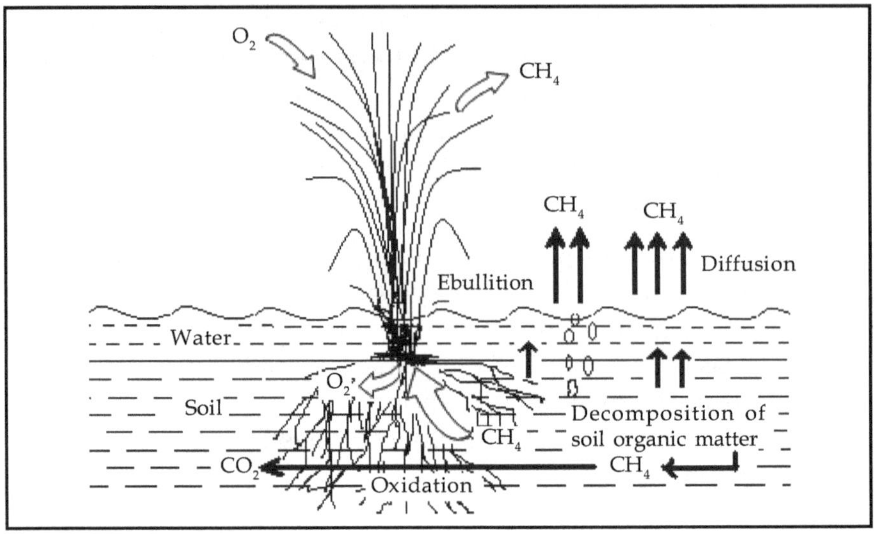

Fig. 14.1. Methane gas escapes into the atmosphere from flooded rice fields.

A growing population has meant more and more mouth to be feed. Because of the land area available for agriculture is limited, high-yielding crop varieties are being grown to increase the agricultural production. These high-yielding varieties require large quantities of nitrogenous fertilizers, which emit nitrous oxide, both from field as well as the industries that make it. Nitrous oxides contributions are also made by leguminous plants that added to the soil.

14.2.2.3. Emission from Motor-vehicles

Cars, buses and trucks are the principal ways by which goods and people are transported from place to place. These are run mainly on petrol or diesel, both are fossil fuels. The burning of fossil fuels releases huge quantities of CO_2 and H_2S into the atmosphere. Carbon dioxide, the principal greenhouse gas generated by man-made emissions, has risen from 280 ppm before Industrial Revaluation to 380 ppm today. It is projected to increase to 500 ppm by the middle of the century. It is worth to note that the last time our planet experienced greenhouse gas levels as high as 500 ppm was some 20-40 million years ago, when the mean sea level were 100 meters higher than today.

14.2.2.4. Use of Weapons of Mass-Destruction

The weapon of mass destruction release enormous energy at the moment of the explosion, causing immediate fire, destructive blast pressure, increase in temperature, and extreme local radiation exposures. In a thermonuclear bomb, an initial fission, such as occurred in the 'atomic' bomb, momentarily creates conditions of enormously high temperature. It blows mountainous volume of dust in air, which may blocks sunlight for several days. The elevated temperature immediately affects the weather parameters of the local areas as well as the adjoining places. Some of the major explosion of the world has been presented in Table 14.2.

Table 14.2. List of weapon of mass-destruction tested and used in the war

Date	Name	Yield (kt)	Country	Significance
16.07.1945	Trinity	19	USA	First fission weapon test
06.08.1945	Little Boy	15	USA	Bombing of Hiroshima, Japan
09.08.1945	Fat Man	21	USA	Bombing of Nagasaki, Japan
29.08.1949	Joe 1	22	USSR	First fission weapon tested by USSR
03.10.1952	Hurricane	25	UK	First fission weapon tested by UK
01.11.1952	Ivy Mike	10400	USA	First 'staged' thermonuclear weapontest (not deployable)
12.08.1953	Joe 4	400	USSR	First fusion weapon tested by USSR (not 'staged' but deployable
01.03.1954	Castle Bravo	15000	USA	First deployable 'staged' thermonuclear weapon; fallout accident
22.11.1955	RDS-37	1600	USSR	First 'staged' thermonuclear weapon tested by USSR (deployable)
08.11.1957	Grapple X	1800	UK	First (successful) 'staged' thermonuclear weapon tested by UK
13.021960	Tsar Bomba	50000	USSR	Largest thermonuclear weapon ever tested
16.10.1964	596	22	China	First fission weapon tested by China
17.06.1967	Test No. 6	3300	China	First 'staged' thermonuclear weapon tested by China
24.08.1964	Canopus	2600	France	First 'staged' thermonuclear weapon tested by France
18.05.1974	Smiling Budha	12	India	First fission nuclear explosive tested by India
11.05.1998	Shakti I	43	India	First potential fusion/boosted weapon tested by India
13.05.1998	Shakti II	12	India	First fission weapon tested by India
28.05.1998	Chagai-I	9-12?	Pakistan	First fission weapon tested by Pakistan
09.10.2006	Hwadae-ri	<1	North Korea	First fission device tested by North Korea

14.3. PHYSICAL EVIDENCE OF CLIMATE CHANGE

Assessing the atmospheric and weather parameters, the scientist from every corner had concluded following evidences of climate change.

- Bubbles ice core from deep in stable Antarctica formations indicate the composition of the pre-industrial atmosphere. The chemical differences between the current atmosphere and that captured in the bubbles reflect human activity.
- Analysis of radiocarbon, which reflect the differing proportions of ^{14}C molecules in samples from different sources, show that emissions from fossil fuel burning have been a major contributor to increased atmospheric concentration of CO_2.
- The temperature difference as record from land and sea since the late 1800s with a variety of instruments, by a variety of institutions.
- Changes in weather parameters, time and duration of monsoon, rain-fall distribution etc.

14.4. IMPACTS OF CLIMATE CHANGE

A team of scientist from every corners of the globe got together under the 'International Panel for Climate Change (IPCC)' has been predicted that the global worming is real, serious and accelerating. Permanent change in weather parameters triggers weather extremes, such as high temperature, floods and droughts. Since the end of the 19th century the earth average surface temperature has increased by 0.3-0.6 ^{0}C. Over the last 40 years, the rise has been 0.2-0.3 ^{0}C. Scientist estimated that man-made emission of greenhouse gases are likely to lead to increase in global average temperatures of between 1.4 ^{0}C and 5.8 ^{0}C by another 100 years (IPCC, 2001). The magnitude of this predicted warming may see negligible, but its rate is faster than any seen in the last 10000 years. The impacts of climate change are many and serious. They include rising sea-level, changes in the availability of drinking water, and an increase in the risk of extreme weather such as floods, droughts and hurricanes; and some of them have been discussed below.

14.4.1. Global Warming and Its Effect on Plant

It is predicted that we are entering in the era of global warming caused by the entrapment of solar energy by atmospheric gases such as CO_2, water vapour, methane, nitrous oxide, choloro-floro-carban and other gases. The planet's temperature has warmed as man-made emission of CO_2 has increased considerably. Since mid-nineteenth century, the temperature has increased by 1.5 ^{0}F. Model of climate change (IPCC, 1990, 1992) predicts an increase in mean global temperature of about 1.5-4.5 ^{0}C in Twenty First century. Rising warmth will lead to an increase in the level of evaporation of surface water; the air also expands and this will increase its capacity to hold moisture. This, in turn, will affect water resources, forests, and other natural ecological systems, agriculture, power generation, infrastructure, tourism, and human health. An increase in the number of cyclones and hurricanes over the last few years has been attributed to change in temperature.

The earth receives energy from sun, which warms the earth's surface. About 30% of this energy is scattered by atmosphere. The rest 70% actually remains behind to heat the earth. Short-wave radiation from the sun passes through the earth's atmosphere. Some of this radiation is reflected back into space, some of its absorbed by the atmosphere. Earth again radiates energy back into the atmosphere through infrared or thermal radiation. The earth emits long wavelength radiation toward the space. However, greenhouse gases form a sort of blanket around the earth and absorb the re-emitted energy increases the temperature (Fig. 14.2). It is because of this greenhouse-like function of the atmosphere that the average global temperature of the earth is 15 ^{0}C, without this blanket of greenhouse gases, the average global temperature would be a frigid -18 ^{0}C and life would not be possible on earth.

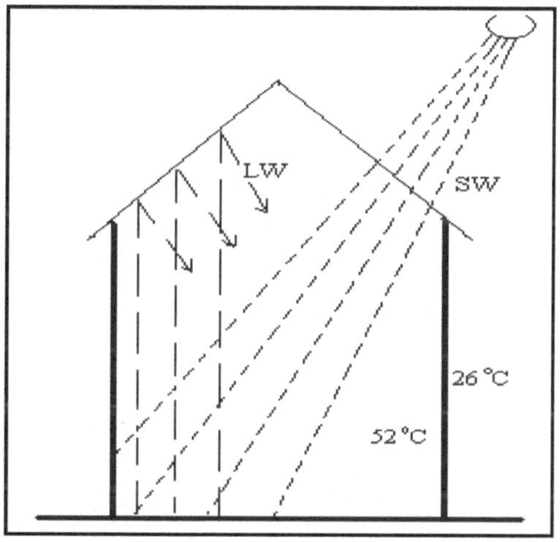

Fig. 14.2. Diagrammatic representation of green house effect. The energy from the sun enters into the greenhouse as short wavelength (SW) rays. The objects inside the greenhouse absorb energy and give it as heat, i.e. long wavelength (LW) rays. The glass does not permit to escape the long wavelength rays, which increase the temperature of the greenhouse.

Industrial revolution and human activities have been releasing more and more of these greenhouse gases into the atmosphere. This leads to the blanket becoming thicker and upsets the natural greenhouse effects. The increased temperature alone will cause the water in the oceans to expand, causing an estimated sea level rise of 20 cm by 2030 (OTA, 1993). Different effects of global warming have been discussed below.

14.4.1.1. Continent-wise consequences of global warming

Asia: Uneven distribution and severity of monsoon rain may increase, causing more floods in the low laying areas, such as Bangladesh, and some other parts of India, whereas, eastern countries expected to receive less rain effecting the agriculture and cropping system. The Siberian *Permaforest* will melt with the increasing temperature affecting the residents nearby. Asian land mass also has influence on ocean current. A huge quantity of cold and heavy water inters into the Artic Ocean through river flow that sinks down and drives currents around the globe. In future, these water sources will warm up causing more light weight, fresh water, thus it may slow down the ocean current.

Africa: The overall rainfall may increase due to global warming, but, due to uneven and untimely distribution of rain will lead drought and flood simultaneously. Africa is the homeland bearing largest desert of the world. The drought around the desert may enhance the expansion of the existing deserts.

Antarctica: It has the highest potentiality to influence the global climate. The increasing temperature would melt the large ice peaks of the world releasing huge quantity of fresh water into the ocean and raising sea level by many meters. There is another possibility, the increase rainfall could increase frozen wastelands grow in size, as more and more snowfalls in the centre, locking up water that would otherwise end up in the ocean.

Australia: Australia is comparatively a dry country. It has no large mountain range. The dry area of this continent may expand more. One of the greatest disasters of warming ocean is the damage of *Great Barrier Reef*, the world largest living structure. Since 1998, the reef has gone two 'bleaching' events on which huge numbers of corals throw of the coloured algae that live alongside them, as a result of stress caused by raising temperature. Climate change also increased frequency of forest fire, which again mount surrounding air temperature and destroys forestlands.

Europe: Global warming changing the climate worldwide. The northern Europe is tipped to gain a more *Mediterranean Climate,* while Mediterranean countries themselves will swelter through increasing frequent droughts.

North America: Agriculture of USA west coast depends on the spring run-off of water from the Rocky Mountains to sustain crop through the parched summer. If this water arrives early as rain, rather than wintering (snowfall) in the mountain tops, then the farmers will affect more.

The melting of Arctic ice is causing concern for ecosystem in the north of the continent. The northern coast of Canada and neighboring Greenland spend much of their year stuck in the polar ice cap or pull away from mainland leaving the polar bears unable to patrol large territories in search of food.

South America: The climate of South America is influenced by the swirling current of the Pacific Ocean. The cool water is replaced by warmer water drifting across from the west, killing fish and causing more rain to fall on the South American coast. Such events are predicted more frequent. But the South American Ocean could have more important effects on the continent climate in a warming world. In 2004, the Brazilian state of Santa Catharina was hit by a strong tropical storm. This was caused by evaporating water in the warmer North Atlantic.

14.4.1.2. Rise in sea level

During last Ice Age the maximum glaciation was found around 16000 BC, when ice sheets covered all of Canada and much of American Midwest and northeast, all of Scandinavia and some regions of Eurasia (Johanstonar-chive.net). The volume of ice was 2-3 times as much as today's volume and the sea level was 120 m lower than what is present. According to Johanston (Johanstonar-chive.net), today's balance between ice caps and the global sea level has been steady since 100 BC, but the climate could change in the next 1000 years. He drives hope from the fact that it took 18000 years to melt 60% of the ice from the last Ice Age.

Mountain glaciers and snow covers have declined on average in both hemispheres (IPCC, 2007). Wider spread in glacier and ice peaks has contributed to sea level rise (Table 14.3). The heating of ocean and melting of glaciers and polar ice sheets, is predicted to raise the mean sea level (MSL) by about 0.5 m over next century. The world two great ice sheets covering Greenland and Antarctica are indeed losing ice to the oceans, and losing it an accelerating pace. Sixty percent of Mount Kilinanjaro of Africa melted; Kivalana Alaska melts eight inches every year. If the unexpectedly rapid shrinkage continues, low lying coasts around the world- including New Orleans, South Florida, and much of Bangladesh could face inundation with a couple of centuries. Greenland and Antarctica have been losing ice over past 5-10 years. In the north, Greenland is shading at least 100 gigatons per year. Current ice sheet loses are not raising sea level faster than 0.1 m per century, but researchers fear that the rates could rise to a meter per century (Anonymous, 2006).

Table 14.3. Observed rate of sea level rise and estimated contributions from different sources (IPCC, 2007)

Source of sea level rise	Rate of sea level rise (m per century)	
	1961-2003	**1993-2003**
Thermal expansion	0.042 ± 0.012	0.16 ± 0.05
Glaciers and ice caps	0.050 ± 0.018	0.077 ± 0.022
Glaciers ice sheets	0.05 ± 0.12	0.21 ± 0.07
Antarctic ice sheets	0.14 ± 0.41	0.21 ± 0.35
Sum of individual climate contribution in sea level rise	0.11 ± 0.05	0.28 ± 0.07
Observed total sea level rise	0.18 ± 0.05	0.31 ± 0.07

Sea-level rise could have a number of physical impacts on coastal areas, including lose of lands due to inundation and erosion, increased flooding, and salt water intrusion and rise in water table. These could adversely affect coastal agriculture, tourism, fresh water resources, fisheries and aquaculture, human settlements, and health. This is particularly crucial in tropical countries such as Bangladesh, with large agricultural regions and high rural population located near sea level.

14.4.2. Impacts of Global Warming on Agriculture

Historical climate change has had a powerful effect on current biogeography. Plants and animals in the natural environment are very sensitive to change in climate. Climate change will affect the agricultural yield directly or because of alternation in temperature and rainfall, and indirectly the change in soil quality, pests and diseases. In particular, the yield of cereals is expected to decline in India, Africa and Middle East. As the temperature rise, conditions will become more favourable for pests such as grasshopper to complete a number of reproduction cycles thereby increasing their population. Global agriculture will face many challenges over the coming decades.

14.4.2.1. *Biophysical impacts of climate change*

Rapid large-scale shift in temperature, precipitation and other climatic parameters could have broad ecological effects, presenting major challenge to the conservation of biodiversity. The rate of plant photosynthesis depends on the amount of the photosynthetically active radiation and level of atmospheric CO_2. Temperature is another important determinant of photosynthesis and subsequently the crop production. The accumulation of biomass also depends on availability of moisture, and nutrients to growing plants.

Temperature: Maximum, minimum and average seasonal temperature has great importance in survival and distribution of plants (detail can be seen in Chapter 5). For example, Palmae/Arecaceae is cold and frost susceptible, whereas, Arctic stress is generally determined by summer warmth. Higher temperature cause heat stress in plants, thus they grow less and produce less crops, in some cases, the plants do not produce at all since excessive heat cause pollen sterility. High temperature in general hasten plant maturity in annual species, thus shortening the growth stages during which pods, seeds, grains or bolls can absorb less photosynthetic products. Increase in temperature will also lengthen the frost-free season in temperate regions, allowing for longer duration crop varieties to be grown and offering the possibility of growing successive crops. Agro-climatic zones are expected to shift pollards as lengthening and warming growing seasons allow new or enhanced crop production (Rosenzweig, 1985).

Extreme temperature in either direction is detrimental for crop production. Prolong hot spells can be specially damaging (Mearns *et al.*, 1984). Critical stage for high temperature injury include seedling emergence in most crops, silking and tasseling in corn (Shaw, 1983), grain filling in wheat (Johnson and Kanemasu, 1983) and flowering in soybeans (Mederski, 1983).

Rainfall: Agriculture is limited by the seasonality and magnitude of moisture availability. Rainfall pattern, intensity, distribution and amount directly affect the plant survival and growth, crop production; and balance to grass and woody plants. Alteration in amount and time of precipitation and evaporation would affect soils and habitats; fresh-water ecosystems are likely to vulnerable to these changes in hydrology (Carpenter *et al.*, 1992). Even minor fluctuation in the availability of water can radically affect habitat suitability for many wetland plant species.

Excessive rainfall results in floods and water logging (see Chapter 3). Water logging soil causes plant roots to rot and heavy rainfall damages tender young plants. Whereas, increased rainfall without flooding may beneficial in very dry areas and allow limited crop growth. So, changes in temperature and precipitation patters as a result of climate change are likely to be bad for large areas of the world but may increase crop production in other regions.

The crop water regime may further be affected by changes in seasonal precipitation, within-season pattern of precipitation, and inter-annual variation of precipitation. Increased convective rainfall is predicted to occur, particularly in the tropics, more moisture in the air, but the distribution and intensity may not be uniform. High rainfall may increase disease infestation in crops, while too little can be detrimental to crop yield. Moisture stress during flowering, pollination, and grain-filling stages is harmful to maize, soybean, wheat and sorghum.

There are precipitations of mid-continental drying the Norther Hemisphere (Manabe and Wetherald, 1986; Kellogg and Zhao, 1988) and the rise in potential evaporation will exceed the rainfall resulting in dryer regions throughout the tropics and low to mid-latitudes (Rind *et al.*, 1990).

Soil: Climate change will have an impact on the soil, a vital element in agricultural ecosystem. Higher atmospheric temperature also cause higher soil temperatures, which may increase solution chemical reaction rates and diffusion-controlled reactions (Buol *et al.*, 1990). The magnitude of soil moisture changes were related to lands use soil texture and topographic conditions (Jasper *et al.*, 2006). Solubility of solid and gaseous components may either increase or decrease depending on the soil temperature range. Higher soil temperature will also accelerates the decay of soil organic matter, resulting in increase of CO_2 to the atmosphere and decrease in C/N ratios.

Soil structure and texture decides the ground vegetation, shrubs and trees. Degrading soils place enormous strain in achieving food security for growing population. With increase in global warming, both evaporation and precipitation will increase. Soil erosion is a major effect of precipitation and its characteristics, viz. quantity, special and temporal variation, intensity, frequency, erodibility, depth, class distribution, rainy days and seasonal rainfall etc. Soil loss also depends on characteristics (erodibility, organic matter content, texture, structure, and infiltration), topography (length, and slope of the land), finally the crops and its management.

Disease and insect pests: Climate change is likely to alter the ranges and distribution of pests. Insects may extend their ranges where warmer winter temperatures allow their over-wintering survival and increase the possible number of generations in every season (Stinner *et al.*, 1989). Insect pest and disease from low altitude regions (where they much more prevalent) may be introduced at higher latitudes. There is also so evidence that the upward expansion of insects and plant diseases will add to the risk of crop losses.

14.4.2.2. Climate change and plant biodiversity

Global warming may destroy the existing plants diversity on the earth. The potential magnitudes of local and global climate change are area of concern. The predicted rate of temperature change that posses the greatest threat to biodiversity. Species of animal and plants are estimated to be going extinct at a rate that is about 100 times faster than the historical record largely due to human activities, which lead to climate change. The availability of species to survive rapid climate changes may partially depends on the rate at which they can migrate to newly suitable areas. Followings are the some of the effects of climate change on plant biodiversity.

1. Plant species with long life cycle and/or slow dispersal are particularly vulnerable to climate change.
2. Artic and alpine species, and Island endemics grow only in some isolated and specific geographical areas, and these are very vulnerable to climate change. Another danger for mangrove which will be squeezed between human settlements and rising sea level. A good and recent example of massive destruction of mangrove in Andaman and Nicober Islands by the *Tsunami* of 26 Dec, 2004 (Fig 14.3).
3. Plant genetic composition may change in response to the selection pressure of climate change. The strong association between distribution of plants species and climate suggests that rapid global climate change could alter plant distributions, resulting in extensive reorganization of natural communities (Graham and Grimm, 1990). Many native plants are also capable of colonizing new urban site. There are called apophytes (Kikle, 1903/04). Kowarik (1992) found 32% of the Berlin's native plants established on urban sites.
4. Increased invasion by alien species may occur.
5. Some plant species are already under threat of climate change. *Aloe dichotoma* of Northern Cape has begun to respond to climate-induced stress. Bluebell woods of UK are threatened by climate change. *Palatathera leucophaea* is a North American threaten orchid that is vulnerable to drought, its numbers has dropped steeply in the last few years as the climate has been warmer and drier.

Fig. 14.3. Massive destruction of mangrove in Andaman Islands by the *Tsunami* of 26 Dec, 2004. **a)** Ariel view of mangrove destruction at Chouldhari, South Andaman; **b)** Mangrove before *Tsunami* at Lohabrak, South Andaman; **c)** View one year after *Tsunami* at Lohabrak, South Andaman. *(For colour version of this figure see page 540)*

14.4.2.3. Plant migration rate

Plant migration rate, as calculated from the fossil pollen record, ranged from about 5-150 km per century (Shugart *et al.*, 1986). Various studies have suggested that rapid climate change would require shifts of plant ranges of up to 500 km shortly (Davis, 1984; Davis and Zabinski, 1992). Plant species individually responds to climate change, migration rates will vary within and among natural communities. Thus, entire community will not move together in response of global warming (Graham and Grimm, 1990). The fossil records proved evidence of decade-or even century-long time lags in species migration (Davis, 1989). Climate change cause more frequent droughts, fires, and pest and pathogen outbreaks are predicted to act in conjugation with climate change to significantly transform the landscape (Peters, 1992).

14.4.2.4. *Dominance of exotic weeds*

Some exotic weeds may become dominant over native species. Many weeds are able to expand quickly, posing serious threats to existing species and overall biodiversity (Schwartz, 1992). Many weed species are widespread, prolific fast-growing annuals capable of establishing in disturbed habitats. Thus climate-induced changes could expose native plants to non-native competitors (Peters, 1992). Exotic weeds may become a problem in the management of many preserves and natural areas. Certain weeds may expand their range into higher altitude habitats depending upon creation of conducive-environment due to climate change.

14.4.2.5. *Physiological changes in plant*

Global warming has both destructive as well as the constructive effects on plant physiological activities. The major physiological effects of global warming on plant have been briefed here.

Water Use Efficiency: The increased global temperature will decrease atmospheric relative humidity affecting the water use potential of crop plants. Under high temperature, plants partially close stomata leading to reduction in stomatal conductance and help in *water saving*. However, slow rate of transpiration will affect evapotranspiration of plants and eventually disturb the cooling system of plant, which accelerates senescence and reduces photosynthesis. High temperature decrease water use efficiency in C_4 plants.

Crop growth and development: Under high temperature speed up the growth and development of crops leading to early flowering and fruiting. Early flowering than the normal duration, reduce potential yield of crops. The crop duration of winter rice and summer rice can be noted, winter rice takes 10-15 days more to reach 50% flowering than the summer rice and have higher yield.

Rate of respiration: Increased in atmospheric temperature increase transpiration rate in plants leading water loss and reduction in photosynthesis.

14.4.3. Adaptation of Plants to Climate Change

Adaptation of plants to climate change exists at the various levels of agricultural organizations (White, 1974; Rosenberg *et al.*, 1989; Waggoner, 1983). In temperate regions, farm-level adoptions include change in planting and harvesting dates, tillage and rotation practices, substitution of crop varieties or species more appropriate to the changing climate regime, increase fertilizer and pesticide applications and improved irrigation and drainage systems.

14.5. PREDICTED BENEFICIAL EFFECTS OF CLIMATE CHANGE

Climate change also has beneficial effects on agriculture. Increase temperature will lengthen the summer in temperate regions, such as Siberia, Northern parts of North America and Northern Europe. Thus, it will lengthen the crop season. Longer growing seasons and increasing rains may boot yields in many temperate regions. Records showed that the season has already lengthened in UK, Scandinavia, Europe and North America.

Beside alteration of climate regime, increasing CO_2 concentration has another beneficial effect on crop production. Higher levels of CO_2 stimulated photosynthesis in certain plants. This is particularly true for so-called C_3 plants because increased CO_2 tends to suppress their photorespiration. This beneficial effect is called *CO_2 fertilization*. The higher photosynthesis rates are then manifested in higher leaf area, dry matter production, and thus yield for many crops (Kimball, 1983; Acock and Allen, 1985; Cure, 1985). Jurik *et al.* (1984) reported upwards shifts in temperature optima for photosynthesis under higher concentration of CO_2 concentration. Higher concentrations of CO_2 and elevated temperature enhanced growth of plants, subsequently increasing the yield of crops (Idso *et al.*, 1987). Experiments based on a 50% increase of current CO_2 concentrations have conformed that CO_2 fertilization can increase crop production by 1.5% under optimal conditions. In contrast, C_4 plants (maize, sugarcane, sorghum, and millets) are less responsive to CO_2 enrichment (Acock and Allen, 1985). However, if trace gas emission continue to grow unchecked, their climate warming effect is projected to continue even up to 2000 ppm (Manabe and Bryan, 1985), but the beneficial boost to photosynthesis appears to level of at about 800 ppm for C_3 crops (Akita and Moss, 1973). Crop specific simulation models predict that doubling the CO_2 concentration from 300 to 600 ppm will produce yield increase by 25-40% for C_3 crop species, but as little as 7% for C_4 crops (Rosenzweig *et al.*, 1992).

High concentrations of CO_2 also tends to close plant stomata, and thus reduces transpiration per unit area of leaf while still enhance photosynthesis. Morison and Gifford (1984) observed an average of 36% reduction in stomatal conductance in an atmosphere enriched by doubled CO_2. However, crop transpiration per ground area may not be reduced commensurately, because decrease in individual leaf conductance tends to be offset by increases in crop leaf area (Allen *et al.*, 1985). Higher CO_2 in atmosphere also increase water use efficiency.

14.6. REFERENCES

Acock B, Allen Jr LH. 1985. Crop responses to elevated carbon dioxide concentrations. In: Direct Effects of Increasing Carbon Dioxide on Vegetation DOE/ER-0238, US Dept of Energy, Washington D.C. pp. 53-97.

Akita S, Moss DN. 1973. Photosynthesis responses to CO_2 and light by maize and wheat levels adjusted for constant stomatal apparatus. Crop Sci. 13: 234-237.

Allen LH, Jones Jr P, Jones JW. 1985. Rising atmospheric CO_2 and evapotranspiration. In: Advances in Evapotranspiration. Proc of the Natl Conf on Advs in Evapotranspiration. Dec 16-17, 1985. Ame Soc Agricl Engineering. St Joshep, Michigan. pp. 13-27.

Anonymous. 2006. Breakthrough of the year. Science. 314(5807): 1850.

Buol SW, Sanchenz PA, Weed SB, Kimble JM. 1990. Predicted impact of climate warming on soil properties and use. In: Impact of Carbon Dioxide. Trace Gase and Climate Change on Global Agriculture, Kimball BA, Rosenberg NJ, Jr Allen LH (eds). ASA Special Publication No. 53, pp. 71-82.

Carpenter SR, Fisher SG, Grimm NB, Kitchchell JF. 1992. Global change and fresh water ecosystem. Annu Rev Eco Systematics. 23: 119-140.

Cure JD. 1985. Carbon dioxide doubling responses: A Crop Survey. In: Direct Effects of Increasing Carbon Dioxide on Vegetation DOE/ER-0238, US Dept of Energy, Washington D.C. pp. 99-116.

Davis MB, Zabinski C. 1992. Changes in geographical range resulting from greenhouse warming effects on biodiversity in forests warming and biological diversity, In: Global Warming and Biological Diversity, Peters RL, Lovejoy TL (eds). Yale University Press, New Haven, CT. pp. 298-308.

Davis MB. 1984. Climate instability, time lags and community diaequlibrium. In: Community Ecology, Dimond J Case TJ (eds.). Harper and Row, New York. pp. 269-284.

Davis MB. 1989. Insights from paleoecology on global change. Ecological Society of America Bull. 70(4): 222-228.

Graham RW, Grimm EC. 1990. Effects of global climate change on the patterns of terrestrial biological communities. Trends in Ecology and Evaluation. 5(9): 289-292.

Idro SB, Kimbal BA, Andersoll MG, Mauney JR. 1987. Effects of atmospheric CO_2 enrichment on plant growth the interactive role of air temperature. Agri Ecosystem Environ. 20: 1-10.

IPCC. 1990. Climate Change: the Intergovernmental Panel on Climate Change scientific assessment. World Meteorological Organisation and United Nations Environment Programme, Geneva. pp. 270.

IPCC. 1992. Intergovernmental Panel for Climate Change Supplement. World Meteorological Organisation and United Nations Environment Programme, Geneva. pp. 70.

IPCC. 2001. Climate Change 2001: The Scientific Basis. In: Contribution of Working Group I to the Third Assesment Report of the Intergovermental Panel on Climate Change, Houghton JT, Ding Y, Griggs DJ, Noguer M, van der Linden PJ, Dai X, Maskell, Johanson CA (eds). Cambridge University Press, Cambridge.

IPCC. 2005. IPCC's Special Report on Safeguarding the Ozone Layers and the Global Climate System.

IPCC. 2007. Climate Change 2007: The Physical Science Basis- Summary for Policymakers. (http://www.ipcc.ch.).

Jasper K, Calanca P, Fuhrer J. 2006. Changes in summertime soil water patterns in complex terrain due to climatic change. J Hydrology Amsterdam. 327(3/4): 550-563.

Johanson RC, Kanemasu ET. 1983. Yield and development of winter wheat at elevated temperature. Agron J. 75: 561-565.

Jurik TW, Weber JA, Gates DM. 1984. Short-term effects of CO_2 on gas exchange of leaves of bigtooth aspen (*Populus grandidentata*) in the field. Plant Physiol. 75: 1022-1026.

Kellogg WW, Zhao ZC. 1988. Sensitivity of soil moisture to doubling of carbon dioxide in climate model experiments, I, North America. J Clim. 1: 348-366.

Kikle M. 1903/04. Die Anthropochoren and der Formenkreis der *Nasturium palustre* DC. Berichte der Zuricher Botanischen Genellschaft. 8: 71-82.

Kimbal BA. 1983. Carbon dioxide and agriculture yield: an assemblage and analysis of 430 prior observations. Agron J. 75: 779-788.

Kowaric I. 1992. Das Besondere der stadtischen Flora and Vegetation. Schriftenreihe der Deutschen Rates fur Landespflege. 61: 33-47.

Manabe S, Bryan Jr K. 1985. CO_2-induced change in a cloud ocean-atmosphere model and its paleoclimatic implication. J Geophysical Research. 90(C6): 11, 689-707.

Manabe S, Wetherald RT. 1986. Reduction in summer soil wetness induced by an increase in atmospheric carbon dioxide. Science. 232: 626-628.

Mearns LO, Katz RW, Schneider SH. 1984. Extreme trmperature events: Changes in their probability with changes in mean temperature. J Climate Appl Meteo. 23: 1601-1613.

Mederski HJ. 1983. Effects of water and temperature stress on soybean plant growth and yield in human temperature climates. In: Crop Reactions to Water and Temperature Stress in Humid, Temperature, Climates, Raper CD, Kramer PJ (eds). Westview Press., Bounder. pp. 35-48.

Morison J. III, Gifford RM. 1984. Plant growth and water use with limited water supply in high CO_2 concentration. I. Leaf area, water use and transpiration. Aust J Plant Physiol. 11: 361-374.

OTA (Office of Technology Assessment). 1993. Preparing for an Uncertain Climate, vol. I. OTA-O-567. US Government Printing Office, Washington D.C.

Peters RL, Lovejoy TL. 1992. Global warming and biological diversity. Yale University Press, New Heaven and London. pp. 386.

Peters RL. 1992. Conservation of biological diversity in the face of climate change. In: Global Warming and Biological Diversity, Peter RL, Lovejoy TL (eds). Yale University Press, New Have, TC. pp. 15-30.

Pielou EC. 1991. After the ice age: the return of life to glaciated North America. University of Chicago Press, Chicago and London. pp. 366

Rind D, Goldberg R, Hasen J, Rosenzweig C, Ruedy R. 1990. Potential evapotranspiration and the likelihood of future drought. J Geophys Res. 95(D7): 9983-10,004.

Rosenberg NJ, Easterling III WE, Crosson PR, Darmstadter J. 1989. Greenhouse Waring: Abatement and Adaptation. Resources for the Future, Washington.

Rosenzweig C. 1985. Potential CO_2-induced climate effect on North American wheat producing regions. Climate Change. 4: 239-254.

Ruddiman WF, Wright HE. 1987. North America and adjacent oceans during the last deglaciation. The Geological Soc America, Boulder, CO. pp. 501.

Schneider SH, Mearns L, Gleick PH. 1992. Climate change senarios for impact assessment. In: Global warming and biological diversity, Peters RL, Lovejoy TL (eds.). Yale University Press, New Haven, CT.

Schwartz MW. 1992. Modeling effects of habitat fragmentation on the ability of trees to respond climatic warming. Biodiversity and Conservation. 2: 51-61.

Shaw RH. 1983. Estimates of yield reductions in corn caused by water and temperature stress. In: Crop Reactions to Water and Temperature Stress in Humid, Temperature, Climates, Raper CD, Kramer PJ (eds). Westview Press., Bounder. pp. 49-66.

Shugart HH, Antonovsky MY, Jarvis PG, Sandford AP. 1986. CO_2 climate change and forest ecosystem. In: The Green House Effect, Climate Change, and Ecosystems, Bolin B, Doos BR, Jager J, Warrick RA (eds). John Wiley and Sons, New York. pp. 475-521.

Stinner RB, Taylor RAJ, Hammond RB, Purrington FF, McCartney DA. 1989. Potential effects of climate change on plant-pest interactions. In: Potential Effects of Global Climate Change in the United States, Smith JB, Tirpak DA (eds). Appendix C-2. US Environmental Protection Agency, Washington D.C. pp. 8-1 to 8-35.

Tibig LV. 1995. *El Nino* ('The Child Chist"). Ang Tagamasid. 23: 18.

Waggoner PE. 1983. Agriculture and climate changed by carbon dioxide. In: Changing Climate National Academy of Sciences Press, Washington D.C. pp. 383-418.

White GF. 1974. Natural Hazards: Local, National, Global. Oxford University Press, New York.

Index

D

E

F

G

H

I

K

L

M

N

O

P

R

S

T

U

V

W

Z

FSC
www.fsc.org
MIX
Papier | Fördert gute Waldnutzung
FSC® C083411

Zeitfracht Medien GmbH
Ferdinand-Jühlke-Straße 7
99095 Erfurt, Deutschland
produktsicherheit@kolibri360.de